World Sustainability Series

Series editor

Walter Leal Filho, Hamburg, Germany

More information about this series at http://www.springer.com/series/13384

Walter Leal Filho · Judy Rogers
Usha Iyer-Raniga
Editors

Sustainable Development Research in the Asia-Pacific Region

Education, Cities, Infrastructure and Buildings

Springer

Editors
Walter Leal Filho
Research and Transfer Centre, Sustainable Development and Climate Change Management
Hamburg University of Applied Sciences
Hamburg
Germany

and

Manchester Metropolitan University
Manchester
UK

Judy Rogers
School of Architecture and Design
RMIT University
Melbourne, VIC
Australia

Usha Iyer-Raniga
School of Property, Construction and Project Management
RMIT University
Melbourne, VIC
Australia

ISSN 2199-7373 ISSN 2199-7381 (electronic)
World Sustainability Series
ISBN 978-3-030-10349-1 ISBN 978-3-319-73293-0 (eBook)
https://doi.org/10.1007/978-3-319-73293-0

Printed on acid-free paper

This Springer imprint is published by Springer Nature
The registered company is Springer International Publishing AG
The registered company address is: Gewerbestrasse 11, 6330 Cham, Switzerland

Preface

This volume introduces many of the papers presented at the Symposium on Sustainable Development Research in Asia Pacific which was held at RMIT University, Australia, in July 2017 under the auspices of the Inter-University Sustainable Development Research Programme. The aim of the Inter-University Sustainable Development Research Programme, established at the World Symposium on Sustainable Development at Universities 2014 in Manchester, is to provide a platform, on which member universities may undertake research on matters related to sustainable development and to assemble interdisciplinary, cross-faculty teams among its member universities, focusing on sustainable development, with a keen interest to engage in bidding for national and international sustainability research projects.

To this end, the symposium on Sustainable Development in Asia Pacific aimed to draw links between research, practice, education for sustainability and the needs of industry, and address the sustainable development goals (SDGs).

Under the broad umbrella of built environment, the symposium brought four themes together:

- Education for Sustainability
- Sustainable Cities
- Sustainable Buildings
- Sustainable Infrastructure

The nature of planning, designing, building and maintaining our built environment called for sharing knowledge, expertise and ideas more so than other sectors. This inaugural symposium brought together industry practitioners, researchers and educators to capture the wealth of experiences on sustainable development research in the Asia-Pacific today.

It is anticipated that this symposium will lead to sharing of ideas and experiences to provide a networked collaborative platform for engagement in the future. The outcomes of this first symposium form a critical path towards the future of the built environment.

We would like to thank all of the authors for their willingness to share their knowledge and experiences as well as the many peer reviewers who contributed their time to ensure the quality of the manuscript.

Hamburg, Germany	Walter Leal Filho
Melbourne, Australia	Usha Iyer-Raniga
Melbourne, Australia	Judy Rogers

Contents

Contributors

Kayani I. Almas Department of Education, Faculty of Social Sciences, PMAS-University of Arid and Agriculture, Rawalpindi, Pakistan

Susan Ang School of Architecture and Built Environment, Deakin University, Geelong, Victoria, Australia

Anurita Bhatnagar School of Construction, RICS School of Built Environment, Amity University, Noida, UP, India

Brijesh Bhatt School of Infrastructure, RICS School of Built Environment, Amity University Noida, Noida, UP, India

Nicole Campbell MLC Center, Sydney, NSW, Australia

Janet Chappell MLC Center, Sydney, NSW, Australia

Andrea Chester RMIT University, Melbourne, Australia

Wes Death Master of Sustainable Practice, RMIT University, Melbourne, Australia

Mark Dewsbury School of Architecture and Design, University of Tasmania, Launceston, Australia

S. T. Garnett Centre for Renewable Energy, Research Institute for the Environment and Livelihoods, Charles Darwin University, DARWIN, Australia

Damien Giurco Institute for Sustainable Futures, University of Technology Sydney, Ultimo, NSW, Australia

E. Halawa Centre for Renewable Energy, Research Institute for the Environment and Livelihoods, Charles Darwin University, DARWIN, Australia

Jung Hoon Han Faculty of Built Environment, The University of New South Wales, UNSW Sydney, Sydney, NSW, Australia

Scott Hawken Faculty of Built Environment, The University of New South Wales, UNSW Sydney, Sydney, NSW, Australia

Colin Hocking Emeritus in School of Education, Formerly La Trobe Learning & Teaching (LTLT), La Trobe University, Melbourne, VIC, Australia

Sarah Holdsworth RMIT University, Melbourne, Australia

J. van Hoof Fontys University of Applied Sciences, Eindhoven, The Netherlands

Edmund Horan Program Director Master of Sustainable Practice, School of Engineering, RMIT University, Melbourne, VIC, Australia

Michael Howes Griffith School of Environment, Cities Research Institute, Griffith University, Southport, QLD, Australia

Usha Iyer-Raniga School of Property, Construction and Project Management, RMIT University, Melbourne, Australia

Margaret Jollands School of Engineering, RMIT University, Melbourne, VIC, Australia

Qudsia Kalsoom Institute of Education, Lahore College for Women University, Lahore, Pakistan

G. Karunasena School of Architecture and Built Environment, Deakin University, Geelong, Victoria, Australia

Afifa Khanam Institute of Education, Lahore College for Women University, Lahore, Pakistan

L. Law Environment Geography and Sustainability Group, College of Marine and Environmental Sciences, James Cook University, Cairns, QLD, Australia

Tim Law School of Architecture and Design, University of Tasmania, Launceston, Australia

Walter Leal Filho Research and Transfer Centre, Sustainable Development and Climate Change Management, Hamburg University of Applied Sciences, Hamburg, Germany; Manchester Metropolitan University, Manchester, UK

Kian Heng Liew SIM-RMIT, Institute for Engineering Leadership, NUS, Liew Strategics, Singapore, Singapore

Liguang Liu School of Government, Central University of Finance and Economics (CUFE), Haidian District, Beijing, China

Alison Lugg School of Education (Bendigo campus), La Trobe University, Melbourne, VIC, Australia

Silvia McCormack College of Arts Social Sciences & Commerce (ASSC), La Trobe University, Melbourne, VIC, Australia

Patricia McLaughlin RMIT University, Melbourne, Australia

Swati Nagpal La Trobe Business School, La Trobe University, Melbourne, VIC, Australia

Triveni Prasad Nanda School of Construction, RICS School of Built Environment, Amity University, Noida, UP, India

Ma Sheilah G. Napalang School of Urban and Regional Planning, University of the Philippines, Diliman, Quezon City, Philippines

Anjula Negi School of Infrastructure, RICS School of Built Environment, Amity University Noida, Noida, UP, India

Tim Nichols School of Architecture and Design, RMIT University, Melbourne, Australia

Campbell Nicole Landcom, Sydney, NSW, Australia

Kristopher Adam Orlowski Department of Infrastructure Engineering, University of Melbourne, Melbourne, VIC, Australia

R. Palliyaguru School of Architecture and Built Environment, Deakin University, Geelong, Victoria, Australia

Sarbeswar Praharaj Faculty of Built Environment, The University of New South Wales, UNSW Sydney, Sydney, NSW, Australia

Budi Prayitno Department of Architecture and Planning, Faculty of Engineering, Gadjah Mada University, Yogyakarta, Indonesia

Naima Qureshi Division of Education, University of Education, Lahore, Pakistan

Samya Rakshit School of Construction, RICS School of Built Environment, Amity University, Noida, UP, India

Jose Regin F. Regidor Institute of Civil Engineering, College of Engineering, University of the Philippines, Diliman, Quezon City, Philippines

Sandra Rodrigues School of Engineering, RMIT University, Melbourne, VIC, Australia

Judy Rogers School of Architecture and Design, RMIT University, Melbourne, Australia

Waseem Saba Department of Education, Faculty of Social Sciences, PMAS-University of Arid and Agriculture, Rawalpindi, Pakistan

S. Safarova Centre for Renewable Energy, Research Institute for the Environment and Livelihoods, Charles Darwin University, DARWIN, Australia

Orana Sandri RMIT University, Melbourne, Australia

Anil Sawhney Department of the Built Environment, Liverpool John Moores University, Liverpool, UK

Samantha Sharpe Institute for Sustainable Futures, University of Technology Sydney, Ultimo, NSW, Australia

Kerry Shephard Higher Education Development Centre, University of Otago, Dunedin, New Zealand

Shilpi Singh School of Construction, RICS School of Built Environment, Amity University, Noida, UP, India

D. T. V. Raghu Rama Swamy RICS School of Built Environment, Amity University, Noida, India

K. K. K. Sylva Department of Engineering Management, Faculty of Engineering, University of Peradeniya, Peradeniya, Sri Lanka

Ian Thomas RMIT University, Melbourne, Australia

J. B. Trombley Centre for Renewable Energy, Research Institute for the Environment and Livelihoods, Charles Darwin University, DARWIN, Australia

Kruti Upadhyay RICS School of Built Environment, Amity University, Noida, India

Alexandra Vella Landcom, Sydney, NSW, Australia

Dexter Villanueva Environmental Engineering, RMIT University, Melbourne, VIC, Australia

Peter S. P. Wong RMIT University, Melbourne, Australia

Simon Wright Institute for Sustainable Futures, University of Technology Sydney, Ultimo, NSW, Australia

Maria Yu Department of Building and Real Estate, The Hong Kong Polytechnic University, Hung Hom, Hong Kong SAR, China

Esther H. K. Yung Department of Building and Real Estate, The Hong Kong Polytechnic University, Hung Hom, Hong Kong SAR, China

Sustainability in Universities in the Asia-Pacific Region: An Introduction

Walter Leal Filho, Usha Iyer-Raniga and Judy Rogers

Abstract In 2014, the Asia and Pacific Region had a population of 4.3 billion, which represented 60% of the world's total.

Keywords Sustainability · Asia-Pacific · Universities · Education

1 Introduction

In 2014, the Asia and Pacific Region had a population of 4.3 billion, which represented 60% of the world's total. The Region is also home to ten of the world's 15 largest cities (United Nations ESCAP and UN Habitat 2015). Between 1980 and 2010, the region's cities grew by around one billion people and are projected to grow by another one billion by 2040. This rapid rate of urbanisation means that half of the region's population will be urban by 2018 (United Nations ESCAP and UN Habitat 2015).

The Asia-Pacific Region is also one of the most disaster prone areas in the world (United Nations ESCAP and UN Habitat 2015). Over 40% of the 3979 disasters that occurred globally between 2005 and 2014 occurred in the region, resulting in

W. Leal Filho (✉)
Research and Transfer Centre, Sustainable Development and Climate Change Management, Hamburg University of Applied Sciences, Hamburg, Germany
e-mail: walter.leal2@haw-hamburg.de

W. Leal Filho
Manchester Metropolitan University, Manchester, UK

U. Iyer-Raniga
School of Property, Construction and Project Management, RMIT University, Melbourne, Australia
e-mail: usha.iyer-raniga@rmit.edu.au

J. Rogers
School of Architecture and Design, RMIT University, Melbourne, Australia
e-mail: judy.rogers@rmit.edu.au

W. Leal Filho et al. (eds.), *Sustainable Development Research in the Asia-Pacific Region*, World Sustainability Series, https://doi.org/10.1007/978-3-319-73293-0_1

the loss of half a million people. More than 1.4 billion people were affected by these disasters, constituting 80% of those affected globally (United Nations ESCAP and UN Habitat 2015). There is a clear link between vulnerability to disaster and the need for sustainable development. Without sustainable urban development and a commitment to the Sustainable Development goals existing inequalities and vulnerability throughout the region will be exacerbated.

In 2015–2016 three significant events occurred: the introduction of the *Sendai Framework for Disaster Risk Reduction* (2015–2030) and the transition from the Millennium development goals to the *Sustainable Development Goals*. In December 2016, the United Nations assembly adopted the *New Urban Agenda* which reaffirmed a:

> global commitment to sustainable urban development as a critical step for realizing sustainable development in an integrated and coordinated manner at the global, regional, national, subnational and local levels, with the participation of all relevant actors. The implementation of the New Urban Agenda contributes to the implementation and localization of the 2030 Agenda for Sustainable Development in an integrated manner, and to the achievement of the Sustainable Development Goals and targets, including Goal 11 of making cities and human settlements inclusive, safe, resilient and sustainable (United Nations General Assembly 2016)

Sustainable Development holds the key to addressing many of the Region's vulnerabilities framed, however, as *transformation* not integration (United Nations ESCAP and UN Habitat 2015). The challenge of transformation does, however, pose many challenges for researchers and Higher Education Institutions that are often still organised around disciplines.

The volume is structured around the four main interlinked themes. Part 1 examines Education for Sustainable Development in the Asia Pacific Region with a particular focus on the Higher Education Sector. Part 2 focuses on cities, Part 3 on infrastructure and Part 4 on Buildings.

2 Part I: Education for Sustainable Development

Higher Education institutions as a key centre of innovation across the region have a key role to play. The 2014 Nagoya Declaration on Higher Education for Sustainable Development recognised the 'essential' role and responsibility of higher education institutions towards creating sustainable societies and called for transformative learning and research. The declaration placed emphasis on 'scaling up and mainstreaming innovative practices in learning and knowledge development, fostering community engagement through the whole-institution approach, ensuring green campus operations, and creating a conducive environment for supporting required transformative processes'. The papers presented in Part I address some of these challenges from a range of different disciplinary perspectives. In Chapter "Researching Higher Education for Sustainable Development: Plan A, Plan B and Moving Beyond Thought-Limiting Clichés" of the volume Shephard asks why so little has changed in higher education for sustainability. He argues that there is a lack of research on the outcomes of education for sustainability and as a result it is unclear what the impact is on graduates perceptions and values. An overemphasis on the need for others to change their practices rather than a focus on the practices themselves is a major barrier to the transformation of sustainability education in the Higher education sector. Shephard identifies two potential plans for education for sustainability Plan A that focuses on persuading University teachers to adopt education for sustainability is, for Shepherd, 'doomed to fail'. Plan B, on the other hand, requires a different approach where university teachers support 'students towards critical and independent thinking and for higher education institutions to research their progress towards any change in student outcome that they have in mind'.

The chapter concludes with a provocation

> that transmission teaching is alive and well throughout higher education and that the broad and progressive construct of student centeredness has passed much of higher education by. It appears to me that for as long as university teachers feel that they can fulfill their role by simply transmitting knowledge, or values, to their students then neither plan A nor Plan B can possibly proceed.

The chapter provides a useful framework to begin to review and interrogate the learning and teaching approaches that follow. In Chapter "Resetting the Compass: Principles for Responsible Urban Built Environment Education (PRUE)" Iyer-Raniga argues that there is a need to 'Reset the compass' to add an additional dimension to education for Sustainability in Higher Education that would involve 'bringing communities, government, public and private organisations to work with educational institutions, particularly universities to engage with the learning content for sustainable development'. In highlighting this shift she poses the question 'What enablers need to be put in place so that responsible built environment education is mainstreamed and implemented effectively, and what other platforms exist for supporting partnerships, collaboration and engagement between academia, industry and research?' Identifying a gap between intent and action the paper proposes the development of Principles for Responsible Urban Built Environment as a vehicle for promoting and developing dialogue between industry, community and educational institutions around education for sustainability for built environment professions. Modelled on the successes of the Principles of Responsible Management Education (PRME) which in 2017 has 660 business and management related higher education institutions involved in 83 countries across the world, PRUE would 'ensure that the objective of developing and delivering the skills for future city design, planning, operation and governance are skills that graduates entering the urban professions of the future will have'.

Chapters "Achieving Both Breadth and Depth: How Sustainability Education Is Being Integrated Across All Undergraduate Courses at La Trobe University, Australia" and "Developing Undergraduate Foundation Courses in Sustainability" shift focus to the institutional and school level, examining issues of implementation. In Chapter "Achieving Both Breadth and Depth: How Sustainability Education Is Being Integrated Across All Undergraduate Courses at La Trobe University,

Australia" Hocking et al. report on the implementation of sustainable thinking as a mandated graduate attribute at La Trobe University Australia. This has depended on defining Education for Sustainability as an educational issue, integrated with other university education development initiatives, and taking a whole-of-institution approach.

Chapter "Developing Undergraduate Foundation Courses in Sustainability" provides an historical overview of the development of a first year foundation course in Sustainability at Griffith University, Australia. In the paper Howes outlines how the development of the course has been refined in response to student feedback before arguing for an approach that requires 'treating sustainability as a design challenge that engages students in interdisciplinary problem analysis and problem solving'. Chapter "A Comparison of Assessment Methods for Engineering Students' Understanding of Sustainability" by Jollands reports on some of the challenges involved in assessing sustainability learning that requires higher order, deep learning.

What happens, however, when students graduate and enter the workforce? How well are they able to embed sustainability education into their everyday practice? Chapter "The Need for the Graduate Attribute Assessment Tool (GAAT)" by Sarah Holdsworth et al. details a research project that evaluated a graduate attribute developed at RMIT to reflect sustainability outcomes. Based on a survey of graduates the research found that while graduates would choose to behave in a way that is consistent with sustainability outcomes if they had 'their way', specific influences act in the workplace to moderate (if not frustrate) this intention. Chapter "University Sustainability Course Creation in China: Experiences from the CUFE" to Chapter "Perceptions of the Research Scholars Regarding Education for Sustainable Development (ESD) in Pakistan" examine some of the challenges in implementing sustainability curriculum in China, India and Pakistan respectively while Chapter "Empowering Women and Girls Through Education: SDG's Vision 2030" by Waseem and Almas focuses on the pivotal role sustainability education and the SDGs can play in empowering women and girls in Pakistan. Chapters Intercultural Dialogue Through Design (iDiDe) as a Platform for Built Environment Education for Sustainability in Rural Developing Contexts: Building Ampara, Sri Lanka and 'Sustainability' After Disaster: Confronting the Complexities of Recovery in the Field: An Educational Experience shift focus once again to discuss curriculum development for sustainability. Chapter "Intercultural Dialogue Through Design (iDiDe) as a Platform for Built Environment Education for Sustainability in Rural Developing Contexts: Building Ampara, Sri Lanka" describes a global mobility study tour model implemented at Deakin University with structured immersive learning that focused upon sustainable rural community development in the Eastern district of Ampara, Sri Lanka. The focus is on sustainable design education, a theme that is also discussed in Chapter "'Sustainability' After Disaster: Confronting the Complexities of Recovery in the Field: An Educational Experience" that explores some of the challenges involved in designing post disaster.

3 Part II: Case Studies: Sustainable Cities

Urban economies revolve around cities as they have become the key nodes of wealth creation and economic growth. While this is the main reason for attracting people into cities for jobs, the down side is that there is still a huge challenge in maintain low costs of production as it is not inclusive nor sustainable in the long term. This leads to issues of shelter, access to potable water, sanitation and infrastructure challenges, and inequalities particularly among women and children. There are some key indicators of the Asia Pacific region that direct the spotlight to cities. The rise of megacities leads to the development of urban corridors and challenge traditional forms of governance as a result of the size and scale (United Nations ESCAP and UN Habitat 2015).

Just as sustainability has been hard to define, however, so also there is difficulty in defining a sustainable city. It is perhaps impossible to achieve a standard definition for a sustainable city but there may be indicators or measures used to understand and move towards sustainable cities. Cities may be considered to be sustainable when they are able to meet all the needs from the three interdependent pillars constituting sustainable development. The World Bank's Urban and Resilience Management Unit currently defines sustainable cities as "urban communities committed to improving the well-being of their current and future residents, while integrating economic, environmental, and social considerations" (cited in (Hoornweg and Freire 2013, p. 9). A traditional engineers' perspective of sustainability being a state of equilibrium where inputs and outputs balance each other has now been replaced by the Sustainable Development Goals, particularly Goal 11, 'Make cities and human settlements inclusive, safe, resilient and sustainable' (Nations 2015, p. 14), and it directly responds to this challenge of sustainable cities.

From a social perspective, other SDGs are equally relevant. While cities are where most of the Asia-Pacific population resides, a fact remains that cities need the rural hinterland to survive. Goal 1 to end poverty in all its forms everywhere; Goal 2 to end hunger, achieve food security and improved nutrition and promote sustainable agriculture; Goal 3 to ensure healthy lives and promote well being for all ages are all essential to ensure life in cities are built on social, ethical and just foundations.

Goal 4 on ensuring healthy lives and promote well-being for all at all ages; Goal 5 on achieving gender equality and empowering all women and girls'; Goal 6 on ensuring availability and sustainable management of water and sanitation for all'; Goal 7 to 'ensure access to affordable, reliable, sustainable and modern energy for all'; Goal 8 on promoting 'sustained, inclusive and sustainable economic growth, full and productive employment and decent work for all'; Goal 9 to 'build resilient infrastructure, promote inclusive and sustainable industrialization and foster innovation'; Goal 10 in reducing 'inequality within and among countries'; Goal 12 on ensuring sustainable consumption and production patterns'; Goal 13 on taking 'urgent action to combat climate change and its impacts'; Goal 14 to 'conserve and sustainably use the oceans, seas and marine resources for sustainable development';

Goal 15 to 'protect, restore and promote sustainable use of terrestrial ecosystems, sustainably manage forests, combat desertification, and halt and reverse land degradation and halt biodiversity loss'; Goal 16 to 'promote peaceful and inclusive societies for sustainable development, provide access to justice for all and build effective, accountable and inclusive institutions at all levels'; and finally, Goal 17 to 'strengthen the means of implementation and revitalize the Global Partnership for Sustainable Development '(Nations 2015, p. 14) all occur within cities that function every day in close partnership with the social, environmental and economic pillars of sustainability. The elements of Goal 11 focus on housing, infrastructure, inclusive and participatory settlement planning, protecting cultural and natural heritage, preparing and protecting from disasters, reducing the environmental impact, universal access to green and public spaces for women and children; recognising the close relationship between urban, peri-urban and rural areas, supporting the development of integrated policies and plans for mitigation, resilience and adaptation for climate change, particularly for vulnerable populations and locations such as least developed countries (United Nations 2015, pp. 21–22).

These recent SDGs are well aligned with the Melbourne Principles of Urban Sustainability resulting from a 2002 meeting (Hoornweg and Freire 2013, p. 11) focusing on vision, economic and society, biodiversity, ecological footprints, modelling cities on eco systems, sense of place, empowerment, partnerships, sustainable production and consumption, and governance and hope. Since many of the Asia Pacific cities are already in existence, they need to be further developed and governed based on principles of eco urbanism. To understand how to take cities onto a path of sustainable development, it is essential to understand how cities produce, consume and transform materials and energy. They need to understand their base line so that they measure changes against these benchmarks. The principle areas to be considered are urban planning and transport, water and biodiversity, energy and materials, and social-cultural features. An equitable, prosperous and sustainable city needs to embrace renewable energy, be connected and well planed, be smart and age friendly, safe and healthy, have a good mix of compact and mixed use, provide opportunities for life long learning, be resource efficient and productive, be regenerative, be resilient and walkable, low carbon and zero waste, be green and be socially inclusive and affordable (United Nations ESCAP and UN Habitat 2015, p. 152).

This section focuses on nine chapters. Villanueva and Horan's paper focuses on the diversity of eco- developments. This is followed by a focus on three case studies, predominantly in Australia. Wright, Sharpe and Giurco discuss the role of government in sustainability transitions. Vella and Campbell discuss creating public value through a collaborative governance process using a case study of the Bays Precinct in Sydney. Chappell and Campbell's paper focus on the Housing Gap in Sydney, Australia. They advocate the use of simple, yet effective communication method to discuss the affordable housing problem. Following these, are two papers focusing on the Indian context. Benchmarking sustainable smart cities in India by Praharaj, Han and Hawkins, followed by the Role of India's smart cities mission in achieving Goal 11 of the SDGs by Bhatnagar, Nanda, Singh, Upadhyay, Sawhney and Raghu Rama Swamy. Syvla's paper focuses on Sri Lanka and considers a circular economic model for a sustainable city in South Asia. Yung and Yu consider urban regeneration in Hong Kong using a case study and finally, Regidor and Napalang discuss another critical feature of mega cities, sustainable transport in Manila. Villanueva and Horan's paper attempts to formulate a multi-dimensional pathway to develop a framework to guide the diversity in eco-developments. Sustainability assessment tools can then aligned to these frameworks in urban areas to ensure that true sustainability outcomes may be achieved or attempts made to be achieved. Lack of clear definitions as already identified makes it very difficult to analyse sustainability performance. They conclude that both qualitative and quantitative measures of sustainability need to be considered. Wright, Sharpe and Giurco use concepts from innovation and transitions theory in a circular economy approach to understand the role of government in support green outcomes. Traditionally, the private sector, rather than the public sector has the lead in innovation solutions whereas governments tend to focus on regulatory measures. Through the use of a particular case study, the paper highlights the that governments can play a lead role in facilitating and leading sustainability transitions.

Vella and Campbell's paper focuses on an urban renewal strategy for a prime location in Sydney; The Bays Precinct. The paper identifies the difficulties with place based leadership, governance, stakeholder engagement and public value creation. By taking a collaborative and organic approach to place making, an engagement approach was tested that underpinned the use of Sustainable Development Goals. Collaborative governance process as demonstrated through the case study showed that a co-created product emerged, leaving a lasting legacy for such complex urban regeneration projects. Chappell and Campbell touch upon an issue that is becoming a key challenge in the mega cities of this world. They focus on the affordability of housing, using Sydney as a case. They discuss the need to take simple and effective measures to fulfil the government's commitment to affordable housing. They support innovative approaches in the private housing market with innovative design and ingenuity in construction, financing and government support in planning and regulation to support overall housing affordability in the market.

Praharaj, Han and Hawken focus on the Indian context. Their paper on evolving an appropriate indicator system for benchmarking sustainable smart cities in India criticises the government lack of indicators to test the success of recent policy launches on smart cities, urban renewal and heritage development. Their paper focuses on closing the gap between intent and actual performance to meet local and global standards. Their study attempts to assess the potential of 98 cities in India to become sustainable smart cities using indicators that are suited to the Indian context to assess and compare the performance of cities.

Bhatnagar, Nanda, Singh, Upadhyay, Sawhney and Raghu Rama Swamy's paper also focus on the growing trend of megacities and increased urbanisation. Their paper explores India's Smart City Mission with SDG 11 and the NUA. Key factors such as affordable housing and provision of basic infrastructure, sustainable and intelligent transport systems, clean cities and sustainable human settlements, resource conservation and efficiency, integrated rural-urban development approach and a range of other factors have been considered but the primary focus of India's smart cities, they analyse has been on ICT, rather than the broader goals of the SDGs including resilience, climate change and adaptation and use of local materials particularly from a built environment perspective. These and a number of other indicators for SDGs have not been included, demonstrating the need to take a more holistic approach to smart cities planning, design and operation in India. Sylva's approach to sustainability questions the traditional approach of using overlapping circles demonstrated in a venn diagram to a nested approach for sustainability. She uses a Sri Lankan case study to adapt the triple bottom line pillars of sustainability to include the circular economy approach. This allows for preservation of indigenous industries, development of appropriate planning mechanisms and indicators for both feed-in and end of pipe solutions that support sustainable development.

Yung and Yu's paper takes us to Hong Kong, where they discuss about urban regenerative approaches. Heritage conservation is traditionally a contested issue where stakeholders and government can often be at loggerheads with each other. Using the case of the Nga Tsin Wai old village, they demonstrate that strong collaboration between urban planners, conservationists and policy makers can support and present solutions where conservation and regeneration may co exist.

The last paper in this section focuses on the Philippines. Regidor and Napalang focus on ride sharing in Metro Manila. They focus on new and innovative models of the use of ridesharing services through the use of Uber and GrabCar as preferred vehicles for transportation over conventional taxis. Their study demonstrates that these options remove congestion due to private cars and also ease the congestion on traditional transport systems. However, regulatory intervention is needed to protect the rights and well being of commuters.

4 Part III: Case Studies: Sustainable Infrastructure

The context for this book is continuing rapid urbanisation in the developing world. In the Asia and Pacific region urbanisation has been taking place at a particularly quick pace. The built environment is a product of a range of complex flows into and out of our built forms, mostly existing in geographical boundaries as villages, towns and cities. By 2050, the Asia Pacific is expected to house over 50% of the world's population (Hoornweg and Freire 2013). In addition to his, is increased urbanisation. As the United Nations ESCAP and UN Habitat (2015, p. 7) note:

> The speed and scope of urbanisation in Asia and the Pacific is unprecedented. Between 1980 and 2010, the region's cities grew by around one billion people. United Nations projections show they will add another one billion by 2040.

> By 2025, cities in Asia, such as Tokyo, Mumbai, Delhi, Dhaka, Calcutta, Shanghai and Karachi are all expected to house populations close to 20 million and in the case of Mumbai, well over 25 million (Hoornweg and Freire 2013, p. 5).

In addition to rapid growth in this region, the status of the country and the support it gets from international donors is also essential. For example, in this region, at least 10 out of 47 countries or 21% are listed as least developed country (United Nations Committee for Development Policy 2017). From the list of developing countries, 19 countries are in the East and South Asian region (Nations 2012). This forms about 18% of the total number of Developing countries comprising the African, Asian (including West Asia), Latin American and Caribbean countries. Asia includes East, West and South Asian regions. South and East Asia form 59% of the total number of developing countries identified here. Therefore, the impact of research outcomes in the region are significant.

The sustainable development goals or SDGs offer a pathway to sustainable development in the built environment. While there are 17 goals with 169 targets, some of these goals relate specifically to infrastructure. Goal 9: Build resilient infrastructure, promote inclusive and sustainable industrialisation and foster innovation includes the development of quality, reliable, sustainable and resilient infrastructure. This is of particular importance in least developed countries where, by 2030, the goal is to raise industry's share of employment and gross domestic product. Both developing and least developed countries are impacted by this goal. In addition, small island developing states in the Oceania region are also going to be impacted. At least 4 of the least developed country are in the pacific region. This goal also includes enhancing scientific research, upgrading technological capabilities of industrial sectors in all countries, encouraging innovation, and public and private research and development spending (UN 2015, Goal 9.5 a,b,c). Information and communications technology (ICT) is an essential part of the process to promote inclusive and sustainable industrialisation. Goal 11 (1–7) is about making cities and human settlements inclusive, safe, resilient and sustainable (UN 2015). This includes housing and sustainable human settlement planning and management in all countries. It also includes expanding public transport, protecting and safeguarding cultural and natural heritage, reduced disasters including water related disasters by protecting poor and people in vulnerable situations, reducing the environmental impact of cities including the quality of air and waste management, provide universal access to safe, inclusive and accessible green and public spaces, improving triple bottom line (TBL) links between urban, peri-urban and rural areas and increasing the number of cities adopting and implementing integrated policies and plans towards resource efficiency, mitigation and adaptation to climate change, resilience and take a holistic approach to disaster management at all levels. A special mention of least developed countries involves financial and technical assistance in building sustainable and resilient building utilising local materials.

Goal 17 (UN 2015) comprises strengthening the means of implementation and revitalising the global partnership for sustainable development through incorporating finance, technology, capacity building, trade and systemic issues. Finance is about strengthening domestic resource mobilisation, including international support from developed countries to developing and least developed countries; mobilising financial resources from multiple sources such as multi lateral banks and implementing investment promotion regimes for least developed countries. Technology in goal 17 focuses on cooperation at both regional and international levels and access to science, technology, innovation and mechanisms such as diffusion to enhance knowledge sharing. Capacity building is about implementing effective and targeted measures to support national plans for all the SDGs. Trade promotes a universal, rules-based, transparent systems such that exports of developing and least developed countries increase. Systemic issues include policy and institutional coherence, where policies for sustainable development support and respect the establishment and implementation of the goals of sustainable development particularly, Goals 1 and 2. It also includes multi-stakeholder partnerships; data monitoring and accountability, where the former enhances global partnerships for sustainable development through mobilisation, sharing of knowledge, expertise, technology and financial resources, and encourages tri partite public, public-private and civil society partnerships. Data monitoring and accountability build on existing initiatives to develop measurements of progress on sustainable development through capacity building. Capacity building foundations of high quality, timely and reliable data with clear measures and data disaggregated to clearly understand and provide transparency of success at national levels needs to be put in place.

This is not to say that other SDGs are not important or relevant in addressing sustainable infrastructure world wide. Goals 1 focusing on no poverty, Goal 2 on zero hunger, Goal 3 on good health and well being, Goal 4 on quality education, Goal 5 on gender equality, Goal 6 on clear water and sanitation, Goal 7 of affordable and clean energy, Goal 8 on decent work and economic growth, Goal 10 on reduced inequalities, Goal 12 on responsible consumption and production, Goal 13 on climate action, Goal 14 on life below water, Goal 15 focusing on life on land, and Goal 16 on peace justice and strong institutions are all relevant and critical to a collaborative approach to the making of sustainable built environments.

To deal with complex challenges in the Asian region, a number of factors that support development in the region is clearly lacking. Rapidly growing cities need to build smartly and efficiently, to avoid locking in inefficiencies, the impact of which will be felt over the next 80–100 years or more.

The impacts of urban spatial forms, operations and governance of cities, resource use and attendant waste generated, transport and other infrastructure are all important. As Hoornweg and Freire (2013) note, cities must adopt sustainable development policies as soon as possible as they are urbanising rapidly. Added to this is also context based research and increased capacities. Higher education for example, is critical as it provides high level skills and research to apply current technologies and to assimilate, adapt and develop new technologies critical to service the market—the two key drivers for productivity (World Bank 2012). Therefore, these may be considered to be drivers of growth. In countries like China, Cambodia and Vietnam quantity of higher education graduates is still too low. Research and higher education go hand in hand. Without the quest for new knowledge and solutions to support the goals of the SDGs, it will be very difficult to solve our current problems with the same old approaches.

Examining the infrastructure levels of countries in 2009, a global report shows the disparities in the low, middle and high income countries (Hoornweg and Freire

2013, p. 8). As a percentage of the total population, high income countries were 77%, whereas in middle level countries it was 48% and in low income countries, 29%. The greenhouse gas emissions per capita, tonnes/year was 23 in high income countries, 4 and 1 in middle and low income countries. Municipal solid waste generation expressed as kilograms per capital per day was 0.4 in low income countries, 1.1 in middle income and 1.6 in high income countries. Energy consumption in KW-hours was 8 for high income countries, 3 for middle income and 0.9 for low income countries. Road density expressed as km per 1000 people was 2, 3.3 and 14.8 for low income, middle income and high income countries respectively. While paved roads, expressed as percentage of total number of roads in high income countries was 87%, in middle and low income it was 30 and 12% respectively. Percentage of population with access to electricity was 100% in high income countries, whereas in middle and low income it was 73 and 30% respectively. Gross capital formation, expressed as dollar per capital was 8374 in high income country, 1086 in middle income and only 137 in low income countries.

The question therefore, is, can infrastructure and particularly urban infrastructure: land for housing and other amenities, water, transport and energy keep up with this growing trends of urbanisation? and what types of solutions may we seek to find for our current complex problems on infrastructure.

This section focuses on four papers. Orlowski's paper on land and sustainable development and seeks to question current paradigms of sustainable and land development. Bhatt and Negi examine the current challenges of providing solar power projects to support India's quest of meeting international agreements for supporting clean power generation. Rodrigues and Horan's paper is on the role of biochar as a model for sustainable agriculture, due to its agronomic potential for crop yields, organic soil carbon, carbon sequestration and climate change mitigation. Death and Horan's paper on using composted kitchen waste for fresh food production in correctional facilities highlights the need to operationalise closed loop systems, and not just focus on the purely greening of correctional facilities from a design perspective. Orlowski's paper on snychronising sustainable development with land development calls attention to the tensions between true sustainability and land development. He argues there are fundamental epistemological differences between the definitions and notions of sustainability and land development. In his quest for identifying the possible ways in which change needs to be made in modern western societies towards sustainable societies, he argues that an integrated conceptual framework and multi scale perspectives are needed to truly understand the exploration of synergies and trade offs between sustainable development and land development to move towards an ideal sustainable land development.

Bhatt and Negi's paper is on the use of rooftop solar photovoltaics as a means to fast track solar power generation. They argue that this is a better option compared to the use of large utility scale, which requires large tracts of land not easily available in India. Policy measures at the national level need state and local government support to improve solar roof top generation to meet the target of 100 GW by the year 2022. Their paper demonstrates the gaps in policy and actual on ground implementation, and a lack of harmonisation between the states in key areas such as consumer tariffs, market readiness, consumer awareness, lack of subsidies, and most importantly on ground and institutional skills effect increased solar capacities on ground.

Rodrigues and Horan's paper focuses on bio-char, a geo-engineered bi product, similar to charcoal. This product can be used for crop yields, as organic soil carbon, carbon sequestration and therefore, climate change mitigation. International case studies in the use of bio char of cocoa orchards, brussels sprouts and soy beans in different countries and climate zones show the potential for mainstreaming these individual cases. The use of bio char in land management practices in Australia at a farm scale supports opportunities for soil carbon sequestration and aligns with government initiatives such as Australia's emissions reduction fund.

Death and Horan's paper on closed loop food production and greening of correctional facilities highlights the use of composted kitchen waste for fresh food production. Using case studies from the US where substantial food waste disposal savings and further savings on fertilizers and fresh produce as a result of food waste composting processes demonstrate benefits that may be support Australian correctional facilities too. In-house agricultural and horticultural production can significantly decrease the prison network's reliance on outsourcing. It can therefore deliver significant food mile and carbon emission reductions along with triple bottom line sustainability benefits.

5 Part IV: Case Studies: Sustainable Buildings

As already indicated previously, cities are the engines of urbanisation in the current high growth regions of the world. Building clean and efficient cities in the developed and developing world requires increased energy efficiency and promote cleaner sources of energy generation to run cities including transport and buildings. Buildings constitute the largest opportunity to improve demand-side energy efficiency (Hoornweg and Freire 2013, p. 28). If buildings are well built, they will reduce operational energy during their life time, therefore reducing the impact of the building on the environment. If older buildings are adapted for reuse, an opportunity exists in reducing the need to build new buildings. Buildings account for approximately 40% of the world's energy use (UNEP 2009a). If emissions as a result of increased building activity are not arrested, by 2030, emissions from buildings alone may reach 15.6 billion t of CO_2 equivalent by 2030. By 2030, the total emissions from buildings in developing countries in Asia Pacific are expected to surpass the rest of the World (Hoornweg and Freire 2013, p. 43).

In some parts of the world, the life cycle of buildings is very low (United Nations ESCAP and UN Habitat 2015). Buildings are rebuilt every 20 or 30 years and existing heritage structures are razed to make way for huge urban buildings such as shopping centres and the like. This has huge environmental impact as the impact of materials used to build needs to be sourced over short spans of time. If buildings are well designed and built, they can last for longer periods of time reducing their impact on the environment. The building sector has great potential to significantly reduce green house gas emissions and estimated emissions may be cut 30–80% whilst making net profits during the building life span (Hoornweg and Freire 2013, p. 44). Recently, the World Green Building Council reported that buildings are currently responsible for 30% of global greenhouse gas emissions (Council 2016) and offer one of the most cost effective ways to reduce emissions through energy efficiency and the use of renewable energy. Therefore, they call for advancing net zero where all buildings, new and existing are net zero emissions by 2050, meaning that buildings will no longer produce emissions. Most of these emissions are expected to be offset through renewable energy generation on site.

To support green building initiatives, several approaches have been used. Green building codes, particularly in the rapidly developing countries of the Asia Pacific have been the focus. Mandatory building energy audits and use of ESCO-Energy saving companies have been tested in various parts of the world. Appliance standards have been supported energy efficiency in personal and building use such as computers and use of appliances in buildings such as lighting and water efficient appliances. Use of solar hot water heaters etc. and use of photo voltaics support energy feedback into the grid or for on site consumption, but this needs to be backed with policies and incentives to encourage adoption. Use of appliances that also rely on renewables such as solar cook stoves, water heaters for domestic hot water also offer other possibilities that have been explored and need to continue to be explored. As indicated, net zero carbon and net zero energy buildings is gathering momentum worldwide.

But focusing on buildings alone is not enough. A full package of opportunities including low carbon life styles are essential. Consumption and production patterns of the users of the building are just as important as the shell that the building provides. No doubt, the shell is an essential step to ensure that building users are given the most opportunities to ensure they follow consumption patterns that support a low carbon life style. The SDGs are all integrated and indivisible (UN 2015, p. 6). Those goals directly linked to building are: Goal 7 access to affordable, reliable, sustainable and modern energy for all; Goal 9 on building resilient infrastructure, promote inclusive and sustainable industrialisation and foster innovation; Goal 11 on making cities and human settlements inclusive, safe, resilient and sustainable, Goal 12 ensuring sustainable consumption and production patterns; Goal 13 to take urgent action to combat climate change and its impacts and Goal 16 to promote peaceful and inclusive societies for sustainable development, provide access to justice for all and build effective, accountable and inclusive institutions at all levels. Goal 4 on ensuring inclusive and equitable quality education and promote lifelong learning opportunities for all is perhaps one that is not directly linked to building at first glance, but very important nevertheless as we need to bring sustainability education front and centre of current educational discourse in the various disciplines comprising the built environment: architecture, engineering, planning, building, surveying and so on.

This section also has four papers. The first paper by Palliyaguru focuses on extending approaches for sustainable development from urban to the rural context. This is followed by Liew's paper on use of new approaches to technology in building refurbishment in Singapore. Prayitno's paper on green modular concept of a sustainable kampong cityblock in Indonesia uses co-habitation architecture concept. Safarova et al's paper is on the impact of sustainable subdivision design on energy consumption of households in hot and humid tropical climate of Darwin in Australia. The final paper in this section by Law and Dewsbury is on understanding the unintended consequences of building regulations focusing purely on energy efficiency and not taking a more holistic view of the implications of tighter buildings. Law and Dewsbury's paper discuss the implications of focusing purely on energy efficiency as the holy grail to reduce energy use. An unintentional consequence of this leads to unhealthy conditions within the home as a result of moisture trapped in buildings in cool climates. Therefore, better understanding of the impacts of tighter buildings particularly in cold climates and better education.

References

Hoornweg, D., & Freire, M. (2013). *Building sustainability in an urbanising world In Urban development series*. Washington: Urban Development; Resilience Unit World Bank.

UNEP (United Nations Environment Programme). (2009). *Buildings and climate change: Summary for decision-makers*. Paris: UNEP, Sustainable Consumption and Production Branch.

United Nations. (2012). World Economic Situation and Prospects (WESP). Statistical annex.

United Nations. (2015). Sustainable development goals. In resolution adopted by the General Assembly on 25 September 2015, A/RES/70/1. United Nations.

United Nations Committee for Development Policy, Development Policy and Analysis Division, Department of Economic and Social Affairs. (2017). List of Least Developed Countries (as of June 2017). In *United Nations ESCAP and UN Habitat. 2015. The State of Asian and Pacific Cities 2015: Urban transformations, Shifting from quantity to quality*. New York: United Nations Human Settlements Programme (UN-Habitat) and The United Nations Economic and Social Commission for Asia and the Pacific (ESCAP).

United Nations ESCAP and UN Habitat. (2015). *The State of Asian and Pacific Cities 2015: Urban transformations, Shifting from quantity to quality*. New York: United Nations Human Settlements Programme (UN-Habitat) and The United Nations Economic and Social Commission for Asia and the Pacific (ESCAP).

United Nations General Assembly. (2016). *New Urban Agenda*. http://habitat3.org/wp-content/uploads/New-Urban-Agenda-GA-Adopted-68th-Plenary-N1646655-E.pdf. Downloaded April 20, 2017.

World Bank. (2012). *Putting higher education to work: Skills and research for growth in East Asia*. The World Bank East Asia and Pacific Regional Report; 246, Washington, DC.

World Green Building Council. (2016). World GBC's Advancing Net Zero project takes step forward as Australia, Canada, Germany, India & US announce plans to recognise zero emissions buildings. In D. Hoornweg & M. Freire (2013), *Building sustainability in an urbanising world*. In Urban development series. Washington.

Part I
Education for Sustainable Development

Researching Higher Education for Sustainable Development: Plan A, Plan B and Moving Beyond Thought-Limiting Clichés

Kerry Shephard

Abstract This paper explores Higher Education for Sustainable Development from a research perspective, and attempts to distinguish progress based on research outcomes, from that based on concerns, commitments, intended outcomes and prizes. The paper's goal is to encourage ESD practitioners to consider the educational nature of the ESD enterprise, to research student outcomes and to envisage alternative ways forward. The paper starts by exploring the considerable achievements of current HESD, described as Plan A, but postulates the presence of some fundamental misunderstandings that may limit its progress in our current higher education systems. An alternative, or contingency, Plan B, is described, that has at its core the aim of enlisting the goodwill of everyone in higher education to do what they want to do, well, rather than to do what they may not want to do. Plan B needs university teachers to support their students towards critical and independent thinking and for higher education institutions to research their progress towards any change in student outcome that they have in mind.

Keywords ESD research · Student learning outcomes · Thought-limiting clichés Researching change · ESD Plan B

1 Introduction

University researchers involved in higher education for sustainable development (HESD[1]) are not necessarily of the 'glass half full' disposition. In general we are concerned that our planet has problems that are mostly produced by people and that,

[1]HESD is used in this paper where the education involved is specifically at degree-level or above. ESD is used more generically as applicable to all education for sustainable development.

K. Shephard (✉)
Higher Education Development Centre, University of Otago,
PO Box 56, Dunedin 9054, New Zealand
e-mail: kerry.shephard@otago.ac.nz

W. Leal Filho et al. (eds.), *Sustainable Development Research in the Asia-Pacific Region*, World Sustainability Series, https://doi.org/10.1007/978-3-319-73293-0_2

for the most part, the people most implicated and most able to do something about it, are those educated by universities. Although we are quite possibly optimistic at heart about much in life, we are also realists and know that solving these problems will require universities to change what they do and how they do what they do (along with much else). After all, we have been working at these issues for quite a few years, as environmental educators for much of the last century, but more collaboratively since the Brundtland report in 1987 (World Commission on Environment and Development 1987) and Rio de Janeiro Summit in 1992 (United Nations Conference on Environment and Development 1992) forced the world to link our environmental concerns more visibly to concerns about people, and cultures, and economies. And we have recently emerged from a bruising decade that focused all that we could muster on education for sustainable development. With respect to student learning in our universities, we have every reason to be wary about too much optimism for dramatic change any time soon.

University researchers involved in higher education for sustainable development are also, by and large, committed to the principle that finding our way through a complex, even wicked, dilemma will need more than stout hearts and strong rhetoric. We need to harness the power and ethos of modern research processes to understand our educational problems and to find research-based educational solutions for them. So, we should first explore the current state of HESD in our institutions and ask ourselves how well this enterprise is working. If we are concerned, then perhaps we should ask some tough questions about Plan A, and consider an alternative, or contingency, Plan B.

2 Plan A

In some respects the essence of Plan A is defined by the broadly based intentions of education for sustainable development. We could, for example, look to Agenda 21 for some key objectives. Agenda 21 suggests, for example, that education is critical for achieving, in general rather than for a sub-set of sustainability-focussed students;

> … environmental and ethical awareness, values and attitudes, skills and behaviour consistent with sustainable development and for effective public participation in decision-making (Agenda 21; United Nations Conference on Environment and Development 1992, Chapter 36, 2)

Similar aspirations have permeated more recent international discourse on change. The European Union, for example, also talks about changing attitudes and concomitant behavioural changes:

> In a continuously changing world, all European citizens should be equipped with the knowledge, skills and attitudes needed to understand and deal with the challenges and complexities of modern day life, whilst taking due account of the environmental, social,

> cultural and economic implications, as well as to assume their global responsibilities. (Council of the European Union 2010, 18).

Similar assertions have become apparent in recent higher education frameworks that seek to take such high-level propositions and to put them into a higher-education teaching and learning context. See, for example, the Future Fit Framework:

> The student learning experience is key: in essence, it is about the full engagement of the learners, including affective, cognitive and active dimensions. (Sterling 2011, 37)

Similarly, broad changes in learners' values and behaviour are characteristic of the kinds of changes expected and promised by those institutions of higher education, or their leaders, actively engaged in HESD. Signatories to the Talloires Declaration, for example, agree to 'Educate for Environmentally Responsible Citizenship' (AULSF 1994). In the USA, the STARS programme (Sustainability Tracking, Assessment & Rating System) of the AASHE (The Association for the Advancement of Sustainability in Higher Education 2017), provides institutions with the means to describe the culture of their institution and awards points to those that claim to promote cultures that align with their sustainability mission.

> Institution conducts an assessment of campus sustainability culture. The cultural assessment focuses on sustainability values, behaviors and beliefs, and may also address awareness of campus sustainability initiatives. (AASHE 2017)

Let there be no doubt that Plan A is creating opportunities for positive, sustainability-focussed change in student learning. The message has in some ways become a movement. Many academics worldwide research in the area and write and speak about their successes in print and at conferences. An HESD field of enquiry has certainly developed, as evidenced by many scholarly journals taking this theme as a substantial element (Environmental Education Research and the Journal of Environmental Education provide two notable examples) or as their main reason for being (for example, the International Journal of Sustainability in Higher Education). Some claim a new discipline of sustainability science is evolving of which ESD is a major component (Barth 2015). Representatives of nearly 500 institutions have signed the Talloires Declaration (AULSF 2016) and other high-level declarations with numerous institutional signatures have been initiated. The STARS webpage (AASHE 2017) is now populated by more than 800 institutions worldwide, many of which have documented their institutional progress towards sustainability. In the UK, People and Planet (2017) have created a league table of higher education institutions, ranking them on sustainability measures. In Europe, substantial effort was focused on researching and supporting the professional development of university academics (UE4SD 2016a). As just one example of many from Australia, Australasian Campuses Towards Sustainability presents annual Green Gown Awards to those institutions taking a proactive approach towards sustainability (ACTS 2016). Progressively, academics are proposing, debating and agreeing the nature of the competencies and capabilities that should result from HESD (see for example Wiek et al. 2011). Plan A is clearly alive in many of our higher education institutions. But is it working and how would we know if it was not?

For a while, and mindful of the challenges involved in demonstrating actual change in student learning, much emphasis was initially placed on indicators of change. See, as examples, Aries (2006) and UNESCO (2007). The latter reference identified a range of useful indicators; mostly relating to inputs into ESD (such as '% of new teachers currently receiving pre-service training in ESD') but also included educational outcomes (such as 'Learners use sustainable practices in daily life'). Closely related to the use of indicators has been the development of funded research projects charged with the responsibility to not only initiate change in universities, through, for example, professional development for university teachers, but also the monitoring of these changes and their consequences. The University Educators for Sustainable Development (UE4SD) project for example, produced a final report documenting extensive development within universities throughout Europe (UE4SD 2016b). One highlight is that there appears to be a substantial student-led push for the promotion of ESD agendas. A special issue of the journal Assessment and Evaluation in Higher Education focused on the broad status of sustainable development in the world's universities and how to assess or evaluate its progress (Shiel et al. 2015).

It is noticeable however, that there are few research-based reports that actually document changes in student perspectives that may or may not result from these extensive developments within institutions. Students come and go, by and large they pass their exams, but so far we have not, in general, managed to monitor, in a research-based manner, substantial changes in student perspectives about sustainability or about their place in the world or in their behaviour after graduation. Indeed, this is a common theme. Substantial changes appear to be occurring within higher education institutions and many commitments are made, but so far we have relatively sparse research evidence that it is having the effect on student learning that we hope it will. There is, of course, some evidence. The highest scoring institution in 2015 in the USA STARS programme was Colorado State University. The institution earned the platinum rating in their category of the programme, scoring 83.5 from a possible 100. The institution scored 36.92/40 in the curriculum section; an important section contributing 40% of the total points possible. Within this section, 4 points accrue for category 'AC-6: Sustainability Literacy Assessment', and Colorado State earned the full 4 points, substantially supported by PhD research. Their score-card is available as is a copy of the survey used and the scoring process adopted by AASHE (2017). Their data suggests that their students know lots about sustainability issues, that they are actually quite interested in sustainability and even that they may be more interested in sustainability having spent time at the University, than they were when they came. (There is not enough information provided about the research methodology to be clear about the research embedded within this data. It is in my nature to be rather pedantic about these things. I do need to know, for example, if the students were anonymous within the survey, as this would potentially influence their responses). Nevertheless this is an important indicator of change. My own research in New Zealand (working with the

University of Otago Education for Sustainability Research Group) does use an anonymous attitude scale to follow students as they progress through their higher education in a range of degree programmes in a longitudinal research project (see for example Shephard et al. 2015). Although we repeatedly measure substantial differences in the environmental attitudes of cohorts of students registered for different degree programmes, we have yet to find substantial changes in these environmental attitudes as our students proceed through their studies. Our data suggest that our students arrive at our institution with a particular set of environmental attitudes and they tend to graduate some years later unscathed by their experiences with us. At least one other research group's approach to monitor changes in students' sustainability attributes is notable (Teisl et al. 2011). If others have researched student outcomes that are altogether more positive with respect to HESD Plan A, our field of enquiry does need to hear about them.

There are other indications that the changes occurring within higher education are not necessarily occurring as rapidly as we may like. Only one university in New Zealand has signed the Talloires Declaration and it seems reasonable to suggest that the majority of universities in the Asian Pacific region are a long way from having carbon-neutral campuses (although notably, as I was writing this paper the University of Melbourne announced that it hoped that it will be by 2030; University of Melbourne 2017). And further afield, I note that although Colorado State University did well in its curriculum rating it did not do so well in other STARS categories, earning, for example, just 1.84/10 for energy. At the end of an extensive project supporting the professional development of university teachers towards ESD in Europe, the UE4SD project reports that; "*ESD is growing fast but is not yet common practice in HE or in the Professional Development of university educators.*" (UE4SD 2016a).

There are, of course, also indications that the world outside of higher education is not changing much either, yet. On a recent visit to Australia I returned dismayed that in a country with so much sunlight, and tertiary education, and so little freshwater, desalination was substantially powered by coal. At home, I am disappointed by New Zealand's limited commitment to emissions control. Further out in the Asia-Pacific area, China is busy importing the coal that Australia is not using for its own energy production. And as I write, Japan's Fukushima nuclear power station is still polluting the Pacific Ocean for all of us. Whatever influence HESD may be having in the Asia-Pacific Region, it is not particularly apparent to me.

And, perhaps, there are indications that our international aspirations for education are also declining. The high hopes for educational outcomes explicit in Agenda 21 are softened somewhat into educational opportunities in the more recent UNESCO Global Action Programme (GAP) on Education for Sustainable Development for some key objectives. The GAP suggests, for example, the need;

> … to reorient education and learning so that everyone has the opportunity to acquire the knowledge, skills, values and attitudes that empower them to contribute to sustainable development (UNESCO 2016)

Rather than pronounce Plan A a failure, naturally it is important to note that changes of the order of the HESD/ESD agenda could not possibly occur quickly. Perhaps with time we shall overcome the barriers to change that occur within higher education and all will be well. In the meantime, just so that we do not waste time waiting, I do think it important to address why Plan A may not be working; not by pursuing the usual barrier analyses, but by asking if there is something fundamentally amiss, educationally speaking, within Plan A approaches.

3 Possible Problems with Plan A

It appears indisputable nowadays that fundamentally ESD is a quest for affective learning outcomes (such as, for example, a willingness to adopt sustainability practices and an emotional aptitude to care), albeit linked to cognitive outcomes (such as knowing how to adopt sustainability practices and how to translate caring into positive actions), rather than cognitive outcomes on their own. This context was well understood in the 1980s as part of the extensive discourse around environmental education, especially in school-based education (see for example Caduto 1983), but it was somewhat overlooked in later years as sustainability developed as a concept and as ESD discussions migrated into higher education (Shephard 2008). Unfortunately, and a significant element of the argument being made here, decisions and behaviours based on emotion tend to be absolute and purposeful rather than partial or neutral. As Fien (1997) points out in relation to this analysis "… the key issue for educators concerned with questions of values and ethics in education should not be to check whether a particular approach to teaching is indoctrination but to ask questions related to the ways, and in accordance with, what values and ends, should schools and teachers 'indoctrinate'." (438). Fien (1997) with others, makes a strong argument that education cannot be neutral. If it does not actively promote change, it is de facto supporting the existing social order.

I see this argument as a fundamental underpinning of current ESD philosophies (as exemplified for example by Agenda 21 and the Future Fit Framework referenced above) but I also see problems in the argument. In some respects it is an 'All or nothing argument' along the lines of 'You are either with us or against us'. Such arguments, particularly those that are affective at heart, are variously described as thought-stoppers, or as thought-limiting or thought-blocking clichés. 'It is simply imperative that learners learn sustainability values and attitudes, and as teachers teach learners it is the teacher's job to achieve this'. Chiras (1992) suggested that "*Thought stoppers elicit an emotional, or gut-level acceptance of an argument, not an intellectual one.*"(467). In some respects, the mission of HESD is to ask university teachers to suspend their own critical judgments and to enact the mission of others. You are either with us or you are against us. Argument over.

I don't think the argument is over for school-based education and I certainly know it's not over for higher education. In my experience, many higher education teachers have responded quite predictably, in choosing to be 'against ESD'.

Many do not identify their role with indoctrination (see for example arguments by Jickling 2009, 2016) and others doubt the sustainability base of either their university role or their personal lifestyle. [In my case, my own university has been burning lignite to heat its buildings and I am looking forward, albeit with a guilty conscience, to flying to Melbourne to deliver this paper].

With respect to my own conceptualization of the viewpoint of proponents of ESD Plan A, I am neither with you nor against you. This analysis suggests to me that Plan A, based essentially on persuading university teachers with diverse sustainability-related values to adopt a particular sustainability-related values-set always was doomed to failure. But I admit that it took me quite a while, and quite a lot of research and thinking, to understand why. The next paragraphs explain Plan B, by detailing my own research journey.

4 Plan B

Plan B starts by asking university teachers what they think about education for sustainable development and their role with respect to their conceptualization of it. Researchers will understand, of course, that the research involved in this task didn't simply 'ask'. Prior research had demonstrated that multiple issues converged around these questions and that it was highly likely that simple questions would elicit diverse responses. Q methodology is a research approach well-suited to determining common patterns of understanding about complex issues-based problems; and we used this approach to learn more about our colleague's viewpoints on HESD (Shephard and Furnari 2013).

We developed 50 statements about the issues involved in HESD and asked 43 university teachers in our own institution to rank them and to record their responses to written questions. Our analysis confirmed four significantly and qualitatively different viewpoints, only one of which actively advocates for sustainability and for education for sustainability. The other three viewpoints all had distinct characteristics that prevent those who own them from using their position within the university to encourage students to act sustainably. Our research did not enable us to say how these four common viewpoint combinations quantitatively distributed amongst the University's teachers, but did help us to understand why some university teachers are willing to use their position to encourage sustainability (along the lines of Plan A) while some were not. We were particularly interested to discover that while many of our research subjects did not wish to formally 'educate for sustainability' themselves, none of them objected to others doing it. They just didn't feel that they wanted to do it, or were able to do it, themselves. A respect for the academic freedom of academic colleagues was likely a major contributor to this perspective. It was also a major element within the mindset of one of the groups that did not advocate for sustainability. That group was highly sustainably-minded themselves. They explored societal values and ethics in their teaching and they contextualised their disciplinary content, perhaps even in interdisciplinary contexts.

But they did much of this within the bounds of neutrality, substantially as discounted as educationally unreasonable, by Fien (1997), in the context of 'learning to care'.

At the time I remember being both encouraged and discouraged by this research data. Encouraged because it helped me to understand better the perspectives of those academic colleagues who were not willing or able to 'green the curriculum' or in other ways advance the causes of HESD. But also discouraged, because it appeared to me at the time to make the task of HESD all the harder. I remember being confronted by combinations of obvious climate-change-denial alongside more philosophically complete conceptualisations representative of highly contrasting ontological and epistemological stances. Here were colleagues, in a range of disciplines, who in particular reacted against what they thought was, in the form of HESD, a highly objective ontology, representing a one-truth realism perspective, alongside a intensely positivist epistemology, providing only one way forwards to environmental and economic stability, and social justice. But there were also colleagues who held almost identical views about the nature of current higher education and its role in maintaining a socially inequitable and environmentally damaging status quo, and their, often unwilling, roles within it.

I remember wondering at the time how Plan A could possibly work within the higher education that I had come to know over many years. Plan A rhetoric that blames higher education for the woes the world and labels university teachers with alternative point of view as 'unwilling to change' seemed, at the time, unlikely to provide a fruitful future strategy.

In 2013 I took a sabbatical year to research my disaffection with the current state of HESD as I understood it. I visited higher education institutions in Australia, Europe and the USA. I employed a constructivist grounded theory research approach (Charmaz 2006). The research attempted a systematic discovery of theory from research data (a grounded theory) with the aim of generating conceptualisations that explain the way the people managed to, or attempt to, resolve the central concerns about ESD and HESD. My data consisted primarily of the viewpoints and stated experiences of university people in a range of formal interviews, group interviews and group discussions. An important element of the research was the constant comparative approach (Corbin and Strauss 2008) which accepts that the researcher's developing understanding is essentially reflexive in nature, incorporating the researcher's own background, situation and development, and operating reflectively and iteratively as new data become available. The research and the resulting model of change were published as a book (Shephard 2015).

My research yielded a discrete number of interacting facets (more technically an integrated set of conceptual hypotheses developed from empirical data) that together created a grounded theory of how sustainability education exists at present in higher education and how it might develop while maintaining the liberal traditions of higher education. Two underpinning liberal ideas or values contributed to this grounded theory. The first focused on minimising the prescription of what university teachers should or should not do by encouraging them to use their academic freedom to decide what to teach and how to teach it. The second promoted the ideal that higher education should be liberating for students in that it should help them find their place in society and not force them into some form of predesigned pattern or mould.

- The first facet that initially dominated my research data was that of barriers to HESD. Barriers were described everywhere and the most pressing for many contributors to my research data were not necessarily system- or financial-barriers, but people who stopped things happening. This first facet interacted strongly with the second; that advocates for sustainability in general did not identify changes that they would make themselves, to their teaching, but called for others to transform. Those with the greatest stake in HESD, particularly those who had academic commitments in the form of research and publishing, were most vocal in calling for others to change their practices in a particular HESD, Plan A direction.
- Next came the discovery about the extent to which hidden curricula dominate many aspects of higher education. Hidden curricula were most obvious when promoted by those advocates for sustainability who were unhappy to, unwilling to, or nervous to tell the world about their sustainability-focused passion. They taught sustainability, but in a hidden context. But my research also noted that hidden curricula were much less obvious when what was hidden seems normal, straightforward or acceptable to society. In particular, many academics teach topics and subjects as if they have no context in the modern world and indeed take a pride in de-contextualising their subject matter. Hidden curricula are not easy to identify and particular skills appear to be necessary to find them. With these skills, hidden curricula can be found almost everywhere in higher education.
- Identifying the means whereby hidden curricula may be identified proved a major breakthrough in this research. For a higher education to be truly liberating for students it must help them to be able to recognise what university teachers are actually teaching them, rather than what they say they teach them. Those who can think critically automatically challenge the assumptions that underpin the statements made by their teachers. Teaching students to be critical thinkers is the most important element of Plan B. Helping university teachers to understand what critical thinking actually is, rather than what they think it is, may be similarly important. My research demonstrated unequivocally to me that most university teachers are passionate in their expectation that their teaching will promote their students to think critically, but that they had diverse views on what thinking critically actually entailed.
- Most universities that I visited have academic or support staff whose key role is to support university teachers to learn how to teach. These people (often identified as academic developers) have a pivotal position in Plan B. Without them, university teachers might continue to misunderstand the nature of critical thinking. My research suggests that, in addition, many university teachers, including many advocates for sustainability, have little personal appreciation of educational theory. Even those who have studied higher education teaching and

learning in a formal programme have often focused on learning in the cognitive domain and know very little about, or have very appreciation of learning in the affective domain.

- Policymakers and administrators and academic leaders also have a pivotal role in Plan B; not necessarily in providing academic leadership, but rather in emphasising and making use of their connectedness within higher education institutions. Change in higher education requires functioning networks. Policymakers and administrators and academic leaders also appear to me to have a passion for evaluating, or assessing the 'quality' of teaching (at least far more so than most university teachers).
- Several other facets came to light in my research. I was deeply impressed by the interdisciplinary focus of some university teachers and it seemed to me that a high proportion of the most interdisciplinary were also advocates for sustainability. Similarly, those university teachers who were voluntarily engaged in their wider community, particularly in a learning and teaching context, appeared to me to be more likely to be advocates for sustainability.
- My last research facet proved to be challenging for me as I developed my grounded theory. I encountered some wonderful role models for HESD and I wondered if these university teachers ought to have a particular place within my grounded theory. However, I also encountered some amazing university teachers who identified with other transformational changes in higher education and who specifically did not advocate for sustainability.

Once the link had been made between critical thinking and the ability to identify hidden meaning in what university teachers tell students (something made possible only by extensive conversation and discussion with academic colleagues during my travels, although it seems so obvious now) the grounded theory came together readily. The grounded theory evolved by setting aside the barriers that advocates for sustainability so readily identified as stopping them doing what they wanted to do and by setting aside any suggestions about what others should be doing rather than what individuals should be doing themselves. The theory took the two core variables of the hidden curriculum and of the critical thinking needed to decipher hidden meanings and made links with the remaining facets identified within the research. The grounded theory, in essence Plan B, suggests that by changing the HESD mission *from* prescribing the sustainability-specific learning outcomes of teaching and learning *to* promoting, and evaluating, quality teaching and learning in higher education, universities and university teachers would find it more straightforward to work towards outcomes compatible with HESD missions. Put simply, university teachers should teach passionately according to their own conscience, but in the process focus on helping their students become independent and critical thinkers. Academic developers or other higher education specialists should help university teachers better understand both how to encourage independent and critical thinking, and higher education pedagogy. University administrators and policy makers should support all university teachers and academic developers do what they want to do, but ensure that change is monitored and assessed or evaluated. In the process university administrators and policy makers will need to work hard to remove restrictions within higher education that limit how university teachers teach and what students are enabled to learn. This research suggests that, in addition, means to encourage multidisciplinary study and community engagement will need to be found, where these are lacking. Above all Plan B is all about enlisting the goodwill of everyone in higher education to do what they want to do, well, rather than to do what they do not want to do, and probably could not do well anyhow.

5 Putting Plan B to the Test

Perhaps Plan B is simply a derivation from Plan A. Perhaps, rather than an alternative, it is indeed a contingency, while we are waiting for Plan A to work. I haven't tried Plan B. Nor to be honest do I think that one researcher could possibly 'try' Plan B. Plan B would for, example, require an institution, or a field of enquiry, or a movement, to redefine its objectives. For as long as the direct objectives of HESD focus on what future generations of students will or will not value, and appreciate, and how they will behave, we are stuck in Plan A mode. We may hope that in future our graduates will make the choices that we currently think are the right choices, but our objectives, under Plan B, should be aimed at facilitating their ability to understand these choices and protecting their freedom to make their own choices. Although Plan B protects the academic freedom of individual university teachers, for the system as a whole it provides no more rights or privileges than to be interested in the choices that our graduates will make.

Clearly Plan B is based on limited evidence and to be useful it needs more research to probe its potential. Some research-based approaches that might help to situate Plan B within our field of enquiry include the following.

1. Evaluation of the impact of change strategies is an essential part of the change process and central to the ethos of a research-based exploration of HESD. It is notable that '*We cannot improve at scale what we cannot measure*' is one of six core principles for educational improvement adopted by the Carnegie Foundation (2017). Evaluating the progress of HESD, while respecting the rights of our students and graduates, allows us, and perhaps obliges us, to be interested in the sustainability-related choices that they make. Within the construct of evaluation, however, we also need to be mindful that, in the wrong hands, evaluation can be seen as managerial or promotional in nature rather than as a research tool. Social and political studies have extensively debated the nature of evaluation and its attempts to be fair and rational. House (1996) in particular identified the potential for evaluation to engage with social justice agendas in hiding or exposing biases and inherent values in evaluative work. The anonymity of subjects in any research-based exploration provides essential

protection of students' and graduates' rights to make their own choices and should become integral to any serious research that we do.

2. Those who research student learning in the context of HESD understand how limiting the research tools at our disposal are proving. Researching changes in values and attitudes, and in constructs such as 'environmental concern', have proved challenging. Although literally hundreds of tools have been developed (Dunlap and Jones 2002) one in particular is most widely used (the NEP or revised New Ecological Paradigm scale). Although deficient in many respects, it has been identified as at present the best available (Hawcroft and Milfont 2010) but surely we can do better? Similarly, clarifying the nature of critical thinking and measuring its development is proving highly challenging for higher education in general. Overall it seems fair to say that Plan B needs better research tools than we have at present if it is to succeed.
3. It appears to me imperative that those who research HESD do need to find some common ground in defining key terms and concepts relevant to our goals. For example, much of what we are researching has been described as 'democratic'; but whether HESD is in fact 'democratic' is contingent on much. My own research suggests that HESD practitioners use the term in a variety of different ways (Shephard and Brown 2016). Similarly, terms such as competency, capability and action-competence are widely used in our field of enquiry, but poorly defined and variously understood.
4. My own research suggests to me that transmission teaching is alive and well throughout higher education and that the broad and progressive construct of student centeredness has passed much of higher education by. It appears to me that for as long as university teachers feel that they can fulfill their role by simply transmitting knowledge, or values, to their students then neither plan A nor Plan B can possibly proceed. For me, a major impediment to any form of HESD is the limited impact that the educational movement of SOTL (the scholarship of teaching and learning) derived substantially from the work of Boyer (1996) is having in higher education. By researching SOTL, we promote an understanding of higher education pedagogy and provide a chance for Plan B and HESD to flourish. Higher education for sustainable development, or higher education for sustainability is, after all, essentially about education.

References

ACTS. (2016). *2016 winners*. http://www.acts.asn.au/initiatives/ggaa/2016ggaa/2016-winners/. Last accessed February 9, 2017.

Aries. (2006). Developing ESD indicators to assess progress during the decade of education for sustainable development. http://aries.mq.edu.au/projects/esdIndicators/. Last accessed February 13, 2017.

Association for the Advancement of Sustainability in Higher Education (AASHE). (2017). The sustainability tracking, assessment & rating system. https://stars.aashe.org/institutions/participants-and-reports/. Last accessed February 13, 2017.

Association of University Leaders for a Sustainable Future. (1994). The talloires declaration 10 point action plan, updated version. www.ulsf.org/pdf/td.pdf. Last accessed February 9, 2017).

AULSF. (2016). Talloires declaration, institutional signatory list. Retrieved February 9, 2017 from http://www.ulsf.org/programs_talloires_signatories.html.

Barth, M. (2015). *Implementing sustainability in higher education: Learning in an age of transformation*. London; New York: Routledge.

Boyer, E. L. (1996). From scholarship reconsidered to scholarship assessed. *Quest, 48*(2), 129–139. https://doi.org/10.1080/00336297.1996.10484184.

Caduto, M. (1983). A review of environmental values education. *Journal of environmental education, 14*(3), 13–21.

Carnegie Foundation. (2017). The six core principles of improvement. https://www.carnegiefoundation.org/our-ideas/six-core-principles-improvement/. Last accessed February 13, 2017.

Charmaz, K. (2006). *Constructing grounded theory*. London: Sage.

Chiras, D. D. (1992). Teaching critical thinking skills in the biology & environmental science classrooms. *The American Biology Teacher, 54*(8), 464–468.

Corbin, J., & Strauss, A. (2008). Basics of qualitative research: Techniques and procedures for developing grounded theory (3rd ed.). Thousand Oaks.

Council of the European Union. (2010). Council conclusions on education for sustainable development. 3,046th Education, youth, culture and sport Council meeting, Brussels, 18 and 19 November. http://www.consilium.europa.eu/uedocs/cms_data/docs/pressdata/en/educ/117839.pdf. Last accessed March 20, 2014.

Dunlap, R. E., & Jones, R. E. (2002). Environmental concern: Conceptual and measurement issues. In R. Dunlap & W. Michelson (Eds.), *Handbook of environmental sociology* (pp. 482–524). Westport, CT: Greenwood Press.

Fien, J. (1997). Learning to care: a focus for values in health and environmental education. *Health Education Research, 12*(4), 437–447.

Hawcroft, L. J., & Milfont, T. (2010). The use (and abuse) of the new environmental paradigm scale over the last 30 years: A meta-analysis. *Journal of Environmental Psychology, 30*, 143–158.

House, E. (1996). The problem of values in evaluation. *Evaluation Journal of Australasia, 8*(1), 3–14.

Jickling, B. (2009). Environmental education research: To what ends? *Environmental Education Research, 15*(2), 209–216. https://doi.org/10.1080/13504620902770345.

Jickling, B. (2016). Losing traction and the art of slip-sliding away: Or, getting over education for sustainable development. *The Journal of Environmental Education, 47*(2), 128–138. https://doi.org/10.1080/00958964.2015.1080653.

People and Planet. (2017). How sustainable is your university? https://peopleandplanet.org/university-league. Last accessed February 9, 2017.

Shephard, K. (2008). Higher education for sustainability: Seeking affective learning outcomes. *International Journal of Sustainability in Higher Education, 9*(1), 87–98. https://doi.org/10.1108/14676370810842201.

Shephard, K. (2015). *Higher education for sustainable development*. London: Palgrave Macmillan UK. https://doi.org/10.1057/9781137548412.

Shephard, K., & Brown, K. (2016). How democratic is higher education for sustainable development? In *Discourse: studies in the cultural politics of education*, (in press, pp. 1–13). https://doi.org/10.1080/01596306.2016.1150254.

Shephard, K., & Furnari, M. (2013). Exploring what university teachers think about education for sustainability. *Studies in Higher Education, 38*(10), 1–14. https://doi.org/10.1080/03075079.2011.644784.

Shephard, K., Harraway, J., Lovelock, B., Mirosa, M., Skeaff, S., Slooten, L., & Deaker, L. (2015). Seeking learning outcomes appropriate for 'education for sustainable development' and for higher education. *Assessment & Evaluation in Higher Education, 40*(6), 855–866. https://doi.org/10.1080/02602938.2015.1009871.

Shiel, C., Leal Filho, W., do Paço, A., & Brandli, L. (2015). Assessing and evaluating sustainable development in higher education. *Assessment & Evaluation in Higher Education, 40*(6), 783–784. https://doi.org/10.1080/02602938.2015.1073028.

Sterling, S. (2011). The future-fit framework. http://www.heacademy.ac.uk/education-for-sustainable-development. Last accessed November 21, 2013.

Teisl, M., Anderson, M., Noblet, C., Criner, G., Rubin, J., & Dalton, T. (2011). Are environmental professors unbalanced? Evidence from the field. *The Journal of Environmental Education, 42*(2), 67–83. https://doi.org/10.1080/00958961003705899.

UE4SD. (2016a). University Educators for Sustainable Development project. Home. http://www.ue4sd.euUS4SD. Last accessed February 13, 2017.

UE4SD. (2016b). UE4SD Innovation report – Year 3 2015 – 2016 September. http://www.ue4sd.eu/images/2016/UE4SD_Innovation_Report_2015-16_final.pdf. Last accessed February 13, 2017.

UNESCO. (2007). Monitoring and assessing progress during the UNDESD in the Asia-Pacific region: A quick guide to developing national ESD indicators. http://unesdoc.unesco.org/images/0015/001553/155304e.pdf. Last accessed February 13, 2017.

UNESCO. (2016). *Goals and objectives - global action programme.* Retrieved February 15, 2017 from http://en.unesco.org/gap/goals-and-objectives.

United Nations Conference on Environment and Development. (1992). Agenda 21. http://www.un.org/geninfo/bp/enviro.html. Last accessed February 13, 2017.

University of Melbourne. (2017). Sustainability plan 2017–2020. https://ourcampus.unimelb.edu.au/application/files/2914/8480/0942/UoM_Sustainability_Plan_2017-2020_40pp.pdf. Last accessed February 9, 2017.

Wiek, A., Withycombe, L., & Redman, C. L. (2011). Key competencies in sustainability: A reference framework for academic program development. *Sustainability Science, 6*(2), 203–218. https://doi.org/10.1007/s11625-011-0132-6.

World Commission on Environment and Development. (1987). *Brundtland report: Our common future.* http://www.un-documents.net/our-common-future.pdf. Last accessed February 14, 2017.

Resetting the Compass: Principles for Responsible Urban Built Environment Education (PRUE)

Usha Iyer-Raniga

Abstract Several international frameworks supporting sustainable development have come into existence since the Rio Conference in 1992. While recent frameworks developed have included education as a key component of sustainability, these frameworks do not offer a clear direction for university educators to embrace sustainability in their curriculum, and wider engagement with co academic and non academic colleagues, student, campus and wider community to work collaboratively. The principal aim of the paper is to develop and support a collaborative networked platform for university educators, particularly in the Asia Pacific to support and enhance their capacities to develop and promote sustainability education. Despite a number of international treaties to support and accelerate global trends towards sustainable development, we are still a long way from living in a low carbon world. Secondary literature review is undertaken to understand the gaps in current treaties to support a way forward to move towards sustainable futures from an educational perspective. The literature demonstrates there are gaps between intent and action. However, there is a long road to be traversed to translate ideas into reality and even more to mainstream industry knowledge, curricula and capacities to support sustainability outcomes in the built environment. There is still a gap between action and intent with regard to achieving sustainability outcomes in the built environment. These gaps are in real danger of becoming even more wider and deeper, and more urgent and critical as we fail to meet our international targets with regard to low carbon futures. PRME: Principles for Responsible Management Education translated to the built environment as PRUE: Principles for Responsible Urban built environment Education offers an opportunity for supporting sustainability into built environment programs in universities.

Keywords Education · University · Built environment · Programs
Sustainable development · Asia Pacific

U. Iyer-Raniga (✉)
School of Property, Construction and Project Management,
GPO Box 2476, Melbourne, VIC 3000, Australia
e-mail: Usha.Iyer-Raniga@rmit.edu.au

W. Leal Filho et al. (eds.), *Sustainable Development Research in the Asia-Pacific Region*, World Sustainability Series, https://doi.org/10.1007/978-3-319-73293-0_3

1 Introduction and Context

A number of recent discussions and targets set at global levels have demonstrated the importance of sustainable development and reaffirmed the importance of concerted efforts and actions for ensuring a way forward for future generations. There is an urgent need to rethink and perhaps arrive at 'out of the box' solutions that continue to use the planet as a home for current and future generations. The role of education is quite critical in driving awareness and change to ensure that individuals and societies are equipped and empowered with the knowledge, skills and values for a sustainable today and tomorrow.

Education for sustainability is a life long process, with the purpose of transforming society. It is a journey where learners are empowered to take informed decisions and a deep seated responsibility for environmental integrity, economic viability and a society that is just and peaceful for current and future generations. Cultural diversity is also celebrated in education for sustainability.

Education for sustainable development (ESD) has four main dimensions (UNESCO 2014a; United Nations 2016):

- Learning content, which informs the curriculum.
- Pedagogy and learning environments, where learners are inspired to act for sustainability. It impinges on action oriented and transformative learning where physical and virtual environments may be used in a learner-centred manner.
- Learning outcomes, focusing on current and future generations where stimulating learning and promoting core competencies is central.
- Societal transformation, empowering learners to transform themselves through enabling a transition to green economies and societies, and empowering people to be global citizens to create a more just, peaceful, tolerant, inclusive, secure and sustainable world.

A fifth dimension also needs to be added. This dimension is about bringing communities, government, public and private organisations to work with educational institutions, particularly universities to engage with the learning content for sustainable development. By adopting this approach, theoretical concepts have a connection, correlation and meaningful association in the real world.

This Chapter examines the historical pulse points of where we have travelled over the last few decades in relation to sustainable development. Despite a number of international treaties to support and accelerate global trends towards sustainable development, we are still a long way from living in a low carbon world. Secondary literature review is undertaken, mainly through reports, internet sources and other grey literature to paint a picture of the current state of play in relation to built environment sustainability education. This provides an understanding of the gaps in the current landscape to support a way forward to move towards sustainable built environment futures from an educational perspective.

The main question guiding this study is: What enablers need to be put in place so that responsible built environment education is mainstreamed and implemented effectively, and what other platforms exist for supporting partnerships, collaboration and engagement between academia, industry and research?

The Chapter commences with examining the key pulse points that have shaped global leadership and trends on sustainable development and education. These are United Nations Decade of Education for Sustainable Development (UN DESD), UNESCO's Global Action Programme (GAP) on Education for Sustainable Development, Inter-University Sustainable Development Research Programme (IUSDRP), Australasian Campus towards Sustainability (ACTS), the 10 Year Framework of Programmes (10YFP), Sustainable Development Goals (SDGs), New Urban Agenda (NUA) and the Principles of Responsible Management for Education (PRME). This is followed by discussions on how the PRME approach may be used for the built environment. The Chapter concludes by calling for action on setting up a parallel PRME for the built environment, called Principles for Responsible Urban built environment Education (PRUE).

The following set of events map the current situation in relation to education and sustainable development, including climate change. Not all the events have been captured, only the key events impacting on the current state of play, particularly in relation to the built environment and education.

2 United Nations Decade for Education in Sustainable Development: UNDESD

The UN Decade for ESD was launched in 2005 to enhance the role of education in supporting and promoting sustainable development. It ended in 2014, running for a full decade. At the Rio+20 (2012) UN Conference on Sustainable Development there was an agreement that the work undertaken to date needs to be spurred further, leading to the development of GAP as the continuation of the UNDESD. The Muscat Agreement (UNESCO 2014b) and the Open Working Group (OWG) working on the development towards the SDGs also supported the development of the GAP.

The role of education as a platform for building a better and more sustainable future has gained increasing attention leading to the declaration of the United Nations Decade of Education for Sustainable Development in 2005, and this has gathered further strength since Rio+20 (UNESCO 2014a). UNESCO was the lead agency in developing the GAP—the Global Action Programme on Education for Sustainable Development. This resulted from extensive consultations with stakeholders, and was formally endorsed by the UNESCO General Conference in 2013 by the 37th session of the General Conference of UNESCO (37C/Resolution 12). The Development Timeline is described in more detail shown in Fig. 1.

Fig. 1 Developmental timeline

3 Global Action Plan (GAP)

GAP seeks to generate and scale-up concrete actions in the UNDESD. It is intended to make a substantial contribution to the post-2015 agenda. The GAP is the follow up to the United Nations Decade of Education for Sustainable Development (2005–2014). The GAP was designed as a new set of sustainable development goals that are action-oriented, global in nature and universally applicable; designed as a concrete tangible contribution to the post-2015 development and action agendas. To assist with implementation, a roadmap has been set up. This roadmap explains the goals, objectives and priority action areas to enable strategic focus and stakeholder commitment along with implementation and monitoring mechanism. It supports various stakeholders—government, civil society organisations, inter governmental organisations, private sector, academic and research communities, the media and other relevant institutions to facilitate and support learning to trainers and individuals.

The vision of GAP is "a world where everybody has the opportunity to benefit from education and learn the values, behaviour and lifestyles required for a sustainable future and for positive societal transformation" (UNESCO 2014a, p. 14). There are two objectives that enable GAP to reach its goals:

Objective 1: "to orient education and learning so that everyone has the opportunity to acquire the knowledge, skills, values and attitudes that empower them to contribute to sustainable development".
Objective 2: "to strengthen education and learning in all agendas, programmes and activities that promote sustainable development" (UNESCO 2014a, p. 14).

The GAP focuses specially on girls and women, small island developing states, and Africa. As girls and women play an important role in sustainable development, there is a special focus on inclusive and sustainable growth and employment of women and particularly, girls. As small island developing states and Africa are dealing with the impact of climate change, education is even more critical to respond to the needs of the local context.

The GAP has identified 5 priority areas to advance the ESD Agenda. These are (UNESCO 2014a, p. 15):

1. Advancing Policy: 'Mainstream ESD into both education and SD policies to create an enabling environment for ESD and to bring about systemic change'. This Priority focuses on alignments to bring about systemic change for real impacts. It supports a collaborative process of engagement with government, private sector, local communities, educators and civil society, who are the main stakeholders in this integrated approach to ESD. The expected outcomes are that ESD is included in 'sub national, national, sub regional, regional and international policy frameworks, plans, strategies, programmes and processes related to education and sustainable development' (UNESCO 2014a, p. 17).
2. Transforming learning and training environments: 'Integrate sustainability principles into education and training settings' (UNESCO 2014a, p. 18). This Priority action is more than preaching and teaching on sustainable development. It focuses on converting thought to action. Whole institution approaches to sustainability are required. It means that the community (consisting of stakeholders from primary and secondary schools, vocational and higher education providers, teachers and professional staff of these institutions, leaders, parents, private, public and small to medium enterprises to name a few), curriculum and university campuses become enabling platforms for sustainability. The outcomes include sustainability plans and strategies.
3. Building capacities of educators and trainers: 'Increase the capacities of educators and trainers to more effectively deliver ESD' (UNESCO 2014a, p. 20). Educators are change agents as far as sustainability education is concerned and they need to acquire the necessary skills, attitude, values and knowledge to change society to become more responsive to sustainability. These educators and trainers, along with key agents from human resource personnel in public and private organisations, civil society and other institutions become the main stakeholders. The expected outcomes are professional development sessions for teachers and key professional staff (to support curricular activities) and capacity building for teachers and trainers.
4. Empowering and mobilizing youth: Multiply ESD actions among youth. The involvement of youth of today as the decision makers of tomorrow have the capacity to make major impacts on sustainable consumption and production patterns. Youth between the ages of 15 and 24 years become the drivers and the beneficiaries of consumption and production patterns. Youth therefore, need to be involved in working with a range of different organisations; government, public and private to drive home the importance of their role today and tomorrow. Youth need to be involved in contributing and advocating ESD across all levels and engaged in bottom-up activities whilst also working on top-down activities.
5. Accelerating sustainable solutions at a local level: At community level, scale up ESD programmes and multi-stakeholder ESD networks. Cities are becoming more important in the journey towards sustainable development. Whilst the problems are complex, rural areas fare no better and whether urban or rural, the importance of engagement with local communities is essential

to finding lasting solutions to our sustainability problems. Networks need to be strengthened to involve a large stakeholder youth population in the discussions and resulting actions. Engagement with local educational institutions and key staff, private and public organisations are key stakeholders. Key outcomes are the integration of ESD programmes and activities into the planning and decision making processes of the community.

4 Inter-University Sustainable Development Research Programme (IUSDRP)

The focus of the IUSDRP is to support the development of the post UNDESD through a new approach to sustainability; Sustainability 2.0. It was launched in 2015. The aim of this programme is to establish a platform on which member universities may undertake more research on matters related to sustainable development, and to consolidate initiatives in the field of sustainable development through (IUSDRP 2017):

- Increasing research income
- Enhancing institutional research profiles in the field of sustainable development
- Increasing the intake of Ph.D. students
- Increasing publication outputs

It is anticipated that through the IUSDRP, committed experts at member universities have a sound basis upon which to undertake research projects, train students, publish in influential journals, and organize events to support and work with each other.

Some of the project areas for research and engagement are energy production and use, production and consumption, water use and management, waste management, transport, climate change, adaptation and resilience, education and curriculum innovation; obviously all in the context of sustainability as well as campus greening. A ‘flagship of the Inter-University Sustainable Development Research Programme is the "World Symposium on Sustainable Development at Universities" (WSSDU) series, which is the leading event in the field. Organised bi-annually, WSSDU congregates the leading organisations and experts on sustainability in higher education, proving a unique opportunity for information exchange, networking and for the documentation of project work and dissemination of results from research projects. This is complemented by the "World Sustainable Development Teach-In Days"' (IUSDRP 2017, p. 8).

5 Australasian Campus Towards Sustainability (ACTS)

ACTS was launched as the Australian Universities Environmental Management Network (AUMEN) before becoming the not-for-profit organization it is today. It follows the UK (The Environmental Association for Universities and Colleges: EUAC) and the US (The Association for the Advancement of Sustainability in Higher Education: AASHE) organisations as a vehicle for facilitating ideas and information exchange across the tertiary sector in Australia and in the region between environmental officers and managers. Since the focus of the book for which this Chapter is written is focused on the Asia Pacific region, so also, the scope of organisations such as ACTS is restricted to this region.

ACTS was officially launched in 2006 and today boasts an international standing operating as the umbrella organization for sustainability within campus operations and learnings for its members. Today, ACTS aims to 'inspire, promote and support change towards best practice sustainability within operations, curriculum and research of the Australasian tertiary education sector' (ACTS 2017). The organization supports the development of resources, sharing of knowledge, developmental and networking opportunities for members and by critically challenging and supporting collaboration with stakeholders to lead sustainability innovation in the sector. ACTS is the convener for the international Green Gown Awards for the Australasian region.

ACTS supports LiFE; Learning in Future Environments (LiFE) launched in 2012. LiFE is designed to help educational institutions manage, measure and improve social and environmental responsibility performance of their own organisations (LiFE Index 2017). It was developed by members working in higher education sectors in UK, Australia and New Zealand. It is a road map to identify gaps, focus resources and set targets to drive change across universities and educational institutions. It covers four main areas: leadership and governance; partnerships and engagement; learning teaching and research; and facilities and operations.

6 10 Year Framework of Programmes (10YFP)

At the United Nations Conference on Sustainable Development (Rio+20) in 2012, Heads of State adopted the 10-Year Framework of Programmes on Sustainable Consumption and Production (SCP) Patterns (10YFP), a global framework for action to accelerate the shift towards SCP in both developed and developing countries. The important transversal role of SCP as an integral part of the 2030 Agenda for Sustainable Development has been clearly affirmed with the inclusion of a stand alone goal (Goal 12) on SCP among the 17 Sustainable Development Goals (UN Environment 2015g).

The period for the 10YFP is 2012–2022. The main objective of the 10YFP is to "accelerate the shift towards sustainable consumption and production to promote social and economic development within the carrying capacity of the ecosystems by addressing, and where appropriate, decoupling economic growth from environmental degradation by improving efficiency and sustainability in the use of resources and production processes and reducing resource degradation, pollution and waste" (United Nations Conference on Sustainable Development 2012, p. 2). The 10YFP accelerates the shift towards SCP, supporting regional and national policies and initiatives; contributes to resource efficiency and decoupling economic growth and environmental degradation and resource use; mainstreams SCP into sustainable development policies, programmes and strategies; supports capacity building and facilitates access to financial and technical assistance and enables all stakeholders to share information and knowledge on SCP tools, initiatives and best practices.

The 10YFP generates collective impact through multi-stakeholder programmes and partnerships, which develop, replicate and scale up SCP policies and initiatives at all levels. The 10YFP fosters knowledge and experience sharing, and facilitates access to technical and financial resources for developing countries. UN Environment serves as the 10YFP Secretariat. The vision of the 10YFP is

> Fundamental changes in the way societies produce and consume are indispensable for achieving global sustainable development. All countries should promote sustainable consumption and production patterns, with the developed countries taking the lead and with all countries benefiting from the process, taking into account the Rio principles… Governments, relevant international organizations, the private sector and all major groups should play an active role in changing unsustainable consumption and production patterns; (United Nations Conference on Sustainable Development 2012, p. 2).

Interested actors from all countries can be involved in the implementation of the 10YFP activities: governments, private sector, civil society, researchers, UN agencies, financial institutions, and other major groups.

The 10YFP is divided into six programmes:

1. Sustainable Public Procurement (SPP): SPP strives to create and maintain a significant hub for exchange and cooperation driving the implementation of SPP worldwide. The programme brings together a variety of stakeholders to foster communication, leverage resources and achieve common objectives (UN Environment 2015b).
2. Sustainable Consumer Information (SCI): This programme directs and supports the provision of quality information on goods and services and the identification and implementation of the most effective strategies to engage consumers in sustainable consumption. It facilitates relevant policies, strategies, projects, partnerships and builds synergies and cooperation between different stakeholders to leverage resources towards mutual goals (UN Environment 2015c).
3. Sustainable Tourism Programme (STP): This programme focuses on transformation of sustainability through efficiency, innovation and adaptability. The programme supports evidence based decision making, adopts a life cycle

approach for continuous improvement, emphasises collaboration among stakeholders and results based project implementation (UN Environment 2015d).
4. Sustainable Lifestyles and Education Programme (SLE): This programme fosters the uptake of sustainable lifestyles as the common norm, building a platform of support for challenges facing us today including resource efficiency and biodiversity conservation, climate change mitigation and adaptation, poverty eradication and social well being (UN Environment 2015e).
5. Sustainable buildings and construction (SBC): The goal of the programme is to shift to SCP patterns in the building and construction sector. The programme supports work on building, upscaling and mainstreaming policies, programmes and actions on promoting knowledge sharing, outreach and awareness raising (UN Environment 2015f).
6. Sustainable Food Systems (SFS): This programmes builds and strengthens multistakeholder initiatives to accelerate the shift towards sustainable food systems. It builds on value on the food chain, from farm to fork. It focuses on promotion of sustainable diets and the reduction of food losses and waste (UN Environment 2015g).

The two programmes that are closely linked to education are the SLE and the SBC programme. By its very nature, SLE is about educating current and future lifestyles to be one that is sustainable. The SBC programme, as part of its remit focuses on knowledge sharing, and raising awareness, although the word 'education' does not appear in its vision or mission statement. There is an implicit understanding that education in all its forms is critical for global building and construction, and particularly in the developing and emerging economies of the world.

In terms of an urban foci, SPP and SBC are the two programmes that are closely linked to Sustainable Development Goal 11 focusing on cities and settlements. The Sustainable Development Goals are described in greater detail below.

7 Sustainable Development Goals (SDGs)

The United Nations General Assembly adopted a resolution on 25th September 2015 (UN 2015) where Transforming our world: the 2030 Agenda for Sustainable Development was adopted, effective from January 1, 2016. The goals and targets are the result of intensive consultation and engagement with civil society and other stakeholders around the world. This work was largely undertaken by the Open Working Group (OWG) of the General Assembly of the SDG and the UN, whose Secretary General eventually endorsed the SDGs (United Nations Environment 2016). There are 17 Sustainable Development Goals and 169 targets. These goals and targets build on the Millenium Development Goals focusing on integrating economic, environmental and social dimensions of sustainable development to stimulate action over the next 15 years in all aspects of critical importance to humanity and the planet. While some progress has been made, there is still a need to progress development in Africa, small island nations, and in vulnerable sections of society, including women and children.

The SDGs build on the Rio Declaration, World summit on Sustainable Development, World Summit for Social Development, the Programme of Action of the International Conference on Population and Development and the United Nations Conference on Sustainable Development. The 17 SDGs with its 169 associated targets are integrated and indivisible.

Using the 5 Ps: People, Planet, Prosperity, Peace and Partnership, these linkages with sustainable development build integrated linkages for realising the goals. People refers to ensuring that all human beings can fulfil their potential in dignity and equality in a healthy environment; Planet refers to enhancing sustainable consumption and production to protect the planet from degradation, sustainably managing natural resources and taking urgent action on climate change to support the needs of present and future generations on the planet, Prosperity ensures fulfilling lives aligned with nature; Peace supports a society that is free from fear and violence; and finally Partnership builds global solidarity particularly considering vulnerable people (Preamble 2/35) (United Nations 2015).

Considering in more detail the built environment, the following SDGs and targets are key touch points as shown and elaborated further in the Appendix.

The Sustainable Development Goals (SDGs)are summarised in Table 1:

Table 1 SDGs 14/35 (United Nations 2015), emphasis (in italics) provided by author where they impact directly on the built environment

Sustainable development goals
Goal 1. End poverty in all its forms everywhere
Goal 2. End hunger, achieve food security and improved nutrition and promote sustainable agriculture
Goal 3. Ensure healthy lives and promote well-being for all at all ages
Goal 4. Ensure inclusive and equitable quality education and promote lifelong learning opportunities for all
Goal 5. Achieve gender equality and empower all women and girls
Goal 6. Ensure availability and sustainable management of water and sanitation for all
Goal 7 Ensure access to affordable, reliable, sustainable and modern energy for all
Goal 8. Promote sustained, inclusive and sustainable economic growth, full and productive employment and decent work for all
Goal 9. Build resilient infrastructure, promote inclusive and sustainable industrialization and foster innovation
Goal 10. Reduce inequality within and among countries
Goal 11. Make cities and human settlements inclusive, safe, resilient and sustainable
Goal 12. Ensure sustainable consumption and production patterns

(continued)

Table 1 (continued)

Sustainable development goals
Goal 13. Take urgent action to combat climate change and its impacts[a]
Goal 14. Conserve and sustainably use the oceans, seas and marine resources for sustainable development
Goal 15. Protect, restore and promote sustainable use of terrestrial ecosystems, sustainably manage forests, combat desertification, and halt and reverse land degradation and halt biodiversity loss
Goal 16. Promote peaceful and inclusive societies for sustainable development, provide access to justice for all and build effective, accountable and inclusive institutions at all levels
Goal 17. Strengthen the means of implementation and revitalize the Global Partnership for Sustainable Development

[a]Acknowledging that the United Nations Framework Convention on Climate Change is the primary international, intergovernmental forum for negotiating the global response to climate change

8 The New Urban Agenda (NUA)

The New Urban Agenda is grounded in the Universal Declaration of Human Rights, international human rights treaties, the Millennium Declaration, and the 2005 World Summit Outcome. It is informed by other instruments such as the Declaration on the Right to Development. The NUA focuses on cities and urban settlements. It therefore, brings the debate back into the context of urbanism. It calls for cities of all sizes, towns and villages to work towards a universal, participatory, people-centred approach to protecting the planet. It supports a long term vision setting out priorities and actions at the global, regional, national, sub-national and local levels that governments and other relevant stakeholders in every country can adopt based on their needs. The NUA calls for particular focus on developing countries, small island developing states, as well as specific challenges facing middle income countries. The NUA calls for all stakeholders to revitalise, strengthen, and create partnerships, enhancing coordination and cooperation to effectively implement the NUA and realise shared vision of sustainable urban development. Cities and human settlements are the drivers of sustainable development in an increasingly urbanised world (Habitat III 2016, pp. 4–5).

The New Urban Agenda can be met within the framework of a revitalized Global Partnership for Sustainable Development, supported by the concrete policies and actions as outlined in the outcome document of the third International Conference on Financing for Development, held in Addis Ababa from 13 to 16 July 2015, an integral part of the 2030 Agenda for Sustainable Development (United Nations 2015, pp. 10/35). The high-level political forum (HLPF) under the auspices of the General Assembly and the Economic and Social Council (EcoSoc) has the central role in overseeing follow-up and review at the global level. Governments as well as 'parliaments, the United Nations system and other international institutions, local authorities, indigenous peoples, civil society, business and the private sector, the scientific and academic community—and all people' are entrusted with implementing the Agenda. It is an Agenda 'of the people, by the people and for the people' (UN 2015, 52, p. 12).

Implementation of the NUA requires enabling environments and access to various resources such as science, technology, innovation and enhanced knowledge sharing platforms. Partnerships with various governments, private sector, civil society, UN systems, and other actors based on the principles of equality, non discrimination, accountability, respect for human rights, solidarity are all critical to achieve the NUA. Building on the Habitat III and the lessons learnt from the process, a collective approach is critical to progress forwards. Inclusion, innovation and integration are seen to be the underling values for the NUA (United Nations Habitat 2017).

With particular reference to education, the NUA under para 155 notes:

> We will promote capacity development initiatives to empower and strengthen skills and abilities of women and girls, children and youth, older persons and persons with disabilities, indigenous peoples and local communities, as well as persons in vulnerable situations for shaping governance processes, engaging in dialogue, and promoting and protecting human rights and anti-discrimination, to ensure their effective participation in urban and territorial development decision-making (Habitat III 2016, p. 20).

Reporting of the outcomes of the NUA is expected every four years, with the first report being submitted during the 72nd session of the General Assembly of the UN through the EcoSoc and will also feed into the HLPF on Sustainable Development.

9 Principles of Responsible Management Education (PRME)

This year marks the first decade of the roll out of PRME. PRME: Principles of Responsible Management Education set out to provide future leaders with the necessary insight and skills to reflect upon, critically analyse and provide leadership with regard to corporate sustainability. PRME was set up by representatives from UN Global Compact, the Association of Advance Collegiate Schools of Business (AACSB), The Aspen Institute's Business and Society Program, the European Foundation for Management Development (EFMD), the Globally Responsible Leadership Initiative (GRLI), Net Impact, a student organisation with more than 13,000 members, Graduate Management Admission Council (GMAC), and the European Academy of Business in Society (EABIA). The first PRME task force, therefore represented international organisations and 60 Deans, University Presidents and Scholars were involved. In July 2007, this outcome was presented to the UN Secretary General Ban Ki-moon and was formally endorsed by him.

The UN Global Compact (UNGC) was the catalysing body for the UN PRME. UNGC was set up in 2000 and has been instrumental in supporting corporate responsibility at a global lelvel. With only 44 companies when it was launched, UNGC became the world's largest corporate responsibility initiative with 6000 business and business participants from over 130 countries in 2010 (Kell and Haertle 2011). In 2014, over 8000 companies in 145 countries have joined UNGC including from the emerging economies of China, India and Brazil (Haertle and Kongstad 2014).

Today, in 2017, the numbers are even higher. Over 660 business and management related higher education institutions are part of the PRME network in 83 countries across the world with 14 regional chapter and 7 issue-area working groups. Over 1200 Sharing Information Progress or SIP reports are available, and over 20 million students have benefited from engagement with PRME in higher education institutions. Key partnerships have been formed with accreditation bodies, regional associations, student and leadership groups, publishers, the UN and businesses (PRME 2017a).

There are 5 main areas of impact. These are:

- "Harmonising a unifying, universal framework of Principles to transform management education
- Expanding the dialogue in higher education to include concepts of sustainability, responsibility and ethics
- Elevating sustainability and purpose as a strategic priority for higher education institutions
- Amplifying global values through local and regional networks
- Communicating UN values and goals as opportunities for higher education institutions" (PRME 2017a, p. 2).

What is really interesting, according to Haertle and Kongstad (2014) are that when businesses were asked to rank their global sustainability challenges, they cited education as the number one challenge at 63%. When asked where companies can have the most positive impact, the survey showed that education again came at 59% in second place, with growth and employment taking the number one position at 83%.

9.1 PRME in Business

The critical link between public sector engagement and private sector is needed to support the full potential of the relationship with the UN and the other organisations (Cogan and Johnstone 2015). In the recent financial crises in 2009, business schools were seen to be responsible for failing to systemically and critically discuss issues of executive incentive programmes, leading to a lack of ethical underpinnings in training students and therefore instilling in graduates these attributes in their professional lives Rasche and Escudero (2010). Other authors such as Horwitz and Grayson call for 'the business of business is sustainable business' (Horwitz and Grayson 2010, p. 29).

Business education often rely on case studies to illustrate and support learning outcomes. A business case study supports education of students and explores the actual situation where a decision was taken, or a challenge or opportunity explored. It also provides a problem or an issue faced by a person (or persons) in an organization. Case studies contain relevant data about the issue available to the key person in the case, plus background information about the organization, the outcomes and the impact of the outcomes. Ivey (2011) notes that case studies, essential for business education, provided 250 cases for study in the ivey website (www.iveycases.com) by December 2011. Today, nearly 6 years on the website boasts 34,000 cases.

9.2 Principles of PRME

"The mission of PRME is to transform management education, research and thought leadership globally by providing the Principles for Responsible Management Education framework, developing learning communities and promoting awareness about the United Nations' Sustainable Development Goals" (PRME 2017b). PRME therefore, focuses on management education. The tag line for PRME is: *Developing the responsible leaders for tomorrow.*

The six principles of PRME are (PRME 2017c):

"Principle 1 | Purpose: We will develop the capabilities of students to be future generators of sustainable value for business and society at large and to work for an inclusive and sustainable global economy.
Principle 2 | Values: We will incorporate into our academic activities and curricula the values of global social responsibility as portrayed in international initiatives such as the United Nations Global Compact.
Principle 3 | Method: We will create educational frameworks, materials, processes and environments that enable effective learning experiences for responsible leadership.
Principle 4 | Research: We will engage in conceptual and empirical research that advances our understanding about the role, dynamics, and impact of corporations in the creation of sustainable social, environmental and economic value.
Principle 5 | Partnership: We will interact with managers of business corporations to extend our knowledge of their challenges in meeting social and environmental responsibilities and to explore jointly effective approaches to meeting these challenges.
Principle 6 | Dialogue: We will facilitate and support dialog and debate among educators, students, business, government, consumers, media, civil society organisations and other interested groups and stakeholders on critical issues related to global social responsibility and sustainability".

PRME provides an opportunity to exchange ideas and best practices, while also being able to stimulate and guide a school's development of responsible management education. While the principles of PRME can be seen as guiding frameworks; they are still open enough to allow educational institutions to demonstrate sustainability outcomes regardless of the context they operate in. PRME acknowledges that the six principles offer contextualisation and innovative meaning, based on reflections and peer-peer discussions for each of the educational institutions. Unlike a membership organisation, PRME provides a platform for innovation, experimentation, learning from mistakes, learning from each other and supporting sustainability outcomes both within the organisation, that is, the universities and the wider socio-economic environment.

Stachowicz-Stanusch (2011) undertook a desk top study of 115 PRME reports by examining their organisational practices presented in the reports under each of the PRME principles. The author's report summarised general approaches to integration of PRME and developing a set of criteria to integrate each of the Principles of PRME into academia. General information included organisational commitments from the senior executives in the form of statements, mission/vision, core values and support into the future. Each of the principles were then examined to see what criteria arose. These included both student initiatives and curricular solutions. Included in student initiatives were clubs and organisations, participation in relevant conferences and through curriculum were case studies, team projects, projects from a company espousing sustainable values for business and society. Principle 2 on values focused on values in strategy and values in code of conduct/ ethics and included criteria focusing on curricular content such as ethics/CSR in curriculum, legal responsibility, fiscal issues, environmental issues, internships/ practicum and special programs for MBAs.

Principle 3 on Methods for effective learning included criteria on institutional solutions with a focus on a person or unit responsible for responsible management education, procedures and norms connected with social responsibility and implemented within educational process; evaluation through surveys, reviews, indicators, etc.; infrastructures through both physical and non physical assets focusing on responsible management; educational practices dealing with conferences, seminars, symposia, discussions as educational practice, e-learning and webinars and study tours.

Principle 4 on research focuses on internal support and research results. Internal support includes research units focused on ethics/responsibility, financial and in-kind support for research teams and research projects and ethics as part of the research agenda/plans. Research results includes publications and conference presentations, awards for research activity and currently conducted research on ethics/responsibility.

Principle 5 on partnerships supports unit or person responsible for contact with stakeholders, partnership with businesses, for example, through joint projects with business, partnership with science and other schools, partnership with local government such as educating officials and joint projects, partnerships with students including student exchange and partnership with other organisations such as projects with NGOs.

Principle 6: Dialogue, include communication channels such as symposia, conferences, seminars, reports, newsletters, websites and such other channels of communication. In addition, examples of best practice may be supported such as use of renewable energy, responsible HR practices, contribution to society and other examples of best practice.

Alongside the development of PRME, HESI also needs to be acknowledged. HESI: Higher Education Sustainability Initiative is a partnership between United Nations Department of Economic and Social Affairs, UNESCO, UN Environment, UN Global Compact's Principles for Responsible Management Education (PRME) initiative, United Nations University, UN Habitat and UNCTAD. It was created in 2012 leading up to Rio+20. HESI commits to teach sustainable development across all disciplines of study, encourage research and dissemination of sustainable development knowledge, green campuses and support local sustainability efforts, and engage and share information with international networks. Through its strong connections with the UN, it was meant to provide a unique interface between higher education, science, and policy making (HESI 2017). HESI includes integrating the SDGs into sustainability strategies, research, teaching, pedagogy, and campus practices to position higher education institutions as key drivers for achieving the SDGs.

10 Discussions

As can be seen from the remit of the various organisations described in the previous sections, there are linkages to education and sustainable development through a number of global initiatives, primarily led by the UN or supported by the UN and its various agencies.

The UNDESD, set up after Rio+20 in 2012 recognised the role of education in sustainable development and focused on setting up sustainability at all levels of education from primary, through to tertiary. It did not have any formal links with any universities. While the GAP has particular areas of focus; girls and women, small island developing states and Africa; and brings in wider partnerships as well as integrating sustainability principles into education and training settings including capacity building for educators and trainers, and involving youth, it does not have any formal links with guiding and supporting curricular change. It also focuses on community engagement and has an urban focus.

IUSDRP, works along similar lines to the GAP but brings in research as an integral part of sustainable development—increasing research in the field of sustainable development through growing research income, publications, Ph.D. students and enhancing institutional profiles. While information exchange and curriculum development is part of their agenda, again, there are no formal links and networks specifically created for supporting curriculum.

ACTS is the equivalent of the American and UK associations supporting the advancement of sustainability in higher education in Australia and the Asia Pacific region. While this organisation also supports teaching and research in higher educational institutions, it does not have any prescribed responsibilities for engaging in curricula or student learning outcomes.

The 10YFP's range of programmes arise from the foundation of sustainable consumption and production patterns, aligned with SDG 12. The SLE programme focuses on formal education across all years of schooling into the higher education sector. The SBC programme while not having formal engagement with the education sector contains research and academic institutions as part of their membership through the lead/co lead and multistakeholder advisory committee membership. The SBC programme supports and encourages knowledge sharing through every available means possible.

The SDGs are a more recent addition to the plethora of global frameworks available. The role of built environment particularly urban environments and education are encapsulated in Goals 4, 7, 9, 11, 12, 13 and 17. In tandem with the SDGs, the NUA brings a greater focus to cities and urban settlements. They are set within the context of partnerships as being the foundation for inclusion, innovation and integration for achieving a shared vision of sustainable development.

PRME is obviously a success story with humble beginings but quite formidable in a decade since its development. While the body that set up and promoted the development of PRME is the UN Global Compact, today, PRME has its own board with its own curricular imperatives, focused on building further and expanding the dialogue in higher education to include the concepts of sustainability, responsibility and ethics in the business world. Taking a leaf from business and management academic education, a parallel for the built environment may be drawn.

Built environment disciplines are many, and most of the disciplines tend to focus on technical competence rather than focusing on ethical discussions in the classroom from a sustainability perspective. As Haertle and Kongstad (2014) note, even though 65% of UNGC signatory companies have developed sustainability strategies at the CEO level, only 35% of the survey respondents train managers to integrate sustainability into strategy and operations and just 21% incorporate sustainability criteria into employee performance assessments. Sixty percent of companies are disclosing their sustainability policies and practices publicly, but very few include details. The same is true with reducing environmental impacts. Sixty six percent of companies have installed and implement environmental management systems (EMS) while 54% monitor their environmental performance and 38% report on their emissions. Therefore, the gap between action and intent is still there. This is even more pronounced in the built environment. There is a gap between design and intent of green buildings and performance of green buildings.

Higher Education Institutions do not exist in isolaton. They are part of a larger network of organisations that interact and develop with each other. From this perspective, PRME can be viewed as a 'well-defined set of concepts and more of a movement to engage students, managers, scholars and teachers in the type of inquiry and learning that aims for a larger vision' (PRME 2017a, p. 4). If this thinking was extended to the built environment, it aligns with both the SDGs and the NUA. Recent developments on the global scale provide a clear structure to engage with the six PRME Principles. SDG 4 is about quality education, and SDGs 7, 8, 9, 11 and 12 are all linked to the built environment. They offer a clear direction to achieve the outcomes in the near future.

The development of student engagement through the PRME SDG student engagement platform creates new avenues for students to critically engage with businesses and the UN around sustainability challenges. This offers a direct parallel opportunity for students in the built environment to also engage and support such platforms.

A parallel process to PRME in the built environment may be called PRUE: Principles for Responsible Urban built environment Education. It may be set up in the same way as PRME was 10 years ago, with support from industry organisations, peak industry bodies, universities and student bodies. This will ensure that the objective of developing and delivering the skills for future city design, planning, operation and governance are skills that graduates entering the urban professions of the future will have. This tri partite partnership will ensure that cutting edge and best practice knowledge will be brought into the classrooms; for industry, the value is in graduates with up to date knowledge and skills so there is no need to retrain the graduates when they enter the work force (and therefore cost savings will be achieved); and for international organisations, there is already a value in a coherent platform to implement/operationalise SDGs and the NUA and support collaboration. PRUE may operate under the tag line of *Developing responsible built environment practitioners for tomorrow*.

Organisations such as RICS (Royal Institute of Chartered Surveyors) who have supported the development of PRME and who operate in the built environment landscape would be critical players. The outputs, should such a partnership emerge would focus initially on preparation of fact sheets, research papers and case studies to be included in teaching and research. Some of these may also be developed as MOOC: Massive Open Online Courses, so that intellectual property is shared and not in the hands of only a few and outcomes of the impact shared by all. It is also essential that the development of in-context case studies for specific countries are also set up.

11 Conclusions

The literature on sustainable development teaching and learning, and outputs for the built environment demonstrates there are gaps between commitment and action. There are no shortages or lack of ideas in the intent to support low carbon futures at national and international levels. In fact, if anything, the timelines for international treaties and agendas on sustainable development have been set up with even shorter time spans over the last few years, and particularly since the Paris Accord in 2015. Urgency towards such international targets have put pressures on various countries to make concerted efforts to meeting the set targets.

The gap between action and intent with regard to achieving sustainability outcomes in the built environment will continue despite the emergence of global agendas such as the SDGs and the NUA. These gaps are in real danger of becoming even more wider and deeper, and more urgent and critical as we fail to meet our international targets with regard to low carbon futures. However, there is a long road to be traversed to translate ideas into reality and even more to mainstream industry knowledge, curricula and capacities to support sustainability outcomes in the built environment. The role of education in bringing the importance of such targets to current and future built environmental professionals is more urgent, now than ever.

PRUE offers an opportunity for supporting sustainability into the built environment programs in universities. It offers practical solution to skill up the built environment professional workforce, especially those working in an urban context and to encourage higher levels of best practice sharing in higher education. It arms these practitioners whether involved in city governance and planning, urban development, or urban design, to be able to cope with rapid economic change, technology change, social change, urban growth, climate change, resilience and adaptation and provide them the nous to be able to practice ethically and sustainably. PRUE will enable these practitioners to develop competences and practices around environmental knowledge, principles of social equity, and good governance.

Appendix

	SDGs directly linked to built environment
	SDGs indirectly linked to built environment
Final list of proposed Sustainable Development Goal indicators[a]	
Sustainable Development Goal indicators should be disaggregated, where relevant, by income, sex, age, race, ethnicity, migratory status, disability and geographic location, or other characteristics, in accordance with the Fundamental Principles of Official Statistics (General Assembly resolution 68/261)	
Goals and targets (from the 2030 Agenda)	*Indicators*
Goal 1. End poverty in all its forms everywhere	
1.1 By 2030, eradicate extreme poverty for all people everywhere, currently measured as people living on less than $1.25 a day	1.1.1 Proportion of population below the international poverty line, by sex, age, employment status and geographical location (urban/rural)
1.2 By 2030, reduce at least by half the proportion of men, women and children of all ages living in poverty in all its dimensions according to national definitions	1.2.1 Proportion of population living below the national poverty line, by sex and age
	1.2.2 Proportion of men, women and children of all ages living in poverty in all its dimensions according to national definitions
1.3 Implement nationally appropriate social protection systems and measures for all, including floors, and by 2030 achieve substantial coverage of the poor and the vulnerable	1.3.1 Proportion of population covered by social protection floors/systems, by sex, distinguishing children, unemployed persons, older persons, persons with disabilities, pregnant women, newborns, work-injury victims and the poor and the vulnerable
1.4 By 2030, ensure that all men and women, in particular the poor and the vulnerable, have	1.4.1 Proportion of population living in households with access to basic services

(continued)

(continued)

equal rights to economic resources, as well as access to basic services, ownership and control over land and other forms of property, inheritance, natural resources, appropriate new technology and financial services, including microfinance	1.4.2 Proportion of total adult population with secure tenure rights to land, with legally recognized documentation and who perceive their rights to land as secure, by sex and by type of tenure
1.5 By 2030, build the resilience of the poor and those in vulnerable situations and reduce their exposure and vulnerability to climate-related extreme events and other economic, social and environmental shocks and disasters	1.5.1 Number of deaths, missing persons and persons affected by disaster per 100,000 people
	1.5.2 Direct disaster economic loss in relation to global gross domestic product (GDP)[b]
	1.5.3 Number of countries with national and local disaster risk reduction strategies
1.a Ensure significant mobilization of resources from a variety of sources, including through enhanced development cooperation, in order to provide adequate and predictable means for developing countries, in particular least developed countries, to implement programmes and policies to end poverty in all its dimensions	1.a.1 Proportion of resources allocated by the government directly to poverty reduction programmes
	1.a.2 Proportion of total government spending on essential services (education, health and social protection)
1.b Create sound policy frameworks at the national, regional and international levels, based on pro-poor and gender-sensitive development strategies, to support accelerated investment in poverty eradication actions	1.b.1 Proportion of government recurrent and capital spending to sectors that disproportionately benefit women, the poor and vulnerable groups
Goal 2. End hunger, achieve food security and improved nutrition and promote sustainable agriculture	
2.1 By 2030, end hunger and ensure access by all people, in particular the poor and people in vulnerable situations, including infants, to safe, nutritious and sufficient food all year round	2.1.1 Prevalence of undernourishment
	2.1.2 Prevalence of moderate or severe food insecurity in the population, based on the Food Insecurity Experience Scale (FIES)
2.2 By 2030, end all forms of malnutrition, including achieving, by 2025, the internationally agreed targets on stunting and wasting in children under 5 years of age, and address the nutritional needs of adolescent girls, pregnant and lactating women and older persons	2.2.1 Prevalence of stunting (height for age < −2 standard deviation from the median of the World Health Organization (WHO) Child Growth Standards) among children under 5 years of age
	2.2.2 Prevalence of malnutrition (weight for height >+2 or <−2 standard deviation from the median of the WHO Child Growth Standards) among children under 5 years of age, by type (wasting and overweight)
2.3 By 2030, double the agricultural productivity and incomes of small-scale food producers, in particular women, indigenous	2.3.1 Volume of production per labour unit by classes of farming/pastoral/forestry enterprise size

(continued)

(continued)

peoples, family farmers, pastoralists and fishers, including through secure and equal access to land, other productive resources and inputs, knowledge, financial services, markets and opportunities for value addition and non-farm employment	2.3.2 Average income of small-scale food producers, by sex and indigenous status
2.4 By 2030, ensure sustainable food production systems and implement resilient agricultural practices that increase productivity and production, that help maintain ecosystems, that strengthen capacity for adaptation to climate change, extreme weather, drought, flooding and other disasters and that progressively improve land and soil quality	2.4.1 Proportion of agricultural area under productive and sustainable agriculture
2.5 By 2020, maintain the genetic diversity of seeds, cultivated plants and farmed and domesticated animals and their related wild species, including through soundly managed and diversified seed and plant banks at the national, regional and international levels, and promote access to and fair and equitable sharing of benefits arising from the utilization of genetic resources and associated traditional knowledge, as internationally agreed	2.5.1 Number of plant and animal genetic resources for food and agriculture secured in either medium or long-term conservation facilities
	2.5.2 Proportion of local breeds classified as being at risk, not-at-risk or at unknown level of risk of extinction
2.a Increase investment, including through enhanced international cooperation, in rural infrastructure, agricultural research and extension services, technology development and plant and livestock gene banks in order to enhance agricultural productive capacity in developing countries, in particular least developed countries	2.a.1 The agriculture orientation index for government expenditures
	2.a.2 Total official flows (official development assistance plus other official flows) to the agriculture sector
2.b Correct and prevent trade restrictions and distortions in world agricultural markets, including through the parallel elimination of all forms of agricultural export subsidies and all export measures with equivalent effect, in accordance with the mandate of the Doha Development Round	2.b.1 Producer Support Estimate
	2.b.2 Agricultural export subsidies
2.c Adopt measures to ensure the proper functioning of food commodity markets and their derivatives and facilitate timely access to market information, including on food reserves, in order to help limit extreme food price volatility	2.c.1 Indicator of food price anomalies

(continued)

(continued)

Goal 3. Ensure healthy lives and promote well-being for all at all ages	
3.1 By 2030, reduce the global maternal mortality ratio to less than 70 per 100,000 live births	3.1.1 Maternal mortality ratio
	3.1.2 Proportion of births attended by skilled health personnel
3.2 By 2030, end preventable deaths of newborns and children under 5 years of age, with all countries aiming to reduce neonatal mortality to at least as low as 12 per 1000 live births and under-5 mortality to at least as low as 25 per 1000 live births	3.2.1 Under-five mortality rate
	3.2.2 Neonatal mortality rate
3.3 By 2030, end the epidemics of AIDS, tuberculosis, malaria and neglected tropical diseases and combat hepatitis, water-borne diseases and other communicable diseases	3.3.1 Number of new HIV infections per 1000 uninfected population, by sex, age and key populations
	3.3.2 Tuberculosis incidence per 1000 population
	3.3.3 Malaria incidence per 1000 population
	3.3.4 Hepatitis B incidence per 100,000 population
	3.3.5 Number of people requiring interventions against neglected tropical diseases
3.4 By 2030, reduce by one third premature mortality from non-communicable diseases through prevention and treatment and promote mental health and well-being	3.4.1 Mortality rate attributed to cardiovascular disease, cancer, diabetes or chronic respiratory disease
	3.4.2 Suicide mortality rate
3.5 Strengthen the prevention and treatment of substance abuse, including narcotic drug abuse and harmful use of alcohol	3.5.1 Coverage of treatment interventions (pharmacological, psychosocial and rehabilitation and aftercare services) for substance use disorders
	3.5.2 Harmful use of alcohol, defined according to the national context as alcohol per capita consumption (aged 15 years and older) within a calendar year in litres of pure alcohol
3.6 By 2020, halve the number of global deaths and injuries from road traffic accidents	3.6.1 Death rate due to road traffic injuries
3.7 By 2030, ensure universal access to sexual and reproductive health-care services, including for family planning, information and education, and the integration of reproductive health into national strategies and programmes	3.7.1 Proportion of women of reproductive age (aged 15–49 years) who have their need for family planning satisfied with modern methods
	3.7.2 Adolescent birth rate (aged 10–14 years; aged 15–19 years) per 1000 women in that age group

(continued)

(continued)

3.8 Achieve universal health coverage, including financial risk protection, access to quality essential health-care services and access to safe, effective, quality and affordable essential medicines and vaccines for all	3.8.1 Coverage of essential health services (defined as the average coverage of essential services based on tracer interventions that include reproductive, maternal, newborn and child health, infectious diseases, non-communicable diseases and service capacity and access, among the general and the most disadvantaged population)
	3.8.2 Number of people covered by health insurance or a public health system per 1000 population
3.9 By 2030, substantially reduce the number of deaths and illnesses from hazardous chemicals and air, water and soil pollution and contamination	3.9.1 Mortality rate attributed to household and ambient air pollution
	3.9.2 Mortality rate attributed to unsafe water, unsafe sanitation and lack of hygiene (exposure to unsafe Water, Sanitation and Hygiene for All (WASH) services)
	3.9.3 Mortality rate attributed to unintentional poisoning
3.a Strengthen the implementation of the World Health Organization Framework Convention on Tobacco Control in all countries, as appropriate	3.a.1 Age-standardized prevalence of current tobacco use among persons aged 15 years and older
3.b Support the research and development of vaccines and medicines for the communicable and non-communicable diseases that primarily affect developing countries, provide access to affordable essential medicines and vaccines, in accordance with the Doha Declaration on the TRIPS Agreement and Public Health, which affirms the right of developing countries to use to the full the provisions in the Agreement on Trade-Related Aspects of Intellectual Property Rights regarding flexibilities to protect public health, and, in particular, provide access to medicines for all	3.b.1 Proportion of the population with access to affordable medicines and vaccines on a sustainable basis
	3.b.2 Total net official development assistance to medical research and basic health sectors
3.c Substantially increase health financing and the recruitment, development, training and retention of the health workforce in developing countries, especially in least developed countries and small island developing States	3.c.1 Health worker density and distribution

(continued)

(continued)

3.d Strengthen the capacity of all countries, in particular developing countries, for early warning, risk reduction and management of national and global health risks	3.d.1 International Health Regulations (IHR) capacity and health emergency preparedness
Goal 4. Ensure inclusive and equitable quality education and promote lifelong learning opportunities for all	
4.1 By 2030, ensure that all girls and boys complete free, equitable and quality primary and secondary education leading to relevant and effective learning outcomes	4.1.1 Proportion of children and young people: (a) in grades 2/3; (b) at the end of primary; and (c) at the end of lower secondary achieving at least a minimum proficiency level in (i) reading and (ii) mathematics, by sex
4.2 By 2030, ensure that all girls and boys have access to quality early childhood development, care and pre-primary education so that they are ready for primary education	4.2.1 Proportion of children under 5 years of age who are developmentally on track in health, learning and psychosocial well-being, by sex
	4.2.2 Participation rate in organized learning (one year before the official primary entry age), by sex
4.3 By 2030, ensure equal access for all women and men to affordable and quality technical, vocational and tertiary education, including university	4.3.1 Participation rate of youth and adults in formal and non-formal education and training in the previous 12 months, by sex
4.4 By 2030, substantially increase the number of youth and adults who have relevant skills, including technical and vocational skills, for employment, decent jobs and entrepreneurship	4.4.1 Proportion of youth and adults with information and communications technology (ICT) skills, by type of skill
4.5 By 2030, eliminate gender disparities in education and ensure equal access to all levels of education and vocational training for the vulnerable, including persons with disabilities, indigenous peoples and children in vulnerable situations	4.5.1 Parity indices (female/male, rural/urban, bottom/top wealth quintile and others such as disability status, indigenous peoples and conflict-affected, as data become available) for all education indicators on this list that can be disaggregated
4.6 By 2030, ensure that all youth and a substantial proportion of adults, both men and women, achieve literacy and numeracy	4.6.1 Percentage of population in a given age group achieving at least a fixed level of proficiency in functional (a) literacy and (b) numeracy skills, by sex

(continued)

(continued)

4.7 By 2030, ensure that all learners acquire the knowledge and skills needed to promote sustainable development, including, among others, through education for sustainable development and sustainable lifestyles, human rights, gender equality, promotion of a culture of peace and non-violence, global citizenship and appreciation of cultural diversity and of culture's contribution to sustainable development	4.7.1 Extent to which (i) global citizenship education and (ii) education for sustainable development, including gender equality and human rights, are mainstreamed at all levels in: (a) national education policies, (b) curricula, (c) teacher education and (d) student assessment
4.a Build and upgrade education facilities that are child, disability and gender sensitive and provide safe, non-violent, inclusive and effective learning environments for all	4.a.1 Proportion of schools with access to: (a) electricity; (b) the Internet for pedagogical purposes; (c) computers for pedagogical purposes; (d) adapted infrastructure and materials for students with disabilities; (e) basic drinking water; (f) single-sex basic sanitation facilities; and (g) basic handwashing facilities (as per the WASH indicator definitions)
4.b By 2020, substantially expand globally the number of scholarships available to developing countries, in particular least developed countries, small island developing States and African countries, for enrolment in higher education, including vocational training and information and communications technology, technical, engineering and scientific programmes, in developed countries and other developing countries	4.b.1 Volume of official development assistance flows for scholarships by sector and type of study
4.c By 2030, substantially increase the supply of qualified teachers, including through international cooperation for teacher training in developing countries, especially least developed countries and small island developing States	4.c.1 Proportion of teachers in: (a) pre-primary; (b) primary; (c) lower secondary; and (d) upper secondary education who have received at least the minimum organized teacher training (e.g. pedagogical training) pre-service or in-service required for teaching at the relevant level in a given country
Goal 5. Achieve gender equality and empower all women and girls	
5.1 End all forms of discrimination against all women and girls everywhere	5.1.1 Whether or not legal frameworks are in place to promote, enforce and monitor equality and non-discrimination on the basis of sex

(continued)

(continued)

5.2 Eliminate all forms of violence against all women and girls in the public and private spheres, including trafficking and sexual and other types of exploitation	5.2.1 Proportion of ever-partnered women and girls aged 15 years and older subjected to physical, sexual or psychological violence by a current or former intimate partner in the previous 12 months, by form of violence and by age
	5.2.2 Proportion of women and girls aged 15 years and older subjected to sexual violence by persons other than an intimate partner in the previous 12 months, by age and place of occurrence
5.3 Eliminate all harmful practices, such as child, early and forced marriage and female genital mutilation	5.3.1 Proportion of women aged 20–24 years who were married or in a union before age 15 and before age 18
	5.3.2 Proportion of girls and women aged 15–49 years who have undergone female genital mutilation/cutting, by age
5.4 Recognize and value unpaid care and domestic work through the provision of public services, infrastructure and social protection policies and the promotion of shared responsibility within the household and the family as nationally appropriate	5.4.1 Proportion of time spent on unpaid domestic and care work, by sex, age and location
5.5 Ensure women's full and effective participation and equal opportunities for leadership at all levels of decision-making in political, economic and public life	5.5.1 Proportion of seats held by women in national parliaments and local governments
	5.5.2 Proportion of women in managerial positions
5.6 Ensure universal access to sexual and reproductive health and reproductive rights as agreed in accordance with the Programme of Action of the International Conference on Population and Development and the Beijing Platform for Action and the outcome documents of their review conferences	5.6.1 Proportion of women aged 15–49 years who make their own informed decisions regarding sexual relations, contraceptive use and reproductive health care
	5.6.2 Number of countries with laws and regulations that guarantee women aged 15–49 years access to sexual and reproductive health care, information and education
5.a Undertake reforms to give women equal rights to economic resources, as well as access to ownership and control over land and other forms of property, financial services, inheritance and natural resources, in accordance with national laws	5.a.1 (a) Proportion of total agricultural population with ownership or secure rights over agricultural land, by sex; and (b) share of women among owners or rights-bearers of agricultural land, by type of tenure
	5.a.2 Proportion of countries where the legal framework (including customary law) guarantees women's equal rights to land ownership and/or control

(continued)

(continued)

5.b Enhance the use of enabling technology, in particular information and communications technology, to promote the empowerment of women	5.b.1 Proportion of individuals who own a mobile telephone, by sex
5.c Adopt and strengthen sound policies and enforceable legislation for the promotion of gender equality and the empowerment of all women and girls at all levels	5.c.1 Proportion of countries with systems to track and make public allocations for gender equality and women's empowerment
Goal 6. Ensure availability and sustainable management of water and sanitation for all	
6.1 By 2030, achieve universal and equitable access to safe and affordable drinking water for all	6.1.1 Proportion of population using safely managed drinking water services
6.2 By 2030, achieve access to adequate and equitable sanitation and hygiene for all and end open defecation, paying special attention to the needs of women and girls and those in vulnerable situations	6.2.1 Proportion of population using safely managed sanitation services, including a hand-washing facility with soap and water
6.3 By 2030, improve water quality by reducing pollution, eliminating dumping and minimizing release of hazardous chemicals and materials, halving the proportion of untreated wastewater and substantially increasing recycling and safe reuse globally	6.3.1 Proportion of wastewater safely treated
	6.3.2 Proportion of bodies of water with good ambient water quality
6.4 By 2030, substantially increase water-use efficiency across all sectors and ensure sustainable withdrawals and supply of freshwater to address water scarcity and substantially reduce the number of people suffering from water scarcity	6.4.1 Change in water-use efficiency over time
	6.4.2 Level of water stress: freshwater withdrawal as a proportion of available freshwater resources
6.5 By 2030, implement integrated water resources management at all levels, including through transboundary cooperation as appropriate	6.5.1 Degree of integrated water resources management implementation (0–100)
	6.5.2 Proportion of transboundary basin area with an operational arrangement for water cooperation
6.6 By 2020, protect and restore water-related ecosystems, including mountains, forests, wetlands, rivers, aquifers and lakes	6.6.1 Change in the extent of water related ecosystems over time
6.a By 2030, expand international cooperation and capacity-building support to developing countries in water- and sanitation-related activities and programmes, including water harvesting, desalination, water efficiency, wastewater treatment, recycling and reuse technologies	6.a.1 Amount of water- and sanitation-related official development assistance that is part of a government-coordinated spending plan

(continued)

(continued)

6.b Support and strengthen the participation of local communities in improving water and sanitation management	6.b.1 Proportion of local administrative units with established and operational policies and procedures for participation of local communities in water and sanitation management
Goal 7. Ensure access to affordable, reliable, sustainable and modern energy for all	
7.1 By 2030, ensure universal access to affordable, reliable and modern energy services	7.1.1 Proportion of population with access to electricity
	7.1.2 Proportion of population with primary reliance on clean fuels and technology
7.2 By 2030, increase substantially the share of renewable energy in the global energy mix	7.2.1 Renewable energy share in the total final energy consumption
7.3 By 2030, double the global rate of improvement in energy efficiency	7.3.1 Energy intensity measured in terms of primary energy and GDP
7.a By 2030, enhance international cooperation to facilitate access to clean energy research and technology, including renewable energy, energy efficiency and advanced and cleaner fossil-fuel technology, and promote investment in energy infrastructure and clean energy technology	7.a.1 Mobilized amount of United States dollars per year starting in 2020 accountable towards the $100 billion commitment
7.b By 2030, expand infrastructure and upgrade technology for supplying modern and sustainable energy services for all in developing countries, in particular least developed countries, small island developing States and landlocked developing countries, in accordance with their respective programmes of support	7.b.1 Investments in energy efficiency as a percentage of GDP and the amount of foreign direct investment in financial transfer for infrastructure and technology to sustainable development services
Goal 8. Promote sustained, inclusive and sustainable economic growth, full and productive employment and decent work for all	
8.1 Sustain per capita economic growth in accordance with national circumstances and, in particular, at least 7% gross domestic product growth per annum in the least developed countries	8.1.1 Annual growth rate of real GDP per capita
8.2 Achieve higher levels of economic productivity through diversification, technological upgrading and innovation, including through a focus on high-value added and labour-intensive sectors	8.2.1 Annual growth rate of real GDP per employed person

(continued)

(continued)

8.3 Promote development-oriented policies that support productive activities, decent job creation, entrepreneurship, creativity and innovation, and encourage the formalization and growth of micro-, small- and medium-sized enterprises, including through access to financial services	8.3.1 Proportion of informal employment in non-agriculture employment, by sex
8.4 Improve progressively, through 2030, global resource efficiency in consumption and production and endeavour to decouple economic growth from environmental degradation, in accordance with the 10-Year Framework of Programmes on Sustainable Consumption and Production, with developed countries taking the lead	8.4.1 Material footprint, material footprint per capita, and material footprint per GDP
	8.4.2 Domestic material consumption, domestic material consumption per capita, and domestic material consumption per GDP
8.5 By 2030, achieve full and productive employment and decent work for all women and men, including for young people and persons with disabilities, and equal pay for work of equal value	8.5.1 Average hourly earnings of female and male employees, by occupation, age and persons with disabilities
	8.5.2 Unemployment rate, by sex, age and persons with disabilities
8.6 By 2020, substantially reduce the proportion of youth not in employment, education or training	8.6.1 Proportion of youth (aged 15–24 years) not in education, employment or training
8.7 Take immediate and effective measures to eradicate forced labour, end modern slavery and human trafficking and secure the prohibition and elimination of the worst forms of child labour, including recruitment and use of child soldiers, and by 2025 end child labour in all its forms	8.7.1 Proportion and number of children aged 5–17 years engaged in child labour, by sex and age
8.8 Protect labour rights and promote safe and secure working environments for all workers, including migrant workers, in particular women migrants, and those in precarious employment	8.8.1 Frequency rates of fatal and non-fatal occupational injuries, by sex and migrant status
	8.8.2 Increase in national compliance of labour rights (freedom of association and collective bargaining) based on International Labour Organization (ILO) textual sources and national legislation, by sex and migrant status
8.9 By 2030, devise and implement policies to promote sustainable tourism that creates jobs and promotes local culture and products	8.9.1 Tourism direct GDP as a proportion of total GDP and in growth rate
	8.9.2 Number of jobs in tourism industries as a proportion of total jobs and growth rate of jobs, by sex

(continued)

(continued)

8.10 Strengthen the capacity of domestic financial institutions to encourage and expand access to banking, insurance and financial services for all	8.10.1 Number of commercial bank branches and automated teller machines (ATMs) per 100,000 adults
	8.10.2 Proportion of adults (15 years and older) with an account at a bank or other financial institution or with a mobile-money-service provider
8.a Increase Aid for Trade support for developing countries, in particular least developed countries, including through the Enhanced Integrated Framework for Trade-related Technical Assistance to Least Developed Countries	8.a.1 Aid for Trade commitments and disbursements
8.b By 2020, develop and operationalize a global strategy for youth employment and implement the Global Jobs Pact of the International Labour Organization	8.b.1 Total government spending in social protection and employment programmes as a proportion of the national budgets and GDP
Goal 9. Build resilient infrastructure, promote inclusive and sustainable industrialization and foster innovation	
9.1 Develop quality, reliable, sustainable and resilient infrastructure, including regional and trans-border infrastructure, to support economic development and human well-being, with a focus on affordable and equitable access for all	9.1.1 Proportion of the rural population who live within 2 km of an all-season road
	9.1.2 Passenger and freight volumes, by mode of transport
9.2 Promote inclusive and sustainable industrialization and, by 2030, significantly raise industry's share of employment and gross domestic product, in line with national circumstances, and double its share in least developed countries	9.2.1 Manufacturing value added as a proportion of GDP and per capita
	9.2.2 Manufacturing employment as a proportion of total employment
9.3 Increase the access of small-scale industrial and other enterprises, in particular in developing countries, to financial services, including affordable credit, and their integration into value chains and markets	9.3.1 Proportion of small-scale industries in total industry value added
	9.3.2 Proportion of small-scale industries with a loan or line of credit
9.4 By 2030, upgrade infrastructure and retrofit industries to make them sustainable, with increased resource-use efficiency and greater adoption of clean and environmentally sound technologies and industrial processes, with all countries taking action in accordance with their respective capabilities	9.4.1 CO_2 emission per unit of value added

(continued)

(continued)

9.5 Enhance scientific research, upgrade the technological capabilities of industrial sectors in all countries, in particular developing countries, including, by 2030, encouraging innovation and substantially increasing the number of research and development workers per 1 million people and public and private research and development spending	9.5.1 Research and development expenditure as a proportion of GDP 9.5.2 Researchers (in full-time equivalent) per million inhabitants
9.a Facilitate sustainable and resilient infrastructure development in developing countries through enhanced financial, technological and technical support to African countries, least developed countries, landlocked developing countries and small island developing States	9.a.1 Total official international support (official development assistance plus other official flows) to infrastructure
9.b Support domestic technology development, research and innovation in developing countries, including by ensuring a conducive policy environment for, inter alia, industrial diversification and value addition to commodities	9.b.1 Proportion of medium and high-tech industry value added in total value added
9.c Significantly increase access to information and communications technology and strive to provide universal and affordable access to the Internet in least developed countries by 2020	9.c.1 Proportion of population covered by a mobile network, by technology
Goal 10. Reduce inequality within and among countries	
10.1 By 2030, progressively achieve and sustain income growth of the bottom 40% of the population at a rate higher than the national average	10.1.1 Growth rates of household expenditure or income per capita among the bottom 40% of the population and the total population
10.2 By 2030, empower and promote the social, economic and political inclusion of all, irrespective of age, sex, disability, race, ethnicity, origin, religion or economic or other status	10.2.1 Proportion of people living below 50% of median income, by age, sex and persons with disabilities
10.3 Ensure equal opportunity and reduce inequalities of outcome, including by eliminating discriminatory laws, policies and practices and promoting appropriate legislation, policies and action in this regard	10.3.1 Proportion of the population reporting having personally felt discriminated against or harassed within the previous 12 months on the basis of a ground of discrimination prohibited under international human rights law
10.4 Adopt policies, especially fiscal, wage and social protection policies, and progressively achieve greater equality	10.4.1 Labour share of GDP, comprising wages and social protection transfers

(continued)

(continued)

10.5 Improve the regulation and monitoring of global financial markets and institutions and strengthen the implementation of such regulations	10.5.1 Financial Soundness Indicators
10.6 Ensure enhanced representation and voice for developing countries in decision-making in global international economic and financial institutions in order to deliver more effective, credible, accountable and legitimate institutions	10.6.1 Proportion of members and voting rights of developing countries in international organizations
10.7 Facilitate orderly, safe, regular and responsible migration and mobility of people, including through the implementation of planned and well-managed migration policies	10.7.1 Recruitment cost borne by employee as a proportion of yearly income earned in country of destination
	10.7.2 Number of countries that have implemented well-managed migration policies
10.a Implement the principle of special and differential treatment for developing countries, in particular least developed countries, in accordance with World Trade Organization agreements	10.a.1 Proportion of tariff lines applied to imports from least developed countries and developing countries with zero-tariff
10.b Encourage official development assistance and financial flows, including foreign direct investment, to States where the need is greatest, in particular least developed countries, African countries, small island developing States and landlocked developing countries, in accordance with their national plans and programmes	10.b.1 Total resource flows for development, by recipient and donor countries and type of flow (e.g. official development assistance, foreign direct investment and other flows)
10.c By 2030, reduce to less than 3% the transaction costs of migrant remittances and eliminate remittance corridors with costs higher than 5%	10.c.1 Remittance costs as a proportion of the amount remitted
Goal 11. Make cities and human settlements inclusive, safe, resilient and sustainable	
11.1 By 2030, ensure access for all to adequate, safe and affordable housing and basic services and upgrade slums	11.1.1 Proportion of urban population living in slums, informal settlements or inadequate housing
11.2 By 2030, provide access to safe, affordable, accessible and sustainable transport systems for all, improving road safety, notably by expanding public transport, with special attention to the needs of those in vulnerable situations, women, children, persons with disabilities and older persons	11.2.1 Proportion of population that has convenient access to public transport, by sex, age and persons with disabilities

(continued)

(continued)

11.3 By 2030, enhance inclusive and sustainable urbanization and capacity for participatory, integrated and sustainable human settlement planning and management in all countries	11.3.1 Ratio of land consumption rate to population growth rate
	11.3.2 Proportion of cities with a direct participation structure of civil society in urban planning and management that operate regularly and democratically
11.4 Strengthen efforts to protect and safeguard the world's cultural and natural heritage	11.4.1 Total expenditure (public and private) per capita spent on the preservation, protection and conservation of all cultural and natural heritage, by type of heritage (cultural, natural, mixed and World Heritage Centre designation), level of government (national, regional and local/municipal), type of expenditure (operating expenditure/investment) and type of private funding (donations in kind, private non-profit sector and sponsorship)
11.5 By 2030, significantly reduce the number of deaths and the number of people affected and substantially decrease the direct economic losses relative to global gross domestic product caused by disasters, including water-related disasters, with a focus on protecting the poor and people in vulnerable situations	11.5.1 Number of deaths, missing persons and persons affected by disaster per 100,000 people
	11.5.2 Direct disaster economic loss in relation to global GDP, including disaster damage to critical infrastructure and disruption of basic services
11.6 By 2030, reduce the adverse per capita environmental impact of cities, including by paying special attention to air quality and municipal and other waste management	11.6.1 Proportion of urban solid waste regularly collected and with adequate final discharge out of total urban solid waste generated, by cities
	11.6.2 Annual mean levels of fine particulate matter (e.g. PM2.5 and PM10) in cities (population weighted)
11.7 By 2030, provide universal access to safe, inclusive and accessible, green and public spaces, in particular for women and children, older persons and persons with disabilities	11.7.1 Average share of the built-up area of cities that is open space for public use for all, by sex, age and persons with disabilities
	11.7.2 Proportion of persons victim of physical or sexual harassment, by sex, age, disability status and place of occurrence, in the previous 12 months
11.a Support positive economic, social and environmental links between urban, peri-urban and rural areas by strengthening national and regional development planning	11.a.1 Proportion of population living in cities that implement urban and regional development plans integrating population projections and resource needs, by size of city
11.b By 2020, substantially increase the number of cities and human settlements adopting and implementing integrated	11.b.1 Proportion of local governments that adopt and implement local disaster risk reduction strategies in line with the Sendai

(continued)

(continued)

policies and plans towards inclusion, resource efficiency, mitigation and adaptation to climate change, resilience to disasters, and develop and implement, in line with the Sendai Framework for Disaster Risk Reduction 2015–2030, holistic disaster risk management at all levels	Framework for Disaster Risk Reduction 2015–2030a
	11.b.2 Number of countries with national and local disaster risk reduction strategies
11.c Support least developed countries, including through financial and technical assistance, in building sustainable and resilient buildings utilizing local materials	11.c.1 Proportion of financial support to the least developed countries that is allocated to the construction and retrofitting of sustainable, resilient and resource-efficient buildings utilizing local materials
Goal 12. Ensure sustainable consumption and production patterns	
12.1 Implement the 10-Year Framework of Programmes on Sustainable Consumption and Production Patterns, all countries taking action, with developed countries taking the lead, taking into account the development and capabilities of developing countries	12.1.1 Number of countries with sustainable consumption and production (SCP) national action plans or SCP mainstreamed as a priority or a target into national policies
12.2 By 2030, achieve the sustainable management and efficient use of natural resources	12.2.1 Material footprint, material footprint per capita, and material footprint per GDP
	12.2.2 Domestic material consumption, domestic material consumption per capita, and domestic material consumption per GDP
12.3 By 2030, halve per capita global food waste at the retail and consumer levels and reduce food losses along production and supply chains, including post-harvest losses	12.3.1 Global food loss index
12.4 By 2020, achieve the environmentally sound management of chemicals and all wastes throughout their life cycle, in accordance with agreed international frameworks, and significantly reduce their release to air, water and soil in order to minimize their adverse impacts on human health and the environment	12.4.1 Number of parties to international multilateral environmental agreements on hazardous waste, and other chemicals that meet their commitments and obligations in transmitting information as required by each relevant agreement
	12.4.2 Hazardous waste generated per capita and proportion of hazardous waste treated, by type of treatment
12.5 By 2030, substantially reduce waste generation through prevention, reduction, recycling and reuse	12.5.1 National recycling rate, tons of material recycled
12.6 Encourage companies, especially large and transnational companies, to adopt sustainable practices and to integrate sustainability information into their reporting cycle	12.6.1 Number of companies publishing sustainability reports

(continued)

(continued)

12.7 Promote public procurement practices that are sustainable, in accordance with national policies and priorities	12.7.1 Number of countries implementing sustainable public procurement policies and action plans
12.8 By 2030, ensure that people everywhere have the relevant information and awareness for sustainable development and lifestyles in harmony with nature	12.8.1 Extent to which (i) global citizenship education and (ii) education for sustainable development (including climate change education) are mainstreamed in (a) national education policies; (b) curricula; (c) teacher education; and (d) student assessment
12.a Support developing countries to strengthen their scientific and technological capacity to move towards more sustainable patterns of consumption and production	12.a.1 Amount of support to developing countries on research and development for sustainable consumption and production and environmentally sound technologies
12.b Develop and implement tools to monitor sustainable development impacts for sustainable tourism that creates jobs and promotes local culture and products	12.b.1 Number of sustainable tourism strategies or policies and implemented action plans with agreed monitoring and evaluation tools
12.c Rationalize inefficient fossil-fuel subsidies that encourage wasteful consumption by removing market distortions, in accordance with national circumstances, including by restructuring taxation and phasing out those harmful subsidies, where they exist, to reflect their environmental impacts, taking fully into account the specific needs and conditions of developing countries and minimizing the possible adverse impacts on their development in a manner that protects the poor and the affected communities	12.c.1 Amount of fossil-fuel subsidies per unit of GDP (production and consumption) and as a proportion of total national expenditure on fossil fuels
Goal 13. Take urgent action to combat climate change and its impacts[c]	
13.1 Strengthen resilience and adaptive capacity to climate-related hazards and natural disasters in all countries	13.1.1 Number of countries with national and local disaster risk reduction strategies
	13.1.2 Number of deaths, missing persons and persons affected by disaster per 100,000 people

(continued)

(continued)

13.2 Integrate climate change measures into national policies, strategies and planning	13.2.1 Number of countries that have communicated the establishment or operationalization of an integrated policy/strategy/plan which increases their ability to adapt to the adverse impacts of climate change, and foster climate resilience and low greenhouse gas emissions development in a manner that does not threaten food production (including a national adaptation plan, nationally determined contribution, national communication, biennial update report or other)
13.3 Improve education, awareness-raising and human and institutional capacity on climate change mitigation, adaptation, impact reduction and early warning	13.3.1 Number of countries that have integrated mitigation, adaptation, impact reduction and early warning into primary, secondary and tertiary curricula
	13.3.2 Number of countries that have communicated the strengthening of institutional, systemic and individual capacity-building to implement adaptation, mitigation and technology transfer, and development actions
13.a Implement the commitment undertaken by developed-country parties to the United Nations Framework Convention on Climate Change to a goal of mobilizing jointly $100 billion annually by 2020 from all sources to address the needs of developing countries in the context of meaningful mitigation actions and transparency on implementation and fully operationalize the Green Climate Fund through its capitalization as soon as possible	13.a.1 Mobilized amount of United States dollars per year starting in 2020 accountable towards the $100 billion commitment
13.b Promote mechanisms for raising capacity for effective climate change-related planning and management in least developed countries and small island developing States, including focusing on women, youth and local and marginalized communities	13.b.1 Number of least developed countries and small island developing States that are receiving specialized support, and amount of support, including finance, technology and capacity-building, for mechanisms for raising capacities for effective climate change-related planning and management, including focusing on women, youth and local and marginalized communities

(continued)

(continued)

Goal 14. Conserve and sustainably use the oceans, seas and marine resources for sustainable development	
14.1 By 2025, prevent and significantly reduce marine pollution of all kinds, in particular from land-based activities, including marine debris and nutrient pollution	14.1.1 Index of coastal eutrophication and floating plastic debris density
14.2 By 2020, sustainably manage and protect marine and coastal ecosystems to avoid significant adverse impacts, including by strengthening their resilience, and take action for their restoration in order to achieve healthy and productive oceans	14.2.1 Proportion of national exclusive economic zones managed using ecosystem-based approaches
14.3 Minimize and address the impacts of enhanced scientific cooperation at all levels	14.3.1 Average marine acidity (pH) measured at agreed suite of representative sampling stations
14.4 By 2020, effectively regulate harvesting and end overfishing, illegal, unreported and unregulated fishing and destructive fishing practices and implement science-based management plans, in order to restore fish stocks in the shortest time feasible, at least to levels that can produce maximum sustainable yield as determined by their biological characteristics	14.4.1 Proportion of fish stocks within biologically sustainable levels
14.5 By 2020, conserve at least 10% of coastal and marine areas, consistent with national and international law and based on the best available scientific information	14.5.1 Coverage of protected areas in relation to marine areas
14.6 By 2020, prohibit certain forms of fisheries subsidies which contribute to overcapacity and overfishing, eliminate subsidies that contribute to illegal, unreported and unregulated fishing and refrain from introducing new such subsidies, recognizing that appropriate and effective special and differential treatment for developing and least developed countries should be an integral part of the World Trade Organization fisheries subsidies negotiation[d]	14.6.1 Progress by countries in the degree of implementation of international instruments aiming to combat illegal, unreported and unregulated fishing
14.7 By 2030, increase the economic benefits to small island developing States and least developed countries from the sustainable use of marine resources, including through sustainable management of fisheries, aquaculture and tourism	14.7.1 Sustainable fisheries as a percentage of GDP in small island developing States, least developed countries and all countries

(continued)

(continued)

14.a Increase scientific knowledge, develop research capacity and transfer marine technology, taking into account the Intergovernmental Oceanographic Commission Criteria and Guidelines on the Transfer of Marine Technology, in order to improve ocean health and to enhance the contribution of marine biodiversity to the development of developing countries, in particular small island developing States and least developed countries	14.a.1 Proportion of total research budget allocated to research in the field of marine technology
14.b Provide access for small-scale artisanal fishers to marine resources and markets	14.b.1 Progress by countries in the degree of application of a legal/regulatory/policy/ institutional framework which recognizes and protects access rights for small-scale fisheries
14.c Enhance the conservation and sustainable use of oceans and their resources by implementing international law as reflected in the United Nations Convention on the Law of the Sea, which provides the legal framework for the conservation and sustainable use of oceans and their resources, as recalled in paragraph 158 of "The future we want"	14.c.1 Number of countries making progress in ratifying, accepting and implementing through legal, policy and institutional frameworks, ocean-related instruments that implement international law, as reflected in the United Nation Convention on the Law of the Sea, for the conservation and sustainable use of the oceans and their resources
Goal 15. Protect, restore and promote sustainable use of terrestrial ecosystems, sustainably manage forests, combat desertification, and halt and reverse land degradation and halt biodiversity loss	
15.1 By 2020, ensure the conservation, restoration and sustainable use of terrestrial and inland freshwater ecosystems and their services, in particular forests, wetlands, mountains and drylands, in line with obligations under international agreements	15.1.1 Forest area as a proportion of total land area
	15.1.2 Proportion of important sites for terrestrial and freshwater biodiversity that are covered by protected areas, by ecosystem type
15.2 By 2020, promote the implementation of sustainable management of all types of forests, halt deforestation, restore degraded forests and substantially increase afforestation and reforestation globally	15.2.1 Progress towards sustainable forest management
15.3 By 2030, combat desertification, restore degraded land and soil, including land affected by desertification, drought and floods, and strive to achieve a land degradation-neutral world	15.3.1 Proportion of land that is degraded over total land area

(continued)

(continued)

15.4 By 2030, ensure the conservation of mountain ecosystems, including their biodiversity, in order to enhance their capacity to provide benefits that are essential for sustainable development	15.4.1 Coverage by protected areas of important sites for mountain biodiversity
	15.4.2 Mountain Green Cover Index
15.5 Take urgent and significant action to reduce the degradation of natural habitats, halt the loss of biodiversity and, by 2020, protect and prevent the extinction of threatened species	15.5.1 Red List Index
15.6 Promote fair and equitable sharing of the benefits arising from the utilization of genetic resources and promote appropriate access to such resources, as internationally agreed	15.6.1 Number of countries that have adopted legislative, administrative and policy frameworks to ensure fair and equitable sharing of benefits
15.7 Take urgent action to end poaching and trafficking of protected species offlora and fauna and address both demand and supply of illegal wildlife products	15.7.1 Proportion of traded wildlife that was poached or illicitly trafficked
15.8 By 2020, introduce measures to prevent the introduction and significantly reduce the impact of invasive alien species on land and water ecosystems and control or eradicate the priority species	15.8.1 Proportion of countries adopting relevant national legislation and adequately resourcing the prevention or control of invasive alien species
15.9 By 2020, integrate ecosystem and biodiversity values into national and local planning, development processes, poverty reduction strategies and accounts	15.9.1 Progress towards national targets established in accordance with Aichi Biodiversity Target 2 of the Strategic Plan for Biodiversity 2011–2020
15.a Mobilize and significantly increase financial resources from all sources to conserve and sustainably use biodiversity and ecosystems	15.a.1 Official development assistance and public expenditure on conservation and sustainable use of biodiversity and ecosystems
15.b Mobilize significant resources from all sources and at all levels to finance sustainable forest management and provide adequate incentives to developing countries to advance such management, including for conservation and reforestation	15.b.1 Official development assistance and public expenditure on conservation and sustainable use of biodiversity and ecosystems
15.c Enhance global support for efforts to combat poaching and trafficking of protected species, including by increasing the capacity of local communities to pursue sustainable livelihood opportunities	15.c.1 Proportion of traded wildlife that was poached or illicitly trafficked

(continued)

(continued)

Goal 16. Promote peaceful and inclusive societies for sustainable development, provide access to justice for all and build effective, accountable and inclusive institutions at all levels	
16.1 Significantly reduce all forms of violence and related death rates everywhere	16.1.1 Number of victims of intentional homicide per 100,000 population, by sex and age
	16.1.2 Conflict-related deaths per 100,000 population, by sex, age and cause
	16.1.3 Proportion of population subjected to physical, psychological or sexual violence in the previous 12 months
	16.1.4 Proportion of population that feel safe walking alone around the area they live
16.2 End abuse, exploitation, trafficking and all forms of violence against and torture of children	16.2.1 Proportion of children aged 1–17 years who experienced any physical punishment and/or psychological aggression by caregivers in the past month
	16.2.2 Number of victims of human trafficking per 100,000 population, by sex, age and form of exploitation
	16.2.3 Proportion of young women and men aged 18–29 years who experienced sexual violence by age 18
16.3 Promote the rule of law at the national and international levels and ensure equal access to justice for all	16.3.1 Proportion of victims of violence in the previous 12 months who reported their victimization to competent authorities or other officially recognized conflict resolution mechanisms
	16.3.2 Unsentenced detainees as a proportion of overall prison population
16.4 By 2030, significantly reduce illicit financial and arms flows, strengthen the recovery and return of stolen assets and combat all forms of organized crime	16.4.1 Total value of inward and outward illicit financial flows (in current United States dollars)
	16.4.2 Proportion of seized small arms and light weapons that are recorded and traced, in accordance with international standards and legal instruments

(continued)

(continued)

16.5 Substantially reduce corruption and bribery in all their forms	16.5.1 Proportion of persons who had at least one contact with a public official and who paid a bribe to a public official, or were asked for a bribe by those public officials, during the previous 12 months
	16.5.2 Proportion of businesses that had at least one contact with a public official and that paid a bribe to a public official, or were asked for a bribe by those public officials during the previous 12 months
16.6 Develop effective, accountable and transparent institutions at all levels	16.6.1 Primary government expenditures as a proportion of original approved budget, by sector (or by budget codes or similar)
	16.6.2 Proportion of the population satisfied with their last experience of public services
16.7 Ensure responsive, inclusive, participatory and representative decision-making at all levels	16.7.1 Proportions of positions (by sex, age, persons with disabilities and population groups) in public institutions (national and local legislatures, public service, and judiciary) compared to national distributions
	16.7.2 Proportion of population who believe decision-making is inclusive and responsive, by sex, age, disability and population group
16.8 Broaden and strengthen the participation of developing countries in the institutions of global governance	16.8.1 Proportion of members and voting rights of developing countries in international organizations
16.9 By 2030, provide legal identity for all, including birth registration	16.9.1 Proportion of children under 5 years of age whose births have been registered with a civil authority, by age
16.10 Ensure public access to information and protect fundamental freedoms, in accordance with national legislation and international agreements	16.10.1 Number of verified cases of killing, kidnapping, enforced disappearance, arbitrary detention and torture of journalists, associated media personnel, trade unionists and human rights advocates in the previous 12 months
	16.10.2 Number of countries that adopt and implement constitutional, statutory and/or policy guarantees for public access to information
16.a Strengthen relevant national institutions, including through international cooperation, for building capacity at all levels, in particular in developing countries, to prevent violence and combat terrorism and crime	16.a.1 Existence of independent national human rights institutions in compliance with the Paris Principles

(continued)

(continued)

16.b Promote and enforce non-discriminatory laws and policies for sustainable development	16.b.1 Proportion of population reporting having personally felt discriminated against or harassed in the previous 12 months on the basis of a ground of discrimination prohibited under international human rights law
Goal 17. Strengthen the means of implementation and revitalize the Global Partnership for Sustainable Development	
Finance	
17.1 Strengthen domestic resource mobilization, including through international support to developing countries, to improve domestic capacity for tax and other revenue collection	17.1.1 Total government revenue as a proportion of GDP, by source
	17.1.2 Proportion of domestic budget funded by domestic taxes
17.2 Developed countries to implement fully their official development assistance commitments, including the commitment by many developed countries to achieve the target of 0.7% of gross national income for official development assistance (ODA/GNI) to developing countries and 0.15–0.20% of ODA/GNI to least developed countries; ODA providers are encouraged to consider setting a target to provide at least 0.20% of ODA/GNI to least developed countries	17.2.1 Net official development assistance, total and to least developed countries, as a proportion of the Organization for Economic Cooperation and Development (OECD) Development Assistance Committee donors' gross national income (GNI)
17.3 Mobilize additional financial resources for developing countries from multiple sources	17.3.1 Foreign direct investments (FDI), official development assistance and South-South Cooperation as a proportion of total domestic budget
	17.3.2 Volume of remittances (in United States dollars) as a proportion of total GDP
17.4 Assist developing countries in attaining long-term debt sustainability through coordinated policies aimed at fostering debt financing, debt relief and debt restructuring, as appropriate, and address the external debt of highly indebted poor countries to reduce debt distress	17.4.1 Debt service as a proportion of exports of goods and services
17.5 Adopt and implement investment promotion regimes for least developed countries	17.5.1 Number of countries that adopt and implement investment promotion regimes for least developed countries
Technology	
17.6 Enhance North-South, South-South and triangular regional and international cooperation on and access to science, technology and innovation and enhance knowledge-sharing on mutually agreed terms,	17.6.1 Number of science and/or technology cooperation agreements and programmes between countries, by type of cooperation

(continued)

(continued)

including through improved coordination among existing mechanisms, in particular at the United Nations level, and through a global technology facilitation mechanism	17.6.2 Fixed Internet broadband subscriptions per 100 inhabitants, by speed
17.7 Promote the development, transfer, dissemination and diffusion of environmentally sound technologies to developing countries on favourable terms, including on concessional and preferential terms, as mutually agreed	17.7.1 Total amount of approved funding for developing countries to promote the development, transfer, dissemination and diffusion of environmentally sound technologies
17.8 Fully operationalize the technology bank and science, technology and innovation capacity-building mechanism for least developed countries by 2017 and enhance the use of enabling technology, in particular information and communications technology	17.8.1 Proportion of individuals using the Internet
Capacity-building	
17.9 Enhance international support for implementing effective and targeted capacity-building in developing countries to support national plans to implement all the Sustainable Development Goals, including through North-South, South-South and triangular cooperation	17.9.1 Dollar value of financial and technical assistance (including through North-South, South-South and triangular cooperation) committed to developing countries
Trade	
17.10 Promote a universal, rules-based, open, non-discriminatory and equitable multilateral trading system under the World Trade Organization, including through the conclusion of negotiations under its Doha Development Agenda	17.10.1 Worldwide weighted tariff-average
17.11 Significantly increase the exports of developing countries, in particular with a view to doubling the least developed countries' share of global exports by 2020	17.11.1 Developing countries' and least developed countries' share of global exports
17.12 Realize timely implementation of duty-free and quota-free market access on a lasting basis for all least developed countries, consistent with World Trade Organization decisions, including by ensuring that preferential rules of origin applicable to imports from least developed countries are transparent and simple, and contribute to facilitating market access	17.12.1 Average tariffs faced by developing countries, least developed countries and small island developing States

(continued)

(continued)

Systemic issues	
Policy and institutional coherence	
17.13 Enhance global macroeconomic stability, including through policy coordination and policy coherence	17.13.1 Macroeconomic Dashboard
17.14 Enhance policy coherence for sustainable development	17.14.1 Number of countries with mechanisms in place to enhance policy coherence of sustainable development
17.15 Respect each country's policy space and leadership to establish and implement policies for poverty eradication and sustainable development	17.15.1 Extent of use of country-owned results frameworks and planning tools by providers of development cooperation
Multi-stakeholder partnerships	
17.16 Enhance the Global Partnership for Sustainable Development, complemented by multi-stakeholder partnerships that mobilize and share knowledge, expertise, technology and financial resources, to support the achievement of the Sustainable Development Goals in all countries, in particular developing countries	17.16.1 Number of countries reporting progress in multi-stakeholder development effectiveness monitoring frameworks that support the achievement of the sustainable development goals
17.17 Encourage and promote effective public, public-private and civil society partnerships, building on the experience and resourcing strategies of partnerships	17.17.1 Amount of United States dollars committed to public-private and civil society partnerships
Data, monitoring and accountability	
17.18 By 2020, enhance capacity-building support to developing countries, including for least developed countries and small island developing States, to increase significantly the availability of high-quality, timely and reliable data disaggregated by income, gender, age, race, ethnicity, migratory status, disability, geographic location and other characteristics relevant in national contexts	17.18.1 Proportion of sustainable development indicators produced at the national level with full disaggregation when relevant to the target, in accordance with the Fundamental Principles of Official Statistics
	17.18.2 Number of countries that have national statistical legislation that complies with the Fundamental Principles of Official Statistics
	17.18.3 Number of countries with a national statistical plan that is fully funded and under implementation, by source of funding

(continued)

(continued)

17.19 By 2030, build on existing initiatives to develop measurements of progress on sustainable development that complement gross domestic product, and support statistical capacity-building in developing countries	17.19.1 Dollar value of all resources made available to strengthen statistical capacity in developing countries
	17.19.2 Proportion of countries that (a) have conducted at least one population and housing census in the last 10 years; and (b) have achieved 100% birth registration and 80% death registration

[a]As contained in **Annex IV** of the **Report of the Inter-Agency and Expert Group on Sustainable Development Goal Indicators (E/CN.3/2016/2/Rev.1)** and agreed upon, as a practical starting point at the 47th session of the United Nations Statistical Commission held in March 2016

[b]An open-ended intergovernmental expert working group on indicators and terminology relating to disaster risk reduction established by the General Assembly (resolution 69/284) is developing a set of indicators to measure global progress in the implementation of the Sendai Framework. These indicators will eventually reflect the agreements on the Sendai Framework indicators

[c]Acknowledging that the United Nations Framework Convention on Climate Change is the primary international, intergovernmental forum for negotiating the global response to climate change

[d]Taking into account ongoing World Trade Organization negotiations, the Doha Development Agenda and the Hong Kong ministerial mandate

References

ACTS. (2017). Australasian Campus Towards Sustainability (ACTS). http://www.acts.asn.au/. Accessed June, 2017.

Cogan, J. K., Hurd, I., & Johnstone, I. (Ed.). (2015). *The UN private sector relationship: A partnership recovered*. Oxford Handbook of International Organizations.

Habitat III: United Nations Conference on Housing and Sustainable Urban Development. (2016). "New Urban Agenda" Quito. http://citiscope.org/sites/default/files/h3/Habitat_III_New_Urban_Agenda_10_September_2016.pdf. Accessed June, 2017.

Haertle, J., & Kongstad, A. (2014). Responsible business—The new normal? *AUC Business Review, 50*.

HESI: Higher Education Sustainability Initiative. (2017). *Higher Education Institutions—Key Driver of the Sustainable Development Goals*. Conference, 19 July 2017, 3–5 pm, Conference room 1, UNHQ, New York.

Horwitz, F., & Grayson, D. (2010). Putting PRME into practice in a business school. *EFMD Global Focus, 4*.

IUSDRP. (2017). Inter-University Sustainable Development Research Programme (IUSDRP). https://www.haw-hamburg.de/en/ftz-nk/programmes/iusdrp.html. Accessed June, 2017.

Ivey, R. (Ed.). (2011). *How to find, produce and integrate case studies that promote PRME values*. Canada: Ivey Publishing.

Kell, G., & Haertle, J. (2011). UN global compact and principles for responsible management education: The next decades. *EFMD Global Focus, 5*, 14–16.

LiFE Index. (2017). LiFE Index. http://life.acts.asn.au/. Accessed June, 2017.

PRME. (2017a). *Impact: A decade of principles for responsible management education*. PRME Global Compact.

PRME. (2017b). Overview. http://www.unprme.org/about-prme/index.php. Accessed June 15, 2017.

PRME. (2017c). Six Principles, Principles for Responsible Management Education. http://www.unprme.org/about-prme/the-six-principles.php. Accessed June 15, 2017.

Rasche, A., & Escudero, M. (2010). Leading change the role of the principles of responsible management education. *Journal of Business and Economic Ethics, 10*.

Stachowicz-Stanusch, A. (2011). The implementation of principles for responsible management education in practice-research results. *Journal of Intercultural Management, 3*, 2.

United Nations (UN). (2015). Sustainable Development Goals. In: *Resolution adopted by the General Assembly* on September 25, 2015.

United Nations (UN). (2016). Transforming our world: The 2030 agenda for sustainable development. In *A/RES/70/1*, United Nations.

United Nations Conference on Sustainable Development. (2012). *A 10-year framework of programmes on sustainable consumption and production patterns*. United Nations, A/CONF.216/5, http://www.scpclearinghouse.org/sites/default/files/10yfp-a-conf.216-5-en.pdf. Accessed June, 2017.

United Nations Educational, Scientific and Cultural Organization (UNESCO). (2014a). *Roadmap for implementing the global action programme on education for sustainable development*. UNESCO, France.

United Nations Educational, Scientific and Cultural Organization (UNESCO). (2014b). Global education for all meeting—The Muscat agreement. In *ED-14/EFA/ME/3* edited by UNESCO.

United Nations Environment. (2015a). UN 10YFP. http://www.unep.org/10yfp/about/what-10yfp. Accessed June, 2017.

United Nations Environment. (2015b). Sustainable Public Procurement. http://www.unep.org/10yfp/programmes/sustainable-public-procurement. Accessed June, 2017.

United Nations Environment. (2015c). Consumer Information Programme for Sustainable Consumption and Production. http://www.unep.org/10yfp/programmes/consumer-information. Accessed June, 2017.

United Nations Environment. (2015d). Sustainable Tourism Programme. http://www.unep.org/10yfp/programmes/sustainable-tourism-programme. Accessed June, 2017.

United Nations Environment. (2015e). Sustainable Lifestyles and Education Programme. http://www.unep.org/10yfp/programmes/sustainable-lifestyles-and-education-programme. Accessed June, 2017.

United Nations Environment. (2015f). Sustainable Buildings and Construction Programme. http://www.unep.org/10yfp/programmes/sustainable-buildings-and-construction-programme. Accessed June, 2017.

United Nations Environment. (2015g). Sustainable Food Systems Programme. http://www.unep.org/10yfp/programmes/sustainable-food-systems-programme. Accessed June, 2017.
United Nations Habitat. (2017). *Action framework for implementation of the New Urban Agenda*. United Nations.

Achieving Both Breadth and Depth: How Sustainability Education Is Being Integrated Across All Undergraduate Courses at La Trobe University, Australia

Colin Hocking, Silvia McCormack, Swati Nagpal and Alison Lugg

Abstract In 2012 La Trobe University decided that every undergraduate student, across every discipline, would have significant and assessed experience of Sustainability Education (SE). The plan for achieving full coverage of SE at La Trobe was reported at a previous World Symposium for Sustainable Development in Higher Education conference (Hocking and Riddle in Aiming for full coverage—integrating sustainability education into all undergraduate courses at La Trobe University, Australia: achievements, lessons learnt and barriers addressed. Springer International Publishing, Cham, pp. 479–493, 2015). Since then, implementing the plan has involved: adopting a university-wide definition of SE; aligned with international definitions, and adaptable within each discipline; establishing a process for tracking subjects and courses that have successfully incorporated SE; and providing education development support for staff, using strategies developed for each discipline context (Hocking in Soc Educ 2015 33(2):26–38, 2015). Across the Arts, Social Sciences and Commerce (ASSC) College (a type of super-faculty), SE has been incorporated in 38 subjects, covering 90% of courses, across all disciplines. Promoting breadth has not compromised depth of development. Two example are: (1) a new common core subject for all Business Degree students called Sustainability; (2) SE is now in five core and elective subjects in the School of Education, increasing the likelihood that students will to encounter SE more than

C. Hocking (✉)
Emeritus in School of Education, Formerly La Trobe Learning & Teaching (LTLT), La Trobe University, Kingsbury Drive, Bundoora, Melbourne, VIC 3086, Australia
e-mail: c.hocking@latrobe.edu.au

S. McCormack
College of Arts Social Sciences & Commerce (ASSC), La Trobe University, Kingsbury Drive, Bundoora, Melbourne, VIC 3086, Australia

S. Nagpal
La Trobe Business School, La Trobe University, Kingsbury Drive, Bundoora, Melbourne, VIC 3086, Australia

A. Lugg
School of Education (Bendigo campus), La Trobe University, Kingsbury Drive, Bundoora, Melbourne, VIC 3086, Australia

W. Leal Filho et al. (eds.), *Sustainable Development Research in the Asia-Pacific Region*, World Sustainability Series, https://doi.org/10.1007/978-3-319-73293-0_4

once in education-related degree programs. These outcomes have depended on framing SE as an educational issue, integrated with other university education development initiatives, and taking whole-of-institution approaches that have involved and provided coordination of stakeholders at all levels.

Keywords Sustainability education · EfS · ESD · Education · Educational development · Sustainable development · Curriculum · Assessment Coverage · Course maps

1 Introduction: How Should Sustainability Education Be Incorporated?

There is discussion in the higher education literature as to the best ways to incorporate sustainability education (SE) in courses and subjects (Ryan and Tilbury 2013; Sterling et al. 2013). Currently most higher education institutions have SE incorporated in some way in 10–20 percent of courses, in Australia (Scott et al. 2012), and the percentage is probably similar internationally (GUNI 2011). This limited incorporation of SE is possibly because employers and countries are not actively seeking this capability from graduates as yet, even though many have sustainability oriented policies and strategies. More countries are becoming aware of the importance of this need. For example, guidelines developed for UK Higher Education institutions in 2014 aim to create a framework for sustainability within the curricula of universities (Targeted News Service 2014).

There is also debate as to whether incorporation of SE across a wider range of courses is possible, or even desirable. Over two decades ago Orr (1994) put a compelling case for broadening the range of courses with SE incorporated:

> The truth is that many of the things on which our future health and prosperity depend are in dire jeopardy… this is not the work of ignorant people. Rather it is largely the results of work by people with BAs, BScs, LLBs, MBAs and PhDs….

2 Purpose: Sustainability Education Across All Courses

In 2012 La Trobe University committed to ensuring that every undergraduate student, across all disciplines, would have significant and assessed experience of sustainability education, whatever their course. Incorporating sustainability into core or large sized elective subjects, in particular, would mean that students would become aware of and reflect on sustainability in the context of their own discipline area. La Trobe University saw this as an opportunity to encourage action for the common good—a more sustainable future for the planet, as well as a way to engage students with the challenging issues of our time (La Trobe University Future Ready 2013).

The plan for incorporating SE at La Trobe was via the introduction into all undergraduate degree courses of three "Essentials". The Essentials are Innovation and Entrepreneurship (I&E), Global Citizenship (GC) and Sustainability Thinking (ST). These are part of a broader strategy to differentiate La Trobe's degrees, and to 'equip all students with the knowledge and skills to solve real world problems' (La Trobe University Future Ready 2013, p. 4). The Essentials intend to "equip students to think and respond beyond conventional boundaries, to foster adaptable thinking and the capacity to apply knowledge and skills in a future as yet unknown to us" (La Trobe University 2015). Although there are significant linkages between the three different areas covered by the Essentials, only Sustainability Thinking (ST) is discussed in this paper. How Sustainability Thinking has been defined, developed, expressed, assessed and monitored, is discussed below.

An outline of the overall plan for achieving the incorporation of ST in all undergraduate courses was reported to the World Symposium on Sustainability Development in Higher Education in Manchester in 2014 (Hocking and Riddle 2015).

3 Approach: Institution-Wide

Once the plan for incorporating ST was put into effect, a number of additional significant processes to those identified in Hocking and Riddle (2015) were required for successful implementation. The final process for implementing the plan involved, in summary:

1. Having a brief, workable University-wide shared definition of what sustainability education meant, called the La Trobe Sustainability Thinking (ST) Essential—see La Trobe University Website (2017). This definition was broadly in line with the Australian Government (2009) National Action Plan for Education for Sustainability (EfS), and with international definitions of Education for Sustainable Development (Tilbury 2011). The definition of Sustainability Thinking for La Trobe University is as follows:

> "Sustainability Thinking is a capacity to engage effectively with social, environmental and economic change and challenges in the contemporary world. These include, for example, climate change, food and water security and human and labour rights.

Sustainability Thinking demands that all La Trobe University students reflect on:

- the complex interactions between natural, economic, social, political and cultural systems;
- our obligations to future generations; and,
- how the choices we make will affect the public good and the wellbeing of future generations.

At La Trobe, Sustainable Thinking is inextricably linked to good global citizenship."

From: Essentials: Quick Guide for Sustainability Thinking (La Trobe University Website 2017)

2. Mandating that ST needed to be evident in each of the intended learning outcomes, student learning activities and subject assessment tasks, for each relevant subject, and at least 25% of the assessment needed to be directly related to the ST Essential.

3. Establishing a process to check that ST was being adequately incorporated in ways that met the University definition. Staff seeking approval of ST in their subject were required to complete a brief approval application document (the Essentials Memorandum) which was reviewed by the appropriate Course Advisory Committee (CAC) of one of the two major Colleges in the University (equivalent to super-faculties). An example of the Essentials Memorandum relevant to Sustainability Thinking is in Appendix 1.
Note While staff coordinating subjects that were incorporating ST were required to meet the minimum standards at least, they were also encouraged to go beyond the minimum, and integrate ST more fully into selected subjects, and across a range of subjects.

4. Developing a course mapping process to identify which subjects were most suitable for incorporation of ST, within each course, and to ensure that each student completed at least one subject with ST incorporated, as part of their course. Appendix 2 shows a generic example of a course map and how the incorporation of ST, and the other two Essentials, were tracked for each course.

5. Appointing an appropriately qualified, University-wide education development officer (Colin Hocking) to work with staff to develop the best ways to incorporate ST in selected subjects, and advise on the best approaches for staff involved in development of ST, and on how best to meet the College-based checking process.

La Trobe University now has two super-faculties, called Colleges, that each oversee about half of the La Trobe Undergraduate courses. At the time of preparing this paper, the College of Arts, Social Sciences and Commerce (ASSC) had achieved close to full coverage for ST in all courses. The other College (Science, Health and Engineering or SHE) is currently in the process of incorporating ST into all courses. For simplicity, the analysis of outcomes for incorporation of ST at La Trobe outlined below focuses on courses and subjects within the College of ASSC.

4 Outcomes: Close to Full Coverage of ST in Courses

According to the course mapping process for Essentials in ASSC (overseen by Silvia McCormack), by mid 2017 ST had been incorporated in 38 subjects, covering 90% of courses and a wide range of disciplines, spanning Media & Journalism, Politics, Languages, Visual Arts, Business, Law, Social Sciences, Education and many more. Examples of how ST has been incorporated in some courses, in ways that go beyond the minimum La Trobe University requirements, are outlined later in this paper.

As a result of the process of incorporating ST in this wide range of subjects across the University, some new perspectives and approaches were developed, around the best ways of aligning the content, learning outcomes and assessment tasks of the subject with those of sustainability. Developing and implementing these approach presented opportunities to explore how to meet the challenge of embedding sustainability within courses with tight discipline boundaries. Some of the ways that this challenge was addressed are reported on in Hocking (2015). They included:

- Having a clear, concise definition of what sustainability thinking means, firmly articulated in University policy, that at the same time allows discipline-appropriate ways of meeting the University requirements.
- Taking an approach that moves backward and forward between the discipline-based academic's expert view of their subject and the experience of the SE education developer, and between the three key requirements for showing successful incorporation of ST, by embedding ST in intended learning outcomes, student learning activities, and assessment tasks (in at least in 25% of assessment, and one major assessment task), until full alignment between these three is achieved (Biggs and Tang 2007).
- Identifying the extent to which elements of ST were already present (if at all) in the chosen subject, and what pathway for development needed to be taken as a result. There appeared to be four main pathways for incorporation of ST in a subject: (1) strengthening ST if it already exists in a subject as a threaded theme: (2) strengthening existing ST that is already a discrete part of the subject; (3) introducing a new section, often linked to a project-based assessment task that could meet the ST requirements; or (4) reframing the focus of part of the subject so that ST could be included (Hocking 2015).
- In instances where it was not clear at the start how ST could be incorporated in a subject (for example, language course subjects, some health subjects, social sciences), a way was sometimes found by identifying an appropriate context for accommodating ST while meeting subject content requirements (Hocking 2015).

For example, in French, the context found for ST was the study of future tenses in the language ('we will be sustainable, we might be sustainable, we can be sustainable, we ought to be sustainable') and then using sustainability as a student learning and assessment context for how to express these various forms of future tense in the language.

The context thinking approach, outlined above, was optimised by using a nested systems model of sustainability, in which the economic is seen as a subset of the societal, and these in turn are seen as a subset of the ecological (Fig. 1). This view of how sustainability relates to subjects helps to inform techniques that develop useful 'context lenses' to identify the best vehicles for embedding sustainability in ways relevant to a course, that also minimise disruption of its disciplinary content. For example, in Orthoptics, the study of eyesight, one useful context identified was diabetes, a major contributor to eyesight problems in communities. A way to embed SE was found by considering the economic (e.g. costs of treatment), social (e.g. community approaches to physical activity) and environmental (e.g. walkable cities, healthy food) dimensions of eyesight problems. Details of the 'context lens' approach are described in Hocking (2015).

In practice, the process of developing ST in subjects, and having this recognized formally in the University, usually proceeded as follows:

1. Subjects identified with potential for incorporation of ST were listed on whole-of-course maps. Subjects deemed appropriate for incorporation of ST were usually chosen for a combination of factors: the suitability of the subject for potential incorporation of ST; the extent of reach of the subject across the course (e.g. whether it was a core subject or other major subject for a course, or even several courses); where the other Essentials were being incorporated into subjects within the course.

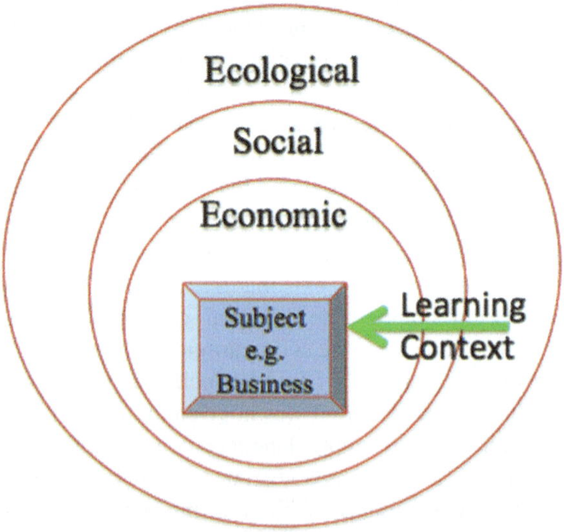

Fig. 1 Sustainability as a set of nested relationships, with the economic nested inside the social that is in turn nested within the ecological. This promotes the identification of a learning context for sustainability that is more complex and inter-disciplinary, and draws on each of these inter-related layers

2. The coordinators of, and other major contributors to, each of the chosen subjects discussed with the ST education developer some potential ways in which ST might be incorporated (using one of the four approaches outlined above)—this might be a rapid process, or it might take several steps. Sometimes external examples or broader ideas for incorporating SE into a particular discipline needed to be identified and followed up. Where the 'context lens' approach was applied, this often involved a type of Course Design Intensive process (Oxford-Brookes University Website 2017).
3. The coordinator of the subject drafted the first version of how ST might look once incorporated into the subject, in terms of the intended learning outcomes, student learning activities and subject assessment tasks (as required by La Trobe).
4. The ST education developer provided feedback on this draft, in terms of curriculum ideas, extent of fit with the University requirements (for each of the three categories) and clarity of expression—and whether the draft would broadly fit with the University requirements or whether additional work on the embedding of ST in the subject was required.
5. Either the final version of the proposal for incorporating ST in the subject was developed by the subject coordinator, or any additional work required was carried out by the subject coordinator and other key staff, supported by the ST education developer.
6. The final version of the proposal for incorporation of ST was framed up within a formal memo for the Curriculum Advisory Committee (CAC) of the appropriate college, outlining how the subject had been incorporated, in terms of subject description for students, intended learning outcomes, student learning activities and subject assessment tasks (Appendix 1).

   Note There was no formal requirement for subject coordinators to consult with the ST education developer, and some staff opted to apply directly to their CAC for confirmation of ST in the subject. This arrangement was valuable in that it placed the ST education developer in the role of advisor, and clearly separated this role from that of the formal recognition process. The distinction between these two roles, and processes used, set up a positive, problem-solving relationship between the subject coordinator and the ST education developer, who was then separate to, but sitting alongside, the approvals process. In most instances where the subject was not successfully approved for ST, the CAC referred the staff member back to the ST education developer for additional work, so that ST could be further developed in ways that would meet university requirements. Over time it became standard for the teaching and learning development and administrative staff of each college to suggest that the staff member check with the ST education developer before formally submitting the subject for approval. In part, this was to make sure that ST in the subject was at an optimum, and in part to reduce the time taken in CAC meetings to consider the incorporation of ST in subjects.

7. Once ST had been confirmed by the CAC for a subject, this was formally recorded on the database for the subject, and in appropriate course maps (Appendix 2). The incorporation of ST in a subject meant that, when a student passed the subject containing ST, they were deemed to have completed ST for their course, and this was linked to the triggering of a mechanism, once the student graduated, for an Australian Higher Education Graduate Statement to be generated and attached to the formal transcript of results issued to the student. This statement was designed to alert potential employers and others to the successful inclusion of ST in the course completed by the student.

5 Reflecting on the Success of Sustainability Thinking at La Trobe

The consistent experience of both the ST Education Developer (CH) and the Academic Coordinator (Coursework) in the College of ASSC (SMc) was that academic staff overwhelmingly embraced the proposal to incorporate ST in appropriately chosen subjects, across each of the courses and disciplines. In only one case was a staff member fully resistant to this idea. In many instances where ST was not already clear within a discipline, the concern of staff was not that they thought it was not worth doing, but rather a lack of clarity about how they would incorporate ST—a positive starting point for considering how such a challenge might be addressed. In many instances, staff became excited by the proposal of making their curriculum and teaching relate to one of the world's greatest challenges—of finding ways to live and work sustainably—and of recasting parts of their discipline outlook and teaching to address this challenge.

In summary, the process of successfully incorporating ST was contingent on several factors:

- mandating at the highest level of the University that ST needed to be embedded;
- having an orderly process for identifying potential subjects, overseen by the overarching learning and teaching committee (CAC) for each college;
- drawing on the support of expertise made available through the designated education developer.

In the absence of any one of these key planks in the process, it is highly likely that the success rate for incorporation of ST would have been much lower. It takes a significant shift in the mindset of many discipline-based academics, combined with expert understanding of what is required for ST, and the diverse ways for it to be incorporated, for the process to be successful. It requires that staff combine their expert knowledge of their discipline and teaching of subjects with an expanded view of their discipline area, sometimes assisted by the ST education developer, in ways that allow the incorporation of ST. Through the development and application of these processes, we have demonstrated conclusively that incorporation of sustainability can be successfully achieved, across a wide range of courses and disciplines.

The successful integration of ST across a wide range of courses and disciplines at La Trobe has vindicated the whole-of-institution approach to Sustainability Education (SE) at La Trobe University, in which SE has been interwoven as a strand of educational development, rather than an initiative of specialist environmental or sustainability interests within the institution (Hocking and Riddle 2015). Central to its success, the endeavour was located in, and coordinated through, the educational infrastructure and processes of the University. This entailed a shift in approach, from SE being pursued either through Operations and Infrastructure, or through a specialist sustainability unit, to one where development of SE was identified primarily as an educational issue (Hocking, Daddow and Ford 2011). As previously pointed out, we would agree with Ryan and Tilbury (2013) who have described this process as follows:

> [In universities there is a] need to access deeper currents of teaching and learning to make ESD (Education for Sustainable Development) a viable education proposition, as well as the potential transfer to other parts of the education and skills sector.

As was hoped, the approach to integrating ST at La Trobe has successfully drawn on the processes of educational development of the institution, and at a more fundamental level has acted to normalise SE as being part of a forward-thinking, innovative approach to Higher Education, one that assumes and promotes SE as part of best educational and institutional practice. In this way, the incorporation of SE becomes aligned with other major shifts in tertiary education development, such as the trend towards more blended modes of delivery, and the incorporation of other major cross-curricular priorities, such as innovation and globalisation. Considering that this is the first instance of a University-wide incorporation of sustainability across all undergraduate courses in Australia, its success is to be celebrated, and bodes well for similar initiatives now under way in other Australian Universities. Rapid change of this type will be needed, and will act as a base for future accelerated development, as our society faces the urgent sustainability challenges that are emerging or already evident.

6 Going Beyond the Minimum—Depth as Well as Breadth

In a number of discipline areas the development of ST went well beyond the expected minimum. Two examples are explored in the sections below.

In Business degree courses at La Trobe, a patchwork of subjects addressing SE in some courses was replaced in 2016 by a core second year subject, entitled 'BUS2SUS—Sustainability', for all students enrolled in any Business degree. More than 1500 students are now enrolled in this compulsory subject. This includes students from a range of business majors including management, marketing, accounting, sport management, finance, events and hospitality. BUS2SUS has been based on a blended learning design that allows for greater scalability across the entire portfolio of majors within Business. With sustainability as the lens or context for change, students are introduced to systems thinking, tools for solving wicked problems, and the role of advocacy in managing change for sustainability.

The process of embedding ST into the core business curriculum presented a number of challenges, including distinguishing sustainability from related streams of corporate social responsibility (CSR) and non-financial measurement and reporting. The curriculum design was ultimately guided by the need for a future set of skills, rather than by identifying disciplinary content that business graduates might require. These skills include critical thinking, creative problem solving, ethical awareness and teamwork. For example, by working in small groups in class, and engaging with 'wicked' global sustainability issues such as climate change, global poverty and renewable energy, students are required to apply a systems lens to examining the true nature of the issues and potential solutions.

The 2017 description of BUS2SUS is as follows:

> This subject introduces you to the concept of sustainability, and a systems approach to understanding the complex interactions between the environmental, economic and social dimensions of sustainability. The subject attracts students from a range of fields, bringing a multidisciplinary team perspective to the researching, analysis and problem-solving aspects of creating positive change for sustainability. In teams, you are required to critique, design and present an action plan aimed at resolving a sustainability issue that has impacts now and for future generations. This subject provides you with the opportunity to enhance, demonstrate and document work-ready skills appropriate to your chosen career path.

Intended Leaning Outcomes for BUS2SUS include the following, that are directly aligned with matching subset assessment tasks that help deliver learning for ST:

- Recognise and reflect on the importance of being a change agent for sustainability, and develop the skills and competencies for creating that change.
- Design and evaluate innovative, systems-based solutions through the application of disciplinary knowledge and skills to researching, analysing and resolving sustainability challenges.
- Identify the interplay between the social, environmental and economic pillars of sustainability, and the implications for ethically complex decision-making.
- Apply the theory and frameworks developed in class to analyse and appraise a specific sustainability-related issue.
- Demonstrate an ability to engage effectively in diverse teams to complete complex team tasks or structured projects in culturally diverse educational settings.
- Demonstrate an ability to write logically focused arguments supported by discipline specific academic referencing and basic rules of grammar.

In the second example of a deeper approach to SE at La Trobe University, in Education and Outdoor Education courses, a patchwork of subjects across two campuses with very different approaches to SE was replaced by a combination of five core and elective subjects containing SE that both met the La Trobe requirements for ST and in most instances developed students' understanding of and engagement with sustainability well beyond the minimum. This means that many students taking a teacher education degree course will encounter SE, in a variety of ways, two or even three times.

Education-related subjects that have incorporated ST, along with their current (2017) subject descriptions, include:

OED1WKN Ways of Knowing Nature

"In this subject students are introduced to the theoretical and practical study of outdoor and environmental studies. Students will examine and explore connections between ways of experiencing the outdoors and knowledge of the outdoors, with an emphasis on implications for environmental sustainability. Students will also develop skills of critical enquiry, academic reading and writing, problem solving, and teamwork. Practical trips provide opportunities for observation and reflection, which are recorded through journal keeping."

OED2ACL Australian Culture and Land

"Students learn about the indigenous and European culture origins and characteristics of land tenure and management of both public and private lands in Australia. They are introduced to the history of our current land management system. Students engage with ideas related to the cultural construction of nature and the way in which current beliefs and practices about land use are embedded in cultural accident and history. In exploring the current institutional and legal structures involved in land use, students are able to understand why differing expectations, regulations and practices apply to a range of land designations."

EDU3PBM Project Based Teaching Method

"This is a Method subject in which students undertake a special project in order to investigate how to teach their chosen Discipline from both a practical and theoretical point of view. They will gather information directly from their school experiences, from Australian teacher education websites, and from other resources to gain in-depth knowledge to inform their practice based upon teaching for an equitable and sustainable future."

EDU3ONP Outdoor and Nature Play

"Students in this subject investigate the notion that unstructured play outdoors (nature play) is fundamental to a healthy childhood in that it provides benefits for health, cognitive, social and emotional development and helps to build resilience and creativity. Students also address the proposition that experience in nature as a child may lead to environmental sensitivity and responsibility later in life. Students design effective outdoor settings for children's nature play and manage aspects of safety while maximising exploration and free play. Students identify and document the learning that emerges from outdoor and nature play."

EDU4SEG Sustainability Education and the Global Teacher

"In this subject students consider their roles as citizens and teachers in a global context. They explore global influences on education, teaching and learning including international trends, policies, political and economic drivers. They critically examine how different cultural factors, perspectives and value systems impact teachers' work, school life and the opportunities and challenges created by globalisation. In this context they explore the connections between globalisation, environmental and social sustainability issues. They investigate concepts of sustainability and sustainability education in the global and Australian context. They enquire about national and local examples of sustainability curriculum, pedagogy and research as it applies to the Australian and Victorian contexts. This work culminates in a project to enact change in their local educational settings."

7 Conclusions

The La Trobe experience of mandating the incorporation of SE centrally, and then setting up the mechanisms and resources for delivering this, has been positively received by nearly all staff, taking SE well beyond the 'usual suspects' of subjects where SE has been traditionally found. Neither has the broadening of SE across courses led to the watering down of SE, but in reality the opposite - conversations have been widened about what sustainable living means, who should be involved, and what contributions can be made by diverse disciplines. In some instances SE has been strengthened and deepened in courses beyond its previous manifestations.

We suggest that these types of step-change approaches to incorporation of SE across diverse courses and disciplines be considered by other institutions, at a time when more sustainability literate graduates in all areas of living and working are very much in need.

Acknowledgements Many staff members at La Trobe University have contributed to the development of Sustainability Education (SE)—we thank them all. Acknowledgements especially relevant to this paper are: La Trobe Learning and Teaching Academic Developers; La Trobe Learning and Teaching Hub Staff; Academic and Administrative Staff from the two La Trobe Colleges; and Academic Staff from the wide range of courses and disciplines who rose to the challenge of incorporating sustainability education into subject they coordinated, or were involved in.

Appendix 1: Example of the Memo Used by Academic Staff to Seek Approval for ST

ESSENTIALS MEMORANDUM

DATE	[Click here to insert date from drop-down menu on right]
TO	College Coursework Committee
SUBJECT	Approval for *<insert subject code>* as a La Trobe Essential

<Insert Subject Code as it appears in CIMS>, *<Insert Subject Name as it appears in CIMS>*

Which Essential is nominated in this subject? ***Please select one and delete other two.***

☐ Global Citizenship ☐ Innovation and Entrepreneurship ☐ Sustainability Thinking

How is The Essential communicated to students in the Subject Description?

Please copy the subject description including information of The Essential addressed being taught. For example, append the following to the end of the Subject Description (customise as needed):

This subject addresses La Trobe's Sustainability Thinking Essential. Sustainability Thinking entails deep appreciation of how the choices we make affects the natural, economic, social, political and cultural systems—now and in the future.

This subject addresses La Trobe's Global Citizenship Essential. Global Citizenship entails deep appreciation of how we live in an interconnected world, being able to recognize the global context of concepts, act across cultures and boundaries, and work with diverse communities—now and in the future.

This subject addresses La Trobe's Innovation and Entrepreneurship Essential. Innovation and Entrepreneurship entails developing the ability to tackle problems creatively, generating new ideas, taking calculated risks and creating change to achieve ambitions—now and in the future.

Where in the subject is the Essential taught?

Please specify the ILOs that incorporate the elements of the Essential and the learning activities that support achievement of the ILOs related to the Essential?

How is the Essential assessed?

At least 25% of the assessment must directly relate to the Essential.

Please provide details of the assessment task(s) and an explanation of how the task(s) assesses the Essential.

Regards,
<Insert name>
<Insert Subject code> Subject Coordinator
School of <insert School name>

Appendix 2

Example of a Generic Course Map showing identification of which subjects Sustainability Thinking (ST) and the two other Essentials at La Trobe have been incorporated.

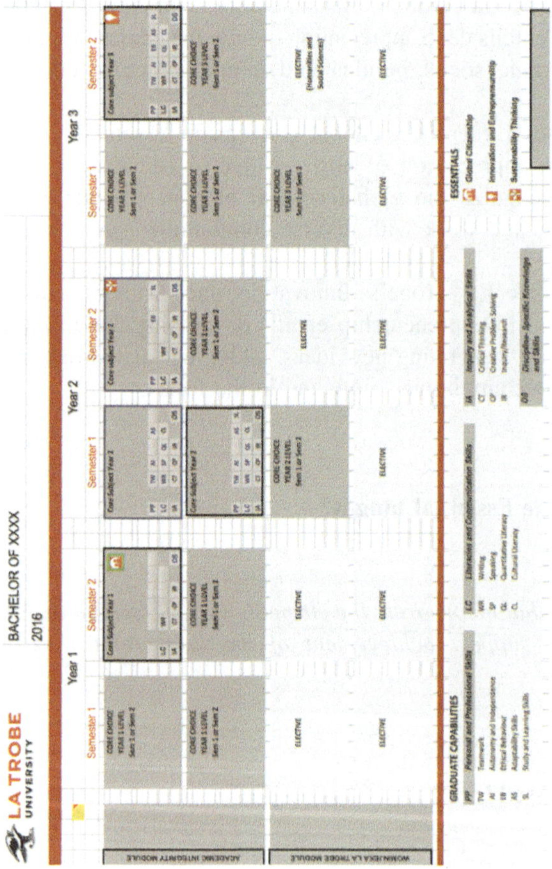

References

Australian Government. (2009). *Living sustainably—The Australian Government's national action plan for education for sustainability.* Department of Environment, Water, Heritage & Arts. http://www.environment.gov.au/education/publications/pubs/national-action-plan.pdf. Last accessed July 24, 2015.

Biggs, J., & Tang, C. (2007). *Teaching for quality learning at university: What the student does* (3rd ed.). Society for Research into Higher Education & Open University Press.

GUNI (2011). Global University Network for Innovation. *Higher education in the world 4: Higher education's commitment to sustainability: From understanding to action.* Palgrave, McMillan.

Hocking, C., & Riddle, M. (2015). Aiming for full coverage—Integrating sustainability education into all undergraduate courses at La Trobe University, Australia: Achievements, lessons learnt and barriers addressed. In W. Leal Filho, L. Brandli, O. Kuznetsova, & A. M. F. D. Paço (Eds.), *Integrative approaches to sustainable development at university level* (pp. 479–493). Cham: Springer International Publishing. https://doi.org/10.1007/978-3-319-10690-8_33.

Hocking, C. (2015). Using social context lenses to assist development of sustainability in a diversity of discipline-based higher education courses. *The Social Educator 2015, 33*(2), 26–38.

Hocking, C., Daddow, A. & Ford, R. (2011). *Building sustainability into the core business of teaching and learning - Victoria University as a case study.* Paper & presentation to Australasian Campuses Towards Sustainability (ACTS) Annual Conference, Adelaide, September.

La Trobe University Future Ready. (2013). Strategic plan 2013–2017—Refreshed 2015. http://www.latrobe.edu.au/about/downloads/La-Trobe-Strategic-Plan-November-2015.pdf. Last accessed March 28, 17.

La Trobe University. (2015). La Trobe essentials procedure policy database document number 114047D. Online resource: http://www.latrobe.edu.au/policy/documents/la-trobe-essentials-procedure.pdf. Last accessed March 26, 2016.

La Trobe University Website. (2017). Essentials: Quick guide for sustainability thinking. http://www.latrobe.edu.au/ltlt/resource-library/sources-archived/essentials-quick-guide-for-sustainability-thinking. Last accessed March 17, 2017.

Orr, D. (1994). *The Earth in mind: On education, environment and the human prospect.* N.Y: Island Press.

Oxford-Brookes University Website. (2017). Course design intensives. https://www.brookes.ac.uk/OCSLD/Courses/Teaching-and-learning/Course-design-intensives/. Last accessed March 20, 2017.

Ryan, A., & Tilbury, D. (2013). Uncharted waters: Voyages for education for sustainable development in the higher education curriculum. *Curriculum Journal, 24*(2), 272–294.

Scott, et al. (2012). *Turnaround leadership for sustainability in higher education.* Canberra: Learning & Teaching Excellence Division, DEEWR, Australian Government.

Sterling, S., Maxey, L. & Luna, H. (2013) *The sustainable university: Progress and prospects.* Earthscan from Routledge.

Targeted News Service. (2014). *UK Universities embrace sustainability in the curriculum with new guidelines.* Targeted News Service, Oct 14. Retrieved from http://ez.library.latrobe.edu.au/login?url, http://search.proquest.com/docview/1616322757?accountid=12001. Last accessed March 20, 2017.

Tilbury, D. (2011). Education for sustainable development. An expert review of the processes and learning. *Report to section for education for sustainable development, division of education for peace and sustainable development.* United Nations, Paris: UNESCO.

Developing Undergraduate Foundation Courses in Sustainability

Michael Howes

Abstract In 1992 the governments of the world committed themselves to pursue sustainable development at the Rio Earth Summit and education was promoted as playing a key role. The commitment was reaffirmed at subsequent summits culminating with the adoption of the United Nations *Sustainable Development Goals* in 2015 which included a specific goal on education. This paper examines how strategic changes can integrate the idea of sustainability into a range of programs at universities. A specific case study is used based on a first year course in environmental sustainability that the author convenes for a large and diverse group of undergraduate students from across all areas of the university. It is argued that in their future professional life, graduates have the potential to be agents of change by helping to transform the state, the private sector, and the community. This analysis is undertaken using the theoretical framework of ecological modernisation that underpins the idea of sustainability, offers a strategic pathway to transform public-private-community interactions, and approaches sustainability as a design challenge. The somewhat ambitious goal is to synthesise the empirical evidence, the practical experience, and theoretical framework into a coherent whole.

Keywords Sustainability · Education · Interdisciplinary · Ecological modernisation

1 Introduction

For a quarter of a century sustainability has been one of the stated goals of governments, businesses and community organisations around the world. Achieving sustainability, however, has proved elusive because the changes required cut across all areas of human endeavour (Howes et al. 2017; Howes 2005). A wide range of

M. Howes (✉)
Griffith School of Environment, Cities Research Institute,
Griffith University, Southport, QLD 4222, Australia
e-mail: m.howes@griffith.edu.au

W. Leal Filho et al. (eds.), *Sustainable Development Research in the Asia-Pacific Region*, World Sustainability Series, https://doi.org/10.1007/978-3-319-73293-0_5

professionals (e.g. scientists, planners, architects, engineers, policymakers, business managers, teachers, health workers, etc.) can play a key role by identifying problems, helping to find solutions, and/or working to overcome the barriers to change. How then can universities best prepare these graduates for the challenges they face? This paper addresses this question, starting with a brief history of sustainability followed by a discussion of the role allotted to education in various transition strategies. After that, an outline of the rise of environmental education at Griffith University is given. This leads to a discussion of a series of courses that were designed to give students in a broad range of disciplines an understanding of sustainability. A case study of a new first year course is then developed, which includes some of the practical lessons learnt with regards to what works and what pitfalls to avoid. The paper finishes by taking a step back and using the lens of ecological modernisation to understand the transformations required in pursuing sustainability, the role of graduates, and the elements that need to be included in their education. The underlying argument is that sustainability should be presented as a design challenge to students training for a wide range of professions and taught by way of interdisciplinary problem analysis/problem solving exercises.

The rise of sustainability to international prominence can be seen as a process of institutional learning by all sectors of society. At the end of the 1960s a rapidly expanding range of new social movements was forcing environmental and social issues onto the political agenda (Howes 2005). This led to new initiatives that expanded the policy focus of governments beyond traditional areas such as security, economic management and welfare (Dryzek et al. 2003). During the 1970s the idea that economic, social and environmental issues were linked began to emerge. This process started with United Nations (UN) *International Development Strategy* in 1970 which linked economic development to a healthy environment. In 1972 concerns about the impacts of industrial development on the environment were the main focus of discussions at the UN Stockholm Conference on the Human Environment and this led to the subsequent creation of UN Environment Programme (UNEP).

A pivotal moment came in 1980 with the release of *The World Conservation Strategy* by UNEP, the International Union for the Conservation of Nature (IUCN) and the World Wildlife Fund (WWF). This introduced the concept of 'sustainable development' as a proposal to utilise ecosystem goods and services for the production of social and economic benefits over the long term (Howes 2005). The concept was elaborated by the World Commission on Environment and Development (WCED) in its final report, *Our Common Future* (WCED 1987) which offered what has become the most commonly used definition: "Sustainable development is development that meets the needs of the present without compromising the ability of future generations to meet their own needs" (WCED 1987, Chap. 2, paragraph 1). Other sections of the report stated that there was a strategic role for education and institutional change in striving for sustainable development (WCED 1987, Chap. 2, paragraph 16; Chap. 4, Sect. 3.2).

The idea was followed by the 1992 UN Conference on Environment and Development (UNCED), otherwise known as the Rio Earth Summit, that produced a strategy for reforms across many sectors of society called *Agenda 21* (UNCED 1992). A commitment that appropriate education and professional training should be a priority appeared multiple times throughout this document (UNCED 1992). National initiatives were developed to supplement these international commitments, including: Australia's *National Strategy for Ecologically Sustainable Development* (1992); the *President's Council on Sustainable Development* (1993–1999) in the USA; and, *Sustainable Development: the UK Strategy* (1994) (Howes 2005). The Australian strategy, for example, had a chapter dedicated education and professional training (Ecologically Sustainable Development Steering Committee 1992, Chap. 26). The principles of this strategy are still enshrined in current legislation, such as the Commonwealth *Environment Protection and Biodiversity Act 1999.*

So by the end of the twentieth century there were international agreements and national policies committing governments to pursue sustainable development. These included recognition that education was a key supporting mechanism that would enable professionals to fulfil a transformative role. These themes were revisited internationally at the 2002 Rio+10 and 2012 Rio+20 summits, as well as the 2015 UN *Sustainable Development Goals*, with Goal 4 focussed on education at all levels (UN 2015).

2 Education and Sustainability

It is important to note from the start that education relating to the environment goes back many generations before the rise of sustainability as an idea. One of the pioneers of environmental education was Patrick Geddes in late-Victorian Britain which preceded the growth of 'nature studies' in the early part of the twentieth century (Palmer 1998). The term 'environmental education' was introduced with the formation of the International Union for the Conservation of Nature and Natural Resources (IUCN) in the late-1940s and promoted at the UNESCO 1968 Biosphere Conference in Paris. There has, however, been some convergence between the two streams of environmental education and sustainability over the last few decades. The 1972 Stockholm conference mentioned previously, for example, generated a principle that stated:

> Education in environmental matters, for the younger generation as well as adults, giving due consideration to the underprivileged, is essential in order to broaden the basis for an enlightened opinion and responsible conduct by individuals, enterprises and communities in protecting and improving the environment in its full human dimension … (UN 1972, Principle 19).

UNEP, another product of the Stockholm conference, co-founded the International Environmental Education Programme (IEEP) and its 1977 conference in Tbilisi set out the principles of environmental education that remain current to this day:

- Life-long learning;
- An interdisciplinary curriculum;
- The interconnectedness of society and the environment;
- The multi-dimensional nature of environmental issues (including the social, political, economic and technological aspects);
- Energy and material resource limits;
- Global and future dimensions;
- Critical thinking and problem solving; and,
- Values and ethics (Palmer 1998).

These Tbilisi principles were picked up in *Agenda 21* (UNCED 1992, Chap. 36, paragraph 36.1).

The UN declared that 2005–14 would be the 'Decade of Education for Sustainable Development'. This entailed the acknowledgement of the need to include sustainability in the training of professionals that would help to transform society (Howlett Ferreira and Blomfield 2016). There was a subsequent shift in emphasis from environmental education to education for sustainability amongst universities that had been developing such programs (Wilensky 2007). With the rise of climate change as one of the major threats to sustainability, there was a further shift in emphasis. The Australian Research Institute in Education for Sustainability (ARIES), for example, published a report in 2007 entitled *Shifting Towards Sustainability: Education for Climate Change Adaptation in the Built Environment Sector*. Disciplines such as planning, architecture and engineering were a particular focus of this report as their graduates can play a key role in climate change mitigation and adaptation for urban areas where the majority of the world's population now live (Lyth et al. 2007).

By the twenty first century education for sustainability was being incorporated into the curriculum for many tertiary students in Australia. The values and principles of sustainability were at the heart of these proposals in line with commitments within *Agenda21,* the *National Strategy for Ecologically Sustainable Development,* and the UN's *Sustainable Development Goals.* Such ideas have been injected into a somewhat turbulent academic environment in Australia of increased competition for scarce public resources, a shift to vocational training, and a growing tension between the demands of research and teaching (Davis 2006; Markwell 2007).

3 Environmental Education at Griffith University

Griffith University opened its doors to students in 1975 to meet the demands of the rapidly growing population and economy of South East Queensland. The teaching programs were deliberately designed to differentiate themselves from more traditional universities by being interdisciplinary and problem-focussed (Quirk 1996). The Australian School of Environmental Studies was one of the four founding schools and took this principle to its core by employing a wide range of both natural scientists (ecologists, chemists, geologists and mathematicians) and social scientists (economists, political scientists, geographers, sociologists and anthropologists). It was the first school of its type in Australia and remains one of the largest in the country to this day (Metcalf 2000). A school of planning was created in 1993 that was then merged into a new and enlarged Griffith School of Environment in 2007. An architecture program was added during another expansion in 2009.

Today the Griffith School of Environment has approximately 53 academic staff divided amongst six discipline groups: planning; architecture; marine science; society and environment; ecology and biodiversity; and, soil and water science. There are several research institutes associated with the school, including: the Cities Research Institute; the Environmental Futures Research Institute; and the Australian Rivers Institute. The author has worked at the school since 2000, was the founding Head of the Society and Environment Discipline, is currently the Deputy Head of School (Learning and Teaching), and has led several research projects within the Cities Research Institute and its forerunner the Urban Research Program.

In terms of teaching, the Griffith School of Environment offers undergraduate, Masters and Ph.D. degrees in urban and environmental planning, architecture, environmental science and marine science. Since the start of 2014 all of the degrees offered by the school require students to take an interdisciplinary first year course that was designed and taught by the author of this paper entitled *1043SCG Introduction to Environmental Sustainability*. This is the focus of the case study that follows and for convenience will be referred to as simply the 'sustainability course' from here onwards.

4 Background to the Case Study: Forerunner Courses

The current sustainability course has a long pedigree. Since its inception, the school of environment has worked through a series of different interdisciplinary first year courses that integrate the natural and social sciences. Three that were taught by the author have had a strong influence on the development of the current course: *1181ENV Sustainable Development*; *1301ENV The Global Environment*; and, *1161ENV Environmental Economics and Policy*.

The first course, *1181ENV Sustainable Development,* was created by the author in 2007 on the Nathan campus as a core for the environmental science and management students as well as an elective for the planning students. The original version of the course took a sector-based approach to sustainability. In the first few weeks the extent of interconnected environmental, social and economic problems was outlined. This was followed by an exploration of the idea of sustainable development as a solution and a set of principles were defined. The remaining two thirds of the course then methodically worked through how these principles could be applied to understanding and addressing issues relating to cities, housing, transport, energy, healthcare and manufacturing. The course was developed further after being taken over by colleagues in 2009 to promote more critical thinking and reflection (Howlett et al. 2016).

The second course, *1301ENV The Global Environment,* had been running on the Gold Coast for more than a decade when the author took it over in 2009. The enrolments grew to 360 students with planning, architecture and environmental science comprising the main cohorts. It took an issue-based approach, looking at land degradation, climate change, the loss of biodiversity, pollution, hazardous waste, and water issues. The approach was based on Earth Systems science and used the drivers-pressures-state-impact-response (DPSIR) analytical framework. This framework is used for State of the Environment reporting by the United Nations, the European Union and the US EPA, as well as both federal and state environment departments in Australia. Although it was interdisciplinary, the natural sciences played a strong role in understanding the issues.

The third course, *1161ENV Environmental Economics and Policy*, was jointly developed in 2009 by the author at the Gold Coast and his colleague, Dr. Peter Daniels, at Nathan. Enrolments on the Gold Coast (where the author was teaching) consisted of 80 students, predominantly from planning. It took a discipline-based approach that was mainly social science but started with a review of key sustainability issues. It also touched on the DPSIR framework in conjunction with key economic and policy tools. Market theory was introduced, then environmental and social problems, such as pollution and climate change, were analysed as evidence of market failures. Economic tools such as contingent valuation and cost benefit analyses were then brought to bear on the issues discussed. This was backed up with a policy analysis of responses such as environmental regulations, taxes, tradable permits and international treaties. The course also considered sustainability as a new goal for macro-economic policy.

In 2014 all three of these courses were replaced by *1043SCG Introduction to Environmental Sustainability* that sought to learn from the experience of these forerunner courses, utilise the best elements of each and integrate them into a functioning whole (Howlett et al. 2016). The development process took 18 months and was funded by a Griffith Learning and Teaching Grant (led by a colleague, Dr. Henry Skates, from architecture). The relevant educational literature was consulted to provide a sound basis for the course design that was grounded on best practice. International resources were used in compiling the course content. A wide range of consultations took place, with experts from all discipline areas of the school represented. External expertise was also brought in, including the university's sustainability officer, the Griffith Institute for Higher Education (now Learning Futures), an information and communication technology expert, and two educational designers. The new sustainability course was launched in Semester 1 of 2014 (that ran February to June).

5 Case Study: The Sustainability Course

The first version of this new course had a modular structure. Module 1 (weeks 1–3) covered the state of the environment, introduced the idea of sustainability, then outlined its key principles. The underlying emphasis of the course was on a values-based approach to sustainability with the commitment to sustainable development at the 1992 Rio Earth Summit viewed as a change in values. Some time was spent explaining the connection between ethics and values using a matrix that ranged from anthropocentric to eco-centric on one axis and deontological to consequentialist on the other. Students were asked to locate themselves on this matrix in order to encourage reflective learning. Module 2 (weeks 4–6) then used the DPSIR framework to explore environmental issues related to the lithosphere, the hydrosphere, the atmosphere, and the biosphere. The approach in this module was largely that of environmental science. In module 3 (weeks 7–9), the focus was on the economy and natural resource management with specific topics covering market theory, market failure and ecological economics. The final module (weeks 10–12) focussed on society and the environment with topics covering culture, indigenous people, settlements and governance. Some conclusions were drawn in week 13 to bring these four modules together. The associated workshops were split between developing basic study skills and reviewing some of the topics covered in the lectures.

The course was completed by 526 students drawn from a broad range disciplines, but the largest cohorts were from planning, architecture, environmental management, science and engineering. There were also students from other groups within the university (including health, business, arts, education and law). All students were surveyed at the end of semester (with a 39% response rate) to evaluate the quality of the course and teaching. This feedback was used to revise the course content and delivery for 2015. In terms of feedback, the students were very happy with the quality of teaching and the overall satisfaction rating placed the convenor in the top quartile compared to other large first year courses at Griffith. The overall satisfaction with the course, however, was mixed. While the students were generally happy, the rating of the statement that "This course engaged me in Learning" was close to the average when compared to similar first year courses. The student comments suggested that this was because the course appeared to be too general, they could not relate much of what was covered directly to their chosen profession, they wanted more of a focus on finding solutions to issues, and the workshops needed to have a stronger link to the lectures. The discipline-based modular approach also appeared to be a problem with such a broad range of students from different cohorts. Students majoring in the environmental and natural science disciplines, for example, enjoyed module 2 but did not engage as well with the economic and social elements of modules 3 and 4. The opposite appeared to be true for the planning and architecture students.

A detailed content analysis of the survey results was undertaken to identify points for improvement. The convenor then went back to the education for sustainability literature to find suitable innovations. A revised course structure was created and put through an internal peer review process with some of the best performing teachers in the school. The resulting new course structure was implemented in semester 1 (February–June), 2015. This new structure was much more integrated and focussed on fewer topics in more detail. The overall approach shifted to treating sustainability as a design challenge (something that had more resonance with the planning, architecture, engineering and science students) rather than a change in values. The principles of sustainability were still taught, but were used in an applied fashion to explain how they can guide the analysis and solution of problems on a week-by-week basis. The underlying idea was to focus on a key problem in the odd weeks, then work through sustainable solutions in the even weeks. A critical thinking component was added with students exposed to different environmental discourses that challenged mainstream views. Each of the environmental science, economics, policy and cultural studies elements were integrated into every lecture to provide a truly interdisciplinary approach to both problem analysis and problem solving.

Emphasising the relevance of the course to each student's chosen profession was addressed in five ways. First, the lectures of week 1 finished with an explicit discussion of how each of the 11 most common professions being pursued by the students could contribute to addressing the state of the environment and striving for sustainability (these professions were: scientists, planners, architects, engineers, health professionals, business managers, policymakers, lawyers, educators, journalists and social scientists). Second, every lecture included elements that were directly connected to these professions either in analysing or solving a specific problem. This was followed up with 'hands-on' workshops that included specific problem solving exercises related to the issues for that part of the course. Third, the students were asked to look up their chosen profession on the *Australian and New Zealand Standard Classification of Occupations* (ABS 2015), identify the key tasks, and design a professional e-portfolio targeting potential employers. Fourth, students were given a written assignment to work on over nine weeks that asked them to outline a role for their chosen profession in addressing climate change and how this would fit in with the broader pursuit of sustainability. Finally, specific case studies were used that involved a range of their professions along with some examples of what Griffith graduates were doing.

In 2015 there were 532 students who took the course and both the evaluations of teaching and the course reached into the top quartile compared to equivalent large first year courses. The course was left largely unchanged for 2016 when the course had 495 enrolments. While the evaluations of teaching for this year remained high, satisfaction with the course slipped into the third quartile, so another revision of the course was undertaken based on the student feedback. The 2017 version of the course (which had 493 enrolled students) had just been completed at the time of finalising this chapter. This version of the course updated the material developed in 2015 but merged coverage of key problems and solutions into single weeks. The workshops were updated and there was a new 'big picture' section added to the last 3 weeks covering scenarios for sustainable economies, governments and societies.

The most recent student evaluation surveys indicated that the course had risen back to the top quartile when compared to equivalent large first year courses.

The last three years of feedback suggests that finding a course that engages with such a broad range of students with very different interest is a challenge. Three innovations have clearly improved the satisfaction with, and effectiveness of, the course: (1) Looking at sustainability as a design challenge; (2) Taking an interdisciplinary approach to problem analysis and problem solving; and, (3) More explicitly relating the topics to the chosen professions the students.

6 The Bigger Picture: Ecological Modernisation

Treating sustainability as a design challenge is the underlying philosophy of the ecological modernisation which provides the theoretical framework that underpins the idea of sustainability (Howes et al. 2010; Howes 2000, 2005). It also offers a strategy for transformation that requires agents for change to operate across all sectors, and the professionals trained by universities can fulfil this role (Howes 2003). Creating interdisciplinary courses in sustainability such as the one discussed in the previous section is therefore important.

Ecological modernisation argues that although the institutions of modernity (i.e. democracy, the state, the market, and industry) have allowed and sometimes even encouraged unsustainable practices, they can be redeemed through a process of transformative change (Christoff 1996; Mol Spaargaren 2000; Dryzek 2005; Howes 2005; Howes et al. 2010). The goal is to decouple economic growth from environmental damage by the application of new eco-efficient technology and the redesign of institutions (Berger et al. 2001; Janicke and Jacob 2004; Dryzek 2005; Howes 2005; Huber 2008; Janicke 2008). This school of thought had its origins in Europe during the early 1980s, and two of its most prominent founders were Martin Janicke and Joseph Huber, although the pedigree of many of its component ideas can be traced back as far as the 1960s (Huber 2000, 2008; Janicke and Jacob 2004; Janicke 2008). Ecological modernisation ranges from the weak techno-corporatist approach that focusses solely on engineering solutions, to the strong reflexive approach that include the redesign of institutions (Christoff 1996). The state is cast in the role of a facilitator of change, with interventions designed to promote economic prosperity, social wellbeing and ecological sustainability (Howes 2005; Howes et al. 2010).

There are five key core themes of transformation in the strong version of ecological modernisation:

1. The redesign of technology to make it more eco-efficient and cleaner (i.e. less material and energy intensive, as well as less polluting);
2. The redesign of economic imperatives to provide incentives for change (e.g. pollution charges and taxes, along with subsidies and tax breaks for sustainable businesses);

3. Political and institutional redesign (i.e. a shift towards a more consensus-based form of governance that encourages partnerships between the state, business and the community);
4. The redesign of the role of social movements to give them more influence in decision making (e.g. engagement with the environmental justice movement); and,
5. A redesign of strategies to constructively engage with public discourse, with greater recognition of social justice and sustainability in decision making (e.g. focussing on green collar jobs as a way to achieve the 'win-win' scenarios) (Howes et al. 2010).

The graduates from the Griffith School of Environment are well placed to engage with each of these transformative themes (Dedekorkut-Howes et al. 2010). They can (re)design products, processes, buildings, urban spaces and infrastructure to take advantage of eco-efficient technology. They can develop policy, planning and economic regimes that reward sustainable development. They can participate in the establishment of more effective community and business partnerships to rethink development activities. They can establish more effective community engagement and consultation processes for decision making. Finally, they can disseminate information and engage in public forums to promote a more sustainable worldview. Training the students to recognise these opportunities and their role as agents of change is therefore an important responsibility for universities and academics. The sustainability course discussed previously has been designed to make a significant contribution towards meeting this responsibility.

7 Conclusions

While there is an international consensus that the pursuit of sustainability is a good idea, developing a strategy that can catalyse the necessary transformations across all sectors of society has proved difficult. Clearly university graduates have an important role to play in making society more sustainable, but how to prepare them? Environmental education and training has been developing for generations, and at institutions like Griffith University it has merged with education for sustainability within a broader interdisciplinary school. By design, trial and error a series of forerunner courses have led to the development of a new first year foundation course in environmental sustainability that is taken by students from many disciplines across the university. Finding ways to engage such a large and heterogeneous set of students has been a steep learning curve. The current version of the course has adopted a number of strategies, such as treating sustainability as a design challenge that engages students in interdisciplinary problem analysis and problem solving. This approach is consistent with the theoretical framework of ecological modernisation and the idea of sustainability that it underpins.

Such a strategy not only guides the education and training of graduates, it also gives them a broader sense of purpose in assisting the pursuit of sustainability.

References

Australian Bureau of Statistics (ABS). (2015). *Australian and New Zealand standard classification of occupations*. Canberra: ABS.

Berger, G., Flynn, A., Hines, F., & Johns, R. (2001) Ecological modernization as a basis for environmental policy: Current environmental discourse and policy and the implications on environmental supply chain management. *Innovation, 14*(1), 55–72.

Christoff, P. (1996). Ecological modernisation, ecological modernities. *Environmental Politics, 5* (3), 476–500.

Davis, G. (2006). The rising phoenix of competition: What future for Australia's public universities? *Griffith Review, 11,* 15–31.

Dedekorkut-Howes, A., Mustelin, J., Howes, M., & Byrne, J. (2010). Tempering growth: Planning for the challenges of climate change and growth management in SEQ. *Australian Planner, 47* (3), 203–215.

Dryzek, J. (2005). *The politics of the earth: Environmental discourses*. Oxford: Oxford University Press.

Dryzek, J., Downes, D., Hunold, C., Schlosberg, D., & Hernes, H.-K. (2003). Green states and social movements: *Environmentalism in the United States, United Kingdom, Germany, and Norway*. Oxford: Oxford University Press.

Ecologically Sustainable Development Steering Committee. (1992). *National strategy for ecologically sustainable development*. Canberra: Australian Government Publishing Service.

Howes, M. (2000). A brief history of commonwealth sustainable development policy discourse. *Policy, Organisation & Society, 19*(1), 65–85.

Howes, M. (2003). The environmental academic. *Spinifex, 24*(4), 6–7. (Queensland Conservation Council, December).

Howes, M. (2005). *Politics and the Environment: Risk and the role of government and industry*. Sydney/Earthscan, London: Allen & Unwin.

Howes, M., McKenzie, M., Gleeson, B., Gray, R., Byrne, J., & Daniels, P. (2010). Adapting the idea of ecological modernisation to the Australian context. *Journal of Integrative Environmental Sciences, 7*(1), 5–22.

Howes, M., Wortley, L., Potts, R., Dedekorkut-Howes, A., Serrao-Neumann, S., Davidson, J., et al. (2017). Environmental sustainability: A case of policy implementation failure? *Sustainability, 9*(2), 165. https://doi.org/10.3390/su9020165. On-line: http://www.mdpi.com/2071-1050/9/2/165/htm.

Howlett, C., Ferreira, J., & Blomfield, J. (2016). Teaching sustainable development in higher education: Building critical, reflective thinkers through an interdisciplinary approach. *International Journal of Sustainability in Higher Education, 17*(3), 305–321.

Huber, J. (2000). Towards industrial ecology: Sustainable development as a concept of ecological modernization. *Journal of Environmental Policy & Planning, 2,* 269–285.

Huber, J. (2008). Pioneering countries and the global diffusion of environmental innovations: These from the viewpoint of ecological modernisation theory. *Global Environmental Change, 18,* 360–367.

Janicke, M. (2008). Ecological modernisation: New perspectives. *Journal of Cleaner Production, 16,* 557–565.

Janicke, M., & Jacob, K. (2004). Lead markets for environmental innovations: A new role for the Nation State. *Global Environmental Politics, 4*(1), 29–46.

Lyth, A., Nichols, S., & Tilbury, D. (2007). *Shifting towards sustainability: Education for climate change in the built environment sector*. Australian Research Institute in Education for Sustainability (ARIES), Macquarie University, Sydney

Markwell, D. (2007). *A large and liberal education: Higher education for the 21st century*. North Melbourne: Australian Scholarly Publishing.

Metcalf, W. (2000). *Evolution of environmental sciences: The first 25 years*. Brisbane: Griffith University.

Mol, A., & Spaargaren, G. (2000). Ecological modernisation. *Environmental Politics, 9*(1), 17–49.

Palmer, J. (1998). *Environmental education in the 21st century: Theory, practice, progress and promise*. London: Routledge.

Quirke, N. (1996). *Preparing for the future: A history of Griffith University*. Brisbane: Boolarong Press.

United Nations (UN). (2015). Sustainable development goals. New York: United Nations. On-line: http://www.un.org/sustainabledevelopment/sustainable-development-goals/#prettyPhoto.

United Nations. (2013). Higher Education Sustainability Initiative (HESI). UN Sustainable Development Knowledge Platform.

United Nations. (1972). Declaration of the United Nations Conference on the Human Environment. New York: United Nations.

UNCED (United Nations Conference on Environment and Development). (1992). Agenda 21. New York: United Nations.

WCED (World Commission on Environment and Development). (1987). Our common future. Final report to the United Nations, New York.

Wilensky, M. (2007). *Tertiary education for sustainability: Four Australian Universities' commitment to sustainability*. St. Louis: Washington University.

A Comparison of Assessment Methods for Engineering Students' Understanding of Sustainability

Margaret Jollands

Abstract Developing the ability of students to think critically and systemically about sustainability issues is assessed in a variety of ways in different disciplines and institutions but there are few reports in the literature about what methods are the most successful. This paper discusses a range of assessment methods and characterises them in terms of Bloom's taxonomy of cognitive processes. Examinations are efficient, standardisable and objective but are not suitable for measuring sustainability learning. They promote surface learning and fail to measure higher order cognitive processes characteristic of sustainability competence. Project based learning (PjBL) is a common approach in engineering disciplines as it mirrors engineering work. PjBL assessment tasks such as reports and presentations can be marked using taxonomies such as the SOLO taxonomy which measure the higher order cognitive processes that are required for sustainability competence. However, markers need a high level of skill to mark reliability with rubrics based on the SOLO taxonomy. Graphical tools such as concept maps are also effective in measuring higher order cognitive processes.

Keywords Project based learning · Bloom's taxonomy · Rubrics
Concept tests · SOLO taxonomy

1 Introduction

The curriculum for sustainability learning is contested, but generally includes critical thinking, systemic thinking, values and ethics (Segalàs et al. 2009). Different disciplines and institutions use a variety of pedagogical approaches to developing students' sustainability competence, ranging from traditional lectures to project-based, community-oriented and constructive-learning pedagogies.

M. Jollands (✉)
School of Engineering, RMIT University, 124 La Trobe
St Melbourne, Melbourne, VIC 3182, Australia
e-mail: margaret.jollands@rmit.edu.au

W. Leal Filho et al. (eds.), *Sustainable Development Research in the Asia-Pacific Region*, World Sustainability Series, https://doi.org/10.1007/978-3-319-73293-0_6

Assessment approaches also vary, from examinations, to reports, presentations and concept maps.

A limited number of small scale studies have reported on the quality of engineering students' learning outcomes in sustainability classes. One study evaluated the curriculum design, the pedagogical approach and the competence of students before and after sustainable development (SD) classes in five leading European technical universities. The authors reported the rather shocking result that students in traditional lecture based sustainability classes had a lower understanding of sustainability at the end of the subject than the beginning. On the other hand, students in classes characterised as community-oriented and constructivist in approach showed an enhanced understanding of sustainability (Segalàs 2009; Segalàs et al. 2009). In another study of engineering graduates from an Australian technical institution, the impact of pedagogy on graduate work readiness was assessed. Graduates from two programs, one traditional and one based on project based learning (PjBL) were compared. PjBL has been used for many years in engineering faculties to develop graduate work readiness (Mills and Treagust 2003; Litzinger et al. 2011; Graham and Crawley 2010). The graduates from the two programs were interviewed about their application of skills in their work places one or two years after graduation. While the study found a number of work ready skills were enhanced in the PjBL cohort, no difference was found in the two cohorts' understanding of sustainability issues (Jollands et al. 2012).

Overall, few studies have been reported and there is no consensus on the best teaching approach to develop sustainability thinking in engineering undergraduates.

The gap is even more pronounced in terms of assessment. This paper reviews the range of assessment approaches used and characterises their strengths and flaws. These are mapped against Bloom's taxonomy and a set of criteria to choose the best assessment for an SD task is proposed.

2 Approaches to Assessment of Sustainability Learning

While a variety of pedagogical approaches to sustainability assessment in engineering classes have been reported, there are very few reports on assessment and no consensus on the best assessment methods. There are no accepted criteria on how to choose the best approach to measure learning for a sustainability. SD classes use a variety of assessments. These include examinations, reports, presentations, concept maps and the SOLO taxonomy. Each assessment type lends itself to different tasks.

University examinations are a generally norm referenced assessment tool that rank candidates quantitatively. Examinations are popular because they are efficient, standardisable and an objective way to measure students' scholastic ability. However, "quantifying performance gives little indication of the quality of the performance, or of what has been learned" (Biggs 1999, p. 145). Examinations promote surface learning by their focus on reproducing facts and memorisation of procedures and formulae (Biggs 1999, p. 167). Students who are good at passing exams are often unable to show that they understand the material (Ramsden 1992, p. 182). Students studying sustainability subjects taught by traditional lectures and assessed by examinations were found to have a lower breadth and complexity of understanding of sustainability at the end of the subject than the beginning (Segalàs 2009).

Problem based learning (PBL) is rarely assessed with examinations. One study of PBL in medical education found "The student assessment methods must challenge problem-solving, … and self-directed learning and not primarily emphasize the recall and recognition of facts" (Barrows 1986). Types of assessment for medical PBL include clinical reasoning, case study, and observation. Students in PBL classes performed significantly better in observation-based clinical performance than their peers in traditional classes (Strobel 2009).

In engineering education, PjBL assessments mimic the everyday experience of practicing engineers, that is, working on projects: assessments include tenders, planning schedules, meeting minutes, progress and final reports, built prototypes and client presentations (Graham 2010). In their seminal report Mills and Treagust (2003) wrote "Almost every task undertaken in professional practice by an engineer will be in relation to a project". These authentic work based assessment tools engage students in more meaningful learning experiences compared to the small fragmented assessments typical of traditional courses (Boud 2010). Quality and consistency of marking can be enhanced by use of rubrics (Stevens and Levi 2005).

However many programs are replacing PBL or PjBL with traditional programs as the high cost of resources are not sustainable (Herreid 2007). Staff report "a lack of knowledge and/or confidence in the design and implementation of assessment process for their PjBL experiences" (Graham 2010). STEM academics and tutors may lack capacity to mark student' sustainability work consistently and accurately especially when part of larger projects with multiple facets. Multiple supervisors may be involved in marking capstone PjBL assessments, so quality and consistency of marking is often an area of concern. If they have not had project based learning experiences and they don't have expertise in sustainability, this will compound the challenge.

A visual quantitative tool that captures the breadth and complexity of student knowledge of sustainability is the concept map. It can be used to teach systematic sustainability thinking as well as assess. A concept map documents an individual's or group's ideas around any topic (Novak 2010). The map records concepts relevant to a question or issue, as well as interlinks between concepts. This allows flexibility to draw concept maps around sustainability questions or issues related to the curriculum. Concept maps have been used in research studies to assess depth of student learning in sustainability (Segalàs 2009; Segalàs et al. 2009, 2010) and in a limited number of cases it has been used to assess student learning (Jollands and Parthasarathy 2013). The method for use of concept maps in assessing students' understanding of sustainability was developed by Segalàs (2009), against taxonomic criteria developed by the Sustainability Portal of the UNESCO Chair of Sustainability at UPC and a benchmark set by evaluating sustainability maps of experts (Segalàs et al. 2009). The method includes calculation of a complexity indicator (CO) and category relevance (CR) for each map, which are then compared against values for experts' maps. Good maps have an even spread of concepts across the taxonomic criteria (CR) and high number of concepts and links between concepts in different categories (CO). Figure 1 shows sustainability concept maps that range in breadth of concepts and complexity of relationships between concepts.

The final assessment method to be considered in this paper is use of the SOLO taxonomy. This tool can assess the sophistication of students' conceptions of sustainability. It characterises the learner's responses to a task in five developmental levels (Biggs and Collis 1982):

- prestructural (no understanding)
- unistructural (only one aspect of data is considered)
- multistructural (considers several aspects of data but fails to link them to the question)
- relational response (draws a simple conclusion based on the given data)
- extended abstract (relates abstract principles to the data and generalises to other contexts without closure).

The taxonomy has had some use in sustainability research studies. Carew and Mitchell (2002) characterised students' understandings of sustainability using the taxonomy and gave examples of a prestructural response as

Before the situation is getting worse we should find a better solution to improve it

and an extended abstract response as

Sustainability is about managing resources for future generations to use. Its about finding a balance between the amount of resources we consume and the amount of time we allow the earth to replenish the resources. It also involves changing current social, economic and political structures so that everything is more connected.

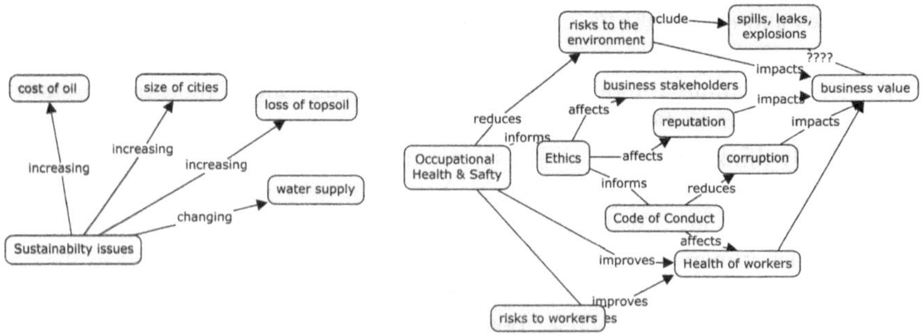

(a) Simple concept map with low complexity

(b) Portion of a complex concept map with high breadth and complexity

Fig. 1 Sample concept maps showing breadth and complexity of sustainability knowledge

Several studies have reported use of the taxonomy in assessment of other kinds of student work. One study in engineering used the taxonomy to mark examination papers comparing closed and open book examinations (Baillie and Toohey 1997). The paper reported that student performance had low correlation in the two exam types, attributed to students changing their approach to learning from surface to deep when assessed by closed rather than open book examinations. Leung (2000) described using the taxonomy to assess students' design and technology (D&T) work. A traditional examination question was posed to students along with a "SOLO" question, and both were marked by three experienced markers using two marking methods: traditional and SOLO taxonomy. The reliability of marking decreased when the SOLO taxonomy was used, attributed to markers' lack of experience using the new marking method. Leung (2000) concluded that markers must be well prepared and trained.

Use of the solo taxonomy in marking requires skills in coding text against a hierarchical taxonomy. This is a qualitative research method analysis skill. Generally, science, technology, engineering and mathematics (STEM) academics have limited experience and low level of skill in qualitative research techniques. Hence STEM academics would need significant professional development and marking moderation to use the SOLO taxonomy reliably. Multiple markers may be involved in marking STEM assessments when part of a PjBL assessment, compounding the challenge of using the SOLO taxonomy to mark PjBL sustainability assessments.

3 Analysis

Bloom's taxonomy was developed to assist teachers to select learning objectives, develop curriculum, provide quality learning activities and aligned assessment. It has six cognitive process dimensions: remember, understand, apply, analyse, evaluate and create and four knowledge dimensions: factual, conceptual, procedural and metacognitive (Anderson et al. 2014). The assessments described in the previous section are characterised in Table 1 in terms of Bloom's cognitive process dimension.

Table 1 shows conventional examinations are a poor assessment method to measure student learning in sustainability. Developing sustainability competence requires more than memorisation, so examinations are a poor measure of and poor motivator for the deep learning approach necessary to develop the capacity for critical thinking, systemic thinking, values and ethics, core competences for sustainability. Examinations encourage students to take a surface approach: the student needs only to remember concepts to pass, without demonstrating any higher order cognitive processes. On the other hand, concept maps or the SOLO taxonomy have more characteristics of effective methods to measure students' learning in sustainability. These methods provide scope to measure higher order cognitive

Table 1 Analysis of sustainability assessment tools in terms of Bloom's taxonomy and marker skill

	Examinations	Concept maps	SOLO taxonomy
Remember	(Conventional) Reproducing facts and memorisation of procedures and formulae	Breadth—number of concepts	Unistructural/ multistructural
Understand		Linkages—number of	Multistructural
Apply		Problem context reflects issue or question	Relational
Analyze		Links between concepts	Relational
Evaluate		Identify critical relationships between concepts	Relational
Create	Open book real work open ended questions with essay based answers		Extended abstract
Marker skill	Low	Low	High

processes that are required for sustainability understanding, especially critical thinking and systemic thinking.

In PjBL the SOLO taxonomy can be used to assess tenders, progress and final reports, prototypes and presentations, although a high level of marker skill is required to achieve reliable results. Concept maps can also be used to show students' understanding graphically. Hence PjBL is a suitable pedagogy for sustainability learning, as reported by Segalàs (2009).

Examinations can be marked with the SOLO taxonomy if they are essay based (Baillie and Toohey 1997). If the examination is open book and questions are open ended and focus on real world sustainability issues then examinations can also be a suitable measure of sustainability learning. Examination answers can also be illustrated with concept maps. Again, reliable results will require a high level of marker skill.

4 Conclusions

Examinations that focus on fact reproduction are not suitable for measuring sustainability learning. They promote surface learning and fail to measure higher order cognitive processes characteristic of sustainability competence. Project based learning (PjBL) is a common approach in engineering disciplines as it mirrors engineering work. PjBL assessment tasks such as reports and presentations can be marked using taxonomies such as the SOLO taxonomy which measure the higher order cognitive processes that are required for sustainability competence. However, markers need a high level of skill to mark reliability with rubrics based on the SOLO taxonomy. Graphical tools such as concept maps are also effective in measuring higher order cognitive processes.

References

Anderson, L. W., Krathwohl, D. R., Airasian, P. W., Cruikshank, K. A., Mayer, R. E., & Pintrich, P. R. (2014). *A taxonomy for learning, teaching, and assessing: a revision of bloom's taxonomy of educational objectives*. Pearson New International: Harlow.

Baillie, C., & Toohey, S. (1997). The 'power test': its impact on student learning in a materials science course for engineering students. *Assessment & Evaluation in Higher Education, 22*(1), 33–48.

Barrows, H. S. (1986) A taxonomy of problem based learning methods. *Medical Education*, Nov 1; *20*(6), 481–486.

Biggs, J. B. (1999). *Teaching for quality learning at university* (p. 9). Bury St Edmunds: St Edmundsbury Press.

Biggs, J. B., & Collis, K. F. (1982). *Evaluation the quality of learning: the SOLO taxonomy (structure of the observed learning outcome)*. New York: Academic Press.

Boud, D. (2010). *Assessment 2020. Seven propositions for assessment reform in higher education*. Sydney: Australian Learning and Teaching Council.

Carew, A. L., & Mitchell, C. A. (2002). Characterising undergraduate engineering students' understanding of sustainability. *European Journal of Engineering Education, 27*(4), 349–361.

Graham, R. (2010). UK approaches to engineering project based learning. *White Paper Sponsored by the Bernard M. Gordon/MIT Engineering Leadership Program*. Available online: http://web.mit.edu/gordonelp/ukpjblwhitepaper2010.pdf.

Graham, R., & Crawley, E. (2010). Making projects work: A review of transferable best practice approaches to engineering project-based learning in the UK. *Engineering Education, 5*, 41–49.

Herreid, C. F. (2007). The death of problem-based learning? In C. F. Herreid (Ed.), *Start with a Story: the case study method of teaching college science* (pp. 153–156). NSTA Press: Arlington.

Jollands, M., & Parthasarathy, R. (2013). Developing engineering students' understanding of sustainability using project based learning. *Sustainability, 5*(12), 5052–5066.

Jollands, M., Jolly, L., & Molyneaux, T. (2012). Project based learning as a contributing factor to graduates' work readiness. *European Journal of Engineering Education, 37*(2), 143–154.

Leung, C. F. (2000). Assessment for learning: using SOLO taxonomy to measure design performance of design & technology students. *International Journal of Technology and Design Education, 10*(2), 149–161.

Litzinger, T., Lattuca, L. R., Hadgraft, R., & Newstetter, W. (2011). Engineering education and the development of expertise. *Journal of Engineering Education, 100*(1), 123–150.

Mills, J. E., & Treagust, D. F. (2003). Engineering education—is problem-based or project-based the answer? *Australasian Journal of Engineering Education*. Available online: http://www.aaee.com.au/journal/2003/mills_treagust03.pdf.

Novak, J. D. (2010). *Learning, creating, and using knowledge: concept maps as facilitative tools in schools and corporations* (2nd ed.). Routledge: New York, NY.

Ramsden, P. (1992). *Learning to Teach in Higher Education*. London: Routledge.

Segalàs, J. (2009). *PhD Dissertation engineering education for a sustainable future*. Barcelona: Universitat Politècnica de Catalunya.

Segalàs, J., Ferrer-Balas, D., Mulder, K. F. (2009). Introducing sustainable development in engineering education: competences, pedagogy and curriculum. In *Proceedings of the*

European Society for Engineering Education (SEFI) Annual Conference, Rotterdam, Netherlands, 1–4 July.

Segalàs, J., Ferrer-Balas, D., & Mulder, K. F. (2010). What do engineering students learn in sustainability courses? The effect of the pedagogical approach. *Journal of Cleaner Production, 18,* 275–284.

Stevens, D. D., & Levi, A. J. (2005). *Introduction to rubrics: an assessment tool to save grading time, convey effective feedback, and promote student learning*. Sterling VA: Stylus Publishing.

Strobel, J., & Van Barneveld, A. (2009). When is PBL more effective? A meta-synthesis of meta-analyses comparing PBL to conventional classrooms. *Interdisciplinary Journal of Problem-based Learning, 3*(1), 4.

The Need for the Graduate Attribute Assessment Tool (GAAT)

Sarah Holdsworth, Ian Thomas, Orana Sandri, Peter S. P. Wong, Andrea Chester and Patricia McLaughlin

Abstract Universities have developed sets of Graduate Attributes (GAs) to be achieved by their graduates across all programs. GAs are often associated with development of 'sustainability' capabilities. However, there is little indication that sustainability GAs are assessed to determine the extent of achievement, application, relevance to professional practice or to provide feed-back into curriculum design. This paper reports on the development of the Graduate Attributes Assessment Tool (GAAT) to determine achievement of the sustainability GA. The research developed, in 2016, a tool to assess the level of graduate attainment, application and use of a sustainability GA in a graduate's workplace. The GAAT is unique as it is founded in the Theory of Planned Behaviour (TPB). The tool was trailed on a sample of RMIT graduates and assessed to determine if it was an appropriate tool. The complex nature of the sustainability GA meant that a modified version of the TPB was required. This resulted in the development of questions that required participants to reflect on their behaviour in their workplace in relation to a specific scenario. Additional questions were included to enable feed-back on curriculum to allow for modification to assist sustainability GA achievement. The GAAT pilot indicated that it provided valid outcomes of the achievement of the sustainability GA and insight for curriculum review. Development of the GAAT and its piloting have provided the basis for a practical, flexible tool for assessing sustainability GAs, and has potential for further development into a tool for assessing the wide range of university GAs.

Keywords Graduate attributes · Sustainability education · Lifelong learning

S. Holdsworth (✉) · I. Thomas · O. Sandri · P. S. P. Wong · A. Chester · P. McLaughlin
RMIT University, Melbourne, Australia
e-mail: sarah.holdsworth@rmit.edu.au

W. Leal Filho et al. (eds.), *Sustainable Development Research in the Asia-Pacific Region*, World Sustainability Series, https://doi.org/10.1007/978-3-319-73293-0_7

1 Introduction

A more sustainable future requires transformative change so that, as empowered and engaged citizens, we can act individually and collectively to critically think and reflect on our current lifestyles and allow for action that results in positive environmental and social change. However, as Fien et al. (2004) note, the paradigms of sustainability, and sustainable development, are complex, nuisance and difficult to define and conceptualise.

Traditionally, sustainability has focussed on compliance with environmental regulations and principles of corporate citizenship. While there is an increased awareness of the broader issues of sustainability and social responsibility there exists a tension between the improvement of environmental and social outcomes and the associated increased costs (as understood over the short-term) (Crocker 2013; Newton 2013).

Sustainability education (SE)[1] is a response to the limits of our knowledge and capability around complex, and serious issues including climate change, environmental decline, food security, poverty (Holdsworth and Hegarty 2016; Sterling and Thomas 2006). These issues are serious because they can produce irreversible effects once certain thresholds are met, and they are complex because they are not easily responded to because they cover many different state, sector, and community boundaries (OECD 2012). Through SE, the development of skills and capabilities to appreciate and respond to complex problems is teleological. Higher education therefore plays a key role in developing knowledge and capacity around sustainability, not just for those studying sustainability and related fields, but for all professions if we are to respond to these 21st challenges (Wals 2014).

There have been many studies over the past decade which attempt to map out learning outcomes for sustainability in the form of skills, attributes, competencies or capabilities depending on the language used (Barth et al. 2007; Barth and Michelsen 2013; Brundiers et al. 2010; Cohen 2007; Dale and Newman 2005; Haan 2006; Hofreitera et al. 2007; Ison 2010; Mochizuki and Fadeeva 2010; Mogensena and Schnack 2010; Parker 2010; Parkin et al. 2004; Scott and Gough 2010; Segalas et al. 2009; Sterling and Thomas 2006; Thomas 2003; Wals 2010; Wiek et al. 2011). There are many papers that present different lists of capabilities with similar themes. A literature review of EfS capabilities (including literature on EfS, ESD and systems education) undertaken by Sandri (2014) and Holdsworth (2010) found

[1]Education for Sustainable Development (ESD) is the term favoured by governments and the UN Decade of ESD. However, others prefer Education for Sustainability (EfS) seeing ESD as too tied to economic development, too instrumentally oriented, and missing cultural, personal and valuative dimensions. Subsequently there has been talk of achieving sustainability curricula, to advance the community's sustainability agenda, morphing to the general term Sustainability Education (e.g. see Jones et al. 2010). Whilst recognising the debate regarding the distinctions of these terms, we will use Sustainability Education in recognition of the broad applicability of the assessment tool.

that the following capabilities are most commonly described as learning outcomes in the literature.

- Sustainability literacy
- Systemic and holistic thinking
- Capability and motivation for lifelong learning
- Critical thinking and reflection
- Reflexivity
- Interdisciplinary skills and ability to work with stakeholders
- Foresighted, anticipatory and futures thinking
- Working with complexity and uncertainty
- Implicit development of values or value of learning and reflection
- Characteristics: empathy, compassion, self-motivation and sense of identity

Experiences in the classroom and consultation with industry suggest that sustainability skills and capabilities are crucial for future employment (Holdsworth and Sandri 2014; Holdsworth and Hegarty 2015; Holdsworth and Thomas 2015; Thomas and Day 2014). Many polices of the Australian government emphasise sustainable development, and many employers seek this expertise (Sterling and Thomas 2006). Sustainability capabilities align with Australian university generic graduate attributes such as life-long learning, problem solving, ethical, social and professional understanding, communication and inquiry (Barrie 2004; Thomas and Day 2014). Furthermore, sustainability capability also aligns with a number of top employability skills demanded by employers including learning, communication, problem solving and initiative and enterprise (Graduate Careers Australia 2014).

It is not surprising therefore that, aspects of sustainability have been integrated into graduate attributes across Australian universities (Thomas and Day 2014) and in various ways, embedded in learning and teaching practice. However questions remain around how to best assess this graduate attribute and ensure it is appropriate for professional practice. Educators are currently in a vacuum as they have few opportunities for knowing whether their curriculum and teaching is effective for graduate activity in sustainability (Sandri et al. 2016). This is not confined to sustainability, is a critical issue that applies to graduate attributes in general. It is important for all universities to be able to measure and capture information about graduates' acquisition of generic learning outcomes as part of their program of study. However, at the beginning of the 21st Century, Barrie et al. (2009, p. 1) were led to observe that 'while all Australian universities make such claims in policy, few can provide convincing evidence of curricula that comprehensively and systematically develop these abilities'. To date there is no clear information to suggest that the situation has changed.

According to Barrie et al. (2009, p. 7) evidence of graduate achievements of graduate attributes are seen as the 'central plank of the next generation of outcomes-based national quality assurance systems'. It is not enough, however, to measure the attainment of these attributes within the curriculum 'we also need to examine the way in which graduates cope with the requirements of their work …'

(van der Velden 2013, p. 211). While a body of literature and theory has developed on sustainability capabilities/attributes, less research has been published on how to evaluate such capabilities and attributes.

2 Graduate Attributes, Sustainability Attributes and Their Measurement

Formal discussion of learning outcomes and generic skills in Higher Education was initiated in the late 1990s, resulting in a framework of generic attributes statements, outlining attributes that university graduates would be required to attain. Consequently individual universities responded with the release of their sets of graduate attributes or capabilities (Bowman 2010). Graduate attributes have now been developed and are being applied by universities across Australia as a way to ensure that graduates achieve learning outcomes that support and promote employability and workplace readiness.

There is, however, 'lack of conceptual clarity' (Green et al. 2009, p. 17) about the exact meaning of terms such as *attributes, skills* and *capabilities* which have been well documented (Barrie 2006). Critically, Sadler (2013) differentiates between *capability, attribute, competencies, competence* and *skills*. He contends that a capability or attribute can be considered as an integrated and large-scale characteristic. A competence may then be described as a smaller-scale identifiable element, a demonstrable skill associated with the completion of a task, which contributes to such a capability/attribute. In particular, capability refers to the ability of an individual to evaluate situational contexts and to appropriately apply skills and competencies. Further, personal abilities and environmental/social contexts influence levels of individual capability. These external conditions enable or limit a person's ability to be capable (Nussbaum 2011).

This lack of clarity is further complicated by the subjective nature of sustainability and it manifestation in sustainability attributes/capabilities. The sustainability attribute at RMIT University articulates as 'Environmentally aware and responsible', and despite reading as a discrete attribute, has several components. According to RMIT University: 'Graduates of RMIT University will have engaged in processes to develop their abilities to recognise environmental and social impacts and to provide leadership on sustainable approaches to complex problems. Examples of how this graduate attribute can be evidenced include:

- Recognise the interrelationship between environmental, social and economic sustainability.
- Appraise and critique context-appropriate sustainability measures.
- Take responsibility for critical decision-making in ensuring sustainable outcomes.
- Appropriately apply their environmental and sustainability literacy in a highly diverse range of contexts.' (RMIT University 2015, p. no page)

In this graduate attribute it can be identified that graduates are expected to achieve the following capabilities to:

- recognise environmental and social impacts;
- provide leadership on sustainable approaches to complex problems;
- recognise the interrelationship between environmental, social and economic sustainability;
- appraise and critique context-appropriate sustainability measures;
- take responsibility for critical decision-making in ensuring sustainable outcomes; and
- apply their environmental and sustainability literacy in a highly diverse range of contexts

Given this cumulative complexity it is not surprising that despite the requirement of Australian Government that all 'universities as part of the AUQA (Australian University Quality Agency) audit process, show how they are embedding such attributes in the teaching of undergraduate degrees' (Barrie et al. 2009, pp. 6–7), 'few can provide convincing evidence of curricula that comprehensively and systematically develop these abilities' (Barrie et al. 2009, p. 1). How graduates then understand and utilise these attributes within the workplace is increasingly important as graduate attributes are also used by accrediting bodies to map out where these are developed in professional degrees in order for the degrees to receive industry accreditation (Barrie et al. 2009). Consequently, it is vital these attributes are evaluated for their individual relevance and attainment.

There is little indication that the situation has changed in more recent years. A small number of studies provide insight into current approaches taken by researchers to assess graduate learning outcomes; post degree-completion, or within workplace contexts. These studies are typically based on graduates' self-reported perceptions of their 'workplace readiness' and satisfaction with their learning outcomes from their degree. An overview of these studies includes:

1. Self-assessment questionnaire/survey

    - Delaney (2004) surveyed 522 graduates of an American University with an aim to determine the degree of satisfaction graduates held in terms of work readiness upon completion of their degree
    - Potts and Kleinpeter (2001) surveyed distance education students with an aim of determining graduates self-perceptions of work—readiness and degree satisfaction.
    - REFLEX graduate study was a large-scale graduate study involving 100,000 graduate participants surveyed in 2005 and 2008, five years after they left higher education, across approximately 20 European countries. The aim was to determine '… the extent to which … competences were required in their current job and the degree to which they actually possessed them' (van der Velden 2013, p. 213).

- Assessment of Higher Education Learning Outcomes (AHELO) (OECD 2012) was an international graduate learning outcomes assessment tool developed and trailed from 2011 to 2013 by the OECD in 17 countries with approximately 10 higher education institutions in each country. The study uses both quantitative and open ended qualitative questions to measure learning outcomes in the disciplines of economics and engineering ensure graduates can 'contribute to economic, scientific and social progress' (OECD 2012).

2. Qualitative interview

   - Martin (2005) conducted a Purposive sample of 16 engineering graduates at a South African university. The aim of the study was to determine engineering graduates' perceptions of how well they were prepared for work in industry.

3. Workplace/work practice assessment

   - University of Graz, Austria (Spiel et al. 2013) triangulated student data with external assessment by combining data from current and recently graduated students, gathered through self-reported surveys, with graduate employer assessment and assessment from current teachers of student performance. An aim was to determine graduate performance.
   - VET (Vocational Education and Training) Practitioner Capability Framework (Innovation & Business Skills Australia 2013) developed a template for the Australian Government for use by training organisations to rate employees' capability. It draws on self-assessment, peer assessment and employer assessment with the aim to determine employee capability.
   - Work integrated learning (WIL) assessment was explained as 'WIL generally uses a range of tools, such as reflective journals and student presentations, to assess learning theory, critical reflection and career management' (McNamara 2013, p. 188). Part of work placement assessment, WIL is being integrated into university learning as a way to give students real workplace learning experience.

Graduate self-reporting (via quantitative surveys, reflection or interviews of perceptions) appears to be the common approach adopted by researchers to assess attribute attainment. While self-reporting can be an efficient way of collecting data (via electronic surveys), it has significant limitations, since it:

- relies on participants having the ability to interpret the questions correctly and be honest in their self-reflection, and
- places responsibility on the respondents to interpret what the question and its context mean, with little or no guidance.

Using a semi-structured interview style can avoid the drawbacks of self-assessment, allowing the researcher to collect supporting data that evidences the graduates' account of their experiences and abilities. The disadvantage being the resources needed to collect even a small data set. Identical comments can be made for assessment of WIL. Learning from these experiences means that the development of an assessment tool to measure graduate capability in professional contexts must consider the advantages and disadvantages of each assessment method identified in this review.

However, a more basic issue is that any assessment undertaken at the end of a program or course/subject to provide a summary of a student's learning is unlikely to provide a viable assessment of a graduate who has entered in the workforce. Firstly, for an employed graduate participation in any university research would be voluntary, rather than associated with a formal learning requirement. Secondly, the form of large scale assessments traditionally used, such as 'written tests, oral presentations, concept maps, problem solving activities, project work, essays, formal assignments and exams' (Queensland Department of Education Training and Employment 2014: no page), are likely to be impractical for graduates to undertake on a voluntary basis. Yet a large-scale tool that provides both a valid assessment of learning and is efficient to apply is desirable, especially in order to compare findings from different programs of study.

An additional consideration is that undertaking an evaluation of a graduate's capability, after the graduate has left university and entered the workforce, increases the difficulty of meaningful and accurate evaluation. This arises because graduates both can be more difficult to contact, and they would have acquired additional learning in the workplace; as noted by van der Velden (2013).

3 Attribute Assessment Consideration and the Theoretical Basis for the Assessment Tool

In order to assess the level of graduate attribute attainment and application, it is important to consider that capabilities must be situated contextually, enabling graduates to demonstrate what they have acquired, these capabilities, and the extent to which capabilities have been acquired. Graduates will then need to draw on their skills, knowledge (competencies) and values, and to be able to use them to critically evaluate the most 'sustainable' options in a given context. Assessment of capabilities may be possible through formal processes during a course/subject where the teacher can observe a student undertaking a task or a student can work on a set task to achieve an outcome. However observation by the teacher may not be practical in research with graduates who have finished their studies. Furthermore, as Lozano et al. (2012) explains capability as when someone is able to undertake some action; which means the person may not necessarily continually demonstrate the ability. In line with the theories of situational learning (Down 2006), it is a graduate's interaction with their environments and social contexts where learning can be demonstrated and assessed. Real life contexts and interaction play a key role and are indeed necessary in order to assess capability. For example, Hager (2006, p. 34) argues that attributes are used in combinations in contexts and thus 'integration in real life practice should never be overlooked'.

Earlier experiences in the development of a tool to assess graduate attributes, using a vignette/scenario approach (Sandri et al. 2016), informed the adoption of Theory of Planned Behaviour (TPB) as the theoretical base of the assessment tool. TPB, developed by Ajzen (1985), is a model that articulates that a person's behaviour is indicative of three critical indicators; a person's attitude towards a behaviour of interest; their perceived social norms related to that behaviour of interest; and a person's perceived behavioural controls over the behaviour. By measuring all three aspects in regards to a specific behaviour, a person's likelihood of performing the behaviour can be indicated.

Based on the TPB model, the 'more favourable the attitude and subjective norm, and the greater the perceived control, the stronger should be the person's intention to perform the behaviour in question … given a sufficient degree of actual control over the behaviour, people are expected to carry out their intentions when the opportunity arises' (Ajzen 2002, p. 1). This relationship is illustrated in Fig. 1.

Given the subjective and diverse nature of sustainable behaviours, the TACT framework (Target, Action, Context and Time) was used to assist in defining the scope of the behaviour (Ajzen 2002). Then the *target* is the graduate application of some key framework that they have learned about in their degree, to fulfil the outcomes drawn from the GA (application of key components, such as to minimise negative impacts to the physical environment) and the *action* is making decisions that use the EAR GA. The *time* was defined as during work-practice in the last 12 months. Additionally, a scenario was developed in order to provide a *context* that was familiar to the participants, to situate the behaviour/s of interest. This approach was guided by Millar and Millar (1996) who argue that attitudes towards

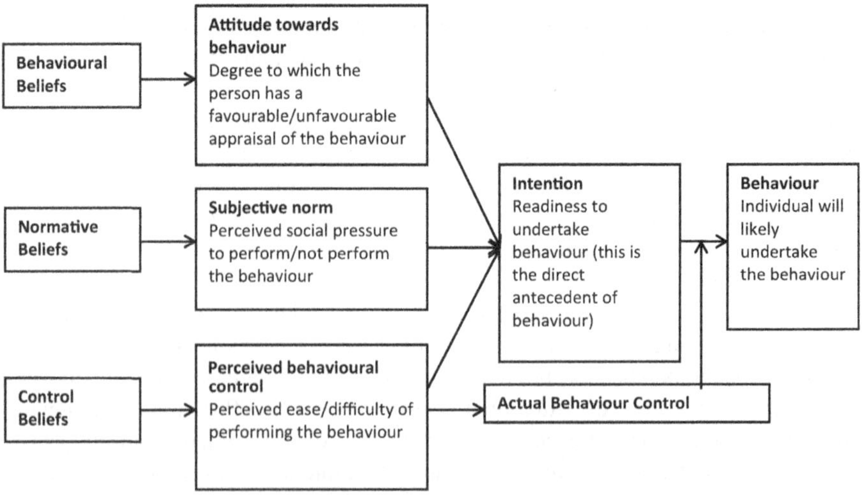

Fig. 1 Components and relationships within Theory of Planned behaviour (TPB)

behaviours are directly influenced by whether or not a person has had prior experience with the behaviour of interest. The scenario required participants to think about their use of a specific piece of environmental legislation (the Commonwealth of Australia's Environment Protection and Bio-diversity Conservation Act, 1999) in making decisions to fulfil the expectations of the GA.

Survey questions were grouped around the five GA sub-behavioural categories (described above), being:

1. Minimise negative impacts to the physical environment
2. Minimise social/community impacts
3. Recognise environmental, social and economic relationships
4. Balance environmental, community and economic outcomes
5. Evaluate a range of alternatives/responses/solutions

From these five sub-behaviours, one question was formulated to measure (separately) each of the following TBP variables:

- Current behaviour typical of the graduate/participant in their workplace
- Attitude of the graduate towards the issue
- Subjective norms operating in society, in relation to the issue, the graduate is aware of
- Perceived behavioural control in the workplace as understood by the graduate
- Leadership that the graduate may take in their workplace.

To cover all aspects arising from these two lists of points, a set of TBP questions was developed. As indicated in Fig. 2, for each of the TBP variables, a question was asked to account for each of the five GA sub-behavioural categories. A draft survey was pre-tested with four fourth year students studying in the School of Property, Construction and Project Management who were completing their honours year while working in the industry; this sample was considered to be similar to program graduates in respect of understanding the survey questions. As a result of the students' feedback, the survey was modified slightly; for example, the opening scenario was repeated in the survey for extra clarity of later questions, and the layout of sections was improved for better readability. Participants in the focus groups were asked to comment on the easy and 'understandability' of the survey, and the role of the scenario.

4 Application of the Tool and Project Findings

Drawing on TPB, a set of questions was developed and trialled through the online survey tool Quatrics. An invitation to participate in the survey was sent to recent RMIT Graduates from the School of Property, Construction and Project Management (involving three degree programs). Overall, invitations were extended to 150 graduates from 2015, 107 graduates from 2014, and 26 graduates from 2013.

TBP variables	*Statement in survey for which respondents are provided with a range of responses*
1. Current behaviour:	I make work-practice decisions that use the EP Act and/or the EPBC Act to [this is in relation to the EAR GA sub-behaviour category, e.g. (1) 'minimise negative impacts to the physical environment'; then (2) 'minimise social/community impacts'; then 3, 4 and 5]:
2. Attitude	If I were free to make my own decisions in my work-practice, I would use the EP Act and/or the EPBC Act to [e.g. in relation to 'minimise negative impacts to the physical environment'; etc] in my work-practice decisions:
3. Subjective norms:	The following stakeholders influence my use of the EP Act and/or the EPBC Act to [e.g. in relation to 'minimise negative impacts to the physical environment'; etc] in my work-practice decisions:
4. Perceived behavioural controls:	The following factors enable me to use the EP Act and/or the EPBC Act to [e.g. in relation to 'minimise negative impacts to the physical environment'; etc] in my work-practice decisions:
5. Leadership:	In using the EP Act and/or the EPBC Act to [e.g. in relation to 'minimise negative impacts to the physical environment'; etc] in my work-practice decisions, I would most likely:

Fig. 2 Framework underlying survey questions

In addition graduates from 2013, from staff LinkedIn contacts, were invited, so that a total of 318 graduates were invited. Along-side the online survey, focus groups with 22 graduates were also undertaken which provided additional qualitative data on both their experiences completing the survey, and their experiences applying the GA in their workplaces.

Analyses of the survey results was designed to both test the validity of the TPB model, and illuminate the degree to which the graduates had achieved RMIT's GA, their preparedness to be leaders for sustainability, and the sources of their understanding of sustainability.

Analysis of the data from the five sub-behaviours demonstrated that the TPB model is a valid approach to assessing graduate attributes. Details of this analysis can be found in Holdsworth et al. (in review), however, these can be summarised, in relation to the five sub-behaviours. For each of the sub-behaviours, results indicate that Behavioural Control has a direct positive impact on Graduates' Actual Behaviour. Despite the suggestion that Graduates' Attitudes and SubNorms be included in the model, their impacts on actual behaviour are not as significant.

In terms of Attitudes (a graduate's inclination to think or behave in a particular way) and Actual Behaviour, the results indicated that the graduates had absorbed the five sub-behaviours of the sustainability GA. Importantly, in their workplace they would seek to behave in a way that is consistent with these elements; although being more constrained than if they were not also influenced by some opposing workplace influences. Broadly, it is clear that their Sustainability Education experiences during their program at RMIT, have been effective; establishing both the knowledge of sustainability and the will to act on these principles in their work-places.

In addition to assessing the respondents' behaviour in terms of knowledge and understanding, the survey also sought insight into how they behaved to promote sustainability outcomes. The results indicated that in all cases the graduates would be *most likely to be directed by employers/managers about what is required.* A moderately close second behaviour was to be informed about the impacts of their work. In essence this demonstrates that in their workplaces the graduates were not seeing themselves as leaders for sustainability. Given the participants demonstrated a strong connection to sustainability, when they had freedom to do so, the indication is that they would be capable of promoting sustainability action to others.

The survey also sought information about the relative importance of some specific sources of information about sustainability. Results from the analysis of data related to the five sub-behaviours indicated that studies at RMIT University were most of the time the important contributor of sustainability related knowledge. The exception was for Physical Environmental knowledge, where Work experience was most important, with RMIT coming a very close second. Additionally, social media and Government or public campaigns, programs or information were much lesser contributors (influential, about half the time), with Activities or information by community groups or programs being only moderately influential in relation to knowledge about social and human communities.

5 Conclusions

Unique to this project is the development and application of a tool for assessing the attainment of graduate attributes (GAs) post degree completion and in the workplace. The measurement tool moves beyond student self-assessment, the dominant approach taken by the small number of existing studies that have attempted similar assessment research (Potts and Kleinpeter 2001; Spiel et al. 2013). Critically the project has both: investigated the role of Theory of Planned Behaviour in 'non self-assessment' of employed graduates; and has provided a mechanism for the collection of information about the extent to which graduates (as students) had absorbed the messages of their university program. The second outcome is especially important, as such information establishes the feed-back loop by which teaching staff and curriculum designers can use the results to review the relevance of their curricular.

As discussed earlier, a key aspect of the paper has been discussion of the development of the assessment tool, based on TPB. Since this approach has not been used previously, it has been important to test the relevance of the approach, for measuring achievement of outcomes, and validate TPB in the context of assessment. As a consequence of the survey and its results, TPB has been shown to be applicable to the assessment of the sustainability Graduate Attribute (GA) at one large Australian university-RMIT. Importantly, the results also show that TPB, and the associated survey, could be applied to the assessment of the sustainability GA in other disciplinary programs at RMIT, and likewise at other tertiary institutions. Further, that TPB is valid for the sustainability GA suggests it could also be valid for assessing other GAs. Such expansion of the application of TPB will be pursued as the next stage for development of the role of TPB.

The starting point for the project was assessment of achievement of the sustainability GA by graduates. By virtue of the survey results it is clear that the sample of graduates have absorbed the five sub-behaviours associated with the RMIT sustainability GA. As a consequence of the framework provided by TPB, it has also been possible to identify that while the graduates would choose to behave in a way that is consistent with these elements, if they had 'their way', specific influences act in the workplace to moderate (if not frustrate) this intention.

Hence, the educators associated with the sustainability curriculum experienced by the survey participants, can have confidence in the general effectiveness of this curriculum (although the results also pointed to some areas where improvements are possible). This has implications for those teaching sustainability in other disciplines, and at other institutions, that their curriculum may be having a powerful effect on their graduates' behaviour, and potentially their work-places.

Likewise, the results related to graduates taking leadership for sustainability, and their source(s) of understanding of sustainability provide important feedback and precise information for curriculum developers and providers. Regarding leadership, that the graduates felt that they would (almost always) be guided by their employers/managers illustrates a lack in the curriculum, to provide strategies for graduates to communicate their knowledge and concerns in their workplaces. Hence, where curriculum changes are possible to assist graduates to reduce work-place influences and restraints, we could anticipate graduates would take a stronger role in leadership for sustainability.

Finally, there are implications for RMIT curriculum, and possibly elsewhere. Within the discipline/program area researched for this project, the basics of the educational experiences provided to students are currently effective in delivering on the sustainability GA. The results demonstrated that the graduates understood what was required for them to act sustainably, and showed preparedness to act appropriately. Yet these actions were clearly moderated by the influence of key stakeholders and contexts associated with their work-places. Hence, if a higher level of the GA elements was to be sought, through class content, assessment tasks and experiences, then students would need to be given curriculum tools to help them resist the opposing influences of the workplace (e.g. Market/end-users, clients, colleagues); and take greater leadership for sustainability. This course of action would apply to RMIT curriculum, and could potentially have substantial benefits for Sustainability Education in other disciplines, and institutions.

Our project to apply TPB to the assessment of the sustainability GA has demonstrated the value of the approach. Specifically it can provide information in relation to the achievement of a sustainability focused GA, and thence highly relevant data for the development or refinement, of curriculum. Equally, the TPB framework potentially provides a valuable approach for the assessment of a range of GAs. Testing this potential provides an opportunity for all who are interested in learning outcomes of graduates, and how the graduates operate in their work-places.

References

Ajzen, I. (1985). From intentions to actions: A theory of planned behavior. In J. Kuhl & J. Beckmann (Eds.), *Action control: From cognition to behavior* (pp. 11–39). Berlin, Heidelber, New York: Springer.

Ajzen, I. (2002). Constructing a TPB questionnaire: Conceptual and methodological considerations, Bielefeld university http://www.uni-bielefeld.de/ikg/zick/ajzen%20construction%20a%20tpb%20questionnaire.pdf

Barrie, S. C. (2004). A research-based approach to generic graduate attributes policy. *Higher Education Research and Development, 23*(3), 261–275.

Barrie, S. C. (2006). Understanding what we mean by the generic attributes of graduates. *Higher Education, 51*(3), 215–241.

Barrie, S., Hughes, C., & Smith, C. (2009). Final Report—the national gap: The national graduate attributes project: Integration and assessment of graduate attributes in curriculum. *Australian Learning and Teaching Council*, [Online] Available: http://www.itl.usyd.edu.au/projects/nationalgap/resources/GAPpdfs/National Graduate Attributes Project Final Report 2009.pdf [Accessed December 2, 2014]

Barth, M., Godemann, J., Rieckmann, M., & Stoltenberg, U. (2007). Developing key competencies for sustainable development in higher education. *International Journal of Sustainability in Higher Education, 8*(4), 416–430.

Barth, M., & Michelsen, G. (2013). Learning for change: An educational contribution to sustainability science. *Sustainability Science, 8*(1), 103–119.

Bowman, K. (2010). Background paper for the AQF council on generic skills. Available http://www.aqf.edu.au/wp-content/uploads/2013/06/Generic-skills-background paper-FINAL.pdf [Accessed January 9, 2015].

Brundiers, K., Wiek, A., & Redman, C. L. (2010). Real-world learning opportunities in sustainability: From classroom into the real world. *International Journal of Sustainability in Higher Education, 11*(4), 308–324.

Cohen, B. (2007). Developing educational indicators that will guide students and institutions toward a sustainable future. *New Directions for Institutional Research, 134,* 83–94.

Crocker, R. (2013). From access to excess: Consumerism, 'compulsory' consumption and behaviour change. In R. Crocker & S. Lehmann (Eds.), *Motivating Change Sustainable design and behaviour in the built environment* (pp. 11–32). Oxon, UK: Routledge.

Dale, A., & Newman, L. (2005). Sustainable development, education and literacy. *International Journal of Sustainability in Higher Education, 6*(1), 351–362.

Delaney, A. M. (2004). Ideas to enhance higher education's impact on graduates' lives; alumni recommendations. *Tertiary Education and Management, 10*(2), 89–105.

Down, C. (2006). Lifelong learning, graduate capabilities and workplace learning. In P. Hager & S. Holland (Eds.), *Graduate attributes* (pp. 187–205). Netherlands: Learning and Employability, Springer.

Fien, J., Guevara, R., Lang, J., & Malone, J. (2004). *Australian country report UNESCO-NIER regional seminar on policy, research and capacity building for education innovation for sustainable development*. Education for sustainable development, Tokyo, Japan: ESCO Australian National Commission.

Graduate Careers Australia (2014) Graduate outlook 2014; employers' perspectives on graduate recruitment in Australia, Melbourne, [Online] Available: http://www.graduatecareers.com.au/research/surveys/graduateoutlooksurvey/ [Accessed November 17, 2015].

Green, W., Hammer, S., & Star, C. (2009). Facing up to the challenge: Why is it so hard to develop graduate attributes? *Higher Education Research & Development, 28*(1), 17–29.

Haan, G. (2006). The Blk '21' programme in Germany: A 'Gestaltungskompetenz'-based model for education for sustainable development. *Environmental Education Research, 12*(1), 19–32.

Hager, P. (2006). *'The nature and development of generic attributes' in graduate attributes* (pp. 17–48). Springer, Dordrecht: Learning and Employability.

Hofreitera, T. D., Monroea, M. C., & Stein, T. V. (2007). Teaching and evaluating critical thinking in an environmental context. *Applied Environmental Education & Communication, 6*(2), 149–157.

Holdsworth, S. (2010). A critique of academic development in sustainability for tertiary educators; Doctoral thesis; RMIT University, Australia.

Holdsworth, S., & Hegarty, K. (2015). *Towards a praxis of sustainability in universities international journal of innovation and sustainable development, 9*(3/4), 205–226. https://doi.org/10.1504/IJISD.2015.071861.

Holdsworth, S., & Hegarty, K. (2016). From praxis to delivery: A higher education learning design framework (HELD). *Journal of Cleaner Production, 122,* 176–185.

Holdsworth, S., & Sandri, O. (2014). Sustainability education and the built environment: Experiences from the classroom. *Journal Education Built Environment, 9*(1), 48–68.

Holdsworth, S.,Thomas, I. (2015). A sustainability education academic development framework (SEAD) In: *Environmental education research* (pp. 1–25).

Ison, R. (2010). *Systems practice: How to act in a climate-change world.* London: Springer and The Open University.

Jones, P., Selby, D., & Sterling, S. (2010). *Sustainability education: Perspectives and practice across higher education*. Earthscan, New York.

Lozano, J. F., Boni, A., Peris, J., & Hueso, A. (2012). Competencies in higher education: A critical analysis from the capabilities approach. *Journal of Philosophy of Education, 46*(1), 132–147.

Martin, S. (2005). Sustainability, systems thinking and professional practice. *Systemic Practice and Action Research, 18*(2), 163–171.

McNamara, J. (2013). The challenge of assessing professional competence in work integrated learning. *Assessment & Evaluation in Higher Education, 38*(2), 183–197.

Millar, M. G., & Millar, K. U. (1996). The effects of direct and indirect experience on affective and cognitive responses on the attitude-behavior relation. *Journal of Experimental Social Psychology, 32,* 561–579.

Mochizuki, Y., & Fadeeva, Z. (2010). Competences for sustainable development and sustainability. *International Journal of Sustainability in Higher Education, 11*(4), 391–403.

Mogensena, F., & Schnack, K. (2010). The action competence approach and the 'new' discourses of education for sustainable development, competence and quality criteria. *Environmental Education Research, 16*(1), 59–74.

Newton, P. (2013). Exploring the role of individual, context and object in sustainable urban consumption. In R. Crocker & S. Lehmann (Eds.), *Motivating change sustainable design and behaviour in the built environment* (pp. 33–55). Oxon, UK Routledge, Oxon, UK: Routledge.

Nussbaum, M. (2011). *Creating capabilities: The human development approach.* Massachusetts: Harvard University Press.

OECD (Organisation for Economic Co-operation and Development) (2012). AHELO: Assessment of higher education learning outcomes. [Online] Available: http://www.oecd.org/edu/skills-beyond-school/skills-beyond-school/AHELO%20Brochure.pdf [Accessed October 30, 2016].

Parker, J. (2010). Competencies for interdisciplinarity higher education. *International Journal of Sustainability in Higher Education, 11*(4), 325–338.

Parkin, S., Johnston, A., Buckland, H., Brookes, F., & White, E. (2004). *Learning and skills for sustainable development: Developing a sustainability literate society*. Higher Education Partnership for Sustainability, London.

Potts, M. K., & Kleinpeter, C. H. (2001). Distance education alumni: How far have they gone? *Journal of Technology in Human Services, 18*(3–4), 85–99.

RMIT University (2015), Graduate attributes: Environmentally aware and responsible. [Online] Available: http://www1.rmit.edu.au/browse;ID=esi2ehsc1kv [Accessed March 1, 2015].

Sadler, R. (2013). Making competent judgments of competence. In Blömeke, S., Zlatkin-Troitschanskaia, O., Kuhn, C. & Fege, J. (Eds.), *Modeling and measuring competencies in higher education: Tasks and challenges* (pp. 13–27). Rotterdam: Sense Publishers, E-book Springer.

Sandri, O. (2014). 'Good practice learning and teaching for sustainability in higher education', PhD, RMIT University [Online] https://researchbank.rmit.edu.au/eserv/rmit:160976/Sandri.pdf [Accessed October 11, 2016].

Sandri, O., Holdsworth, S., & Thomas, I. (2016). Vignette question design for the assessment of graduate sustainability learning outcomes. *Journal of Environmental Education Research*. http://dx.doi.org/10.1080/13504622.2016.1263280

Scott, W., & Gough, S. (2010). Sustainability, learning and capability: Exploring questions of balance.*Sustainability, 2,* 3735–3746.

Segalas, J., Ferrer-Balas, D., Svanstrom, M., Lundqvist, U., & Mulder, K. F. (2009). 'What has to be learnt for sustainability? *A Comparison of Bachelor Engineering Education Competences at Three European Universities' Sustainability Science, 4,* 17–27.

Spiel, C., Schober, B., & Reimann, R. (2013). Modeling and measurement of competencies in higher education: The contribution of scientific evaluation. In S. Weber F. Achtenhagen & F. Oser (Eds.), *Modeling and measuring competencies in higher education: Tasks and Challenges* (195–206). Sense Publishers. [Online] Available: http://link.springer.com.ezproxy.lib.rmit.edu.au/book/10.1007/978-94-6091-867-4 [Accessed December 17, 2014].

Sterling, S., & Thomas, I. (2006). Education for sustainability: The role of capabilities in guiding university curricula. *International Journal of Innovation and Sustainable Development, 1*(4), 349–370.

Thomas, I. (2003). Employers expectations of graduates of environmental programs: An australian experience. *Applied Environmental Education & Communication, 2*(1), 49–59.

Thomas, I., & Day, T. (2014). Sustainability capabilities. *Graduate Capabilities, and Australian Universities, International Journal of Sustainability in Higher Education, 15*(2), 208–227. https://doi.org/10.1108/ijshe-05-2012-0046.

van der Velden, R. (2013). Measuring Competences in Higher Education: What Next? In S. Weber, F. Achtenhagen & F. Oser (Eds.), *Modeling and measuring competencies in higher education: Tasks and challenges* (pp. 207–216). Sense Publishers [Online] Available: http://link.springer.com.ezproxy.lib.rmit.edu.au/book/10.1007/978-94-6091-867-4 [Accessed 11/10/2016].

Wals, A. E. J. (2010). Mirroring, gestaltswitching and transformative social learning: Stepping stones for developing sustainability competence. *International Journal of Sustainability in Higher Education, 11*(4), 380–390.

Wals, A. E. J. (2014). Sustainability in higher education in the context of the UN DESD: a review of learning and institutionalization processes. *Journal of Cleaner Prodduction, 62,* 8–15.

Wiek, A., Withycombe, L., & Redman, C. L. (2011). Key competencies in sustainability: A reference framework for academic program development. *Sustainability Science, 6,* 203–218.

University Sustainability Course Creation in China: Experiences from the CUFE

Liguang Liu

Abstract Higher education for sustainable education (HESD) has earned more attention in China rhetorically. Reorientation of the curriculum is taken as the essential vehicle for the implementation of HESD. However, there is less in-depth and systematic research on how a stand-alone course is designed for undergraduates of all majors in China and what perceptions college students have on the sustainability education. Taking Central University of Finance and Economics as a case, the study investigates the creation of a sustainability course and provides an initial survey of students' perceptions on the course contents and arrangement. Students recognized their deficiency on the global sustainable development action and accepted the importance of sustainability curricula. They also made suggestions on course contents and pedagogies. The study argues that current HESD initiatives essentially manipulated by the government may miss out on reaching a majority of universities, therefore, to promote sustainability in universities in China, the importance of governance and politics should not be ignored. The research findings provide a useful reference to instructors to improve teaching and further pedagogical research, as well as educational authorities to transform HESD policy making and implementation.

Keywords Higher education · Course design · Sustainable development China

1 Introduction

Higher education plays a vital role in achieving strategic development in sustainability. Since 1992, the need of Education for Sustainable Development (ESD) has begun to emerge and integrate with the earlier environmental education in China

L. Liu (✉)
School of Government, Central University of Finance and Economics (CUFE),
39 South College Road, Haidian District, Beijing 10081, China
e-mail: liuliguang@cufe.edu.cn

W. Leal Filho et al. (eds.), *Sustainable Development Research in the Asia-Pacific Region*, World Sustainability Series, https://doi.org/10.1007/978-3-319-73293-0_8

(Niu et al. 2010). The Decade of Education for Sustainable Development (DESD), ranging from 2005 to 2014, also encouraged higher educational institutions (HEIs) to integrate sustainability into innovations in teaching, research, social engagement, campus administration, and student development. With the promulgation of the 2030 Agenda for Sustainable Development and adoption of the Sustainable Development Goals (SDGs), higher educational institutions confront challenges to deepen governance reforms towards a more sustainable future.

In its National Medium and Long-Term Education Reform and Development Plan (2010–2020), Chinese government identified the ESD as an important strategic theme for future educational reform. Although environmental education, to some extent, has been involved in the courses for primary and secondary education in China, there are no clear regulations for universities to develop such curricula and initiate sustainability programs. On the one hand, the effectiveness of sustainable campus construction highly relied on the voluntariness and capabilities of individual universities; on the other hand, government leans toward supporting some piloted universities to accumulate experiences. The higher education for sustainable development (HESD) has optimistic prospects, though the current result is not optimized.

The aspects that HEIs can contribute to sustainable development include green campus operations, sustainable research, networks among institutions, curricular innovation, public outreach and community engagement. Over the past decade, there have been more studies on university sustainability initiatives in China from various perspectives, such as university administration and motivations (Zhao and Zou 2015; Lo 2015), ESD policy development (Wang 2014), awareness status of campus sustainability (Yuan and Zuo 2013). There are also normative studies on significance and approaches to promote higher education for sustainable development (Wang 2007; Wang and Li 2011). Case studies, interviews, questionnaires, and document analysis are the most employed methods.

Since ESD is, in essence, interdisciplinary and university students and faculties are recruited and categorized by specific majors, it has always been challenging to initiate and foster an integrated sustainability curricular system to students. Some pioneer green universities in China have been highlighted and the progress proved to be evident (Zhao and Zou 2015), however, the efforts of the majority of universities, some of whom also seek to enhance their competence through updating its ideas and reforming its institutional governance, have received less attention. Although a case of one university may seem unique and particularly individual, under a rather centralized regime in China, it reveals representative elements and features that peer universities share.

The Central University of Finance and Economics, which is a public university located in Beijing, China, is taken as a case. A survey was conducted to 229 undergraduate students of different majors, who were taking the optional sustainability course in the fall semester of 2016. The purpose of this research is to investigate students' perceptions on sustainability course learning, in order to better understand the Chinese students' interests and views on sustainability course study and to improve curriculum design and pedagogical research thereafter.

2 HESD and Its Development in China

This section provides a contextual background of how China has launched the HESD initiatives and how Chinese universities have incorporated more sustainability programs into their development agenda.

2.1 Higher Education Governance in China

Notwithstanding its sheer size and diversity, China has been a unitary and centralized state. The main organs of state power are the National People's Congress (NPC, the highest legislative branch) and the State Council (the highest executive branch). Since the Communist Party of China (CPC) is the founding and ruling political party, the Politburo and its Standing Committee play most significant roles in policy decision. The subnational (provincial) government in mainland China exercises jurisdiction over 22 provinces, 5 autonomous regions, and 4 municipalities under the central government (i.e. Beijing, Shanghai, Tianjin and Chongqing).

The Chinese Ministry of Education is a cabinet-level agency under the State Council that takes charge of the overall planning, coordination, and management of all educational system in mainland China, including compulsory basic education, vocational education, and higher education. In China, the ways of supervising higher education institutions are diversified. In 2015, there were as many as 2560 universities and colleges and 292 institutions for adult education. The number of enrolled undergraduate and graduate students in higher educational institutions reached 28.16 million, and 7.36 million college students entered the job market. Both figures have been increasing dramatically (Statistical Yearbook 2016). According to Lo (2015), the HEIs in China can be generally categorized into four types based on the administrative structure. The first type is central-level HEIs, which are directly affiliated with the Ministry of Education. These HEIs are normally well-funded and considered as top-tier universities. The second type is provincial-level HEIs. These HEIs are under the control of provincial governments. This is the largest group of HEIs, but is smaller in scale, less prestigious and less funded. The third group is private HEIs, which are privately owned and are operated for profit. The fourth group is independent HEIs, managed through partnerships between a public university and private sectors. Private and independent HEIs do not receive funding from the government and face serious challenges of policy inconsistencies, perceived low quality and low reputation (Lei 2012; Li and Morgan 2011).

Although the majority of Chinese universities are state-owned, the funding varies greatly, even within the same group. To promote the educational competitiveness and build world-class universities, Ministry of Education adopted the differentiation strategy, initiating the Project 211 in the 1990s and the Project 985 in 1998. The formal project includes 112 HEIs till the mid-2000s and the latter one includes 39 universities from the list of Project 211 universities. It is reported that from 2009 to 2013, the research funding for Project 985 HEIs, Project 211 HEIs (73 universities), and other 670 common HEIs were 139.5 billion RMB (52.7%), 51.1 billion RMB (19.3%), and 72.4 billion RMB (28%) respectively (Xu 2014). The Project 211 and Project 985 have been widely regarded as aggravating the imbalance of competition among universities and local development (Xu 2014; Serger et al. 2015; Wang 2013).

2.2 ESD Governance and University Initiatives

The education for sustainable development (ESD) was first raised by the United Nations Educational, Scientific and Cultural Organization (UNESCO) in 1988 and proposed at Rio Conference in 1992. It was the time that the Chinese government began to incorporate the idea of sustainability into development agenda. In China's Agenda 21, education and capacity-building were officially identified as fundamental strategies to sustainable development. The Chinese government also established the Administrative Center for China's Agenda 21 (ACCA21) under the Ministry of Science and Technology to coordinate relevant sustainable development programs and strengthen the training of researchers. With sustainable development being enriched to cover economic, social and environmental aspects, the National Development and Reform Commission (NDRC) gradually played a dominant role in drafting and integrating general policies and guiding the domestic sustainable development practices. Ministry of Education is responsible for the relevant initiatives in all levels of education institutions. According to Chinese National Sustainable Report (NDRC 2012), China made great achievements in enlarging investment in education and achieved a full-scale free compulsory education in both urban and rural areas, as well as promoting the equitable development of education, especially in underdeveloped regions.

In China's Agenda 21, China has raised tasks of strengthening environmental education in primary and secondary schools and developing curricula and materials in universities and introducing post-graduate courses on sustainable development. Within this program, various efforts have been made to reorient teaching and research in Chinese universities and to bring sustainability practice into campuses. Since 2003, with support from international agencies such as UNESCO, UNICEF, and UNEP, environmental education initiatives flourished in China. The Chinese government and UNESCO have enhanced their partnership in the field of education. The Chinese Flagship Project on Environment and Population Education for Development (EDP) was promoted within the framework of UNESCO's program on Education for Sustainable Development (ESD). China later actively participated in the UN Decade of Education for Sustainable Development (DESD) during 2005–2014, which aims to the enforcement of integration of ESD into education, public awareness and training, and community development. Since 2015, the Global Action Program (GAP) on ESD has become the follow-up to the DESD and seeks to scale-up concrete actions that contribute to the achievements of SDGs. The

Chinese ESD program, initiated by Ministry of Education and the UNESCO, chose 13 provinces and municipalities to set institutions for the operation.

Several global sustainability declarations, e.g. Talloires (1990), Copernicus (1994), and Lunenburg (2001) provide useful publicity to encourage universities' involvement in sustainability progress, but they are not sufficient to change institutional and disciplinary practices in universities (Bekessy et al. 2007). In China, the demand for sustainable development, in tandem with rapid economic growth over the past decades has provided an essential impetus for some HEIs to change strategies to engage sustainable education in teaching, researching, outreach and service to society. According to Niu et al. (2010), approximately 50 percent of all the Chinese universities have launched general courses linked to the sustainable development and almost all universities have begun to reoriented curriculum in line with ESD. Some prestigious universities, e.g. Tongji University, Tsinghua University, and China Renmin University, have become leaders in sustainability education and green campus practices. The participation and engagement in the international sustainability network, such as APRUC (UNEP—Asia-Pacific Regional University Consortium), ProSPER.Net (Promotion of Sustainability in Postgraduate Education and Research Network) and BELL Project, contributed to the capacity building and visibility elevation and earned them prominence in the world.

3 Sustainability Course Design and Survey at CUFE

This section takes Central University of Finance and Economics (CUFE) in China as a case to investigate how it has initiated its sustainability course creation and how the undergraduate students enrolled in the class have responded to the course study. The selection of CUFE case is not just to simply highlight its uniqueness, it is used as an instrumental case to understand how the sustainability course is practiced generally in China.

3.1 CUFE Introduction

Central University of Finance and Economics (CUFE) is located in Beijing. It was first established in 1949 with the name Central School of Taxation. Later, it went through several stages of development, from Central Institute of Finance, Central Institute of Finance and Economics, to Central Institute of Finance and Banking. It had long been attached to the Ministry of Finance. In 1996, the institute was officially renamed as Central University of Finance and Economics (CUFE) under the direct leadership of the Ministry of Education and in 2005, it became one of the national Project 211 universities. The university is a well-known university especially in the fields of finance and economics. If measured in size and comprehensive capacity, CUFE ranks about the 80 to 120th; however, in terms of undergraduate student quality measured by scores to get admitted, it can be listed in the group of top-20 most competitive universities in China.

CUFE is a medium-scaled university, with 10,036 undergraduates, 4289 master students, and 673 doctoral students. It hosts 29 schools, including 50 undergraduate programs, 76 master degree programs, and 31 doctoral programs. There are 1157 faculties and 544 staffs working with it (CUFE 2015). To accommodate the growing development of the institution, in 2007, CUFE began to build its second campus—Shahe suburban campus, which is 6 times larger and located 30 km north of the Downtown campus. When the main buildings and facilities finished in 2009, the Shahe Campus was put into use. Currently, more than 7500 students, mainly freshmen, sophomores and junior undergraduates study and live on Shahe campus, while seniors, master's and doctoral students, and international students stay at the downtown campus. The university provides regular daily commuting buses between two campuses. CUFE announced its goal and vision of sustainable development, but does not have an official sustainability mission and coordination office at this stage.

3.2 Sustainability Course Arrangement

The current curricula for undergraduates at CUFE include compulsory and optional courses. Besides optional courses within involved disciplines, general education courses are open for undergraduate students of all majors to choose. These courses are approved by the Office of Teaching Affairs and set by the principle of "commitment of socialist core values with Chinese characteristics and adherence to excellent national tradition and culture". General educational courses shall not be repeatedly created and are divided into 5 categories: namely, Culture, Literature and Art; History, Politics and Society; Philosophy, Psychology and Living; Science, technology and Environment; Life, Earth and Universe. Each course is counted as 2 credits, and the student is required to earn at least 10 credits from no less than 3 categories within three academic years.

The general education courses have been updating each semester. Currently, there are about 170 courses on the list. In the fall semester of the 2016–17 academic year, Sustainable Development and the Global Challenge course was launched by a lecturer from the School of Government. The School, a relatively new school compared with others, was formed in 2007 from a single department of public administration. It later established departments of public affairs, city management, and international politics. In 2013, a new department of resource and environmental management was founded. The department has 7 faculties on duty but its degree programs have not been approved. The faculties are encouraged to teach courses for other majors and create general educational courses.

The course of Sustainable Development and the Global Challenge is a 36 contact-hour course. The course was scheduled 3 h every week for 12 weeks in

Shahe Campus. The instructor has a decade of working experience in environmental and development research, and 5 years experience in teaching courses in public administration, public policy, resource and environmental management, and public program management.

3.3 *Learning Objectives and Contents*

The curriculum setting was under the contextual background of the formulation and unanimous adoption of the sustainable development agenda by the United Nations in 2015 and a series of economic, social and environmental challenges that the world is new facing. The United Nations Summit in 2015 has adopted 17 Sustainable Development Goals (SDGs) in the 2030 Agenda for Sustainable Development. These new goals, built on the experiences of the Millennium Development Goals (MDGs), aims to mobilize global efforts to end all forms of poverty, fight inequalities, tackle climate change, and protect the environment. In addition, how countries, regions and the world can integrate policies and efforts is crucial to the achievements of SDGs.

The creation of the optional course was to complement students' study in their specialized majors. It intended to provide a basic introduction to the interdisciplinary field of sustainable development, describes the complex interactions of the world development and the Earth's environment. It introduced the recent theoretical and practical development in sustainable development research and provides a broad overview of the key challenges and potential solutions. The contact time per week was designed as 2 h for lecturing, and 1 h for questions or discussion. The contents of the 12-week course are shown in Table 1.

The lecture materials, including slides used in class and suggested readings were posted to a course mailbox that was open to all registered students and stayed up for the duration of the course. There were two assignments and one essay to complete after class. The assignments were questions that students can find answers from the reading materials. The essay, accounting for 60% of the final grade, asked students to use what they have learned in class to analyze and comment on a real development case.

4 Questionnaire Survey and Results

Since the course is an optional stand-alone module for undergraduates, and it was a newly created course, a small-scale questionnaire survey is conducted halfway to gain some understanding of the registered students' prior knowledge of sustainable development and their perceptions of the course. The questionnaire consisted of 4 sections. The first section was designed to collect demographic information about the respondents; the second section examined students' preference of topics

Table 1 Contents of the optional sustainable development course

Schedule	Content	Modules
Week 1	Introduction, planetary boundary, research methods	Module 1: overview of sustainable development
Week 2	History of sustainable development (Environmental history, Brundtland Report, Rio Earth Summit)	
Week 3	History of sustainable development (from MDGs to SDGs)	
Week 4	Human population and health	Module 2: specific topics of environmental concerns
Week 5	Economic development	
Week 6	Poverty and inequality	
Week 7	Energy and environmental management	
Week 8	Addressing climate change, questionnaires*	
Week 9	Policy, government, and politics	
Week 10	Green campus and community development	
Week 11	Carbon market development in China and the US; environmental ethics; air quality issues	Module 3: topics of interests
Week 12	The role of government and governance; policy making process; NIMBY issues; gender equality	

*The questionnaires was delivered to collect information of students' prior knowledge of sustainable development and their perceptions on the course contents

lectured in class. The third section is designed to investigate students' understanding of the political dimension of the policy process, especially the role of government. The fourth section consisted of non-structured questions, asking students whether the course has reached their expectations and what possible suggestions they want to make. The survey was conducted in week 8 and week 9 (to those who were absent in week 8) when important concepts and events of sustainable development had been lectured.

4.1 Questionnaire Results

In total, a big class of 258 graduate students chose to take this newly created course. The students were invited to fill out the questionnaire and 229 questionnaires were returned with a response rate of 88.8%. Of 229 respondents, there are 105 male students (45.9%), 124 female students (54.1%). The students are from different grades, 71 students as freshmen (31%), 103 as sophomores (45%), and 55 as juniors (24%). Students come from 21 schools and 43 undergraduate programs out of 29 schools and 50 undergraduate programs of CUFE.

The students were asked whether they have had stand-alone ESD courses or relevant environment education courses in their previous studies at school. 191 students gave negative responses (83.4%). Some responded that they had received some environmental education in their elementary years, but not at high school. Students were asked to assign a number from 1 (meaning no knowledge at all) to 5 (full understanding) to indicate their prior knowledge about key historical events of sustainable development before they had taken the course. Five historical events were selected as indicators: 1972 Human Environment Conference; Brundtland's definition of sustainable development; Rio Earth Summit and Agenda 21; MDGs adopted in 2000; and SDGs in 2015. Figure 1 shows the survey results. In general, students had insufficient knowledge on these events (No knowledge responses occupied the most), but the situation gets improved to the more recent events in 2000 and 2005.

The students were asked to choose topics they are more interested or less interested. These topics were what had been proposed in the syllabus and some had been taught. The purpose was to get to know students' preference of the topics. The question was designed to be multiple response questions, students were advised to choose 3–5 items that they felt most appropriate. Figure 2 shows the students' responses, sequenced by the order of percentage of cases that indicates a higher degree of interest.

Since the 2030 Sustainable Development Agenda has adopted 17 goals for global efforts, among these goals, what areas of SDGs are students more interested? The results showed a generally stable distribution, with higher interests in Goal 4 (quality education, 44.1%), Goal 5 (gender equality, 39.3%), Goal 8 (good jobs and economic growth, 31.9%); and lower interests in Goal 15 (life on land, 7.4%), Goal 17 (partnerships for the goals, 13.5%) and Goal 14 (life below water, 14.4%). Besides, students responded that beyond the case in China (47.7% out of 464 responses), they are also interested in the sustainable development practices in developed countries (68.2%), other developing countries (57.7%) and less developed countries (37.3%).

The course is provided by the instructor from the School of Government, who in some occasions, stressed the roles played by government, especially to the centralized country like China. 63.8% of respondents thought it is necessary to take

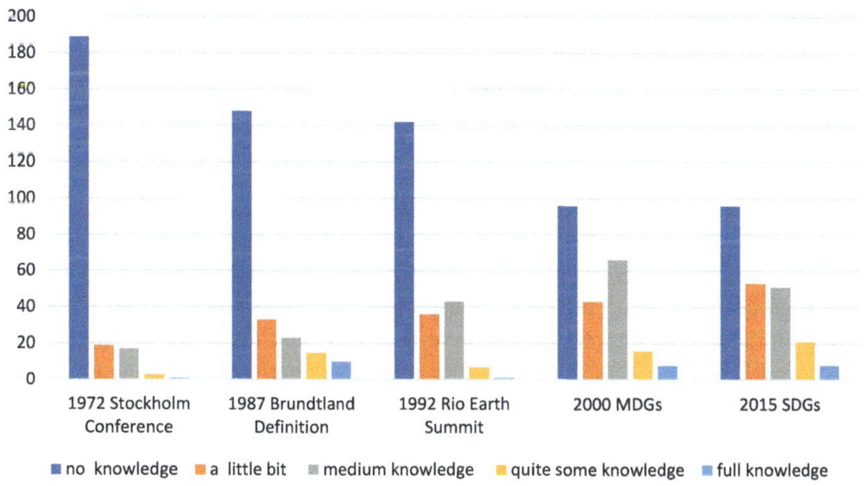

Fig. 1 Students' prior knowledge about SD events

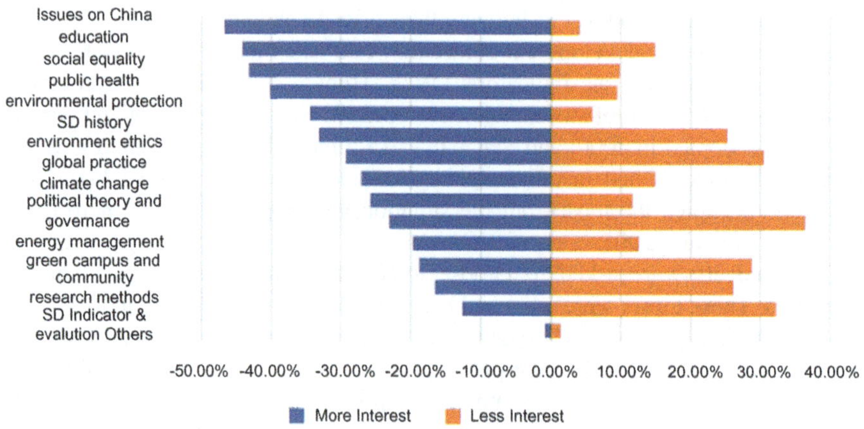

Fig. 2 Survey responses to topics of more interest and less interest

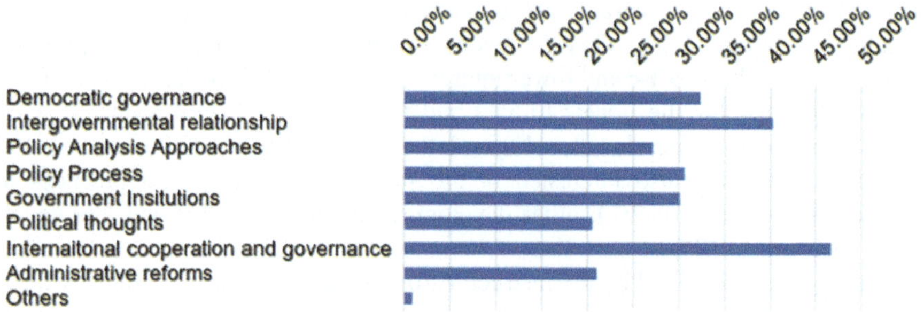

Fig. 3 Survey responses to the issues of interest about government

additional courses to study government and how it works. The students were asked to make 2–3 multiple response choices among some topics concerning government and politics that they want to deepen their studies, the results showed that international cooperation and governance, inter-governmental relationship, and democratic governance are highlighted as the top three topics (Fig. 3).

A number of foreign and domestic policy actions, in areas of economic and social development, as well as environmental protection were presented as cases in class. Although most students had never taken public policy courses before, students showed great interests in discussing concrete cases and connecting them with policy processes. Taking a recent local protest on a waste incineration project in south China as a case, respondents were asked to rank the order of relative significance of various factors that influence policy making and contribute to conflict resolution.

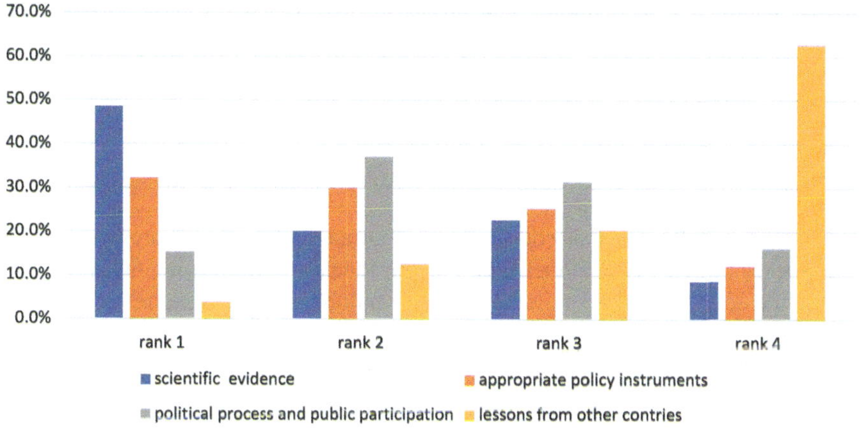

Fig. 4 Survey response to the priorities of formulating policy solutions

As Fig. 4 shows, scientific evidence is taken as the most important factor in policy making, and lessons from other countries ranked the last. Students suggested some other factors, such as citizen mentality, capacity to tackle policy side-effects.

The class generally reported a positive appraisal to the course. 214 students (93.4%) claimed it has reached their expectations; 11 students (4.8%) thought it basically live up to the objectives, and 3 gave negative recognition. A majority of students reflected that the course was informative and necessary, and it helped them to broaden the views and follow the sustainable development issues in China and abroad.

Students provided suggestions to the improvement of the course: less theory instruction and more practical cases related to ordinary life; opportunities to arrange some sightseeing or field trip; more class discussions and interactions; more videos or movies related to relevant topics; more book and website recommendation for after-class reading. Some students complained that the course time occasionally conflicted with their personal and school activities; or the class had bypassed some important topics, such as the Belt and Road Initiatives; land use in rural areas in China etc.

5 Discussion

The discussion on what we can learn from sustainability curricular development at CUFE is generally framed by the three dimensions of ESD, namely learning content, pedagogy and learning environments, learning outcomes (UNESCO 2014). In addition, the outcomes of ESD in general, and HESD, in particular, are greatly influenced by the factor of ESD governance. By putting together some findings from the survey and subjective observations, this section discusses why political factors should not be ignored when discussing the development of sustainability curricula and initiatives in higher institutions in China.

5.1 Learning Content

The creation of the optional course acted as an attempt to contribute to curricular improvement at CUFE. It attracts 258 undergraduates of various majors and grades to take the course. To reflect the holistic and multidisciplinary features of ESD, the course has tried to incorporate critical issues, such as the evolution of sustainable development, climate change, social equality, sustainable community development, public health, quality education, into the curriculum. Such arrangement has generally met the needs of students of diverse needs within 12-week duration.

Although ESD has been widely reported to be incorporated into curricula at all levels of education in China, the students' knowledge about the sustainability historical evolution is insufficient, even for these highly-qualified students that had stood out in the college entrance exam. The students have shown their diverse interests and concerns related to sustainable development. The class has addressed some of these topics to the best of the instructor's possibility.

In general, the students' perceptions on course contents were roughly the same across different groups by gender or by grades. Some minor differences did exist and are worth being tested in the future. Comparatively, male students are more interested in sustainability history (5.5% vs. 10.7% within a gender), while female students have less interest in politics coverage (17.3% vs. 10.1%). Concerning the sustainable development goals, female students are more interested in gender equality (12.1% vs. 6.7%), and male students in health and well-being (8.4% vs. 4.9%).

There are some interesting findings when analyzing statistical outputs of different grades. Students of freshmen, sophomores, and juniors all ranked issues of China, sustainability history, social equality and environmental protection as top

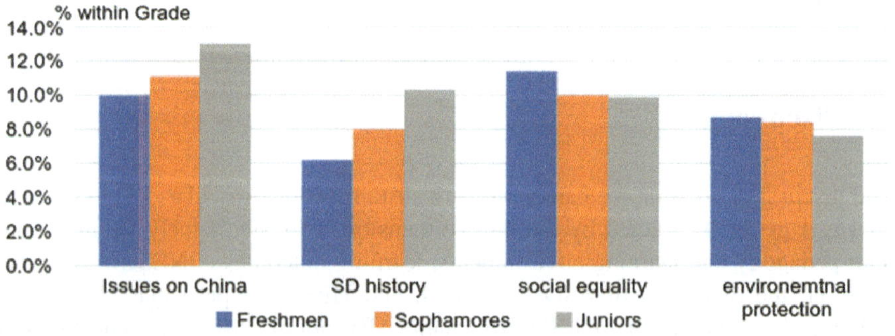

Fig. 5 The perception difference of top interesting topics by grades

topics of interest; but with school year accumulating, the interests in the Chinese issue and the history increased, while the interests of the social equality and environmental protection reduced (Fig. 5). This might be explained by the proposition that the study of their specialized majors may help formulate their new concerns and cool down their passionate sensitiveness to problems in development. However, this warrants further investigation.

5.2 *Pedagogy and Learning Environment*

Pedagogy and learning environments influence how effective the teaching is in terms of student learning. To optimize teaching and learning, educators should conduct a systematic assessment of whether and how teaching approaches and learners' behavior influence the learning effectiveness. This means instructors should not stop at sincere teaching but transform up towards scholarly teaching (Gurung and Schwartz 2013).

The questionnaire survey gave more focus on students' perceptions of learning content of the sustainability course, not on the pedagogy. The class status and some students' comments and suggestion did reflect a lot of unmet expectations existed. In any case, teaching should be taken in an interactive, learner-centered way that enables more engagement of students. Without a better understanding of students' interests and expectations, the course loses students and could not reach the goals set by the designer. A challenge is to get to know the generalized and unique interests of the students from different majors and to connect learning goals with teaching techniques. Although the optional course was proved to be a fruitful start, further pedagogical research is required.

The current learning environment is basically fine. Instructor and students met weekly in a classroom with a capacity of 300 people. The online course has not been planned yet, but to arrange a short campus visit to gain an understanding of CUFE greening practice is in progress. Besides, the Department of Resource and Environment Management is developing a new course of "Management of Green Campus" to attract students that may take interests.

The direct output was that 246 students (95.3%) passed the course (the optional course is graded in "pass" or "fail"), based on a calculation of student presence (20%), assignments (20%), final essay (60%). As a matter of fact, final essay weighed the most and all submitted essays were above the acceptable level. There were some excellent essays that students apply what they had learned in class and in their major study and wrote in a clear and coherent manner with solid and rich information.

In subsequent weeks after finishing the course, students voluntarily formed groups writing proposals for National College Student Innovation Training Competition, a country-wide program sponsored by the Ministry of Education and encourage students to conduct theoretical or practical research. Some students took advantage of personal networks built from the sustainability class and the instructor received emails soliciting suggestions or being invited to advise their projects. Some ideas were come up from the class, and research groups wanted to go deeper.

A complete assessment of universities' sustainability performance can be processed from other instruments, such as GASU (Graphical Assessment of Sustainability in Universities), which facilitates analysis and longitudinal comparison with other HEIs (Lozano 2006). Sustainable campus construction in HEIs takes a great deal of time and effort. Curricular development, as one pillar of HESD, signifies a crucial step forward to institutional reform. Without an institutional culture in which such type of grass-roots sustainability effort is recognized and valued, the time and energy will be spent elsewhere and sustainable development will always be a rhetoric.

5.3 Governance Performance and Political Influences

The finding from the survey indicates that though students have diverse interests of sustainable development practices among different country types, they gave special attention to issues of China. Of topics related to government, higher grade students tend to choose democratic governance as topic of interest (16.3% in juniors vs. 12.9% in sophomores vs. 10.6% in freshmen), meanwhile, juniors also led in percentage of accepting necessity of in-depth study on government (9.1% in juniors vs. 1.9% in sophomores vs. 0% in freshmen). Moreover, students have a universal resistance towards political theory instruction (15.2% in juniors vs. 13.1% in sophomores vs. 13.9% in freshmen, aggregated result in Fig. 2).

The inconsistency seems understandable. The political courses in Chinese HEIs are compulsory with political theories often full of Party rhetoric, opportunistic and undebatable. Students have been persistently trained to accept its sociopolitical construct from childhood and devote themselves to serving the great cause. However, there is a complex interplay between increasingly self-centered ambition and social justification (Jost and Hunyday 2005), the students may have realized the implication and importance of politics with the age and knowledge increasing, but lacking strong motivation to be independent in thinking and doing, they either stop by commending highly of China's economic achievements and be tolerant of its drawbacks, or they would make compromise, working hard for a wealthy personal life and be indifferent to the public affairs. There is no surprise that the government and school administration in China maintain overt and covert control over student behavior, under such environment, the individual students with passion and sympathy will be often feeling isolated and helpless. This possibly explains the reasons why students did not show accompanying interests in having a further study on government, and they were hesitant to discuss government in class.

A separate lecture on "policy, governance, and politics" was delivered in week 9. The contents included relevant concepts, he Goal 17 of SDGs and its targets, in particular, the systematic issues (UN 2015); and World Bank's global governance Indictors, which provides a governance framework for longitudinal and cross-countries comparisons. Moreover, influential political thoughts were introduced, and relationships among state capacity, democratic accountability, and rule of law were analyzed. The main purpose of such arrangement was to stimulate students' further thinking of current challenges in China. There is no doubt that government plays leading roles in establishing and implementing policies for sustainable development, but it is also true that if we want better government, we had better talk politics.

6 Conclusions

Higher education for sustainable education has earned more attention in China. The curricular development provides an opportunity for HEIs to get involved in campus sociability activities, and for students to equip with skills for sustainable lifestyles. This study illustrates a questionnaire survey which was conducted at an optional course in a university in Beijing, China, and investigates undergraduates' perceptions on the course arrangement and their interests. The results showed that students diversified in their interests of topics about sustainable development. Most students had limited knowledge about the sustainable development history, but they generally accepted that the course was very useful and helpful, and had reached their expectations. The survey showed some preference differences of the relative importance of sustainability topics by grade and gender, to some extent reflecting the unique characteristics of each category. Students are more concerned about China's development and issues that happening around, and they proposed diversified teaching approaches and practices for course improvement. It seems that students recognized the importance of government in promoting sustainable development, but are not ready to give their focus on the workings of government. It is fair to say that by creating such a course, students would like to accept a framework for sustainable development to reconsider China's achievements and reflect on the path of the world.

Some concluding remarks are still tentative. There have been a growing number of publications and directives drawing attention to the importance and approaches of integrating sustainability content into the curriculum, of best pedagogical practices for optimizing teaching and learning; and of sharing findings through rigorous performance measurement and analysis. It requires more substantial efforts in pedagogical research to understand the interactions of teaching, learning, and learning environment for HESD with a view of the integration of curriculum development, research, community service within and beyond HEIs.

In China's governance of HESD, although universities have a higher degree of discretion to design and initiate its own green campus activities, Ministry of Education plays dominant roles in determining which institutions will be selected as a demonstration site and be provided with funds and resources. The government administrators and universities, who have been accustomed to the rules of the game, follow the rules routinely. Only a few lucky universities in China can take this advantage and pioneer their sustainability education, for a majority of universities like CUFE, the institutional motivation and capacity for HESD could not be guaranteed and they have to muddle through the process.

China has its unique political regime and governance system, the tradition of centralized top-down approach potentially builds a country of strong state and weak society, and creates a culture of subordination over autonomy, obedience over democracy. If the country is confident in realizing a transition towards green economy and society, it is doubtable that it can realize another goal of ESD transformation, this is "empowering people to be 'global citizens' who engage and assume active roles, both locally and globally, to face and to resolve global challenges and ultimately to become proactive contributors" (UNESCO 2014). This broader issue also relates to politics and has impacts on curricular development and green campus initiatives. However, it is beyond the scope of this paper.

References

Bekessy, S. A., Samson, K., & Clarkson, R. E. (2007). The failure of non-binding declarations to achieve university sustainability: A need for accountability. *International Journal of Sustainability in Higher Education, 8*(3), 301–316.

CUFE (2015). Basic information of CUFE. http://en.cufe.edu.cn/aboutcufe/statistics/index.htm. Accessed April 27, 2017.

Gurung, R., & Schwartz, B. (2013). *Optimizing teaching and learning: Practicing pedagogical research*. West Sussex: Wiley-Blackwell, UK.

Jost, J. T., & Hunyday, O. (2005). Antecedents and consequences of system-justifying ideologies. *American Psychological Society, 14,* 260–265.

Lei, J. (2012). Striving for survival and success: Chinese private higher education in the twenty-first century. *On the Horizon, 20*(4), 274–283.

Li, F., & Morgan, W. J. (2011). Private higher education in China: Problems and possibilities. In J. W. Morgan & B. Wu (Eds.), *Higher education reform in China* (pp. 66–78). New York, NY: Routledge.

Lo, K. (2015). Campus sustainability in Chinese higher education institutions: Focuses, motivations and challenges. *International Journal of Sustainability in Higher Education, 16* (1), 34–43.

Lozano, R. (2006). A tool for a graphical assessment of sustainability in universities (GASU). *Journal of Cleaner Production, 14*(9–11), 963–972.

NDRC (2012). China's national report on sustainable development. http://www.china-un.org/eng/zt/sdreng/P020120608816288649663.pdf. Accessed June 08, 2012.

Niu, D., Jiang, D., & Li, F. (2010). Higher education for sustainable development in China. *International Journal of Sustainability in Higher Education, 11*(2), 153–162.

UN (2015). Goal 17: Revitalize the global partnership for sustainable development. http://www.un.org/sustainabledevelopment/globalpartnerships/. Accessed May 13, 2017.

UNESCO (2014). UNESCO Roadmap for implementing the global action Programme on education for sustainable development. http://unesdoc.unesco.org/images/0023/002305/230514e.pdf. Accessed May 14, 2017.

Serger, S., Benner, M., & Liu, L. (2015). Chinese university governance: Tensions and reforms. *Science and Public Policy, 42*(6), 871–886.

Statistical Yearbook (of China) (2016). National bureau of statistics of China. http://www.stats.gov.cn/tjsj/ndsj/2016/indexch.htm.

Wang, Q., Yuanping, L. (2011). Theoretical features and practical implications of China's education for sustainable development (in Chinese). *Education Theory and Practice*, 31.
Wang, S. (2013). The economic analysis of imbalance in the regional development of higher education in China. *Education Science of Journal of Hunan Normal University, 3,* 111–114.
Wang, T. (2014). Education for sustainable development in china. *Journal of Sustainability Education*, 6. http://www.susted.com/wordpress/content/education-for-sustainable-development-in-china_2014_06/.
Wang, X. (2007). Problems and solutions of higher education for sustainable development (in Chinese). http://www.cnsaes.org/homepage/saesmag/jyfzyj/2007/1A/gj070116.htm.
Xu, Y. (2014). 70% research funding from the government was taken by universities of Project 211 and Project 985 (in Chinese). Changsha Evening Daily. http://www.china.com.cn/news/2014-11/19/content_34089357.htm. Accessed November 19, 2014.
Yuan, X., & Jian, Z. (2013). A critical assessment of the higher education for sustainable development from students' perspectives: A Chinese study. *Journal of Cleaner Production, 48,* 108–115.
Zhao, W., & Zou, Y. (2015). Green university initiatives in China: A case of Tsinghua. *International Journal of Sustainability in Higher Education, 16*(4), 491–506.

Sustainability in Construction Management Education: An Indian Perspective

Shilpi Singh, Samya Rakshit, Triveni Prasad Nanda, Anurita Bhatnagar and Anil Sawhney

Abstract With increasing impact of human survival on the environment and perpetual debate of our responsibility towards future generations, sustainability has come a long way. With urbanization and sustained economic development in India, greater building activity is expected, as India gears up to build 70–80% of the expected future building stock. As developing countries fight for a greater role in changing the environmental scene globally, role of sustainability in construction education has gained ample momentum. A systemic integration in education is required to sensitize students and future leaders to think, operate and apply sustainability concepts. As educators, it is our responsibility to condition young minds to view sustainability as an indispensable attitude and empower them to steer business decisions in alignment with sustainable goals. The construction management pedagogical approach steers postgraduate programs towards a holistic understanding of sustainability and inculcating built environment professionals who are the stewards of our sustainable future. This paper aimed at analysing one such construction management program at Royal Institution of Chartered Surveyors School of Built Environment, to study the efficacy of this pedagogical approach which goes on to show knowledge evolution of sustainability concepts over a period of three semesters. Using quantitative methods, the statistical significance of the change in student perception through course work has been analysed with additional insights and comparison with the alumnus cohort.

Keywords Sustainability · Construction education · Sustainable construction India · Construction management curriculum

S. Singh (✉) · S. Rakshit · T. P. Nanda · A. Bhatnagar
School of Construction, RICS School of Built Environment,
Amity University, Block–F2, Fifth Floor, Sector–125, Noida 201313, UP, India
e-mail: shilpimago@gmail.com

A. Sawhney
Department of the Built Environment, Liverpool John Moores University, Liverpool, UK

W. Leal Filho et al. (eds.), *Sustainable Development Research in the Asia-Pacific Region*, World Sustainability Series, https://doi.org/10.1007/978-3-319-73293-0_9

1 Introduction and Background

The long term impact of human consumption patterns are evident through climate change and overutilization of natural resources globally. The contribution of the construction industry towards overutilization of natural resources and high energy consumption both during construction and operations of buildings is huge (Iyer-Raniga et al. 2010). The construction sector globally accounts for one-fifth of the energy consumption (Architecture 2030 2017). The Indian construction industry is not far behind, accounting for close to 40% of total energy consumption in the country. It is expected to grow to USD 1 trillion a year market by 2025, according to a study by Global Construction Perspectives and Oxford Economics (Fallis 2015). The reason for this growth spurt is attributed to the thriving urbanization rate, with 10.4 million extra inhabitants just in the Indian capital alone (Fallis 2015). The report further predicts that India will overtake Japan by 2030, to be the third largest global construction market. The major emphasis in the building stock is to be in the areas of housing and infrastructure, with an estimated housing demand of close to 170 million to be built by 2030 alone (Fallis 2015). With this demand, it is crucial that steps are taken to ensure that the construction industry embraces sustainability as a core ethos. This research aims at assessing RICS SBE's efforts at empowering future construction professionals to comprehend the meaning of being truly sustainable. The following sections discuss the extant literature pertinent to the current discourse on sustainability in construction management education.

2 Literature Review

With developing countries fighting for a greater role in this changing economic and environmental global scenario, the role of education in promoting sustainable development among the youth has emerged to be of prime importance (UN 1992). The Brundtland Commission gave the world a broad definition of Sustainable Development in 1987 as "Sustainable development is development that meets the needs of the present without compromising the ability of future generations to meet their own needs" (Brundtland 1987).

Tinker And Burt have derived a separate definition from this for sustainable construction as "those materials and methods used to construct and maintain a structure that meets the needs of the present without compromising the ability of future generations to meet their own needs" (Tinker and Burt 2003). The earth and all its resources are a shared reserve that we as humans have used and continue to deploy beyond its holding capacity. The building and construction industry, within the domain of built environment, forms the largest consumer, having a far reaching impact on the environment and natural resources. It also has a greater responsibility and a larger role to play in contributing towards the Sustainable agenda (Murray and Cotgrave 2007). Consequently, with increasing population, global economic development, the current global consumption patterns and dependence on non-renewable resources, the need to include Sustainability in educational programs for the built environment is indispensable (Holdsworth and Sandri 2014).

As Raufflet notes, the overall intent is to achieve a systemic integration to build connections and sensitize students and managers to think, operate and feel differently regarding the concept of sustainability (Raufflet 2013). It is important to assess how the current generation of graduates in construction management understand the importance of sustainable development. The role of educators is not limited to ensuring that the course curriculum is as per university requirements but also, how well the young minds are being conditioned to face the challenges and steer business decisions towards or in alignment with sustainable development (Murray and Cotgrave 2007; Chau 2007). Therefore, the overarching goal of construction educators is to impart responsive education, one that inculcates sustainability as an inherent project driver and not just as an add-on objective.

At the Royal Institution of Chartered Surveyors School of Built Environment (RICS SBE), the pedagogical approach is centred towards a holistic understanding of sustainable development as it applies to the Built Environment. As RICS SBE prepares its students in the field of construction management and business practices, it is vital that we assess, how our education is impacting the potential built environment professionals who represent the future decision makers.

3 Sustainability in Construction Management Education

The early models of business education primarily focused on educating potential managers in order to maximize returns for owners and investors (Sharma and Hart 2014). The business education has focussed on 'core concepts like finance, accounting with minimal inclusion of education on Ethics, Corporate Social Responsibility, and Sustainability' (Haski-Leventhal and Concato 2016). The Brundtland Commission Report changed this perception as it argued for a major role of business in achieving sustainable development. The report marked a turning point in business education and was followed by a demand on "businesses to operationalize the concept of sustainable development to address their negative social and environmental impacts" (Sharma and Hart 2014). With the turn of the century, the need for incorporating sustainability in management education has been further emphasised: the United Nations Global Compact has made an endeavour to align the Sustainable Development Goals with the Principles of Responsible Management Education (Haertle 2016). Construction Management is a course at the confluence of the engineering and the management disciplines, the need for a sustainable perspective in this field is indispensable (Lim et al. 2015).

In a study done by Sharma and Hart in 2014, it was found that most of the major business schools in the world have added some kind of initiative focusing on sustainability (Sharma and Hart 2014). However, many of such initiatives are nothing but appendages to the core course curriculum. These initiatives are mere "saddle bag" issues which are siloed away from the core curriculum, to be addressed when the 'meat' or the real content of the coursework is over (Sharma and Hart 2014). Students in the built environment disciplines go on to become construction professionals, lecturers, and researchers amongst assuming other roles. It is imperative that these prospective professionals are prepared to deal with the provisioning and upkeep of the built environment and make decisions for a more sustainable future (Hayles and Holdsworth 2008). This conditioning of these young minds depends on the awareness, knowledge, skills, and values they acquire during their requisite period of study in the university (Ameh et al. 2010; Ahn et al. 2009).

Stephen et al. (2008) emphasised that "institutions of higher education are important places of production, perpetuation, and dissemination of knowledge". These institutions globally recognize that they have a unique responsibility towards the goal of achieving sustainability and have adopted practices and course curricula to this end (Stephens et al. 2008; Chau 2007). The university is a rational environment for bringing together diverse and discrete ideas on sustainability and integrating the concept into relevant educational curricula (Ameh et al. 2010).

4 Imparting Sustainability Education—Indian Context

While India was recently classified as "lower-middle income" country by World Bank, the economic development in the last few years has definitely pushed it to the global frontiers (World Bank 2017). In early 2015, India announced an ambitious climate change policy to reduce the intensity of carbon emissions by 33–35% by 2030 from 2005 levels and switch to 40% non-fossil fuel in the energy mix by 2030 (Accenture 2015). In addition, India plans to quadruple its renewable energy capacity and bridge the energy supply and demand gap by 2022 (Accenture 2015). With such energy efficiency targets, the Indian business environment is gradually shifting towards a mind-set that appreciates and understands the benefits of greater resource efficiency. Furthermore, in the CDP report for 2015, most of the participating private and public sector corporations have reported a reduction in emissions and thus their impact on climate, by use of energy efficiency measures (Carbon Disclosure Project 2015). Thus, the business environment in India is gradually remoulding itself to be inclusive of sustainable endeavours and initiatives. This transformation centred on India's commitment towards sustainability is indeed a daunting challenge, one that requires future business decisions to be taken with sustainability as an integral stakeholder, but also one that requires that our future business leaders are adequately conditioned to understand and appreciate sustainability and can also innovate beyond its goals. The opportunities being created within this transforming business environment are plenty and the need to address how our education system has indeed stepped up to prepare the business professionals becomes essential (Vijayakumar Bharathi et al. 2014). According to research conducted by Imam et al. in (2011), the need to include environmental sustainability, ethical decision-making and social responsibility in management education within the Indian context is centred on two factors. First is the external pressure from the newfound societal relevance of concepts of social, economic and environmental responsibility and second due to a greater internal pressure by faculty who have focussed their research on business and natural environment (Imam et al. 2011).

As clearly documented (Accenture 2015), the construction sector has a greater role and responsibility in overall contribution towards sustainability in the built environment, as it bears the maximum accountability to improvise energy and resource efficiency (Murray and Cotgrave 2007; Ekundayo et al. 2011). Additionally, with greater knowledge dissemination of benefits of sustainable development, there is a need to prepare management graduates with concepts of sustainability (Haski-Leventhal and Concato 2016). Along with construction education, the management education has an equal role to play in leading the path towards achieving sustainable growth. Together, these two sectors have a profound contribution in India achieving its sustainable goals. The RICS SBE serves these two specific industry sectors as it provides a business administration degree in construction project management, it caters to two critical domains of construction industry and business management sector.

5 Research Aim and Objectives

This research aims to evaluate the impact of construction management coursework on the perceived understanding of postgraduate students in concepts of sustainability as it applies to the built environment. In view of this aim the research objectives are as follows:

1. Understand the current state of knowledge of construction management students in the field of sustainable construction
2. Conduct a peer review of sustainability focussed postgraduate coursework with other schools offering similar programs to assess the relevance of imparting sustainable knowledge in construction management education
3. Evaluate the role of course curriculum in shaping the knowledge base of student group in the realm of sustainable development/sustainability
4. Analyse knowledge progression in students as they gradually move towards their course culmination
5. Evaluate application and relevance of knowledge pertaining to sustainable construction, acquired during coursework by recent graduates in business/project scenarios.

6 Research Methodology

The methodology followed in this research is explained in Fig. 1. The first step in the research method consisted of literature review on various aspects of sustainability, sustainability in construction management education and in the Indian context. This was followed by a comparison of coursework of RICS SBE with five other selected schools and culminated in a survey questionnaire of students at RICS SBE. The comparison of coursework helped the researchers to achieve objective 2 and to fulfil objectives 1, 3, 4 and 5, the questionnaire survey was carried out. The literature review was highly instrumental in developing the survey questionnaire.

7 Survey Design and Specifics

The survey respondents were divided into three cohorts as shown in Table 1.

The survey was open for seven weeks from November 27, 2016, to January 15, 2017, to the student cohorts at RICS SBE. A total of 271 responses were recorded and included in the student survey.

The survey consisted of three parts: the first part contained questions testing general awareness regarding sustainability, the second part comprised of questions testing specific knowledge of sustainability within the domain of the construction industry; and the third part dealt with the perception of the importance given to sustainability in the Indian construction industry. Part 1 and 2 were relevant in

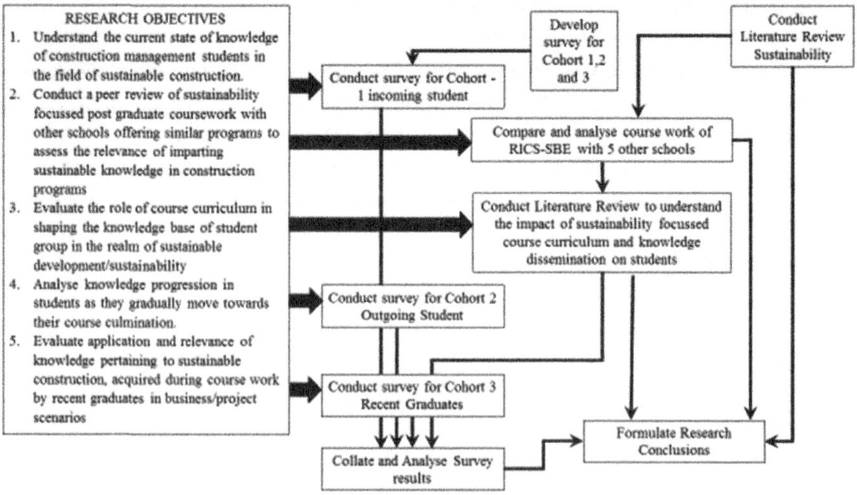

Fig. 1 Research methodology

Table 1 Survey respondent composition and characteristics

Cohort 1	Incoming students in the post-graduate program at RICS SBE	Second semester students
Cohort 2	Outgoing students in the post-graduate program at RICS SBE	Fourth semester students
Cohort 3	Recently placed graduates	Recently placed students

gauging information from the perspective of research objectives 1, 3 and 4; while part 3 was helpful in reaching research objective 5.

Cohort 1 was only required to take the first two parts of the survey, while Cohort 2 and 3, who had been exposed to the construction industry through internship and/or professional experience, answered all three parts. Questions were targeted towards specific areas, to assess topics understood to be relevant from the literature survey.

> Part 1 and 2 of the survey were designed considering the issue that in many courses, sustainability is taught only as an introductory course and not as an in-depth one (McDonald 2004) and that students often lack awareness about the specific meaning of concepts related to sustainability. Questions were also targeted to understand whether sustainability-education was integrated with the rest of the course (Iyer-Raniga et al. 2010; Raufflet 2013; Sharma and Hart 2014), whether the knowledge is only theoretical or practical applications were emphasized upon (McDonald 2004), and whether students got an opportunity to probe into research during the coursework (Ganah et al. 2008). Part 3 was designed to examine the market demand, awareness, and acceptability of sustainability in the Indian market (Economic Policy Forum 2014). The focus of some questions was also to understand how well the classroom knowledge is addressing the industry needs and practices (Iyer-Raniga et al. 2010). As eventually, classroom knowledge seeks solutions through the implementation of concepts.

8 Program Comparison

In order to assess the robustness of the course curriculum at RICS SBE, a comparison was conducted with institutions offering similar programs. The selected institutions were, School of Planning and Architecture (SPA) Delhi, Indian Institute of Technology Madras (IIT-M), Indian Institute of Technology Delhi (IIT-D), National Institute of Construction Management and Research, Pune (NICMAR) and Centre for Environmental Planning and Technology, Ahmadabad (CEPT). All the institutions mentioned above conduct a post graduate program in construction management/Building Engineering and Management/Construction Engineering and Management. The authors conducted an analysis of construction management coursework at the above mentioned institutions to compare the inclusion of sustainability concepts in built environment sector.

The program comparison revealed that RICS SBE has a more definitive bend towards inclusion of sustainability in the program structure by providing sustainability focussed courses such as, 'Sustainable Practices for Construction' and 'Advanced Energy Management'. In addition to these mandatory courses, several sections of general courses also cover sustainability aspects such as, 'BIM for Built Environment: Sustainability Modelling and Environmental Analysis' and 'Lean Practices in Construction Projects: Linkages of Lean and Green Concepts in Design of Built Environment', etc. It was also observed that IIT-D has no requisite course that offers a holistic knowledge in sustainable construction practice, however, an elective course titled 'Sustainable Materials and Green Buildings' is available for students. IIT-M follows a similar pattern for requisite courses but offers a host of courses targeting environmental studies related to all aspects of the built environment. Additionally, with a focused group of environmental program within the civil engineering program, both IIT-D and IIT-M have professional experts with vast experience in green building consultancy and environmental impact assessment for infrastructure projects. Similarly, CEPT has no requisite course but an elective named 'Indian culture and sustainable construction practices' which is open to both undergraduate and postgraduate students.

SPA Delhi has a requisite course named, 'Environmental Clearances' which enumerates Environmental Impact Assessment and similar statutory bye-laws required for project approvals. CEPT and SPA Delhi, both primarily being design schools, have a strong base of architectural and planning professionals that inculcate a strong pedagogy which is inclusive of sustainable design practices in construction and infrastructure sectors. Thus, rather than having explicit exposure to subjects related to sustainability or green building design, the students at CEPT and SPA Delhi receive more implicit exposure to these concepts.

NICMAR has also opted for an elective course called 'Green construction management'. Thus, it can be safely assumed that the students who do not opt for this elective have no way of being exposed to concepts on green building practices and sustainability in built environment.

9 Survey Key Findings

The survey revealed consistencies with the literature survey. Responses from students helped to affirm that course curriculum at RICS SBE has a considerable impact on increasing student knowledge and awareness on sustainability concepts. Also, there were several events and competitions in addition to coursework that inculcated the need for self-learning of such concepts by students. One such example is the annual event of "Green Building Week". The event is a conglomeration of Dialogues with Industry Experts on Sustainable Construction, Workshops from Design and Construction Companies implementing sustainability practices, showcasing hands-on ideas and experiences and student competitions on sustainability practices in the built environment. Such events have been instrumental in creating an environment of passive and self-learning for students at RICS SBE.

Comparison between the responses from the first and second parts of the survey revealed a considerable knowledge evolution in students' understanding of sustainability concepts as they progressed through the program at RICS SBE.

Both the cohorts of 4th semester and alumni students performed better than the 2nd semester students. This was assessed through *T*-tests. The *p*-value for one-tailed *t*-test for 4th semester students was 0.00019 and 0.04514 for alumni respectively. Figures 2 and 3 indicate a significantly better performance of fourth semester students and can be attributed to requisite and elective courses in sustainability that are offered during the 3rd semester of the program at RICS SBE.

It was also found that introduction of sustainability at undergraduate education is essential in forming a strong foundation that helps in developing a more inclusive perspective for construction professionals. It was observed that undergraduate course learning or self-learning in concepts of sustainability by students helped to

Fig. 2 Performance in Part 1 of the survey: basic questions

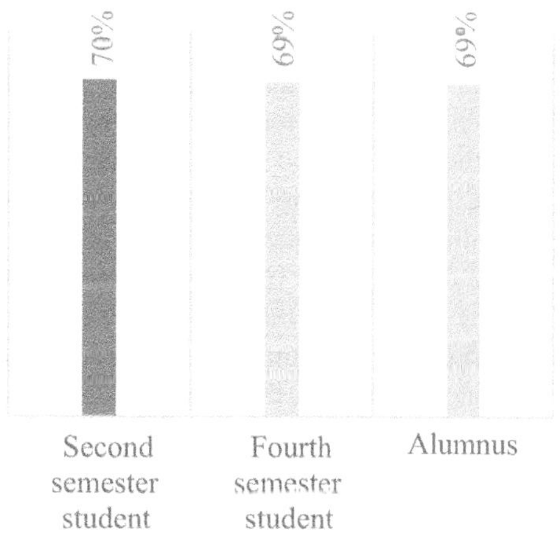

Fig. 3 Performance in Part 2 of the survey: advanced question

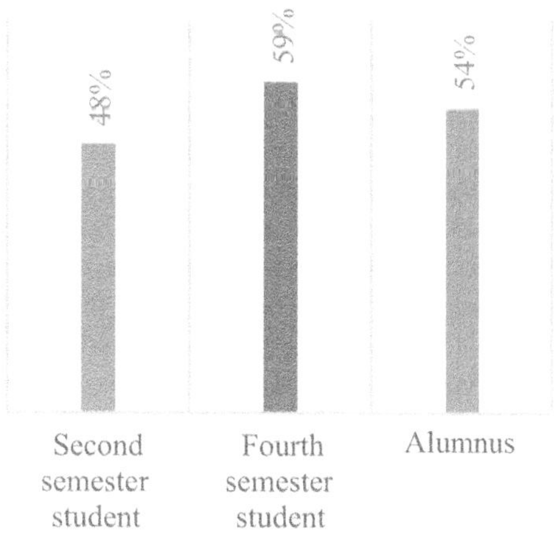

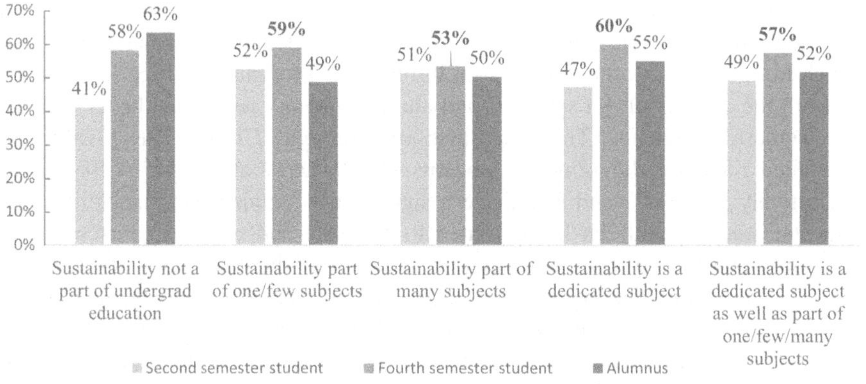

Fig. 4 Undergraduate learning and prowess of advanced concepts of sustainability

establish a strong knowledge base and develop a perspective to comprehend and appreciate sustainability practices in construction. As shown in Fig. 4, this was apparent in survey responses from second semester students, as they scored less than their peers who were familiar with Green Buildings in their undergraduate course in some form or other. *t*-tests were performed to investigate whether the second-semester student-group unfamiliar with Green Buildings was outperformed by the other groups of second semester students at a statistically significant level, and the *p*-values of the one-tailed *t*-test were, 0.024461 (Significant) for unfamiliar vs. those who had come across the term 'Green Building', in spite of there being no dedicated course on the issue; 0.022965 (Significant) for unfamiliar vs. those who had a dedicated course/chapter(s) on green buildings and 0.027732 (Significant) for unfamiliar versus those who learnt about Green Buildings on their own. Thus, for second semester students, there was a statistically significant gap between students who had been unfamiliar with green buildings at the undergraduate level and students who had been familiar, based on measuring the performance in the basic questions.

Program structure and course curriculum at RICS SBE has a significant impact in filling gaps in sustainability knowledge from the students' undergraduate education and successfully bringing all students to a level platform.

This was an important observation and learning for RICS SBE, as the students entering the program come from a diverse spectrum of undergraduate courses.

In order to assess this, *t*-tests were performed to investigate whether the performance gaps between the fourth-semester student group unfamiliar with green buildings and each of the other three groups, familiar with green buildings on various levels during their undergraduate studies are statistically significant. Among the *p*-values of the one-tailed *t*-tests, the only significant one is for the student group unfamiliar with green buildings vs. those who learned about green buildings on their own. This proves that the gap between students is adequately bridged by their fourth semester regardless of their exposure to green buildings in their

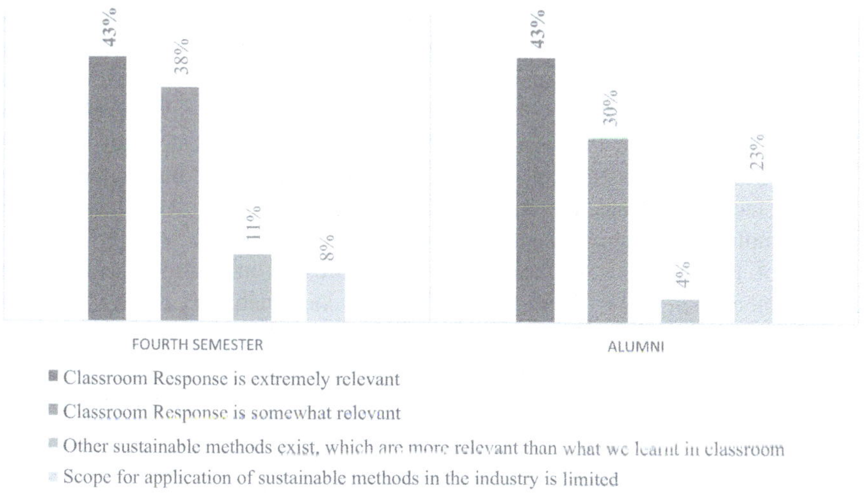

Fig. 5 Relevance of classroom knowledge in industry

undergraduate coursework. Thus, reinforcing that the RICS SBE curriculum acts effectively in imparting knowledge about sustainability to the students.

One of the survey questions addressed to both fourth semester students and alumni attempted to gauge student perception regarding the relevance of classroom knowledge in sustainability. The intent of this question was to recognize whether the students understood why they are studying sustainability concepts, and also act as a sense check for staying abreast with current industry practices. This is extremely relevant as the sustainability practices are ever evolving. As shown in Fig. 5, the survey revealed that 43% of both, fourth semester students and alumni think that classroom knowledge is extremely relevant, 38% of fourth semester students and 30% alumni think that classroom knowledge is somewhat relevant, 11% of fourth semester students and 4% alumni think that "Other sustainable methods" exist which are more relevant than what is being taught in classroom.

Additionally, 8% of fourth semester students and 23% alumni reported that scope for application of sustainable methods in the industry is limited. These figures represent that fourth semester students are more optimistic about the application of sustainable techniques in the construction industry and the alumni, based on industry interaction after their course completion have come across newer sustainability concepts and practices, in the realm of built environment.

Close to 46% of fourth semester students and 40% of the alumni surveyed felt confident that they have inculcated knowledge about sustainability, sustainable construction guidelines and capability to solve real life problems from a sustainable perspective. This implies that the percentage of students receiving the full spectrum of sustainability-related education has increased from the alumni to the fourth semester students.

However, the percentage of students who reported to have learnt about sustainability and guidelines for sustainable construction is 29% for fourth semester students and 39% for the alumni. The percentage of students reported to have learned about sustainability only without guidelines or real life problems has increased from 21% in the alumni to 25% in the fourth semester students. This points towards the need to create greater knowledge depth for final semester students by inclusion of elective courses that provide focused vertical knowledge of more complex sustainability concepts in the built environment.

One of the questions in the survey also addressed inclusion of concepts in courses other than those whose primary focus is sustainability, which is an issue highlighted in several texts of the literature survey (Iyer-Raniga et al. 2010; Raufflet 2013; Sharma and Hart 2014). The response received as shown in Fig. 6, reinforces that while the RICS SBE curriculum has a considerable focus of including general concepts of sustainability, there is a large percentage of students for whom greater knowledge penetration is still necessary.

The presence of a strong research focus in a construction management program provides greater learning opportunities for students. This is all the more relevant when it comes to sustainability aspects in construction management (Ganah et al. 2008). As observed in the survey, 17% of students reported that they have been made familiar with the sustainability-related research, 47% have looked for research once/few times, 25% many times, while 11% refer to sustainability research on a regular basis. Therefore, a large percentage i.e., 83% of students have been familiar with sustainability-related research during their coursework at RICS SBE.

Fourth semester students were asked to rank the perceived importance of sustainability in the Indian Construction Industry, results of which showed 44%

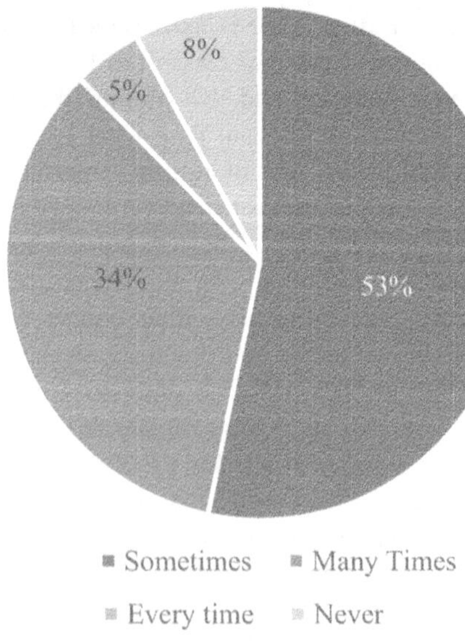

Fig. 6 Response to the question "Did general subjects taught at RICS SBE include a Sustainable Perspective?"

students and 32% alumni reporting that "Extreme Importance" is given to sustainability in the Indian construction industry, 24% students and 30% alumni think it is of "Moderate Importance"; 18% students and 21% alumni think it is of "Occasional Importance"; 13% students and 15% alumni think it is of "Extremely Rare Importance", while 1% students and 2% alumni think it is of "No Importance". The students seem to have a more optimistic view regarding importance given to sustainability in the industry: giving the rating scores (like in a Likert scale) from 1 to 5, the mean score of the students is 3.97 while that of the alumni is 3.75. However, there is no statistically significant difference in the *t*-test values.

10 Conclusion

The survey indicated that RICS SBE's curriculum is effective in imparting sustainability concepts in the built environment domain. This can be attributed to two key strengths of the coursework, one being the overall curriculum interspersed with sustainability concepts as opposed to siloed sustainability courses. The second aspect being the transdisciplinary approach towards inclusion of sustainability into the construction management curriculum. RICS SBE has a comprehensive approach of including sustainability concepts into core construction courses during early semesters of the program and then exposing the students to core sustainable construction concepts during the later semesters. This helps not to force fit sustainability, but exposes the students towards overall resource efficiency and builds critical thinking as it relates to sustainability concepts as shown in Fig. 7.

This helps the students to form an inclusive perspective that they build on through the program and also once they become part of the industry. As Conte points out, sustainability should permeate into construction curricula as an integral component, similar to conventional parameters of cost, time and quality (Conte 2016; Wright et al. 2015).

In addition, the value generated by a transdisciplinary approach in developing a curriculum focused towards sustainable development is also amplified at RICS SBE, by way of specialized courses that focus on sustainable concepts that are offered across different disciplines of the built environment as shown in Fig. 6.

The recent graduates of the course appeared to have achieved a good level of understanding of both general and specific aspects of sustainability and its applications to the built environment. In addition, the course work is successful in imparting the exact meanings and significances of technical terms relating to sustainability as outlined in the literature review (Iyer-Raniga et al. 2010). The knowledge gap that existed between students due to different levels of exposure to sustainability in their undergraduate coursework, is adequately bridged towards the culmination of the program at RICS SBE.

Furthermore, the program at RICS SBE creates an environment comprising of both active and passive knowledge means for student learning. The survey also highlighted a few areas that require greater focus, such as the need to create greater

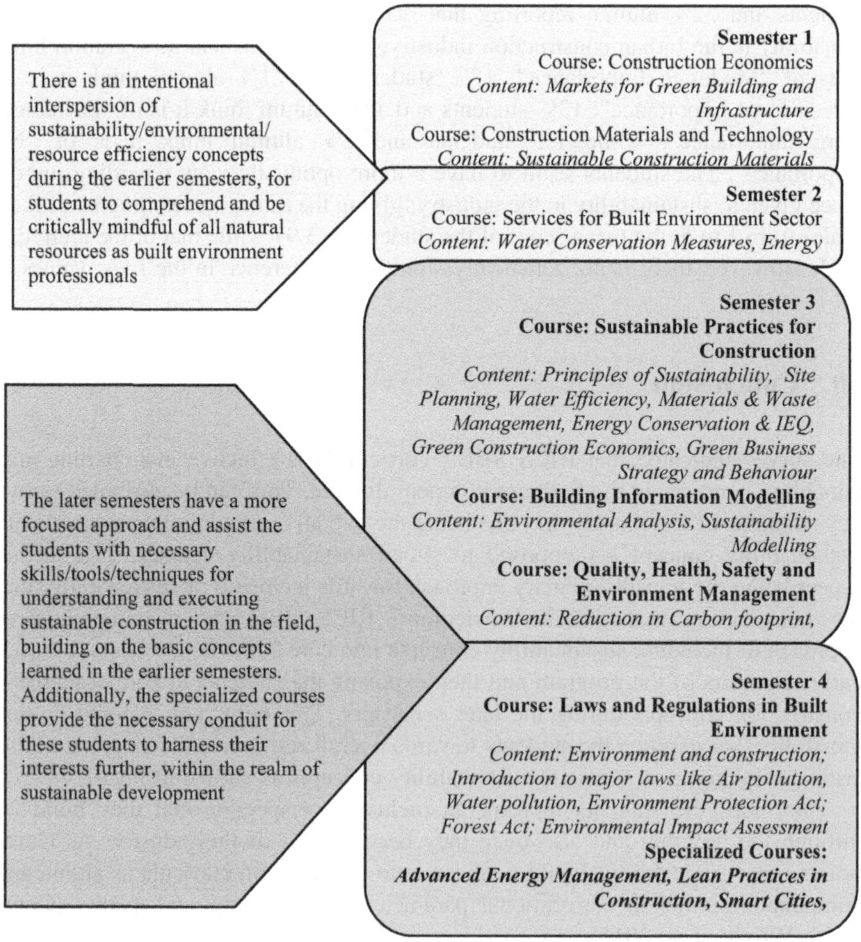

Fig. 7 Inclusion of Sustainability concepts in core construction courses in early semesters and focus on Sustainable construction course in later semesters at RICS SBE

depth of knowledge in application of sustainability concepts. While a majority of students undergo research work in their course of study but it is important for the students to see the implementation of these learned concepts (Ganah et al. 2008). Focussed sustainability training and student competitions at RICS SBE have been instrumental in achieving this end. Greater participation from industry mentors in sustainable practices coupled with events such as the "Green Building Week" have also been an ongoing effort at RICS SBE.

Regarding the existence of sustainable practices in the Indian construction industry, the survey indicates that students, who have been less exposed to the industry, seem to be more optimistic. The demand for sustainable construction practices in India is on an upward trend due to mandatory Environmental regulations being initiated by the Government of India and an increased demand for green buildings fuelled by potential benefits of energy efficiency and resource intensiveness (Carbon Disclosure Project 2015; Accenture 2015). However, as this faction of the Indian construction industry grows, the need to intervene as educators is both imminent and timely. As Murray and Cotgrave rightly note in their research that "*Construction has the potential to enhance rather than degrade the environment and to promote rather than exasperate social and economic equity. If this potential is to be realised, everyone within the industry will need to attain some level of sustainability literacy*" (Murray and Cotgrave 2007). In the light of economic and urban development in India, there is a definite opportunity to make an inclusive and sustainable intervention in the talent that enters the Indian construction industry. There is a definite demand for adequately trained professionals that can steer the development in India in a more sustainable manner (Economic Policy Forum 2014). At RICS SBE, creation of the right environment for student learning and development is ensured that builds on the construction knowledge and cultivate their critical abilities to prepare them to be the stewards of development that is truly inclusive and sustainable.

References

Accenture. (2015). *Energy efficiency bears fruits for India Inc.*, New Delhi.

Ahn, Y. H., Kwon, H., & Pearce, A. R. (2009). Sustainable education for construction students. In *Proceedings of Associated Schools of Construction Conference*.

Ameh, S., et al. (2010). Sustainable construction education: Assessing the adequacy of built environment professional's training. In *West Africa Built Environment Research (WABER) Conference, 27–28 July 2010*. Ghana, pp. 27–28.

Architecture 2030. (2017). Why the building sector?|Architecture 2030. http://architecture2030.org/buildings_problem_why/. Accessed June 12, 2017.

Brundtland, G. H. (1987). Our common future: Report of the world commission on environment and development. *Medicine, Conflict and Survival, 4*(1), 300.

Carbon Disclosure Project. (2015). *CDP Climate Change Report 2015, India edition*,

Chau, K. W. (2007). Incorporation of Sustainability Concepts into a Civil Engineering Curriculum. *Journal of Professional Issues in Engineering Education and Practice, 133*(3), 188–191. http://ascelibrary.org/doi/10.1061/(ASCE)1052-3928(2007)133:3(188).

Conte, E. (2016). Sustainability and built environment . The role of higher education in architecture and building engineering. *European Journal of Sustainable Development, 5*(3), 1 10.

Economic Policy Forum. (20140. *Promoting Sustainable and Inclusive Growth in Emerging Economies: Green Buildings*. Available at: https://economic-policy-forum.org/wp-content/uploads/2016/02/Sustainable-and-Inclusive-Growth-Green-Buildings.pdf.

Ekundayo, Da, et al. (2011). Mapping of sustainability education to construction related curricula: A case study of quantity surveying (QS) degree programme. *In RICS COBRA Conference, 2011*, 12–13.

Fallis, A. (2015). *Global Construction 2030*, London.

Ganah, A., Pye, A., & Hall, G. (2008). The role of knowledge transfer in sustainability research in the built environment discipline. In *24th Annual ARCOM Conference* (pp. 299–307). Cardiff, UK: UK Association of Researchers in Construction Management.

Haertle, J. (2016). Realising the sustainable development goals through responsible management education AGENDA 2030—A plan for action for people, planet and prosperity. In *PRME DACH Conference*. FH Krems, Austria: PRME, UN Global Compact.

Haski-Leventhal, D., & Concato, J. (2016). *The State of CSR and RME in Business Schools and the Attitudes of their Students.*

Hayles, C. S., & Holdsworth, S. E. (2008). Curriculum change for sustainability curriculum change for sustainability. *Journal for Education in the Built Environment, 4205*(March), 25–48.

Holdsworth, S., & Sandri, O. (2014). Sustainability education and the built environment: Experiences from the classroom. *Journal for Education in the Built Environment, 9*(1), 48–68. http://journals.heacademy.ac.uk/doi/abs/10.11120/jebe.2014.00011.

Imam, A., Ashraf Ali, M., & Zadeh, M. N. (2011). Environmental ethics and integrating sustainability into management education. *Purushartha: A Journal of Management Ethics and Spirituality, IV*(1), 15.

Iyer-Raniga, U., Arcari, P., & Wong, J. P. C. (2010). Education for sustainability in the built environment: What are students telling us? In *Proceedings 26th Annual ARCOM Conference*, (September, pp. 1447–1456).

Lim, Y. S., et al. (2015). Education for sustainability in construction management curricula. *International Journal of Construction Management, 15*(4), 321–331. Available at: http://dx.doi.org/10.1080/15623599.2015.1066569.

McDonald, M. (2004). *Sustainable Environmental Design Education (SEDE)*. USA: California.

Murray, P. E., & Cotgrave, A. J. (2007). Sustainability literacy: The future paradigm for construction education. *Structural Survey, 25*(1), 7–23. Available at: http://www.emeraldinsight.com/doi/10.1108/02630800710740949.

Raufflet, E. (2013). Integrating sustainability in management education. *Humanities, 2*(4), 439–448. Available at: http://www.mdpi.com/2076-0787/2/4/439/.

Sharma, S., & Hart, S. (2014). Beyond "Saddle Bag" sustainability for business education. *Organization & Environment, 27*(1), 10–15. http://oae.sagepub.com/cgi/doi/10.1177/1086026614520713.

Stephens, J. C., et al. (2008). Higher education as a change agent for sustainability in different cultures and contexts. *International Journal of Sustainability in Higher Education, 9*(3), 317–338. http://www.emeraldinsight.com/doi/10.1108/14676370810885916.

Tinker, A. K., & Burt, R. (2003). "Greening" the construction curriculum. In *ASC Proceedings of the 39th Annual Conference*, (April, pp. 113–118).

UN. (1992). United Nations Conference on Environment & Development Rio de Janerio, Brazil, 3 to 14 June 1992. In *United Nations Conference on Environment & Development* (p. 351). Rio de Janerio.

Vijayakumar Bharathi, S., et al. (2014). Inclusion of sustainability education in business schools-An Indian B-school case study of MBA-ITBM curriculum. *International Journal of Applied Engineering Research, 9*(23), 22703–22725.

World Bank. (2017). World bank country and lending groups. Available at: https://datahelpdesk.worldbank.org/knowledgebase/articles/906519-world-bank-country-and-lending-groups. Accessed June 12, 2017.

Wright, M. F., et al. (2015). Beyond sustainability: A context for transformative curriculum development. *Transformative Dialouges, Teaching & Learning Journal, 8*(2), 1–19.

Perceptions of the Research Scholars Regarding Education for Sustainable Development (ESD) in Pakistan

Qudsia Kalsoom, Naima Qureshi and Afifa Khanam

Abstract Education for Sustainable Development (ESD) is not a mainstream theme either in school education or in higher education in Pakistan. In November, 2016 the Institute of Education, Lahore College for Women University, Pakistan organized an international conference on "Building Knowledge Competencies for Sustainable Development (SD) in Asia: Achieving the Goals of Life-Long Learning". The conference received ninety two paper abstracts. However, only a few articles directly addressed research problems related to ESD. In response, the study detailed here was conducted to investigate the perceptions of research scholars regarding SD and ESD to understand the reasons for lack of research in the field of ESD in Pakistan. The study employed survey method to collect data from the research scholars who presented their papers at the conference. The respondents were from departments of education of five Pakistani universities. Forty two scholars participated in the study. The data were coded and categorized under pre-determined categories. Data indicated that majority of the research scholars were not familiar with the concepts of SD and ESD and their importance. They generally perceived ESD as a set of programmes which sustain over time; and sustainability competencies as those abilities which would help the students in their future employment. The study findings indicate a need of mainstreaming ESD in Pakistan so that research scholars become aware of the concepts of SD and ESD. This awareness might help them identify ESD relevant research problems.

Keywords Education for Sustainable Development · ESD in Pakistan ESD-based research · ESD perceptions

---

Q. Kalsoom (✉) · A. Khanam
Institute of Education, Lahore College for Women University, Jail Road, Lahore, Pakistan
e-mail: qudsia_kalsoom@yahoo.com

N. Qureshi
Division of Education, University of Education, College Road Township Lahore, Lahore, Pakistan

W. Leal Filho et al. (eds.), *Sustainable Development Research in the Asia-Pacific Region*, World Sustainability Series, https://doi.org/10.1007/978-3-319-73293-0_10

1 Introduction: ESD and Teacher Education

Education for Sustainable Development (ESD) has emerged as a new field in education to address the problems of unsustainability. The United Nations Conference on Environment and Development (UNCED) held in Rio de Janeiro in 1992 coined the term 'Education for Sustainable Development (ESD)'. Since then sustainable development has been consistently addressed in various UN conferences (UNESCO 2005). To implement ESD globally, the United Nations declared the years 2005–2014 as UN Decade of Education of Sustainable Development (DESD). UNESCO was assigned a leading role in the implementation of DESD. It worked with ministries of education and other concerned departments to implement the ESD agenda. DESD helped in re-orienting education from early childhood education through universities across the world (Buckler and Creech 2014; McKeown and USTESD 2013).

Teacher education is a key space to implement ESD (Buckler and Creech 2014; McKeown 2014; Tilbury 2011) and, therefore, "teacher education institutions need to reorient preservice teacher education to include ESD" (McKeown 2002, p. 33). Considering the key role of teacher education in ESD implementation, teacher education programmes have been re-oriented in many countries and states like Sweden, Scotland, Australia (Buckler and Creech 2014; Ferreira et al. 2007). In Pakistan, ESD has not been mainstreamed (Kalsoom et al. 2017). However, some fragmented initiatives regarding the implementation of ESD have been taken during DESD. Some of the initiatives are: UNESCO's Associated Schools Project Network (ASPnet); a course on ESD in the Master programme of the University of Education (Buckler and Creech 2014); and participation of the University of the Punjab in 'The International Network of Teacher Education Institutions' (McKeown, n.d). Over sixty schools were part of ASPnet in Pakistan and they developed a 'Peace and ESD Education Programme' to share contextual knowledge on respect and cultural diversity with the teachers, students, parents and wider community (KNCU 2009).

One pedagogy-based initiative/intervention regarding ESD was recently undertaken at the Institute of Education, Lahore College for Women University, Pakistan. The initiative aimed to expose the preservice teachers to the concepts of sustainable development (SD) and ESD to enhance their sustainability consciousness (Kalsoom and Khanam 2017). As part of the initiative, preservice teachers carried out small-scale undergraduate research projects in 11 weeks. These projects involved empirical inquiries into local sustainability issues (effect of socio-economic status on education, challenges faced by women students in higher education, environmental consciousness of the preservice teachers, culture of teacher education institutes etc.). Besides conducting empirical inquiries, the preservice teachers participated in weekly 'research tutored' (Healey and Jenkins 2009) sessions for 11 weeks. In these sessions, participants discussed sustainability issues as well as shared their research work. The initiative was designed by the teacher as part of a compulsory course 'Research Methods in Education'. In this study, on one hand, participants were conventional researchers and on the other hand, they learnt concepts of SD, sustainability, sustainability competencies. Along with cognitive learning, preservice teachers developed attitudes and behaviours favourable towards sustainability (Kalsoom and Khanam 2017).

Though there are some individual or institutional efforts related to ESD implementation in teacher education, ESD is still outside education policy in Pakistan. This can be witnessed from the analysis of different documents like: education policy (Government of Pakistan 2009), provincial education sector plans (Government of Balochistan 2013; Government of KPK 2012; Government of Punjab 2013–2017. Government of Sindh 2014), Curriculum of B.Ed. Honours Programme (Higher Education Commission 2012), national professional standards for teachers (Ministry of Education 2009) and National Standards for Accrediting Teacher Education Programmes (NACTE 2009). None of the documents have included or even mentioned Education for Sustainable Development (ESD) as a priority. Some content related to environmental education, economy, human rights and citizenship has been included in the B.Ed. (Honours) curriculum. Civic responsibility, social cohesion and tolerance have been mentioned in education policy, education sector plans and in the national professional standards. However, no document has mentioned the concept of ESD as a discrete focus. Such a weak emphasis on ESD has resulted into two problems: (i) preservice teachers' knowledge of sustainability issues and their attitudes and behaviours towards sustainability are not up to expectations (found in another study by Kalsoom et al. 2017); (ii) lack of research on ESD based issues.

The current study originated from researchers' observations in the international conference on "Building Knowledge Competencies for Sustainable Development (SD) in Asia: Achieving the Goals of Life-Long Learning" in November 2016. The conference was organized by the Institute of Education, Lahore College for Women University, Pakistan. The conference received 92 abstracts altogether written by 142 researchers. Out of 142 researchers, 121 were from education departments of the universities and colleges of Pakistan. Significantly, few articles directly addressed research problems related to ESD. Among the ten abstracts which were directly related to ESD themes, six were written by preservice teachers who had participated in the pedagogical intervention (Kalsoom and Khanam 2017), as mentioned earlier.

The observation regarding the lack of research on ESD, reflected on a prior database search about ESD-based articles written by Pakistani authors. The databases searched were Springer; Taylor & Francis; and four national journals on education. We found over 2500 articles on ESD from the data bases of Springer and Taylor & Francis. Our search indicated that no empirical study on ESD had been done by Pakistani authors. Moreover, we analysed 353 articles which had been published in national journals from June 2004 to June 2016. We noticed that no article had been published on the subject of ESD in national journals. Potential reasons for this lack of research seemed no or little exposure of the researchers to the concepts of SD and ESD because ESD has not been prioritized at the national level. Current study was an attempt to further understand the problem of lack of research on ESD in Pakistan. The research question addressed in this study is:

What are the perceptions of Pakistani research scholars regarding SD and ESD?

Literature on sustainable development, education for sustainable development and sustainability competencies (Ciegis et al. 2009; McKeown 2002; UNESCO 2005; Tilbury 2011; UNECE 2012; Wiek et al. 2011; De Haan 2006) constitute framework of the study.

2 Sustainable Development

The concept of sustainable development has been defined in different ways in literature signifying the complexity of the concept. "Since the path-breaking deliberations of the Brundtland Commission, the expression 'sustainable development' has been used in a variety of ways, depending on whether it is employed in an academic context or that of planning, business or environmental policy" (Redcliff 2005, p. 213). The Brundtland report defined sustainable development as the "development that meets the needs of the present without compromising the ability of future generations to meet their own needs" (WCED 1987). Society, environment and economy are key areas of sustainable development (Harris 2000; UNESCO 2005; Ciegis et al. 2009; Sachs 2012). Besides three individual dimensions of SD i.e. society, economy and environment, a system perspective is also an essential feature of SD. Bossel (1999, p. 6) asserts that "sustainable development is possible only if component systems as well as the total system are viable". Similarly, Ciegis et al. (2009) also argue that sustainable development is based not on economic, social, ecological, or institutional dimensions, rather on their system as an integrated whole. McKeown and Nolet (2013, p. 5) view sustainable development as an "overarching paradigm at the United Nations to address numerous interrelated problems (e.g. poverty reduction, environmental protection, social justice, etc.)". They further assert that the related term *sustainability* is a thinking paradigm about the future in which economic, social and environmental concerns are balanced. This thinking paradigm is based upon the values associated with human dignity and human rights (McKeown and Nolet 2013, p. 6).

3 Education for Sustainable Development (ESD)

ESD is one of the programmes of sustainable development (McKeown 2002). This programme comprises of four strands: understanding and addressing sustainability issues; skills; perspectives; and values that "guide and motivate people to seek sustainable livelihoods, participate in a democratic society, and live in a sustainable manner" (McKeown 2002, p. 13). ESD may include: "values education, civic and citizenship education, health education, education for HIV and AIDS prevention, human rights education, ICT, gender equality, and environmental education" (Tedesco et al. 2011). DESD monitoring and evaluation report elaborated ESD as a broader concept involving knowledge about sustainable development and the key processes. Tilbury (2011) maintains that ESD involves processes of collaboration and dialogue; processes which engage the 'whole system'; processes which innovate curriculum as well as teaching and learning experiences; processes of active and participatory learning". ESD processes aim at ESD learning which involves learning to ask critical questions; learning to clarify one's own values; learning to envision more positive and sustainable futures; learning to think systemically; learning to respond through applied learning; and, learning to explore the dialectic between tradition and innovation" (Tilbury 2011). ESD learning can also be viewed in terms of sustainability competencies.

4 Sustainability Competencies

Researchers have tried to address the issue of 'ESD outcomes' by proposing a set of key competencies. Wiek et al. (2011), in their literature review, categorized the key competencies in five groups. They are: Systems thinking, normative competence, strategic competence, anticipatory competence and interpersonal competence.

System thinking refers to student's "ability to identify and critically reflect on the values pertinent to a specific sustainability challenge, as well as students' skills in identifying and prioritizing challenges across the three sustainability domains" (Remington-Doucette and Musgrove 2015, p. 539).

Normative competence is "the ability to collectively map, specify, apply, reconcile, and negotiate sustainability values, principles, goals, and targets… This capacity is based on acquired normative knowledge including concepts of justice, equity, social-ecological integrity, and ethics" (Wiek et al. 2011, p. 209).

Anticipatory competence refers to "the ability to collectively analyze, evaluate, and craft rich 'pictures' of the future" (Wiek et al. 2011, p. 207) regarding the issues of sustainability.

Strategic competence is the ability to "collectively design and implement interventions, transitions, and transformative governance strategies toward sustainability" (Wiek et al. 2011, p. 210).

Interpersonal competence is "the ability to motivate, enable, and facilitate collaborative and participatory sustainability research and problem solving" (Wiek et al. 2011, p. 211). Interpersonal competence comprises of advanced skills in communicating, deliberating and negotiating, collaborating, leadership, pluralistic and trans-cultural thinking, and empathy (Wiek et al. 2011).

In this study, participants' perceptions regarding Education for Sustainable Development have been analysed to explore if they fall under any of the concepts i.e. SD, ESD or sustainability competencies.

5 Method

The study detailed here employed survey method to collect data from the research scholars who presented their papers at the international conference: "Building Knowledge Competencies for Sustainable Development (SD) in Asia: Achieving the Goals of Life-Long Learning". The conference was held in Pakistan in November 2016. Survey was used as a research method because of its relevance for the study. It helps to "gather data at a particular point in time with the intention of describing the nature of existing conditions [situation]" (Cohen et al. 2007, p. 205).

5.1 *Respondents*

A relatively smaller number of respondents participated in the study. Cohen et al. (2007) noticed that there was no minimum number of respondents who could participate in a survey. Some researchers had carried out survey with lowest number of 28 participants (Cohen et al. 2007). As there is no minimum limit of the number of participants in a survey, therefore, forty two participants were included in this study. Participants were selected through purposive sampling. The sample consisted of more 'knowledgeable people' (Cohen et al. 2007), the ones who presented papers in the international conference on sustainable development.

Forty two scholars who participated in this study were from departments of education of five Pakistani universities. Twenty five of them were M.Phil. students, eleven were Ph.D. students and six were faculty members. The majority of the paper presenters in the conference were from Lahore College for Women University (LCWU). This affected the study sample. In our study, data came from 27 scholars from LCWU. Fifteen scholars were from four other universities. As the study aimed at investigating the perceptions of scholars who presented their work in a conference on 'sustainable development' therefore, we decided to proceed with data collection with uneven representation of universities. Among the study respondents five were those who had experienced pedagogical intervention (Kalsoom and Khanam 2017).

5.2 *Procedure for Data Collection*

Presenters from different universities at the conference registration desk were asked to participate in a short interview of 4–5 min. They were told that their responses would be included in a study about "Perceptions of the Research Scholars regarding ESD". Twelve scholars agreed to be the interview. Six of them were interviewed before the conference while six were interviewed during tea break. Three more participants were interviewed during lunch break and at the end of second day of the conference. Paper presenters from authors own university were interviewed on day 3 of the conference and afterwards.

5.3 *Interview*

Data were gathered through short interview with each scholar. Interviews were not audio-taped. Interviewers jotted down respondents' points in their diaries. Each interview lasted for 4–6 min. The interview included only five questions.

- Have you ever studied a stand-alone course on SD or ESD?
- How would you describe sustainable development?
- How would you describe Education for Sustainable Development?
- Which processes or activities are necessary for ESD in classroom and in the university?
- What are the expected outcomes of ESD in terms of students' competencies?

Probes like 'why do you think so', or 'how can it be' or 'what else' were used. The data were analysed quantitatively as well as qualitatively. Qualitative data were coded and categorized under pre-determined categories i.e. sustainability competencies, ESD processes, and ESD Content. Two additional categories also emerged from the data.

6 Findings

Data indicated that the respondents had not studied a stand-alone course on ESD. Other study findings have been summarized in the following five categories.

1. Sustainability competencies
2. ESD processes
3. ESD content i.e. society, economy, and environment
4. Education that sustains or continues
5. Education that aims at sustainability

6.1 *Sustainability Competencies*

Sustainability competencies identified by the scholars have been presented in Tables 1 and 2.

The majority of the competencies mentioned by the respondents (communication, collaboration, problem-solving, leadership, conflict resolution) fall under Wiek

Table 1 Frequency of the sustainability competencies

Competence	Frequency
Communication	31
Planning for future career	20
Critical thinking	16
Problem-solving	14
Team work/collaboration	10
Decision making	9
Creativity	8
Leadership and vision	7
Environmental care	4
Social justice	4
Conflict resolution	2
System thinking	2
Empathy	1
Lifelong learning	1
Honesty	1

Table 2 Number of competencies mentioned by individual respondents

Six competencies or more	2 respondents
5 competencies	3 respondents
4 competencies	8 respondents
3 competencies	16 respondents
2 competencies	10 respondents
1 competence	3 respondents

et al. (2011) category of interpersonal competence. Respondents mostly did not mention normative, anticipatory and strategic competencies (Wiek et al. 2011) in their responses. Only one respondent mentioned that ESD should aim at making people honest while four mentioned social justice and environmental care. A total of 26 out of 42 respondents (62%) mentioned 2–3 competencies as outcomes of the ESD. Communication was the most mentioned competence followed by planning for future career. This indicates that the research scholars view ESD as a means of preparing students for better jobs.

Respondents also mentioned critical thinking, problem-solving skills, team work and creativity as outcomes of ESD. Below are some representative responses:

> Education should promote critical thinking and creativity. It should empower students to take decisions. If it does not promote them [these skills], it is not Education for Sustainable Development.

> Students should learn to plan for their careers. There is no career counselling in our schools and universities. Students must also learn to communicate well and collaborate. These are essential life skills.

> Problem solving, decision making and communication… In fact communication is most important to sustain at work place.

ESD should promote thinking skills like problem-solving and creativity. It should also expose students to different professions so that they make sustainable decisions.

The responses indicate that the research scholars' perceptions of sustainability competencies were those competencies which are goals of any quality education programme. They viewed 'sustainability' as an ability of the individuals to sustain their employment. Likewise, '*making sustainable decisions about a profession*' also indicates that the study respondents perceived ESD in terms of sustainable employments.

7 ESD Processes

Table 3 presents a summary of responses regarding ESD processes mentioned by the respondents.

The data indicate that the majority of respondents perceived active participatory learning as an essential process in implementing ESD. Some respondents also viewed collaboration and curriculum innovation as important processes. However, majority of the respondents who talked about curriculum change referred to 'career counselling' instead of three dimensions of SD. None of the respondents mentioned the processes of 'engaging the whole system'.

7.1 ESD Content

Participants' responses in terms of their perceptions regarding ESD content have been summarized in Table 4.

Table 3 Perceptions regarding ESD processes

Processes	Key words mentioned	Frequency of responses
Collaboration	Cooperation, collaboration, working together, team work	13
Engaging with the 'whole system'		0
Curriculum innovation	Courses on career counselling	7
	Environmental education	4
Innovation in pedagogy	Field trips	1
Active and participatory learning	Activity based learning	19
	Experiential learning	3
	Student-centred learning	5

Table 4 Perceptions regarding ESD content

ESD content	Frequency of Responses
Content related to society, economy, environment and their relationships	3
Content on citizenship	2
Content on social justice	1
Content on economy	0
Content on population and resources	0
Content on environment	6

Table 4 indicates that research scholars are generally not aware of the ESD-based content. Only three respondents mentioned content on all three dimensions of sustainable development as part of ESD. Some respondents mentioned content on environment as part of ESD. This indicates that a limited number of respondents had awareness of the environmental dimension of the sustainability.

8 Education that Sustains or Continues

Study respondents interpreted ESD in a general way as 'an education that sustains or continues'. To nearly half of the respondents (20), Education for Sustainable Development was an education that provides 'sustained' learning instead of passing examination. For example:

> Education which is long-term. It should prepare students for future careers.
>
> ESD does not teach for examination. It is about that learning which is permanent.
>
> Education which gives opportunities for real learning instead of not rote memorization.

The above responses indicate that research scholars perceived ESD as that education which helps students to learn something that would be needed and beneficial in future.

9 Education that Aims at Sustainability

Two respondents viewed ESD in terms of environment education. They mentioned that ESD aims at conserving nature. For this, they maintained that environmental issues and ways to protect the environment should be a core component of the school and university curriculum. Respondents who had participated in the pedagogical intervention (Kalsoom and Khanam 2017) described Education for Sustainable Development more holistically and as a means of achieving sustainable development. They mentioned that ESD makes students familiar with socio-economic and environmental issues. Moreover, it also helps the learners to develop attitudes, skills and behaviours which can contribute towards sustainable development. A respondent mentioned:

> ESD is different from simple education. ESD does not make people good technicians but make them good citizens. ESD makes students learn about local and global issues of society and environment. It changes students' values and behaviours and life styles. After becoming aware of inequity, injustice, environmental degradation, waste of water, we are trying to modify our life styles. ESD should be compulsory.

The above response indicates that some research scholars who had an exposure to SD, perceived ESD as a way of changing lifestyles that can contribute towards broader aim of SD.

10 Discussion

This small study was undertaken in a context where ESD is not a priority at the national or provincial level. There are fragmented efforts at individual and institutional level to integrate ESD into existing courses.

The findings of the current study indicate that research scholars in Pakistan are generally not aware of the concepts of SD and ESD. Similarly, they are not aware of ESD processes and ESD competences. This seems a potential reason for lack of research in the field of ESD in Pakistan. The majority of the study respondents perceived ESD as a strategy to help the learners to get better employment in the future. Though they viewed ESD as an education for future, their description of future was 'future employment'. This seems a reason that majority of the respondents mentioned communication skills and other interpersonal skills (Wiek et al. 2011) as the outcome of ESD. As participants' perception of 'future' was not about future of the planet and human kind, therefore their description of ESD outcomes did not include normative, anticipatory and strategic competencies (Wiek et al. 2011). It is also notable that 62% of the respondents mentioned only 2–3 competencies while literature reports a range of different sustainability competencies (De Haan 2006; Wiek et al. 2011; Mochizuki and Fadeeva 2010). The lack of awareness of ESD among the research scholars is also supported by their description of ESD.

Scholars' perceptions of ESD were mostly not in line with its description in literature (UNESCO 2005; Tilbury 2011; McKeown 2002). Majority of the research scholars perceived ESD generally instead of conceptually. With this general perception, they viewed ESD as a means of 'sustained learning' by developing students' thinking skills. However, they did not elaborate thinking skills. To them ESD discourages rote learning and short-term learning to pass examination. This generic perception of the research scholars reflects the reason for their research papers which addressed general education and learning. Some examples of the paper titles are:

- Relationship of Pedagogical Beliefs and Classroom Practices of Secondary School Teachers
- Teaching English in an Age of Globalization: Some Reflections on Pedagogy and Curricula in Pakistan
- Effects' of Different Learning Styles on Students' Performance

Although research scholars are generally not aware of the concept of ESD, they are familiar with different ESD processes except 'engaging the whole system'. The reason could be that most of the ESD processes are key features of any quality education programme. As the respondents in this study belonged to the discipline of education, therefore they were aware of the importance of collaborative and participatory learning. The process of engaging the whole system is peculiar to ESD, so it did not appear in the study responses. Even those respondents who perceived ESD as a tool for SD, did not mention this ESD process. This indicates that although integration of ESD in an existing course (Kalsoom and Khanam 2017) was helpful in making participants perceive ESD as a tool for SD but it was not enough to make the participants fully aware of the need of the 'processes which engage the whole system'.

11 Conclusion

Study findings indicate that current programmes of education and teacher education in Pakistan are not building students' (future teachers) awareness of ESD. Their perceptions of ESD are naïve and not aligned with the agenda of SD. This seems a potential reason for sparse research on ESD in Pakistan. This finding also indicates a strong need for re-orienting the 'whole system' to develop people's (teachers, students, staff) sustainability competences. Education departments or education institutes are one component of the larger university. ESD should be reflected in university policy, curriculum, pedagogy, assessment practices, research, university culture, student services, campus operations (physical), university-community partnership, and ethics. Leal Filho (2010) suggests six criteria for universities to achieve excellence in the field of SD. It includes: core staff with formal qualification in the field of SD; strong publication profile on SD; robust research programme on SD; the implementation of principles of SD as intrinsic part of university; the integration of SD in university activities and operations (e.g. campus greening); the existence of a programme of extension (outreach) on SD topics, for the non-students public (pp. 280–281).

In the light of current situation, Pakistan seems far behind in meeting the mentioned criteria. However, teacher education institutes and departments may start the change by introducing stand-alone courses on ESD as well as integrating ESD across existing courses. Hegarty et al. (2011) noticed that stand-alone courses help in achieving sustainability goals of universities. "Undergraduate research is the pedagogy for the 21st century" (Council on Undergraduate Research and National Conference on Undergraduate Research 2005). Undergraduate research especially "research based" and "research tutored" (Healey and Jenkins 2009) activities focusing on sustainability issues may be one direction to facilitate undergraduate students to become aware of the concept of SD as well as contribute towards research on SD. Education for Sustainable Development is a key field in the post-2015 development agenda, and the findings of the current study may be used by the policy makers and UNESCO commission for Pakistan to take initiatives to mainstream ESD in universities.

Acknowledgements We are grateful to Sibte Hasan for his guidance in conducting this research.

References

Bossel, H. (1999). *Indicators for sustainable development: Theory, method, applications: A Report to the Balaton Group*. Manitoba: International Institute for Sustainable Development.

Buckler, C., & Creech, H. (2014). *Shaping the future we want: UN decade of education for sustainable development (2005–2014)*. Paris: UNESCO.

Ciegis, R., Ramanauskiene, J., & Martinkus, B. (2009). The concept of sustainable development and its use for sustainability scenarios. *Engineering Economics, 62*(2), 28–37.

Cohen, L., Manion, L., & Morrison, K. (2007). *Research methods in education* (6th ed.). London: Routledge.

Council on Undergraduate Research and National Conference on Undergraduate Research. (2005). *Joint statement of principles in support of undergraduate research, scholarship, and creative activities*. http://www.cur.org/about_cur/history/joint_statement_of_cur_and_ncur/. Last accessed September 15, 2015

De Haan, G. (2006). The BLK '21' programme in Germany: A 'Gestaltungskompetenz'-based model for Education for Sustainable Development. *Environmental Education Research, 12*(1), 19–32.

Ferreira, J.-A., Ryan, L., & Tilbury, D. (2007). Mainstreaming Education for Sustainable Development in initial teacher education in Australia: A review of existing professional development models. *Journal of Education for Teaching: International Research and Pedagogy, 33*(2), 225–239.

Government of Balochistan. (2013). *Balochistan Education Sector Plan 2013–2017*. Balochistan: Education Department.

Government of KPK. (2012). *Education Sector Plan 2010–2015*. KPK: Elementary and Secondary Education Department.

Government of Pakistan. (2009). National Education Policy. Islamabad

Government of Punjab. (2013–2017). Punjab Education Sector Plan. School Education Department

Government of Sindh. (2014). *Sindh Education Sector Plan 2014–2018*. Sindh: Education & Literacy Department.

Harris, J. M. (2000). *Basic principles of sustainable development*. Global Development and Environment Institute, Tufts University. http://www.ase.tufts.edu/gdae/publications/working_papers/Sustainable%20Development.PDF. Last accessed March 25, 2016

Healey, M., & Jenkins, A. (2009). *Developing undergraduate research and inquiry*. York: The Higher Education Academy.

Hegarty, K., Thomas, I., Kriewaldt, C., Holdsworth, S., & Bekessy, S. (2011). Insights into the value of a 'stand-alone' course for sustainability education. *Environmental Education Research, 17*(4), 451–469.

Higher Education Commission. (2012). *Curriculum of education*. Islamabad: Higher Education Commission.

Kalsoom, Q., & Khanam, A. (2017). Inquiry into sustainability issues by preservice teachers: A pedagogy to enhance sustainability consciousness. *Journal of Cleaner Production, 164,* 1301–1311.

Kalsoom, Q., Khanam, A., & Quraishi, U. (2017). Sustainability consciousness of the preservice teachers in Pakistan. *International Journal of Sustainability in Higher Education, 18*(7), 1090–1107. https://doi.org/10.1108/IJSHE-11-2016-0218.

KNCU. (2009). *Regional collection of good practices in achieving MDGs through ESD in Asia and the Pacific Region.* Seoul: Korean National Commission for UNESCO.

Leal Filho, W. (2010). Teaching Sustainable Development at University Level: Current trends and future needs. *Journal of Baltic Science Education, 9*(4), 273–284.

McKeown, R. (2002). *Education for sustainable development toolkit.* http://www.esdtoolkit.org. Last accessed June 24, 2014

McKeown, R., & USTESD-Network. (2013). Reorienting Teacher Education to Address Sustainability: The U.S. Context. White Paper Series, No. 1. Indianapolis, IN: United States Teacher Education for Sustainable Development Network

McKeown, R. (2014). The leading edge of teacher education and ESD. *Journal of Education for Sustainable Development, 8*(2), 127–131.

McKeown, R. (n.d.). *International network of institutions of teacher education: Five years of work on reorienting teacher education to address sustainability.* UNESCO. http://www.unescobkk.org/fileadmin/user_upload/esd/documents/workshops/kanchanburi/mckeown_reorienting.pdf. Last accessed March 9, 2015

McKeown, R., & Nolet, V. (2013). Education for sustainable development in Canada and the United States. In R. McKeown & V. Nolet (Eds.), *Schooling for sustainable development in Canada and the United States* (3p). Berlin: Springer,

Ministry of Education. (2009). *National professional standards for teachers.* Islamabad: Government of Pakistan.

Mochizuki, Y., & Fadeeva, Z. (2010). Competences for sustainable development and sustainability. *International Journal of Sustainability in Higher Education, 11*(4), 391–403.

NACTE. (2009). *National standards for accreditation of teacher education program.* Islamabad: National Accreditation Council for Teacher Education.

Redclif, M. (2005). Sustainable development (1987–2005): An oxymoron comes of age. *Sustainable Development, 13,* 212–227.

Remington-Doucette, S., & Musgrove, S. (2015). Variation in sustainability competency development according to age, gender, and disciplinary affiliation. *International Journal of Sustainability in Higher Education, 16*(4), 537–575.

Sachs, J. D. (2012). From millennium development goals to sustainable development goals. *Viewpoint, 379,* 2206–2211.

Tedesco, J. C., Opertti, R., & Amadio, M. (2011). *The curriculum debate: Why it is important today.* UNESCO-IBE. http://www.ibe.unesco.org/fileadmin/user_upload/Publications/Working_Papers/curr_debate_ibewpci_10_en.pdf. Last accessed December 10, 2014.

Tilbury, D. (2011). *Education for sustainable development: An expert review of processes and learning.* Paris: UNESCO.

UNECE. (2012). *Learning for the future: Competences in education for sustainable development.* https://www.unece.org/fileadmin/DAM/env/esd/ESD_Publications/Competences_Publication.pdf. Last accessed September 18, 2015

UNESCO. (2005). *Guidelines and recommendations for reorienting teacher education to address sustainability. UNESCO education for sustainable in action.* Paris: UNESCO.

Wiek, A., Withycombe, L., & Redman, C. L. (2011). Key competencies in sustainability: A reference framework for academic program development. *Sustainability Science, 6*(2), 203–218.

WCED. (1987). *Report of the world commission on environment and development: Our common future.* http://www.un-documents.net/our-common-future.pdf. Last accessed March 15, 2016

Author Biographies

Qudsia Kalsoom has recently completed her doctoral thesis on Education for Sustainable Development from the Institute of Education, Lahore College for Women University, Pakistan. Qudsia has been serving as a teacher and a teacher educator for past nineteen years. Her research interests are: Education for Sustainable Development; Action Research; and Social Justice Education.

Naima Qureshi is working as an Assistant Professor in the University of Education, Lahore, Pakistan. She obtained her Ph.D. from the University of Warwick, UK. She has been involved in teacher education since 2002. Naima has contributed as a teacher educator, project manager and a researcher in different projects. Her major research interest is teachers' professional development.

Afifa Khanam is working as an Assistant Professor and Incharge Department of Research and Evaluation in Lahore College for Women University (LCWU), Pakistan. Besides, she coordinates Ph.D. programme in the Institute of Education, LCWU. She is engaged in research projects on sustainability education, moral education and teachers' development.

Empowering Women and Girls Through Education: SDG's Vision 2030

Waseem Saba and Kayani I. Almas

Abstract The Sustainable Development goals (SDGs) and Vision 2030 is a thought provoking issue after un-accomplishment of MDGs 2015. This study was conducted by considering goal 5 of SDG vision 2030 about Empowering Women and girls in the perspectives of Education. The design of the study was descriptive in nature. A sample was selected in three different regions of Punjab (south, central and upper). A stratified random sampling technique was used to collect a proportionate sample size of 600 educated and uneducated women to collect data for the study. Data was collected by questionnaire, and focus group discussions. The results were collected using SPSS version 16 and inferential conclusions were made by content analysis. The main objective of the study was to identify the factors which play a pivotal role in empowering women and girls through education. The findings of the study revealed that the factors which riveted educated and uneducated women were different and varies by the area of Punjab they belong. Factors which affected the women of Rural South Punjab were ignorance and poverty, patriarchal thinking and the caste system. In Rural Central Punjab and the Rural Upper Punjab women were confronting economic status and gender discrimination in their inclination towards Education and Empowerment. Therefore it was concluded that Education may act as a vector to awareness, equity and equality for girls and women in their transformative process of empowerment and sustainable development of a country. The beneficiaries of the study are educated and uneducated women, stake holders, Policy makers and the future researcher.

Keywords Women empowerment · Education · Rural Punjab · Sustainable development goals (SDGs) 2030

W. Saba (✉) · K. I. Almas
Department of Education, Faculty of Social Sciences,
PMAS-University of Arid and Agriculture, Rawalpindi, Pakistan
e-mail: sabawaseem1999@gmail.com

W. Leal Filho et al. (eds.), *Sustainable Development Research in the Asia-Pacific Region*, World Sustainability Series, https://doi.org/10.1007/978-3-319-73293-0_11

1 Introduction: Women Empowerment Through Education and Sustainable Development

Education is an essential part of human progress and a basic right of every citizen. Education has a strong association with social and economic growth of a country. Countries like Pakistan can only attain sustainable socio-economic growth by an increase in female literacy rates as they comprise half of its population. Education enhances people's understanding and enhances their technical capabilities for exploring new ideas. It increases the eminence of their lives and leads to extensive social assistances to people and society. As explained by Kongolo and Bamgose (2002, 82) education and empowerment go hand in hand and "there has been dearth of political drive and sustained commitment to fulfill the monetary requirements and wellbeing of most rural women by local authorities and governments". According to a UNDP 2010 report, Pakistan ranked 120 in 146 countries in relations to Gender-related Development Index (GDI), and in terms of Gender Empowerment Measurement (GEM) position, it ranked 92 in 94 countries. According to United Nations (2015) Pakistan intends to accomplish the Sustainable Development Goals and also intends to eradicate gender disparity at all levels of education by the year 2030.

The sustainable development Goals (SDGs) and vision 2030 is a global vision. After ratifying the SDGs and vision 2030 Pakistan considered it as the substantial milestone in achieving sustainability.

Pakistan is a country where women make up half of its population and it is vital to explore the status of women. Goal 5 of the SDGs is the focus in this study which emphasizes women empowerment and gender equality. It highlights equal status of women in society, equivalent human rights, equal esteem, decision-making power, access to law and justice, equal in economic empowerment opportunities, worth and value. This entails that a woman's empowerment is only possible through education and awareness which will leads to sustainable development. It also challenges illiteracy, ill-health, poverty, ethnicity, prejudice, patriarchal thinking, feudalism, caste system, discrimination, rural-urban inequities, dispossession and banishing the weak.

2 Education for Sustainable Development

Education is considered important for setting a country on the path of development but unfortunately the educational sector as a whole and female education as a fragment has been ignored by the succeeding governments therefore the literacy rate of Pakistani women is among the lowest in the world. For women empowerment education has to play an instrumental role. In the rural areas of Punjab, Pakistan education for girls is considered trivial and unimportant as compared to boys, and this dispossession of girl's education is serious issue. The female literacy rate in Pakistan is only 58% according to the Economic survey of Pakistan, (Islamabad: Ministry of Women Development 2015). Female Literacy rate of (15 year and above) population of Punjab, Pakistan comprises of 51%, female urban literacy rate is 70% and female rural literacy rate is 38% (PSLM 2015–16). There might be many reasons for the reduced female literacy rate in rural areas but cultural constraints are the greatest logjam. Insufficient efforts have been made to educate women living in a male prejudiced society. The cause for this is often claimed to be a male dominated conservative society, but the study detailed here reveals that this indifference is basically due to the cultural constraints which restrict women to come out for formal schooling, leaving behind their conventional role in the home and family (Mahmood 2012). There are many deterrents faced by women in their way to empowerment through education.

3 Empowerment as a Cyclic Continuous Process in Sustainable Growth

Stromquist (2002) describes empowerment as a process that needs to regulate the power both in organizations and interpersonal relations. At the individual level, empowerment can be observed as the skill to direct and regulate one's life. It is the practice by which women can develop a grip on their lives by understanding and claiming their rights at every stage of society. Knowledge and understanding is a fragment of women empowerment, the knowledge of self-awareness, getting the perceptible tools of command and an affirmative self-concept cannot be achieved over night. Empowerment is an incessant process. This means that it is a cyclic process one stage is connected to the proceeding stage. Rawland (1997) explains that there is a need to be aware that power can be defined in diverse forms like power over, power to, power within, and power from within.

Obanya (2004) confirms that the basic skills that may be a prerequisite for empowerment are functional literacy which comprise basic literacy proficiency, knowledge, expertise, and abilities which consist of awareness improvement, logical skills, decision-making skills, managerial skills and methodical skills. As far as socio-economic skills are concerned these include vocational skills achievement, vocational skills development, sustainable income making skills. He further argued that empowerment is a lifelong learning process for social, economic support and continuous learning.

4 Education as a Catalyst for Change in Women Empowerment and Sustainability

Kabeer (2001) explains that education is a reagent for empowerment; access to education may in itself not be as appropriate a measure of empowerment as that education has to enrich women's abilities and autonomy. She argues that all forms of education may not be empowering resources because the outcome is reliant on context and that a slightest inception level of education may be required for education to be actually empowering. Kabeer (1994) has introduced another self-motivated explanation of empowerment. She favors a notion that theory pertaining to empowerment needs to transform into practicality rather than be a mere slogan of empowerment. Kabeer (1994) consider that self-respect and emotion of being as active agent are the vital values of empowerment. Naz (2006) expands Kabeer's description of empowerment by suggesting that "empowerment should be taken as a concept of recognizing oneself as an dynamic representative capable of making decisions". Young (1993), also advocates the same idea that "empowerment allows women to regulate their life, on their own, set their own strategy, be able to help each on, and propose demands on the government for sustenance and on the people itself for transformation". For Young, empowerment requires the transformation of the progress and formations accountable for women's poor standing in the society. It is based on a concept that a woman's position can be transformed in such a way that the development will be sustained. Improved understanding, self-awareness and cognizance of gender impartiality are scales of the empowerment process (Murphy-Graham 2008). There are indications that these components are commonly established during and as a outcome of higher education (Maslak and Singhal 2008). Educated women turn out to be more assertive (Maslak and Singhal 2008) and they have enriched communication abilities and can protect their identity in an operative and subtle way and are able to make decisions for themselves. Educated women generally have a better understanding of human rights, valuing difference of thoughts, and keep good care of social standards and try to solve issues and problems and inculcate peace build peace and coherence in society. Education is termed as a vector for empowerment and mobilization in the transformative process of empowerment. It has been carved for the ability to build capacity, boost self-confidence and increase collaboration/contribution in social and economic changes (Acker and Gasperini 2009). Srilatha (1993) has rightly observed, "The term Empowerment is associated with the right of women to get equality and integrity in society".

5 Eradication of Poverty and Employment Opportunities for Sustainability

According to Knowles et al. (2002) gender inequalities in education significantly affect economic growth and sustainable development. Chaudhry (2007) explored the impact of gender disparity in education on financial progression in Pakistan. In an empirical study Chaudhry (2009) explored factors that affects rural poverty in South Punjab (Pakistan), and he concluded that the graph of poverty can be slowed down by increasing job opportunities for females, giving access to education, reduced family sizes and dependence percentage. Education has the considerable association with employment opportunities. Chaudhry (2009) further reveals that skilled and educated women employees will not only affirm women's welfare, but will also raise the general output of the female employees due to more competitiveness. Hence, feminist economists discuss that it is necessary for government to consign more resources for women's education, as it will advantage the entire society. Moreover, social and cultural taboos that confine women to their homes only and prohibits them to compete for resources outside home need to be challenged to transform the social and economic dependence of women on men. The nature and degree of women's oppression and subservience can be different from areas to areas and the rural and urban gap in Pakistan. Khan (2007) observed that patriarchal thinking are comparatively more noticeable in the rural and ancestral setting where native traditions and aboriginal laws create stronger male control and power over women.

6 Sustainable Development Goals Vision 2030

The Sustainable Development Goals (SDGs), officially known as transforming our world: the 2030 Plan for Sustainable Development is a transnational set of 17 aspiring Goals comprising 169 targets. The Goals are confined in paragraph 54 United Nations Resolution A/RES/70/1 of 25 September 2015. The Resolution is a wider transnational agreement that, although acting as the Post 2015 Development Outline builds on the Values agreed upon under Resolution A/RES/66/288, generally known as the future of our choice. World leaders from 193 countries agreed to the 2030 Outline for Sustainable Development. It will apply valuable experiences acquired from recently completed strategies of the fifteen-year Millennium Development Goals. This study will focus on the fifth Sustainable Development Goal—"Gender equality and women's empowerment." It illustrates how the 2030 Agenda implementation is aided by shaping substantial, atypical, intricate, multi-dimensional and necessary partnerships, comprising donor communities; countrywide, local, national and native governments; individuals, nuclear and extended families, public leaders, heads of social institutions; civil society organizations, and most critically, women and girls everywhere. Women have a critical role to play in all of the SDGs, with many targets specifically identifying women's equality and empowerment as both the objective, and as part of the solution. Deep legal and legislative changes are needed to ensure women's rights around the world. In many nations, gender discrimination is still woven through legal and social norms. The SDGs ambition is human rights of all and to achieve gender equality and the empowerment of all women and girls; "a world in which every single woman and girl have gender equality and equal legal, social and economic hurdles to their empowerment have been removed". The sustainable development goals seek to change the course of the 21st century, addressing key challenges such as:

Goal 1. Reduce poverty
Goal 2. Eliminate starvation
Goal 3. Safeguard healthy lives
Goal 4. Certify quality education (inclusive and equitable)
Goal 5. Attain gender equality and empower all women and girls
Goal 6. Make sure accessibility and sustainable management of water
Goal 7. Confirm right to use economical, sustainable energy for all
Goal 8. Encourage sustained employments for all
Goal 9. Stimulate comprehensive and sustainable industrialization
Goal 10. Decrease disparity within and among countries
Goal 11. Ensure safety of humans and cities for sustainability
Goal 12. Confirm sustainable consumption and production patterns
Goal 13. Take imperative action to fight climate change
Goal 14. Preserve marine assets for sustainable use
Goal 15. Safeguard, reinstate and stimulate terrestrial ecosystems for sustainability
Goal 16. Encourage peace and harmony in societies for sustainable development
Goal 17. Support and rejuvenate the worldwide enterprise for sustainable development

7 Comparison and Reasons That SDGs Have Improved Outcomes

The MGDs have been criticized for being too narrow and applicable only to the poor countries and to be achieved through the support of the wealthy nations. On the other hand, the SDGs are broader in scope and are applicable to all the countries The MGDs were expected to take the nations halfway to the goals, however, the SDGs aim at reaching to the statistical zero and finish the task at hand. The private sector played a substantial role in SDGs, through many initiatives of United Nations (UN) while MDGs remained underrated under-explored within the under developed countries. SDGs will be another evolutionary step in getting people around the world to think a little bit more as global citizens and think about poverty, inequality, sustainability, consumption and discrimination. Certainly The SDGs are accelerative in nature that if development is meant for every one so that no one is left behind and safeguard the rights of all human rights and principles. Seven SDG targets explicitly refer to persons with disabilities; an additional six targets refer to people in vulnerable situations, while seven targets are universal and two refer to non-discrimination. Inequality is not just measured in terms of growth but in terms of making sure the most excluded can exercise their human rights. Goal 5 of SDGs vision 2030 is about women empowerment.

8 Goal 5: Achieve Gender Equality and Empower All Women and Girls

- Eliminate discrimination in all forms against women and girls universally
- Eradicate violence against all women and girls in the communal and remote areas
- Maintain women's active involvement and equal participation at all levels of decision-making in political, economic and public life
- Confirm and protect worldwide women's reproductive rights to sexual and reproductive health and human rights in agreement with International Conference on Population and Development and the Beijing Platform for Action
- Carry out reforms to give women equivalent rights to financial resources, accession of proprietorship and control over land and other forms of property, monetary services, share in inheritance and in hereditary property, in agreement with general laws
- Boost the use of enabling technology particularly ICT information and communications technology to support women empowerment
- Implement and reinforce comprehensive policies for the elevation of girls and women the empowerment all levels

9 The Role of Education in Empowering Women and Girls for Sustainable Development in Pakistan

The study detailed here aimed to identify the views of Educated and uneducated women about role of Education in empowering women for the sustainable development in Pakistan. A sample of 600 female in the age cohorts of 15–60 years was randomly drawn from rural areas of three regions of Punjab (upper, central and southern Punjab). Both qualitative and quantitative methods were used for data collection. The analysis of the study is carried out at two levels. First a statistical analysis of sample 600 females: 300 educated and 300 uneducated women was carried out. Data was collected through questionnaire designed for both educated and uneducated women. The questionnaire comprised of 52 items with 10 themes for the collection of relevant data. Secondly, interviews with women from the target population (women living in different marginalized rural areas of Punjab representing educated and uneducated women) were conducted using focus group discussion. Focus group discussion was conducted with 100 women in 4 groups (each group comprises of 25 women) from different social class and occupations. The results of the questionnaire were analyzed and interpreted through variable driven data and were triangulated with the results of focus group discussions.

10 Results and Finding

Descriptive statistics in the form of graphical representation were computed from the data driven variables assessed for this Survey. Data collected was organized, tabulated and analyzed. The Results are presented in the respective pie graphs.

Table 1 indicates that educated women strongly agreed that they have an easy access to their schools to get education while uneducated women strongly disagreed that easy access to schools was accessible to them.

Figure 1 indicates that 59% of women from upper Punjab strongly agreed that accessibility to schools without hurdles helps them get educated and empowered,

Table 1 Accessibilty to schools and its effect on the status of women

Variable	Status of women		Total
I find easy access to education	Educated women	Uneducated women	
Strongly agreed	6	32	38
Agreed	58	31	89
Disagreed	66	112	178
Strongly disagreed	82	200	282
No response	5	8	13
Total	300	300	600

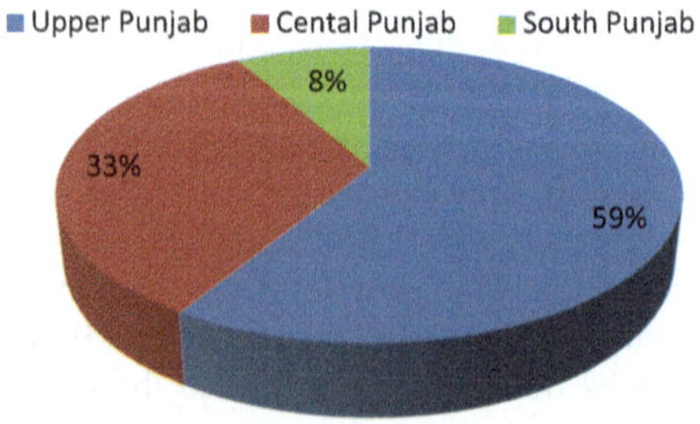

Fig. 1 Access to schools and its effects on Education

Table 2 Caste system and its effect on the status of women

Variable	Status of women		Total
Caste system is a deterrent in getting education	Educated women	Uneducated women	
Strongly agreed	65	128	193
Agreed	82	145	227
Disagreed	77	8	85
Strongly disagreed	85	5	90
No response	5	0	5
Total	300	300	600

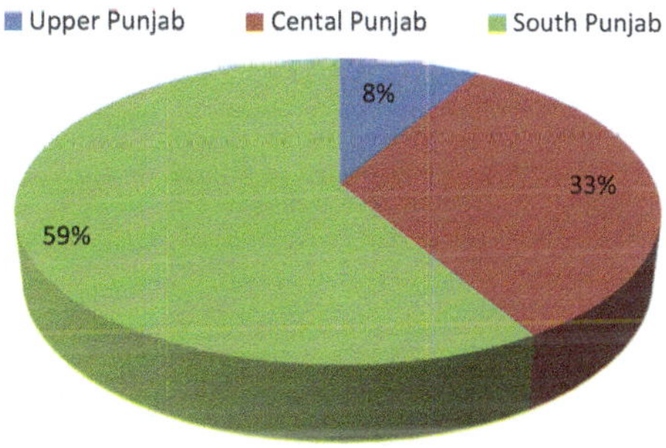

Fig. 2 Caste system is hurdle to get education

33% of women from Central Punjab agreed with their access to schools and 8% of women from South Punjab agreed to have access to schools. The results revealed that in South Punjab there is less accessibility for women to schools and hence women are less educated and less empowered.

Table 2 indicates that most of the uneducated women agreed that caste system is the deterrent in their way of getting educated and empowered, the perception of educated women is vice versa which disagree that caste system is a not an obstruction in their way of education.

Figure 2 indicates that 59% of women from South Punjab agreed that caste system is main constraint towards getting education and getting empowered, 33% of women from central Punjab also agreed that caste system refrain them to get Education and 8% of women from upper Punjab agreed that caste system is an obstacle in getting them educated and empowered.

Table 3 indicates that uneducated women agreed strongly that poverty was the economic constraint due to which it is difficult for them to pursue education while educated women strongly disagreed that that poverty is not a constraint in getting them educated.

Figure 3 indicates that 77% of rural women in South Punjab are poverty stricken and agreed that poverty is hindrance in their way of getting Education and make them empowered, 17% of women agreed from central Punjab and 6% of women

Table 3 Poverty and its effect on the status of women

Variable	Status of women		Total
I face poverty as a constraint in getting education	Educated women	Uneducated women	
Strongly agreed	21	221	242
Agreed	25	114	139
Disagreed	117	22	139
Strongly disagreed	26	36	62
No response	11	7	18
Total	300	300	600

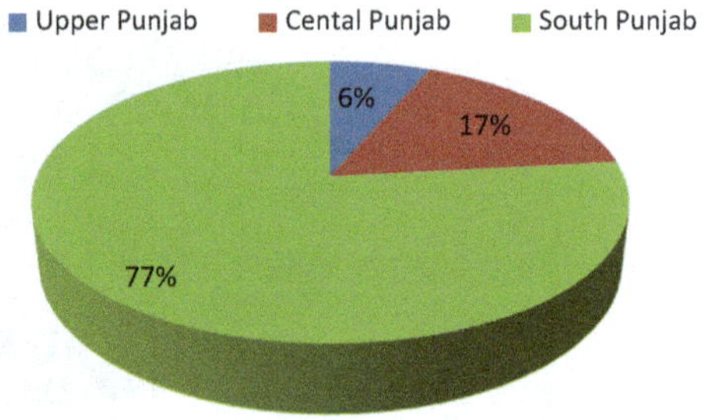

Fig. 3 Poverty is hindrance in the way of getting education

Table 4 Inadequate resources and its effect on the status of women

Variable	Status of women		Total
Inadequate resources was a bottleneck to get education	Educated women	Uneducated women	
Strongly agreed	49	160	209
Agreed	29	141	170
Disagreed	54	69	98
Strongly disagreed	67	31	98
No response	10	15	25
Total	300	300	600

from upper Punjab agreed that poverty is the problem in their way of getting Education.

Table 4 indicates that uneducated women agreed strongly that inadequate resources was the main bottleneck due to which it is difficult for them to pursue education while educated women strongly disagreed that inadequate resource is not a restriction in getting them educated.

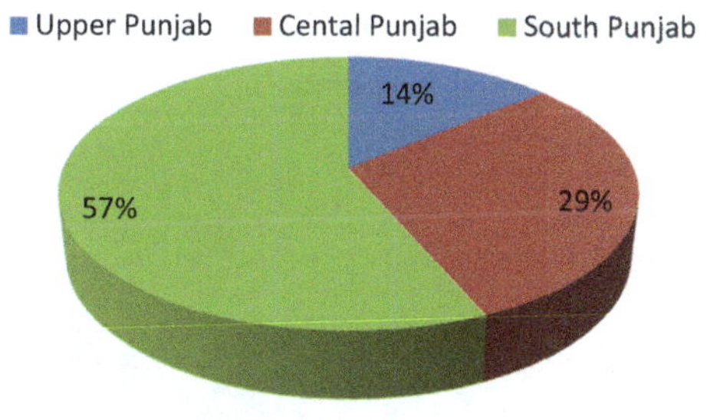

Fig. 4 Inadequate resources creates restriction in getting education

Table 5 Socio-cultural norms and its effect on the status of women

Variable	Status of women		Total
Social and cultural norms are the bottlenecks to get education	Educated women	Uneducated women	
Strongly agreed	68	40	108
Agreed	73	170	243
Disagreed	140	68	208
Strongly disagreed	10	21	31
No response	9	1	9
Total	300	300	600

Figure 4 indicates that 57% of women belong to Rural Lower Punjab agreed that inadequate resources prohibit them in getting education, 29% of women from central Punjab agreed that inadequate resources restricts them to get education. 14% of women in upper Punjab agreed that inadequate resources was an obstacle for them in getting educated.

Table 5 depicts that uneducated women agreed that social and cultural norms are the bottlenecks in their way to get education while the educated women disagreed with this notion.

Figure 5 indicates that 60% of women belong to rural South Punjab agreed that they find social and cultural norms as an obstacle in their way to get educated and be empowered, 28% of women in Rural Central Punjab agreed that social and cultural norms is a hurdle in their way to get education and become empowered, 12% of women belong to Upper Punjab agreed the social and cultural norms do cause hurdles for being educated and empowered.

Table 6 reveals that uneducated women agreed to an extent that skill training centers is helpful in getting them skilled educated and empowered while educated women strongly agreed that skill training inevitably help in making them empowered.

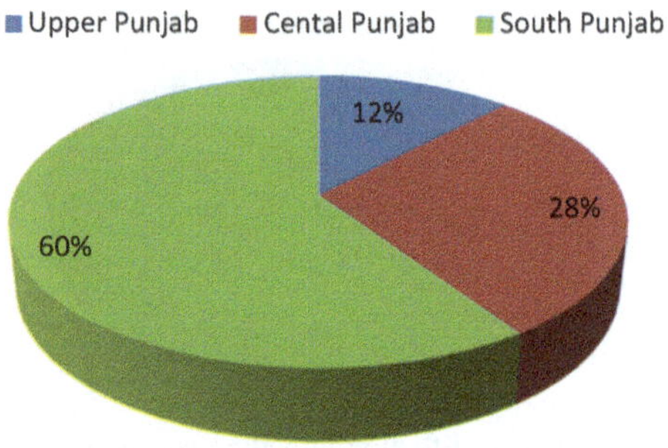

Fig. 5 Social, cultural norms cause hurdles in the way to education

Table 6 Skill training and its effect on the status of women

Variable	Status of women		Total
Skill training helps me in education	Educated women	Uneducated women	
Strongly agreed	157	150	307
Agreed	81	84	165
Disagreed	3	34	37
Strongly disagreed	49	27	76
No response	9	6	15
Total	299	300	599

Fig. 6 Inadequate skill training programs

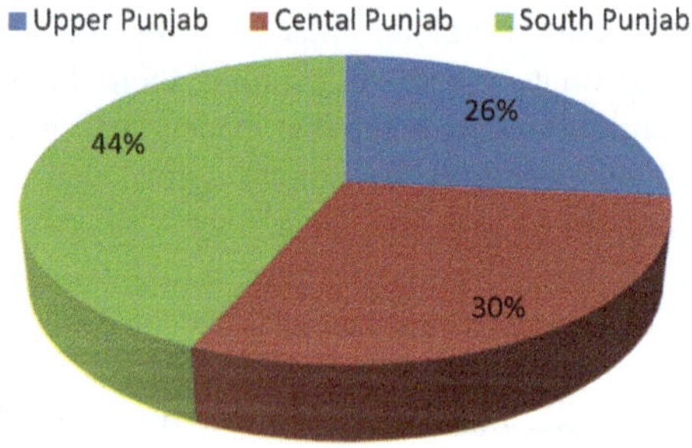

Figure 6 indicates that 44% of women belong to South Punjab agreed that there are inadequate skill training programs which help them to educate and empowered, 30% of women in Central Punjab agreed the inadequacy of skill training which can help them in rising their financial condition to empower and 26% of women in upper Punjab agreed that there are inadequate number of skill training programs in upper Punjab.

Table 7 Discrimination at home and its effect on the status of women

Variable	Status of women		Total
Discrimination at home	Educated women	Uneducated women	
Strongly agreed	50	218	268
Agreed	11	159	170
Disagreed	24	30	54
Strongly disagreed	67	31	98
No response	0	5	10
Total	300	300	600

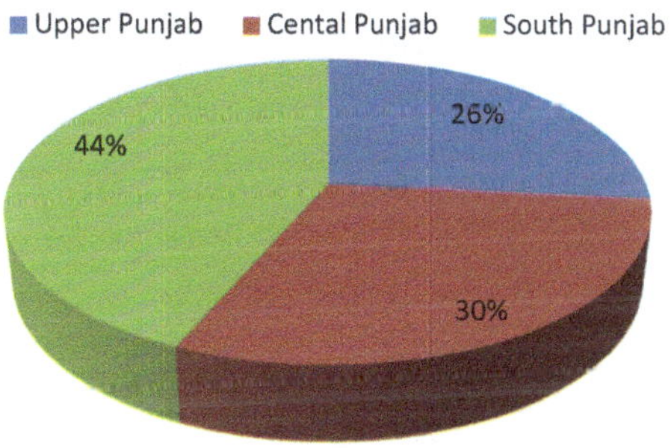

Fig. 7 Discrimination at home is a barrier towards education

Table 7 indicates that uneducated women strongly agreed that they remain uneducated due to discrimination at their home on the other hand educated women disagreed with this opinion.

Figure 7 indicates that 44% of women belong to South Punjab agreed that discrimination within the family and society creates barrier for them from getting educated and empowered, 30% of women belong to Central Punjab agreed that they discriminatory behavior is a deterrent for them to be educated and empowered, 26% of women in Upper Punjab agreed that they also face discrimination as a hurdle in getting education and empowerment.

Table 8 indicates that educated women strongly agreed with the opinion that their parents support helps them to be educated while uneducated women disagreed that their parents support helps them in getting educated.

Figure 8 indicates that 77% of women belong to Upper Punjab agreed that parental support help them to be educated and empowered, 19% of women in central Punjab agreed that parental support is with them to get education, 4% of women in south Punjab agreed with the parental support in getting them educated and empowered.

Table 9 indicate the opinion of educated women that they strongly disagreed with the patriarchal thinking of family is a hurdle in getting them educated on the other hand uneducated women strongly agreed that patriarchal thinking of family is the key hurdle in their way to education.

Table 8 Parents support and its effect on the status of women

Variable	Status of women		Total
Parents support help their daughters to get education	Educated women	Uneducated women	
Strongly agreed	108	6	114
Agreed	106	26	132
Disagreed	70	158	228
Strongly disagreed	10	107	113
No response	7	6	13
Total	300	300	600

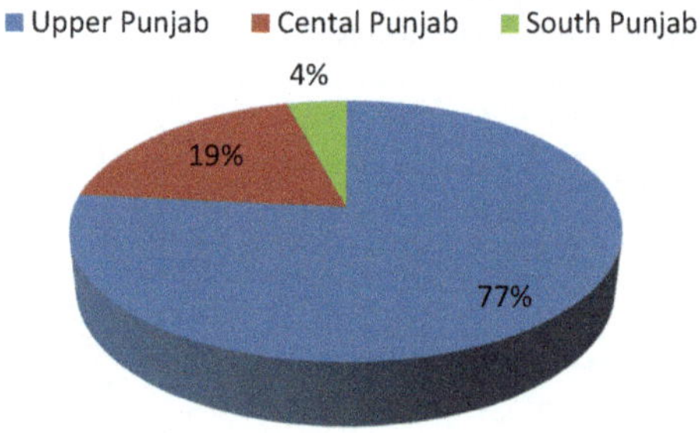

Fig. 8 Parent's support and its effect on education and empowerment

Table 9 Patriarchal thinking and its effect on the status of women

Variable	Status of women		Total
Patriarchal thinking of family is hurdle towards education	Educated women	Uneducated women	
Strongly agreed	2	132	134
Agreed	23	132	155
Disagreed	176	22	198
Strongly disagreed	96	13	109
No response	3	1	4
Total	300	300	600

Figure 9 indicates that 65% of women in south Punjab agreed that patriarchal thinking is a hurdle for them to be educated. 27% of women belong to central Punjab agreed that patriarchal is an obstacle to make them educated and 8% of women in upper Punjab agreed that patriarchal thinking resists them to be educated and empowered.

Table 10 indicates that educated women agreed that she is well aware of her basic rights but uneducated women disagreed with the opinion that she is aware of her basic human rights.

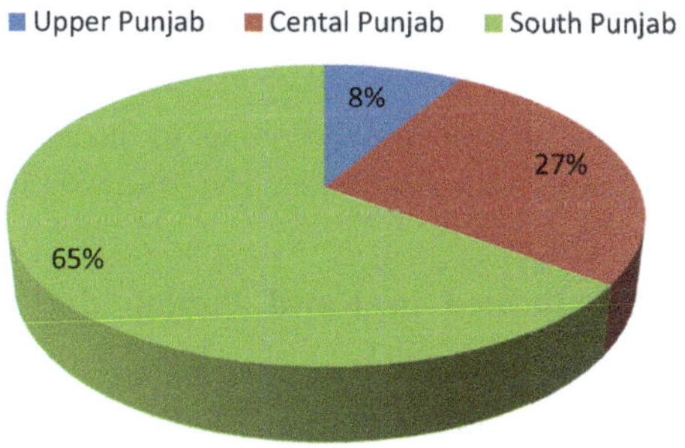

Fig. 9 Patriarchal thinking is a deterrent towards education

Table 10 Awarness of rights and its effect on the status of women

Variable	Status of women		Total
I have awareness of my basic human rights	Educated women	Uneducated women	
Strongly agreed	49	39	88
Agreed	137	61	198
Disagreed	70	69	139
Strongly disagreed	27	128	155
No response	15	5	20
Total	300	300	600

Fig. 10 Awarness of rights and its effect on education and empowerment

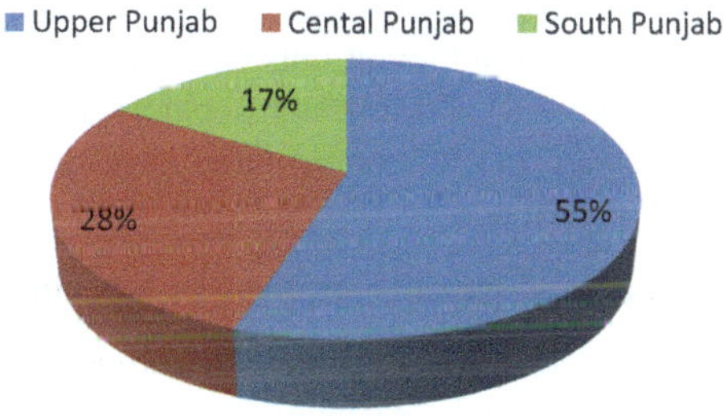

Figure 10 indicates that 55% of women belong to Upper Punjab agreed that they are aware of their rights and laws to get empowered and educated, 28% of women in central Punjab agreed to the fact that they are well aware of their rights, 17% of women in south Punjab agreed that they are aware of their rights.

11 Findings of Focus Group Discussion

Themes are made for discussion and decoding of these themes was done for inferential findings of focus group discussion. The following themes were identified in this study.

- **Women remain uneducated in Rural Punjab due to inaccessibility to schools**

Participants of focus group discussion revealed that due to the remoteness of marginalized rural areas of Punjab the inhabitants do not have easy access to formal schools. They are confined to their homes or can have madrassa education (religious education). Educated women in this area have not graduated from high school and can only be able to read and write. The reason given by the majority of participants was inaccessibility of schools was the far off distance to schools.

- **Lack of awareness is an obstacle for women in Rural Punjab**

The majority of participants of focus group discussions exposed that females are unaware of their basic human rights. Whether educated or uneducated women are ignorant about their rights regarding land ownership, productive heath rights, inheritance rights and socio economic rights. It was also unveiled by the participants that lack of knowledge of information technology like internet and other ICT devices for girls and women in remote areas is another bottleneck.

- **Discrimination in family and society is barrier towards education**

It was a view that in Rural Punjab women are supposed to be at their best as mothers, daughters, sisters and wives and an indoor inhabitant. They have nothing to be concerned about outside of home especially at community level. Women are not motivated and inspired to elucidate their views in public. If they are allowed to participate in outdoor activities this could contribute to the sustainable growth of their societies.

- **Lack of Parents support in the home restricts girls education**

Participants articulated that there is a lack of parent support for girls and women in South Punjab. Women are not able to display their self-confidence and exhibit their views because parents never seek their consent in any matter. They are never a part of decision making process at home or in a family. Even educated women belong to rural Punjab expressed that they have no right to make decisions on their behalf therefore they lack independence and confidence. Above all it was found that cordiality among the family and community members results in a great friction. Women were not allowed to attain education in south Punjab due to the absence of parent support.

- **Inadequate skill training programs make women reliant and redundant**

Focus group participants expressed their views that women are not gaining sufficient skills due to inadequate skill training centers. There was an identified lack of well-designed skill training programs for the girls and women in order to develop their skills and abilities even though skills will further improve the women's approach to communal and educational information.

- **Inadequate resources are a deterrent in the way towards education for rural women**

Participants of focus group discussions conveyed their views that most of them did not get education due to inadequate resources like lack of transport facilities to reach schools, lack of materials like (books, bags, stationary), lack of monetary resources, and lack of infrastructure.

- **Patriarchal thinking is a main obstacle for women in rural Punjab to get education**

Most of the women belonging to rural South Punjab communicated in the focus group discussions that effort is needed to change the patriarchal thinking of the society. They are confronting patriarchal thinking in their families and in society at large which is the main obstacle in rural Punjab for women to get into schooling for formal education. Both educated and uneducated women emphasized the fact that they are striving in a male controlled society which creates many hurdles for them to find their way for education.

- **Poverty is a key factor for restraining women towards education**

In focus group discussions women belonging to Rural Punjab especially South Punjab conveyed that the families they belong to are poverty stricken and for them it was more important to earn bread and butter than to send their daughters to schools for formal education. In the region of rural Punjab women are meant to stay at home for household chores.

- **Caste system is a hindrance in the way to education for rural women**

Focus group discussions revealed that the caste system is an ultimate and foremost hindrance for rural women to go to schools. Rural women voiced that since they reached the age of puberty they are confined to their homes for indoor madrassa education and refrained from going out. Feudalism has also a great hold in rural south Punjab and rural central Punjab, which make rules and laws for the society.

- **Social and cultural norms are the challenges towards getting education**

Women from Rural Punjab openly spoke about stringent social and cultural norms of society. According to them it is a male controlled and male chauvinistic society where women cannot make decisions. Women cannot get equal opportunities for getting education or getting a job. Women gave a stance that they can work in fields with men but they are not permitted to work on their own. Social taboos characteristic of rural society put women in a difficult situation.

12 Discussion

Results of the study indicated that education and women empowerment is considerably influenced by many factors, such as he traditional socio-cultural norms and taboos of the community, distance away from schools which are not present in the rural vicinity and its inaccessibility for girls is another hurdle in getting education. 65% of women in south Punjab agreed that patriarchal thinking is a hurdle for them and leads to loss of self-confidence and self-reliance. Lack of available resources is another bottleneck faced by rural women, 77% of women agreed that poverty is the main barrier for women belonging to rural Southern Punjab which has to be eradicated. The caste system is the most challenging factor which is blended with feudalism discriminating against women particularly in South Punjab as compared to women in central and upper Punjab. Discrimination is, however faced by all the three regions but its ratio is more in South Punjab as compared to the upper and central Punjab., Findings of the questionnaire reflects that 83% of women from South Punjab are ignorant about their rights and only 17% agreed that they have awareness of their basic rights. Parent support is a vital variable identified in this study. 4% of women from South Punjab agreed that they have parents support to get education. Finally the results of the study revealed that women empowerment can be improved by working on the identified significant factors in remote regions of rural Punjab by elevation of education and awareness.

In summary it was found that social and cultural norms, discrimination, patriarchal thinking, poverty, inadequate resources, lack of accessibility, caste system, lack of family support, and lack of awareness, all combine to impact on opportunities for women in Rural Punjab. The rural women from South Punjab are impacted by these social barriers more than in other regions. According to Pakistan Bureau of statistics 2015, 74% of the Rural Women from 7 Districts of South Punjab (D. G. khan, Bahkar, Rahimyar khan, khanewal, Rajanpur, Muzaffargargh, Vehari) have no formal education at all.

Women, however, are key agents of change in the shift towards sustainable development but due to lack of functional literacy programs to upgrade the reading and writing skills of girls and women in the marginalized areas of Rural Punjab the literacy rate is touching 38% in average according to PSLM (2015–16). Women's capacity building and human development activities are not enriched in these areas due to lack of awareness of these literacy programs and lack of accessibility of public information to make them more substantial and progressive. There is an imperative need for future policy development on Education for Sustainable Development to integrate a gender perspective, promoting a transformative and universal approach to women's empowerment.

13 Conclusion and Recommendations

Empowerment of women aims at striving towards acquiring higher literacy level and education for the sustainable development of Pakistanrefer to SDGs. The female population is almost half of the total population of the country and the literacy rate is still alarming in the Rural Punjab. Addressing the sustainable development Goals (SDGs) 2030 will help to ensure that all girls and women may have equal ownership of productive resources, increased participation in decision making, awareness of their rights, improved standards of living and productive health, self-confidence, self-reliance in women. The present study highlighted that poverty, inequality, traditional cultural norms and patriarchal thinking, discrimination, and lack of resources, cannot be eradicated alone until Government policies and programs are properly executed for women in these Rural areas of Punjab. Without education women are unable to understand their rights and their importance. Detailed analysis of the data revealed that almost all participants believed that education can bring remarkable transformation in women's lives resulting in social, cultural, economic, political sustainability in the long run by implementing the following:

- Enhance their confidence by giving them family support.
- Raise their status in the family and society by confronting the social cultural norms.
- Bring awareness about their rights by eliminating inequality.
- Boosting their self-esteem by evading discrimination between boys and girls
- Reduce their economic dependency by giving equal opportunities for employment.
- Better upbringing of their children with the help of education.
- Enhance their mobility by encouraging their accessibility to schools, colleges and offices.
- Open career opportunities by giving them an equal chance of being selected.
- The paper concludes with the following recommendations that are more directive than the above—these need to be rephrased.
- Imparting education to girls should be encouraged at all levels especially in rural remote areas of Punjab. Teams should be made by the Government to provide incentives to the families which will be ready to send their girl child to schools.
- It is a dire need to open skill training centers for girls and women because if women will be well educated and skilled, they can play their role effectively to enhance entrepreneurship.
- The ICT centers should be introduced for girls and women in the form of rural ICT educational schemes for rural areas.
- Distance learning programs should be introduced for rural marginalized female population to enhance learning.
- Orientation workshops should be arranged by the counseling team in order to discourage the Patriarchal thinking.

- Facilitate women by providing micro financing and monetary support to cope up with their inadequate resources to get education.
- Conventional authorities of Marginalized areas should manage to remove the socio cultural inhibitors which preclude women from receiving equivalent admittance to these schools and skill training centers.
- It is recommended that women should be supported by their family and community at all levels of decision-making and governance to achieve leadership skills.
- The women should be encouraged with the introduction of rural scholarship reform schemes for rural women so they can go through formal education and attain higher level of education.
- Functional literacy programs for the female adults should be introduced to develop their reading and writing skills. Apparently such skills will further enrich the women's access to community and educational information.
- Allocation of female budget by Government bodies should be allotted for areas of Rural Punjab, for the development and training for women to be able to attain skills to boost their careers.

References

Acker, D., & Gasperini, L. (2009). *Education for rural people. The role of education, training and capacity development in poverty reduction and food security*. Rome, Italy: Food and Agriculture Organization of the United Nations (FAO).

Chaudhry, I. (2007). Impact of gender inequality in education on economic growth: An empirical evidence from Pakistan. *The Pakistan Horizon, 60*(4).

Chaudhry, I. (2009). Poverty alleviation in Southern Punjab (Pakistan): An empirical evidence from the project area of Asian Development Bank. *International Journal of Finance and Economics.*

General Assembly Resolution. (2015). 70/1, Transforming our world: The 2030 Agenda for Sustainable Development, A/70/1. Available from undocs.org/A/RES/70/1.

Kabeer, N. (1994). *Reversed realities: Gender hierarchies in development thought*. London: Verso.

Kabeer, N. (2001). Reflection on the measurement of women's empowerment in SIDA Studies No. 3. *Discussing women's empowerment: theory and practice*. Stockholm.

Khan, S. (2007). Patriarchy, gender violence and poverty amongst Pakistani women: A social work inquiry. *International Journal of Social Work and Human Services Practice, 2*, 56–6.

Knowles, Stephen, Lorgelly, Paula K., & Dorian Owen, P. (2002). Are educational gender gaps a brake on economic development? Some cross-country empirical evidence. *Oxford Economic Papers, 54*(1), 118–149.

Kongolo, M. B. (2002). Participation of rural women in development: A case study of Tsheseng, Thintwa and Makhalaneng villages, South Africa. *International Women Studies*, 79–92.

Maslak, M. A., & Singhal, G. (2008). The identity of educated women in India: Confluence or divergence. *Gender and Education, 20*(5), 481–493.

Murphy, G. E. (2008). *Opening minds, improving lives: Education and women's empowerment in honduras*. Nashville: Vanderbilt University Press.

Naz, F. (2006). *Pathways to women's empowerment in bangladesh*. AH development Publishing House. Dhaka, Bangladesh.

Obanya, P. (2004). *The dilemma of education in Africa*. Ibadan: Heinemann Educational Books.
Pakistan Social and Living Standard Measurement Survey. (2015–16). Govt. of Pakistan, Islamabad.
Rowland, J. (1997). *Questioning empowerment*. Oxford: Oxfam.
Sandhu, A. M. (2012). Politics of Women's education in Pakistan and government policies, 1947-2008, History & Culture. *A Research Journal of National Institute of Historical Research, 33*. (Quaid-i-Azam University Islamabad, Pakistan).
Srilatha, B. (1993). The meaning of women's empowerment: New concepts from action. In G. Sen, A. Germain, & L. Chen (Eds.), *Population policies reconsidered. Health, empowerment and rights* (pp. 127–138). Cambridge: Harvard University Press.
Stromquist, N. P. (2002). Education as a means for empowering women. In J. Parpart, S. Rai, & K. Staudt (Eds.), *Rethinking empowerment: Gender and development in a globalized/local world* (pp. 22–38). London: Routledge.
UNDP, (2010). *Human development report*. New York: Oxford University Press.
United Nations. (2015). *Partnerships for sustainable development Goals-17 Partnerships*. United Nations Department of Economic and Social Affairs Division for Sustainable Development.
Young, K. (1993). *Planning development with women: Making a world of difference*. London: Macmillan.

Intercultural Dialogue Through Design (iDiDe) as a Platform for Built Environment Education for Sustainability in Rural Developing Contexts: Building Ampara, Sri Lanka

Susan Ang, G. Karunasena and R. Palliyaguru

Abstract The 2016 Pritzker Prize awarded to socially conscious Chilean architect Alejandro Aravena for dedication towards improvement of social and humanitarian issues around the globe heralded a firm spotlight on socially responsible architecture and a holistic view of sustainability. Deakin University's iDiDe (pronounced "i-dee-dee") delivered a global mobility study tour model with structured immersive learning that focused upon sustainable rural community development in the Eastern district of Ampara, Sri Lanka. The program facilitated student learning of sustainability across the realms of environmental, cultural and social dimensions. Deakin University partnered with a number of institutions and organizations and pursued community participatory approaches to the design process. Mixed teams of students engaged collaboratively in context analysis and research for sustainable design. Three prototype projects being a Community Based Organisation (CBO) community facility, an adaptable classroom and low-cost teacher's house, alongside the introduction of bamboo as an integral sustainable building material were initiated in 2016. This paper evaluated student assessment outcomes and reflected upon partnerships in the context of the collective response to community needs. It concluded that there is academic merit in an immersion unit such as iDiDe to act as a global platform for education for sustainability in Asia. The adoption for an integral sustainable design and construction framework appropriate for rural contexts has been recognized as a way forward and a direction for future research.

Keywords iDiDe · Built environment education · Rural community Sustainability · Intercultural collaboration

S. Ang (✉) · G. Karunasena · R. Palliyaguru
School of Architecture and Built Environment, Deakin University, Gheringhap Street, Geelong, Victoria 3220, Australia
e-mail: susan.ang@deakin.edu.au

W. Leal Filho et al. (eds.), *Sustainable Development Research in the Asia-Pacific Region*, World Sustainability Series, https://doi.org/10.1007/978-3-319-73293-0_12

1 Introduction

iDiDe was established in 2010 in Deakin University as a global mobility study tour that provided architecture and built environment students immersive authentic learning through intercultural collaboration. Strategic partners included non-Australian universities from Malaysia, Thailand, India, Indonesia and Sri Lanka. Over 30 Australian students have been successful recipients of the Australian Government's New Colombo Plan and between 2010 and 2017, over 200 students have gained an iDiDe experience. Feedback from participants indicated deep and transformative learning outcomes. High impact transformative themes identified by Ang (2017) included increased confidence gained first hand through immersion resulting in developed ability to initiate and conduct effective collaboration and communication across diverse cultures and high motivation for the pursuit of further opportunity for travel, work and study in internationalized environments. Ang (2017) identified the organic nature of iDiDe as a defining characteristic that exemplifies the essence of intercultural dialogue and intercultural understanding. Over a period of seven years, the iDiDe model demonstrated student learning outcomes including preparation for professional practice for future graduates of design and built environment. A key value was of enriched and enhanced learning experience of real world scenarios.

This paper explores embedded curriculum and transformative learning about sustainability through intercultural collaboration. The lens reported has been that of a subject for architecture design and built environment students. Firstly, the paper illustrates iDiDe as a pedagogical model for sustainable built environment education. A distinctive feature is the collaborative learning gained from working alongside peers from diverse cultures to achieve a holistic understanding of sustainability. Secondly, it reports on the Building Ampara rural community project and learning objectives and outcomes in the context of iDiDe as a case study of curriculum innovation. At Deakin University, the iDiDe unit is offered for credit towards coursework with assessable tasks. The Unit Learning outcomes (ULO) and Graduate Learning Outcomes (GLO) were achieved through structured activities focused upon issues of sustainable rural community empowerment and building improvements. The research identifies student learning outcomes that embedded lessons in sustainability. Finally, the paper reflects upon the contribution of iDiDe partnerships as part of the collective and concluded that there is academic merit in an immersion unit such as iDiDe to act as a global platform for education for sustainability in Asia.

2 Background and Context of the Building Ampara Project

After the civil war in Sri Lanka ended in 2009, several community development projects were initiated by international agencies. In rural areas, progress and re-development were slower than in the cities due to disparities between community needs and top down policy driven initiatives. Aspirations Education Foundation (AEF), a Sri Lankan NGO organisation delivered sustained effort over a period of nine years to improve conditions for the communities in the Ampara District. AEF's vision and goals closely mirrored issues of empowerment of women, education for children and income generation, identified more roundly in Sustainable Development Goals (SDGs) (2030 Agenda for Sustainable Development, United Nations 2015).

Deakin University's partnership with AEF Building Ampara saw efforts achieved in multi-disciplinary, community development collaboration with several social enterprise initiatives successfully implemented in the rural communities of Ampara (AEF The Ampara Project 2015a, b). The established trust and growing empowerment within the remote and marginalised communities allowed prevalent issues of under development and lack of resources to be well identified. iDiDe Building Ampara considered a whole of project and longer-term vision and recognised limitations and constraints related to local legislation, local government infrastructure policy and governance, legalities of land ownership, funding models, architecture design, community involvement, realisation processes and post construction evaluation. The use of sustainable building materials and innovation of traditional building techniques in the implementation of a model of self-help within the communities acted to enable capacity building in the longer term as a priority. iDiDe Building Ampara aligned itself wherever practical with the vision and mission of AEF and those of SDGs relevant to the context of Ampara and served as the premise of the context of learning for students towards understanding sustainability.

3 Methodology

The Deakin iDiDe study tour program was offered to both undergraduate and postgraduate students in the architecture degree as an elective (optional enrolment through faculty approval) unit for credit towards coursework. Students were from second year level to fourth year level as well as from related inter-disciplinary courses in Architecture, Planning, Landscape Architecture, Building Economics, Quantity Surveying, Facilities Management, and Integrated Design. The unit was delivered over eleven weeks with three assessment tasks worth 20%, 50% and 30% respectively. The four-week duration for the first task acted as a "pre-travel" phase. The three-week immersive study tour travel phase comprised intensive collaborative activities. The unit concluded with a four-week "post-travel" assessment dedicated to reflection. During the immersive collaborative phase, Deakin students worked under intensive study tour conditions (on the move) and on the premises of the Faculty of Architecture at the University of Moratuwa under design studio workshop conditions alongside students from other partner universities (Malaysia and India) to propose the design outcomes. The mix of student participants from multidisciplinary backgrounds added another dimension to the cross-cultural learning experience although the impact of this aspect has not been covered in this paper.

Students were introduced to the philosophy of integrated sustainable eco-design described by Kishnani (2012) and undertook three major assessments: (1) discipline related research report; (2) community engagement in design and development, and (3) self-reflective journal. Students considered integrated sustainable eco-design principles and applications in Building Ampara through the context of sustainable development goals of empowerment of women and education for children, proposing design ideas that would offer the means for long term capacity building for the community, participatory community design and co-building approaches to bridge the gap between top down and bottom up initiatives and ultimately propose built environment solutions that would result in adaptive, flexible, regenerative and resilient building opportunities. The scope of authentic learning included research and context analysis, topographical mapping, identification of local materials, vernacular building typology study, local construction techniques, documentation, collation, reporting, intercultural dialogue, interpersonal communication and cross-cultural and multi-leveled collaboration alongside supervised participation in community engagement activities.

4 Scope of the Paper

The research paper is about university student learning outcomes that provided an evidence based case study of a learning and teaching project that introduced students to the principles of sustainable design through immersion. The scope is an evaluation of Deakin assessment outcomes and a reflection of the overall project outcomes for the community. Non-Deakin students and student experience were not part of the scope of this research. The research outcomes provided evidence that student learning included understanding and awareness of issues relating to sustainability. Immersion and active learning were conducted over two iDiDe study tours to Sri Lanka conducted in 2016 and 2017. In 2016, Deakin's School of Architecture and Built Environment collaborated with Sri Lankan strategic university partner, the Faculty of Architecture at the University of Moratuwa, Aspirations Education Foundation (AEF), Melbourne-based Art Building Children's' Dreams, a Melbourne-based NGO, and the International Islamic University of Malaysia (IIUM). In the second study tour, Manipal University, India, replaced IIUM as a fourth university partner. A total of ninety students from four countries participated in the two iDiDe study tours to Sri Lanka and to the overall Building Ampara project in Sri Lanka.

5 iDiDe Conceptual Pedagogical Model for Sustainable Design Education

Altomonte (2013) explained how sustainable design education could be understood and translated into responsible practice.

> Sustainable design education can be understood in three stages: sensitization, validation and reflection. This can be delivered across undergraduate, graduate and postgraduate degrees, or be condensed in a single cycle of higher or post professional education. Without prescribing an ideal curriculum, these three different stages of learning—from the exploratory, through a propositive onto a critical approach—suggest the progression of abilities in sustainable design that should be attained for responsible practice (Altomonte 2013).

Strategies for interdisciplinary, early implementation and strong student action were identified for education in sustainability for architects (Domenica et al. 2013; Gürel 2010; Altomonte et al. 2014). Studio experiences developed a heightened awareness of sustainability as a multidimensional concept that requires critical thought processes, and greatly influenced the recognition of environmentally responsible design as an imperative in education (Gürel 2010). Actively engaging students in interdisciplinary, service-oriented projects through a design charrette were valuable in teaching concepts of sustainability in higher education (Walker and Seymour 2008). The education of future designers must include opportunities for constructing knowledge through an iterative problem-solving process of reading, assimilating, reasoning, and reflection (Betrabet Gulwadi 2009). The learning context offered is one that recognizes human and non-human interdependence. Universities advocate education for sustainability.

> The process of education would emphasize active, experiential and collaborative learning and real-world problem solving on the campus and in the larger community. For example, as part of the curriculum, the learning experience for students would include working on actual, real-world problems facing communities, government and industry. It would also increase group work and learning so students would be able to effectively collaborate as future manages and leaders on complex problems (Kaur 2016).

Designing and learning from the environment was described as a three-dimensional textbook (Taylor 2009). The iDiDe experience crystalized impacts of higher education forming partnerships with local and regional communities to help make them socially vibrant, economically secure and environmentally sustainable. The iDiDe model offered an interdisciplinary design studio combined with elements of an international study tour. Refer Fig. 1 showing iDiDe as a central platform of learning for collaboration alongside academics, community members, government agencies and industry professionals to actively engage in a process of active and authentic learning. The architecture curriculum at Deakin

University does not explicitly singularise sustainability as subject. Sustainability in architecture and built environment curriculum is a matter of integration (Wright 2003; Bradley et al. 2010).

Educating for sustainability may require a whole of university approach (McMillan and Dyball 2009). Students did not yet understand the tools available to assist their sustainability goals (Iyer-Raniga et al. 2010). Sustainability did not constitute a prescriptive body of knowledge, but instead required the application of broad principles and critical thinking to support a process of continuous learning where the practitioner bears a responsibility to build on and apply their learning from one project to the next (Iyer-Raniga et al. 2010). A transformation meaning a complete re-design of education based on sustainability principles required a paradigm change so that education would be built on learning as change and education as sustainability. In practice, this would mean that the goal of all education would be sustainable development, and the different disciplines and subjects would all contribute to it (Sammalisto and Lindhqvist 2008).

> No country is sustainable or has come close to becoming sustainable. There is no proven recipe for success. This has led to the realization that sustainability is essentially an ongoing learning by doing process that actively involves stakeholders in creating their vision, acting and reviewing changes (UNESCO 2002) … [Hence, there is a] need to challenge mental models rather than merely inform society (Tilbury 2006).

Current literature endorses the critical role universities play in facilitating the active relationships between sustainable knowledge principles and the application of them in practice.

Fig. 1 iDiDe pedagogical model developed and implemented by S. Ang 2016

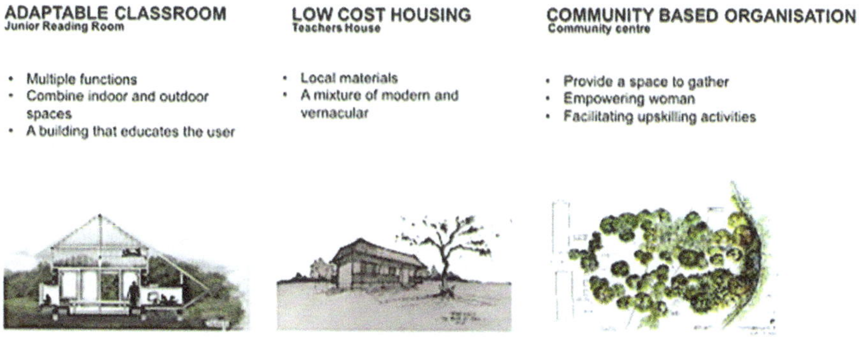

Fig. 2 Building Ampara design concept prototypes produced by iDiDe students 2016

6 Evaluation and Outcomes

Assessment Task 1: Individual Research Report (Pre-travel Phase)
As part of the first assessment, students completed a research task that demonstrated evidence of discipline knowledge, personal skills and experience that contributed towards a team approach for the project. Individual reports defined a challenge/problem within a prescribed research study context theme of Building Ampara and demonstrated understanding of relevant key issues and challenges informed and supported by research. Five major research themes were offered. Refer Table 1 for research themes and topics that students chose to expand upon.
A sample of student initiated topics were tabled (refer Table 2). Research reports prepared under these topics showed clear evidence of enhanced learning of sustainable development through their literature review, bibliographies and discussions of issues.

Students evidenced self-appraisal to determine contributions at both individual and group levels through goal setting that enabled their reflection upon learning and experiences at the conclusion of the unit. The report comprised two sections: (1) Program of investigation that defined the problem; and (2) A personal learning plan that comprises of evaluation of the student's skills and experience that could be applied to the project, and identified personal goals. This assessment scaffolded research in awareness of Sri Lankan social, cultural, environmental and political contexts. This prepared students for field work in rural sites. The outcomes evaluated student achievement of the following unit learning outcomes: (1) Organize and undertake a program of investigation appropriate to the discipline chosen area that responds to the study tour context; (2) Set personal achievement goals and reflect upon action in relation to personal skills and experience goals throughout the

Table 1 Suite of major research topics offered for students to choose and expand upon

Theme 1 Sustainable rural development	Theme 2 Capacity building	Theme 3 Social inclusion in the built environment	Theme 4 Integrated systems	Theme 5 Built environment education
Building methods and practices	Social and cultural dimensions	Accessibility and inclusive design	Integrated design approaches	Collaboration and student engagement in built environment education
Sri Lankan rural contexts Comparative analysis study	Community design and co-building	Equity and diversity	Environmental agendas	Evaluating the iDiDe model for collaboration and student engagement
Participatory approaches	Role of architects/built environment professionals/ architecture	Gender equality	Alternative building materials (e.g. bamboo)	An architectural class's efforts towards social architecture Student involvement in a live community project
	Grass roots activism		Adaptive building design strategies; Resilient and responsive systems	Power of partnerships—academia and community

Table 2 Student research topics related to sustainable development

Student	Research topic
1	A literature review on the origins of capacity building and the approaches to practice in rural communities
2	Understanding capacity building: How do we design communities to build capacity?
3	Holistic sustainable design paradigms to empower sustainable development in rural communities: the case of Building Ampara, Sri Lanka
4	Implications of integrating environmentally adaptable and resilient systems into rural community architecture
5	How impersonal architecture and the deficient client relationship are impeding social and economic progress in Ampara, Sri Lanka
6	The context of social inclusion in the built environment of Sri Lanka
7	Instigating positive environmental change through education in the modern era
8	Integrative, adaptive and sustainable design for Building Ampara
9	How best for NGOs to provide non-biased aid to make delivery of aid more effective
10	Leadership, communication, collaboration and reflection-in-action
11	Bamboo: an alternative to steel reinforcement in concrete
12	Exploring capacity building practices: Sri Lankan rural contexts
13	Instigating positive environmental change through education in the modern era: Issues, responsibilities, education, roles and pathways
14	Integrated adaptable design systems

study tour experience; and (3) Demonstrate ability to enter the debate surrounding engagement with diverse rural communities and cultures through ethical and sustainable approaches. Summative assessment 1 assessed student achievement of the following graduate learning outcomes. GLO8—through student ability to demonstrate understanding and definition of challenges within diverse urban or rural communities and cultures; GLO6—through student ability to evaluate student's own skills and experience and to utilize them to set personal goals. A sample of personal learning goals related to sustainable development identified by individual students in the unit highlighted how these goals responded directly to the aims and objectives of Building Ampara community development project. Refer Table 3 and Image 1 which show an exemplar example of one student outcome for Assessment Task 1.

Assessment Task 2: Group Project (In-Country Travel Phase)

In the second assessment task, students engaged in collaborative group activities undertaken in cross cultural groups in Sri Lanka through active involvement and contribution to study tour project (Building Ampara) over the three-week travel

Table 3 Personal learning goals related to sustainable development identified by individual student

Student	Personal learning goals related to sustainable development identified by individual students
A	• Develop understanding of Sri Lankan culture, history and traditions • Develop communication and understanding between the team and the local community (Ampara) • Develop awareness and exposure to Sri Lanka and the broader Asia Pacific region • Develop flexible leadership skills and applying them in various situations
B	• Building cultural awareness and learning collaborative skills • Improve communication skills beyond Australia • Increasing construction knowledge
C	• Develop interpersonal and teamwork skills • Increase communication techniques and skills • Improve self-reliance and global awareness
D	• Focus on the issues that are critical to the design of the community centre in a manner derived from passion and not from a need to achieve high grades • Find the balance between the future aspirations of architecture and the pre-existing ideals in Ampara, and analyze the importance of both realms in creating social impact • Create a space where others feel entirely comfortable to express their opinion, and always consider the relevancy within all words
E	• Develop skills in inquiry, self-learning and collaboration
F	• Cultural sensitivity • Develop communication skills across diverse stakeholder-ships • Develop design skills related to community needs

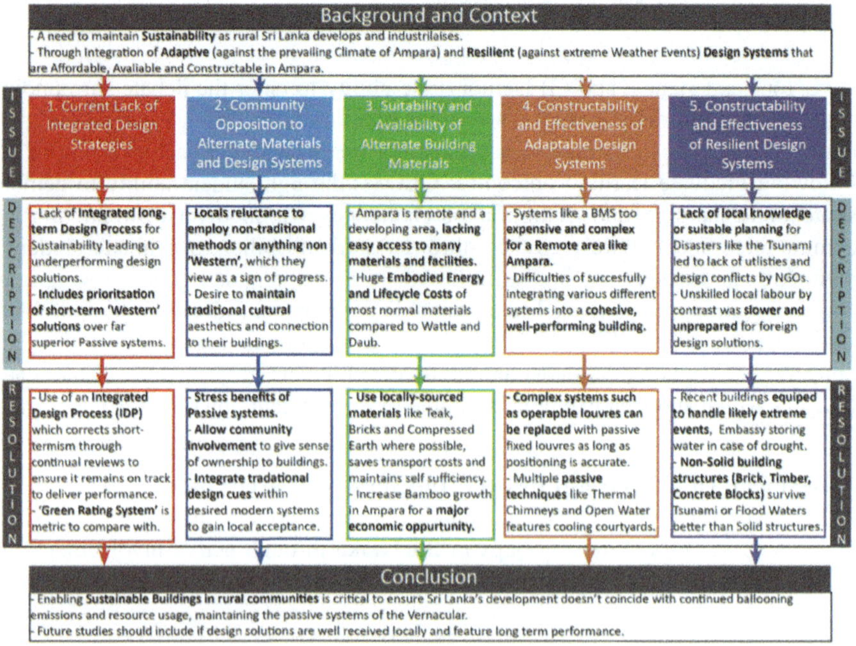

Image 1 Diagrammatic representation of approach to enabling sustainable building in rural communities by J. Argent, Deakin student 2016

period of the study tour. The assessment required students to participate in the following activities through intercultural collaboration and immersive authentic learning:

1. Icebreaker activities that helped scaffold intercultural collaboration.
2. Individual short oral presentation on assigned building of significance that represents the culture, heritage, history, architecture and society of Sri Lanka.
3. Journaling on study tour experience capturing cultural immersive experiences such as local sights of significance and interest.
4. Active involvement to assigned tasks that contribute to community data collection, project development and documentation.
5. Active participation and contribution to overall iDiDe experience.
6. Contribution to self and peer critique.
7. Active participation and contribution to iDiDe Exhibition in Sri Lanka.

Unit learning outcomes for Assessment 2 included: (1) Set personal achievement goals and reflect upon action in relation to personal skills and experience goals throughout the study tour experience; (2) Demonstrate ability to enter the debate surrounding engagement with diverse rural communities and cultures through ethical and sustainable approaches. Summative Assessment 2 assessed student achievement of the following graduate learning outcomes: (1) GLO7—through student ability to engage effectively in teamwork and collaborative learning with others from different disciplines and diverse backgrounds; and (2) GLO8—through student ability to engage collaboratively to demonstrate understanding of challenges and to provide sustainable solutions within diverse urban or rural communities. Image 2 is of students from Deakin University, International Islamic University and University of Moratuwa engaged in a mothers group community forum during the community design process conducted in January 2016. Images 3, 4 and 5 are exhibition outcomes of the group design project produced by Deakin students, Moratuwa students and University of Manipal students. Student responses to Assessment Task 2 were observed to be embedded in submission of Assessment Task 3 in the form of reflection of the summative study tour experience and also upon personal learning goals set out in Assessment 1. Students reported their learning goals through responding to the assessment tasks criteria in the formats of research reports, posters and artefacts as part of an iDiDe exhibition presentation.

Assessment Task 3: Individual Reflective Journal (Post-travel Phase)
In this assessment, the unit learning outcomes were to review how personal achievement goals set in Assessment Task 1 were achieved and a reflection of the study tour experience and project. Students collated and presented learning experiences gained from the study tour. This required students to critically reflect, self-appraise and self-critique their own involvement, contribution and performance in all individual and group activities. Discourse and dialogue that evidenced heightened awareness of self-development across iDiDe philosophies and principles of intercultural competence and designing in diverse contexts were strongly encouraged. Students reflected upon their own engagement within the context of

Image 2 Community forum with mothers of Okkampitiya Primary School, iDiDe student teams, 2016. Photo by Deakin University

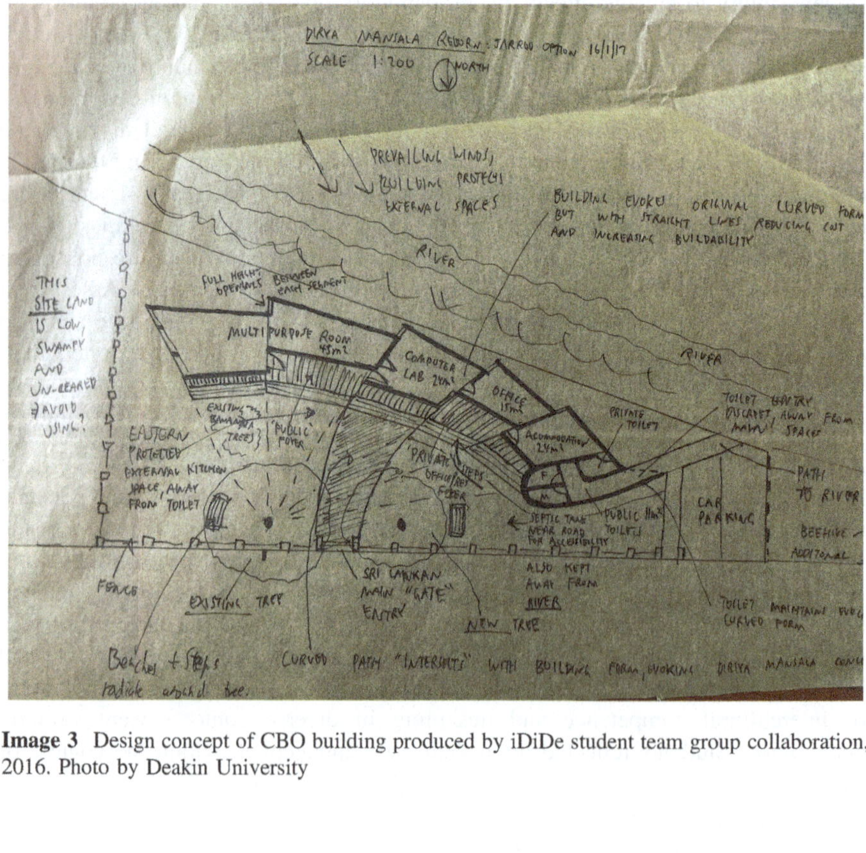

Image 3 Design concept of CBO building produced by iDiDe student team group collaboration, 2016. Photo by Deakin University

Image 4 Render of conceptualized CBO prototype building for Aranalu Community, Polwatta, Ampara, Sri Lanka, assessment outcome 2, iDiDe student team group collaboration, 2016

Image 5 Physical model of conceptualized CBO prototype building for Aranalu Community, Polwatta, Ampara, Sri Lanka, assessment outcome 2, iDiDe student team group collaboration, 2016

their peers. Assessment Task 3 assessed student achievement of these graduate learning outcomes: (1) GLO6—through student ability to demonstrate understanding and definition of challenges within diverse urban or rural communities and cultures; and GLO8—through student ability to engage collaboratively to demonstrate understanding of challenges and to provide sustainable solutions within diverse urban or rural communities.

One student reflection showed strong consideration of sustainability related to the Building Ampara project:

> Designers need to disengage themselves with physical aesthetics, and opt for architecture inspired by the human potential surrounding them. If this connection is made, and the architect works communally with the people, entire groups of Ampara will be positively influenced. Socially responsible architecture has the potential to engage local professionals, teach people new skills and ignite self-worth. It is imperative that architecture reconnects with humanity and utilizes participative measures to ensure the needs of the community are met. This is turn will generate the sense of personal accomplishment and encourage Ampara to work towards a strong social and economic future. (Extract from Deakin student G. Spencer, Deakin student, 2016)

7 Reflection of Outcomes for the Community and the Role of iDiDe Partners

Outcomes of the first study tour included a proposal for integrated sustainable vision for community development in Ampara, and a vision to implement an integrated approach to sustainable building that could facilitate capacity building. Four key principles were identified:

- Participatory community based feasibility studies and concept design development.
- Co-design and co-building techniques with community input and involvement.
- Use of local materials, local techniques, references of vernacular with new vision and modern applications as first option.
- Introduce the use of bamboo as a building material with longer term intentions of instigating higher levels of government agency for a bamboo industry. (Currently, the use of bamboo in most of Sri Lanka was recognized to be limited to craft purposes).

The design outcomes of the community co-design process included three prototype community buildings—Adaptable Classroom, Low Cost House and a Community based organization facility (CBO building). Refer Fig. 2.

This was underpinned by on-ground research conducted that indicated how design and construction processes were experiencing substantive disintegration of collaboration between stakeholder groups and community needs, resulting in non-sustainable processes and outcomes. All university partners endorsed inter-disciplinary and integrated framework for acquiring "truly" sustainable buildings in the rural contexts as a sustainable approach. Community forums with stakeholders conducted with several communities resulted in successful adoption of co-design and co-building approaches. The full scope of iDiDe experiences cannot be covered in this paper. However, a small translated vignette of a report of a visit to Okkampitiya Wijayabahu Primary School in January 2016, highlighted issues uppermost on the community's agenda.

> The School Principal and the parents of the school children expressed their thoughts on the roles and possibilities for their CBO building, "Diriliya Mansala". The community came up with the word "liya" in "Diriliya" as reference to empowering women. Diriya Mansala, was the Sinhala meaning for "a place of strength and courage to gather". The community spoke of a variety of activities that could empower young mothers as well as perform many economic, social and cultural activities. Many mothers hung around and outside the school premises for long hours waiting for their children to finish school. The community expressed desire to conduct social and educational activities to empower and initiate income generation. "We want to learn anything that will help us earn income", "We can sell the things we make" and "we want to teach younger mothers to be independent". The centre could promote health and nutritional value of village based traditions for children and adults in line with the concept of a community kitchen; prepare mid- day meal for school children with nutritional vegetables.

All partners subscribed to collaborative, cohesive and open planning processes. Each partner was integral to the success of the iDiDe event and crucial to facilitating education for sustainability. Universities forming partnerships with local and regional communities help make them socially vibrant, economically secure, and environmentally sustainable (Cortese 2003). Equally, input from academics who lead the study tour and the project cannot be underestimated. "Capacity-building of educators" must be considered the cornerstone of transforming universities to better empower students to become change agents for sustainability (Edwards 2004). Sri Lankan partners were instrumental for their intrinsic knowledge of local customs, culture, practice and language. In the months leading up to Ampara visit, Deakin University staff and students raised funds towards the procurement of building materials and community labor costs towards the proposed CBO Building for the Arunalu community in Polwatta. Future involvement will see a lead role played by local community in the implementation of the project. New theories of holistic sustainable design have empowered stakeholders to readdress their perspectives towards ecological outcomes. Universities have a critical role in improved environmental and sustainable literacy (Cortese 2003; Vanegas 2003), and to make a contribution towards improving architecture education in sustainability. Building Ampara saw the successful adoption of a design proposal by the CBO Community leadership group, site survey and context analysis, material study and recommendation for sustainable innovated applications of a local material palette along with environmentally appropriate construction techniques. Adjunct projects successfully delivered by the students included design build of several childrens' swing projects gifted and installed in five schools and building consultant appraisals for future building improvements and facilities management.

8 Conclusion and Further Research

This paper evaluated student assessed learning of sustainability through engagement in a collaborated event that involved all levels of local government and local community stakeholders (schools, teachers, parents, students and families). It positioned academia as an objective agency to facilitate local and international partnerships focused upon promoting and delivering a sustainable education and research agenda. The adoption of an integral sustainable design and construction (ISDC) framework appropriate for rural contexts has been recognized as a direction for the way forward. iDiDe Sri Lanka Building Ampara initiated stakeholders to buy-into the project and band together to achieve sustainable community development. In coherence with the concept of 'Spatial Agency', architects must recognize that "the most valuable commodities are the giving of time, the making of time and the not giving up" (Awan et al. 2013). Architecture and the subsequent construction process have the potential to unite the factions of the Ampara community through localized involvement. The potential to eradicate the inequality experienced by religious, gender and socio-economic minorities, can be initiated through prototype buildings that highlight the communal needs of regions alike Ampara. There is clear evidence to conclude that iDiDe offered a robust sustainable learning platform for education in sustainability. Further research that will be helpful will be evaluation of non-assessable student outcomes and participant experiences.

Acknowledgements of Contribution by Partners and Well-Wishers The authors wish to acknowledge and thank all students, academic colleagues, stakeholders and well-wishers who contributed to iDiDe since 2010 and specifically the Building Ampara Project (January 2016–January 2017), all of whom have provided the inspiration and purpose for this research.

References

AEF. (2015a). Aspirations Education Foundation. http://aspirations.edu.lk/foundation/deakin-university-australia/. Last accessed January 10, 2017.

AEF. (2015b). Aspirations Education Foundation. http://aspirations.edu.lk/foundation/who-we-are/. Last accessed August 29, 2016.

AEF. (2016). Aspirations Education Foundation. https://www.facebook.com/permalink.php?story_fbid=1277237768959264&id=1058264917523218. Last accessed April 12, 2017.

Altomonte, S. (2013). In the Architects Journal, "Five views on sustainability in architecture education". https://www.architectsjournal.co.uk/news/students/five-views-on-sustainability-in-architectural-education/8651398.article. Last accessed April 03, 2017.

Altomonte, S., Rutherford, P., & Wilson, R. (2014). Mapping the way forward: Education for sustainability in architecture and urban design. *Corporate Social Responsibility and Environmental Management, 21*(3), 143–154.

Ang, S. (2017). Intercultural dialogue through design (iDiDe): A model of intercultural collaboration and student engagement. In R. Tucker (Ed.), *Collaboration and student engagement in design education* (Vol. 11, pp. 230–256). IGI Global.

Awan, N., Schneider, T. & Till, J. (2013). *Spatial agency: Other ways of doing architecture.* Routledge.

Betrabet Gulwadi, G. (2009). Using reflective journals in a sustainable design studio. *International Journal of Sustainability in Higher Education, 10*(1), 43–53.

Bradley, J. F., Sayce, S., & Lewis, A. (2010). Sustainability and built environment professionals: A shifting paradigm. In *Sustainability education: Perspectives and practice across higher education* (pp. 257–272). London: Earthscan Publications.

Cortese, A. D. (2003). The critical role of higher education in creating a sustainable future. *Planning for Higher Education, 31*(3), 15–22.
Domenica, I. L., Gorby, C., Poerschke, U., Nickolas Kalisperis, L., & Woollen, M. (2013). Environmentally conscious design–educating future architects. *International Journal of Sustainability in Higher Education, 14*(4), 434–448.
Edwards, B. (2004). Sustainability and education in the built environment. *The Sustainability Curriculum: The Challenge for Higher Education*, 129–140.
Gürel, M. Ö. (2010). Explorations in teaching sustainable design: A studio experience in interior design/architecture. *International Journal of Art & Design Education, 29*(2), 184–199.
Iyer-Raniga, U., Arcari, P., & Wong, J. (2010). Education for sustainability in the built environment: What are students telling us? In C. Egbu (Ed.), *Proceedings of 26th Annual ARCOM Conference* (pp. 1–10), Leeds, UK.
Kaur, R. (2016). Education for sustainability. *ACADEMICIA: An International Multidisciplinary Research Journal, 6*(3), 169–173.
Kishnani, N. (2012). *Greening Asia: Emerging principles for sustainable architecture*. Singapore BCI Asia Construction Information Pte Ltd.
McMillan, J., & Dyball, R. (2009). Developing a whole-of-university approach to educating for sustainability. *Journal of Education for Sustainable Development, 3*(1), 55–64.
Sammalisto, K., & Lindhqvist, T. (2008). Integration of sustainability in higher education: A study with international perspectives. *Innovative Higher Education, 32*(4), 221–233.
Taylor, A. (2009). *Linking architecture and education: Sustainable design for learning environments*. UNM Press.
Tilbury, D. (2006) *Environmental education in Australia*. Paper prepared for the 2006 Australian State of the Environment Committee, Department of the Environment and Heritage, Canberra. http://www.deh.gov.au/soe/2006/emerging/education/index.html. Last accessed December 09, 2016.
United Nations Sustainable Development Agenda. (2015). https://www.weforum.org/agenda/2015/09/what-are-the-sustainable-development-goals/. Last accessed April 04, 2017.
Vanegas, J. A. (2003). Road map and principles for built environment sustainability. *Environmental Science & Technology, 37*(23), 5363–5372.
Walker, J. B., & Seymour, M. W. (2008). Utilizing the design charrette for teaching sustainability. *International Journal of Sustainability in Higher Education, 9*(2), 157–169.
Wright, J. (2003). Introducing sustainability into the architecture curriculum in the United States. *International Journal of Sustainability in Higher Education, 4*(2), 100–105.

Author Biography

Susan Ang is a Senior Lecturer and Founding academic leader of iDiDe at the School of Architecture and Built Environment, Deakin University Australia. Susan has extensive experience in teaching and researching in architecture and built environment education. Prior to academia Susan was a registered practicing architect for 15 years. Since 2010, Susan has led the establishment of academic partnerships with strategic university partners between Australia and Asia. In 2011, Susan's Master of Architecture by research thesis was published as a book, titled, "Valuing Practical Experience: Enhancing student learning outcomes through work integrated learning in combination with intercultural experiences" by Lambert Academic Publishing, GmBH & Co., Saarbrucken, Germany. Susan's research interests resides in the intersection of architecture and social health and development. Susan has been involved in collaborative research in inclusive design and social inclusion in built environment design and education for more than 7 years. Susan was part of the team awarded "SMART Researcher Award in 2012 by the City of Greater Geelong for outstanding contribution to this field. Susan is a lead collaborative researcher that seeks to evaluate best practice applications of collaboration in design education, universal design in the built environment and the development of an integrated sustainable design framework for sustainable rural contexts.

'Sustainability' After Disaster: Confronting the Complexities of Recovery in the Field: An Educational Experience

Tim Nichols and Judy Rogers

Abstract This paper focuses on the learning and teaching approach adopted in an undergraduate landscape architecture design studio at RMIT University that reveals some of the challenges involved when design students are asked to confront and work within the broad area of disaster response and recovery. Departing somewhat from the traditional approach often employed in design studios where students work individually or in small groups on a 'problem' with the aim of designing a 'solution' the studio approached the problem(s) of disaster recovery by focusing on collaborative 'problem(s)' definition. The focus of the studio was on the small township of Marysville in the state of Victoria, Australia that was almost completely destroyed in the Black Saturday fires of February 2009. Eight years on visitors to the town are delighted by the' appearance' of a town restored, however underneath the town remains vulnerable on a range of different fronts. The case study begins to reveal many of the pitfalls for designers and other professionals who are called on to work in post disaster environments, and the, often, narrow perspective about what constitutes 'sustainability' and 'recovery' that can lead to an overemphasis on 'things' at the expense of long-term processes of change towards what is an increasingly complex and unknown future.

Keywords Disaster recovery · Sustainability · Design after disaster Vulnerability · Curriculum development

1 Introduction: Black Saturday

On 7th February 2009 the State of Victoria, Australia experienced a disaster that was quickly to be named 'Black Saturday' (see Fig. 1). On that day fires erupted across the state, fuelled by high temperatures, strong winds, low humidity and the consequences of a 10-year drought that resulted in very high and very dry fuel

T. Nichols · J. Rogers (✉)
School of Architecture and Design, RMIT University, Melbourne, Australia
e-mail: judy.rogers@rmit.edu.au

W. Leal Filho et al. (eds.), *Sustainable Development Research in the Asia-Pacific Region*, World Sustainability Series, https://doi.org/10.1007/978-3-319-73293-0_13

loads. Given the conditions, the fires travelling at speeds of up to 50 km per hour, a situation that caught many communities by surprise (Teague et al. 2010). The fires resulted in 173 fatalities, 2133 homes destroyed, 78 townships affected and over 4500 km^2 burned. The fires left an estimated 7500 people homeless (Rural Australia Institute Australia (RAI) 2013). The scale and the impact of the fires were unprecedented (Teague et al. 2010), presenting enormous challenges for Government, emergency services and Relief Agencies.

On the 8th February, 2009 the Australian Government's Disaster Response Plan was activated and the Victorian Bushfire Reconstruction and Recovery Authority (VBRRA) was established two days later on 10 February 2009 'to oversee and coordinate the largest recovery and rebuilding effort Victoria has ever faced' (VBRRA 2010). Significantly, the slogan adopted by the Authority from the outset was 'we will rebuild'.

Not long after, on the 16 February, the Victorian Bushfires Royal Commission was established to investigate the circumstances that led to the enormous loss of life and property. The Commission's terms of reference included causes and consequences, the preparedness and the response of government, emergency services and households; measures taken to prevent disruption to essential services; and any other matters deemed relevant by the Commissioners (Teague et al. 2010). The Commission heard evidence from 434 experts and lay witnesses and received more than 1200 public submissions. It handed down 67 recommendations to government in its final report, including policy recommendations for bushfire safety, ignition prevention, emergency and incident management, land use planning and building, land and fuel management, and organisational structures (Teague et al. 2010).

In 2012 three years after the fires class actions were launched in the Supreme Court that alleged that the fire was caused by faulty power lines. These class actions

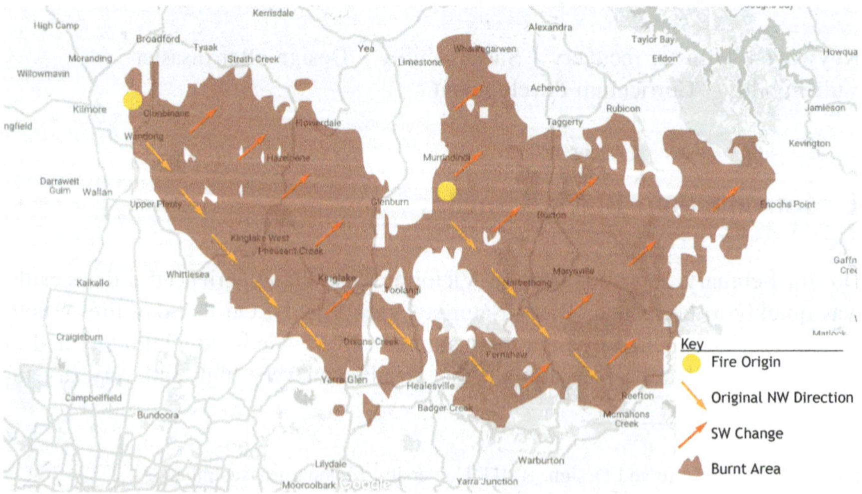

Fig. 1 The case study area

were settled out of court in 2017. Many issues however remain unresolved and recovery is on going.

2 Marysville: A Case Study

Amongst the towns devastated by the fires the small rural township of Marysville was almost totally destroyed. The fires killed 39 people in the township and surrounding area, and destroyed or damaged more than five hundred buildings. In addition, the commercial Centre of the town was destroyed, as was the core of the town's economic activity in tourism and hospitality. Public infrastructure, including the police station, primary school, kindergarten and health clinic, was also destroyed. The fires also led to the extensive displacement of the local population and 8 years on only 56% of houses have been rebuilt. The economy of Marysville and the Triangle was, and continues to be based on the natural attractions of the area and tourism. However, the fire caused the closure of key tourist sites including Lake Mountain a nearby Ski resort that was extensively damaged. There was also the loss of accommodation for tourists.

In the immediate aftermath of the fires action was taken quickly to clear the town so that rebuilding could begin. Recovery after disaster is not, however, a seamless, linear process and the consequences of a singular focus on rebuilding exacerbated many of the issues facing the town. As Leadbeater (2013) so clearly notes "input rebuilding' does not equal 'output recovery'. Amongst the many issues facing the town is what the RAI labels the reconstruction 'mirage' (RAI 2013) where reconstruction offered a short lived economic boom for the town and where the focus on 'things' led to overinvestment in infrastructure that is difficult to maintain and does not serve the long term needs of the community (RAI 2013). This tension to provide tangible outcomes or things was clearly identified in the VBRRA (2011) Legacy Report that acknowledged:

> Tight timelines were imposed by government objectives, including the rapid allocation of funds held in the Victorian Bushfire Appeal Fund and the desire to show early and visible progress in reconstruction. However, for fire-affected communities, the complexity and bureaucratic nature of the processes imposed by government, the difficulty in determining priorities and gaining broad community support for the plans were serious issues.

One example of the haste to rebuild is the Urban Design Framework for Marysville and the Green Triangle that was released on the 11th November 2009, 10 months after the fires (Roberts 2009). The Framework identified a 'unique opportunity to rebuild Marysville as a safer, more sustainable town: a town with a distinctive character and a high quality environment, creating a memorable place to visit" (Roberts 2009). Based on what was claimed to be 'extensive discussions and consultations' which included a celebration day three months after the fire, a 3 day workshop, three community meetings and opportunities to provide feedback, the objectives of the UDF were to:

- Strengthen the economic, social and environmental sustainability of Marysville and Triangle communities
- Support the rebuilding of Marysville as an attractive, memorable place with a distinctive character
- Facilitate a viable business and community services core in Marysville to underpin a vibrant local economy
- Improve accessibility and connections within the Triangle and within Marysville
- Protect and enhance the natural beauty of the environment and the high quality of water
- Build resilience to bushfire threat
- Facilitate a mix of housing options (Roberts 2009, 15).

Future looking in its intent the Urban Design Framework gives scant attention to the enduring impact of the fires on the township and focused instead on the physical rebuild. Yet eight years on while some essential infrastructure and buildings have been rebuilt the town remains vulnerable on a range of different fronts.

The 'tension between the need for people to achieve a level of personal recovery and the urgency to provide a framework for rebuilding a town centre that is largely destroyed' (VBRRA 2011, 180) was also identified in the VBRRA legacy report. While the authority adopted a community led approach to reconstruction clearly some members of the community were more able to 'lead' than others. It would seem that pressure from outside to rebuild was also a, if not the, major factor setting the tone of urgency that led to the rebuilding of some things at the expense of others.

One of the key ambitions of the Urban Design Framework was to ensure that Marysville would be 'sustainable'. What this entailed in practice was the construction of 'things' or what were then current fads in the field of sustainable design—particularly water sensitive urban design. Marysville is replete with rain gardens, soft landscaping, solar panels—the 'what' of sustainability. These are all travelled 'things' brought from outside that signal environmental sustainability at the expense of tackling the more difficult questions of sustainability; how does the recovery process promote and support long term sustainability, and sustainability of what and for whom?

When viewed as a whole and considering the broader understandings of sustainability that are missing from the framework, these 'things' are solutions to the 'low-hanging fruit' problems of urban design. Speaking of the architectural profession, Till (2010) states that the issue is not so much with the utility or merit of these 'things' themselves, but the framing of the problem that leads to their realisation as the solution:

> I have argued elsewhere that is it a value system that is still in thrall to the basic tenets of the modern – the will to order, to control, to stamp out contingency. The construction of a better world is thus too often associated with the construction of an ordered world. While these tenets may have a 'self-evident' logic in their initiation (self-evident because self-determined and so self-fulfilling), their consequences are perhaps less edifying; indeed violent. Think of Captain Ahab, that harbinger of modernity in Melville's Moby Dick. - "all

> my means are sane, my motive and my object mad" – and then think of an architect. My means – let's say the drains in the planned quarter – are indeed sane, but my objects – those segregated and segregating buildings – are quite mad. Who can argue with drainage? But that invisible logic masks the madness of the effect (Till 2010).

Without wishing to single out rain-gardens and solar panels too much as they are well intentioned on their own, the question is whether the environmental sustainability that these features may achieve is worth the expense of an investment in money, time and energy that could have otherwise focused on how to build greater *sustainment* in the town. They also raise questions about whether these 'things' merely represent 'sustainability' without people and ecologies (because these seem to be largely absent) and what is the relationship between sustainability and recovery? As Proudly (2013) suggests

> How is recovery measured and who judges what constitutes a complete or successful recovery? When lives are disrupted so profoundly it takes time to process the trauma and weave it into our life story. The event then forms, to varying degrees, part of the person's identity in the present and into their future. Recent anniversaries of two severe bushfires in Australia – 30 years since Ash Wednesday, 10 years since the Canberra fires – remind us that recovery (or healing) is an ongoing process.

Indeed there is growing evidence in Marysville of the reconstruction 'mirage' described by RIA, with Council highlighting growing financial stress in ongoing maintenance and upkeep of costly investments, including town centre landscaping works (RIA 2013). Investment in the underutilized and costly to maintain infrastructure is one consequence of a problem/solution approach to design without adequate definition of the 'problem' itself. What is also left out is an understanding of the memory of the place and the length of time needed to recover and move on. As Leadbeater 2013 suggests

> it is imperative that recovery agencies and government at all levels acknowledge that disasters do not happen in a vacuum. Prior to being impacted, every community has existing values, networks, projects, relationships, knowledge, and capacity that underpin its day to day operations and indeed, it's very identity. But the imposition of externally constituted and 'templated' recovery models can seriously undermine inherent community resilience (Leadbeater 2013, 46).

The impacts of the Black Saturday bushfires on the township of Marysville are therefore many, varied and deeply complex. In the haste to rebuild and to make things 'right' again key considerations that were acknowledged by VBRA but left out included the capacity of the 'community' to contribute to and be involved in the 'community consultation' processes that were so quickly established, an understanding that a local 'community' is not necessarily uniform and nor will they necessarily 'agree', and that community led recovery is far from straightforward (RIA 2013, 41). The somewhat narrow perspective adopted of what constitutes 'sustainability' also led to a lack of integration with an overemphasis on 'things' at the expense of long-term processes of change towards what is an increasingly complex and unknown future. Also set aside was an acknowledgement of the need for long term thinking (recovery) as well as short term (response) and perhaps an acknowledgement that the pre-existing conditions in Marysville could not be 'restored'. Somewhat ironically, it appears that the haste to rebuild 'brick by brick, school by school' deprived the communities of taking the time to reflect on and document what the character of the town was.

3 Sustainability and Recovery as Process

Within the context of Sustainability and Recovery in post disaster contexts, an approach to design can be described across a spectrum between two extremes: the rational or modern approach ('problem/solution') and the 'liquid' model as identified by Till (2010).

In the rational model the student starts with the identification of a problem and then proceeds through a number of pre-determined problem solving steps to achieve and document a solution. These typically involve analysis, research, conceptualisation, design development, and presentation. In this case the 'problem' is predetermined. This model is useful when a clear problem can be identified that requires a largely functional response. Further along the spectrum is a variation of the rational model where the problem is identified as part of the pre-determined problem-solving steps, usually during analysis and research phases. Whilst this places a greater emphasis on the definition of the problem, there is the risk that this process tends to identification of simpler 'problems' that lead to 'self-evident' solutions rather than the consideration of more complex 'problems'.

These two processes promote arrival at 'pre-established truths', which is of utility in functional applications where a problem is pre-established and universally understood. However, with greater complexity there is need for a model that puts greater emphasis on the definition of the problem, and requiring the designer to hold off on their natural tendency toward problem solving.

Manzini et al. (2008) see the designer's role as working with the fluidity of social systems to enable realisation of projects and deliverables through social innovation rather than focusing on the actual production of objects and 'things'. In this model the designer's objective is to empower others through the sharing of design knowledge. The designer must immerse themselves in the social context in order to 'liberate the desires and enable the needs of others' (Till 2010).

This process of social innovation and empowerment is a radical departure from the notion of the designer operating as the expert, and also represents a shift in thinking from notions of participatory design where communities are invited to work *with* designers on problem-solving. Till takes this approach of working with the fluidity to define 'liquid design', which instead positions the designer as a 'catalyst to release the latent spatial intelligence of others' (2010).

To empower others to develop design solutions, the problem cannot be re-determined but instead must arise from the same processes of social innovation. The social innovation Till refers to is the actions taken by communities navigating within the socio-political and economic structures, with the designer acting as a facilitator or 'catalyst'. In this sense the designer is positioned as a 'trusted outsider' (Edmunds 2012). However, this level of trust is not established simply through expertise, but is earned through a process of listening that Anderson et al. (2012) describe through their work with recipients of international aid:

> Listening is challenging. It takes time and energy, it demands attention and receptiveness, and it requires choices. Listening at both the interpersonal level and the broader, societal level is a discipline that involves setting aside expectations of what someone will say and opening up, instead, to the multiple levels at which humans communicate with each other.
>
> At the interpersonal level, one needs first to be quiet long enough to let the other person talk (a practice that is difficult for some of us!). Then one needs to ask questions and probe the ideas offered rather than interject one's own opinions and analyses or jump to quick conclusions about what the other person means. A listening conversation is distinct from an interview. It opens space for dialogue on issues of importance to both parties. The act of listening is a way of showing respect.

Anderson et al. (2012, p. 142) conclude that working in this manner requires early investment (in terms of both time and funds) in listening on a deeper level. This act of listening allows the sharing of knowledges, which at first is an opportunity for the designer to immerse themselves in the situational specificities and adopt local knowledges, before finally sharing design knowledges to assist communities to achieve their goals. It is only once the specific goals or projects are established (by others) designers are able employ problem solving skills to assist in the delivery of projects.

However, the wider policy, polity and economic settings within which designers operate in the west rarely allow for the timeframes and discourses required at the far end of this spectrum (De Weaver and Lloyd 2005). Political promises, such as Kevin Rudd's commitment to rebuild 'brick by brick' (Nicholson and Rood 2009), and the need to be seen to be 'doing something' imparts short timeframes. Additionally, the long timeframes required with an approach such as the liquid model rarely fit within consultancy fee budgets. One tactic designers can employ is to leverage off (or stitch on to) shorter term design processes to build trust, relationship and capacity within communities to feed this into future design processes, so that over time the design responsibility is led by the same communities that it intends to serve.

So how does the liquid model of design fit within the disaster recovery process? Immediate disaster response typically involves urgent timeframes that are not amenable to the processes of deep listening and knowledge sharing required in order to drive design response under this model. These initial responses are typically functional applications for lower order needs (in terms of Maslow's hierarchy) such as emergency or temporary shelter. In these cases the role of design can be purely functional and employ rational models of problem/solution. Though as the response and recovery process progresses, the priorities shift from lower order and more urgent needs to higher order needs such as rebuilding homes or the town's character, but also to the bigger questions surrounding the ongoing security and resilience of the town and its residents. It is in this ongoing process of recovery that genuine listening and facilitation skills of a designer utilising the liquid model are required.

Therefore we posit that a designer employed at the initial response and recovery stages of the recovery cycle must act with a long-term focus and commitment. Involvement in initial projects offers an opportunity to initiate conversations that are absolutely crucial to establishing the relationships required to transition to the position of a 'trusted outsider' able to facilitate greater empowerment through ongoing design scenarios. Once involved in these areas and recovery processes a designer must commit to ongoing involvement in that place, to ongoing conversations and development of knowledges, and crucially to learning from their mistakes as they continue, and involvement that is iterative and leverages of smaller and more urgent projects to foster relationships and capacity to empower communities to play a greater role in recovery through design projects.

So if designers need to adopt a longer term commitment and approach to the disaster recovery cycle, how can this approach be taught in design studios in a single semester? To us, the first step was to highlight the short-comings of the problem/solution approach as evident in the Urban Design Framework and to encourage students to have to confidence to hold off on the impulse to problem solve a design solution.

4 The Design Studio

Amaratunga and Haigh (2011) have argued that the post disaster period offers a window of opportunities to address vulnerability in a community but that it is an opportunity often missed because well planned reconstruction takes time. They argue

> …there is a window of opportunity, but it is beset with challenges. A pragmatic approach to the development of a resilient built environment needs to include an understanding of these difficulties and their implications for what can actually be done, at least in the short term. While the humanitarian efforts are frequently a rushed process, effective rebuilding for resilience will require reflection, discussion and consensus building. This should not undermine the importance of starting this process early in the recovery phase; indeed, a failure to consider long-term reconstruction goals early in the recovery can lead to wasted or misguided effort, as well as undermine efforts for future resilience.

It is this missed opportunity that the studio set out to grapple with and to work with the complexities involved. However, such a level of complexity and the need to take time presents a range of challenges for a design studio where the focus is often on developing a solution to a 'problem'. A different approach is clearly required that demands a pedagogy that shifts the role of the teacher away from 'expert' to that of facilitator (Manzini 2008) or as Davis and Sumura (2007) suggest 'what teaching is can never be reduced to or understood in terms of what the teacher does or intends. Rather, teaching must be understood in terms of its complex contributions to new, as-yet-unimaginable collective possibilities'. They argue

> Teaching here is more about a conscientious participation in expanding the space of the possible by creating the conditions for the emergence of the not-yet-imaginable, rather than about perpetuating entrenched habits of interpretation (or even exploring the limits of current imagination). Teaching, like learning, is not about convergence onto a pre-established truth, but about divergence – about broadening what can be known and done. In other words, the emphasis is not on what is, but on what might be brought forth. Teaching thus comes to be a participation in a recursively elaborative process of opening up new spaces of possibility while exploring current spaces.

The studio began with two field trips to Marysville. However, rather than provide students with background information to understand the impacts of the disaster students were asked to firstly curate day one of the field trip and secondly to become involved in a working bee at a makeshift museum in Marysville. Curation of the field trip involved allocating groups of students to different locations in fire affected towns (see Fig. 1) and asking then to tell the story of the place at each stop. For their site each group was asked to consider existing condition through the lense of the Disaster management cycle from preparedness—or pre-fire conditions through to recovery. They were specifically asked to address the following questions:

- How prepared was the community pre-fire?
- What were the prevailing conditions that led to the fire? (consider here vegetation, available resources, access, slope.
- How was the site and the community impacted by the fire?
- What steps were taken in the immediate aftermath of the fire?
- How was the recovery process managed/how did it unfold?
- 8 years on is the community less vulnerable? More resilient?

This allowed each group to understand the conditions of the disaster management cycle specific to their locality, and through the process of curation to develop a personal connection to the locality. Additionally, the sharing of these findings allowed for the comparison between other groups findings to draw out common threads, and those that are situationally specific.

The second key aspect of the field trips involved two working bees at the Phoenix Museum, currently housed in the old Marysville Bowls Club at the bottom of town (Fig. 2). The museum was developed as a discrete opportunity for visitors to connect with the stories of the fires, whilst at the same time taking the stress off local traders and the Visitor Information Centre about the 2009 fires. It opened on

Fig. 2 Working bee at the Phoenix Museum

the 6th March 2010 in a portable building, and is a repository of memories, visual stories, objects left from the fires. The ambition for the museum is bushfire awareness and is currently run by volunteers, with growing interest from bushfire education school groups. Some but by no means all of the objects salvaged from the fires are on display in the museum or have or will be returned to their owners when they are ready to receive them. Through the working bees at the museum students were fortunate to be exposed to some of these stories not through direct oral engagement but through engagement with the objects and the memories themselves. This provided for the development of the students' personal understandings of the impact of the fires.

One final aspect of the studio is an acknowledgement that in Disaster scenarios no one person acts alone, nor can a single designer, working alone come up with a solution to a complex set of problems. The studio therefore initially adopted a collaborative approach to problem identification where all 37 students worked as a group to research the context of the fires. Students were divided into 6 groups to prepare a joint background report for mid semester (Fig. 3). The report titled—*What is the present context of Marysville and how did we get here?* The report explored the impact of Black Saturday on the township of Marysville through all phases of the disaster management cycle. This was followed by a class discussion on the 'problems' and possible ways of intervening using small catalectic moments. In order to facilitate this the studio drew on the metaphor of urban weaving (Till

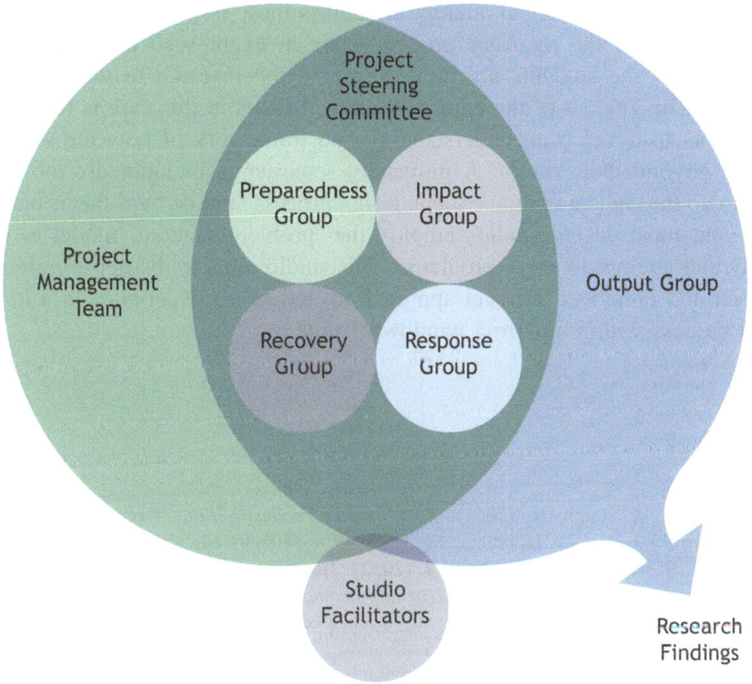

Fig. 3 Structure of the studio

2017) to allow places for intervention that did not involve large-scale visions of what Marysville could be in the future. And so while the primary goal of the studio was to contribute to greater understanding of the role of problem identification and definition in design, a secondary aim is to begin to 'stitch' the urban design framework into the study area through small, locally specific interventions.

5 Conclusion

There is no doubt that designers will increasingly be called upon to work in disaster response and recovery into the future. Within this context 'sustainability' needs to become more than a leitmotif but rather as a framework for thinking about and intervening to address the long-term processes of recovery and resilience. How then do we as educators equip students with the necessary skills to work in such environments in a way that acknowledges, engages with and honors what has gone before? How do we foster the confidence to step back from problem/solution approach and instead take the time necessary to develop both a nuanced understanding of the problem as well as a design approach that empowers those involved?

This studio was one small attempt to address these issues. First it demonstrates the complexity of the recovery process from an event such as black Saturday. Following this the students are invited to lead the research to identify a/or the 'problem'. The key act of the educators as facilitators in this case is to work with the students to develop an understanding and framework of how the project can progress beyond their studio. A university semester is undoubtedly too short in which to go through an empowerment design process, and perhaps this is part of the reason that most design studios employ the 'problem/solution' model to rapidly identify and attempt to solve problems. This studio aims to disrupt this short-term approach and employed a novel approach to deliver an experience in a long and patient process within a narrow window of time.

References

Amaratunga, D., & Haigh, R. (2011). *Post-disaster reconstruction of the built environment: Rebuilding for resilience*. Hoboken, NJ, USA: Wiley-Blackwell.

Anderson, M. B., Brown, D., & Isabella, J. (2012). *Time to listen: Hearing people on the end of international aid CDA collaborative learning projects*. Cambridge: Massachusetts.

Davis, B., & Sumura, D. (2007). Complexity science education: Reconceptualising the teacher's role in learning. *Interchange, 38*(1), 53–67.

De Weaver, L., & Lloyd, D. (2005). The language of community engagement in a regional and indigenous context. In *Proceedings of International Conference on Engaging Communities* (pp. 1–15). Brisbane: Queensland Department of Main Roads.

Edmunds, M. (2012). A new story – Roebourne: A case study", Chapter 12. In B. Walker (Ed.), *The challenge, conversation, commissioned papers and regional studies of remote Australia* (pp. 152–180). Desert Knowledge Australia, Alice Springs.

Jeremy, T. (2010). Urban Weaving' in *Collected Writings* First published as Till, Jeremy 2010. In B. Lilet, B. Lexter & T. Steven van (Eds.), Urban Weaving', in 6(0) Ways: Artistic Practice In Culturally Diverse Times (pp. 52–57). Rotterdam: NAI Publishers.

Leadbeater, A. (2013). Community leadership in disaster recovery: A case study. *Australian Journal of Emergency Management, 28*(3), 41–47.

Manzini. E., Walker, S., & Wylant, B. (2008). *Enabling solutions for sustainable living: A Workshop*. Manitoba: University of Calgary Press.

Nicholson, B., & Rood, D. (2017). We'll rebuild brick by brick. *Sydney Morning Herald* 11th February 2009. 10th April 2017.

Proudley, M. (2013, April). Place matters [online]. *Australian Journal of Emergency Management, 28*(2), 11–16. Availability: http://search.informit.com.au/documentSummary;dn=364072181428846;res=IELHSS>ISSN:1324-1540.

Roberts, D. (2009). *Marysville and triangle urban design framework report*. Melbourne: Prepared by Roberts Day Pty Ltd.

Rural Australia Institute (RAI). (2013). Natural disasters report: Case Studies RAI, Australia.

Teague et al., (2010). In R. Teague & S. McLeod. *Pascoe 2009 Victorian Bushfires Royal Commission final report*. http://www.royalcommission.vic.gov.au/Commission-Reports/Final-Report.html. Accessed February 15th, 2017.

Victorian Bushfire and Reconstruction and Recovery Authority (VBRRA). (2010). Anecdote report: In their own words. Available https://www.rdv.vic.gov.au/__data/assets/pdf_file/0004/1203475/In_their_own_words_-_VBRRA_Anecdote_Report.pdf. Downloaded February 2017.

Victorian Bushfire and Reconstruction and Recovery Authority (VBRRA). (2011). Legacy Report available https://www.rdv.vic.gov.au/__data/assets/pdf_file/0017/1151090/VBRRA-Legacy-Report-Summary.pdf. Accessed February 2017.

Author Biography

Judy Rogers Judy Rogers is Program Manager of the Master of Disaster Design and Development at RMIT University Australia. She has published widely in the field of Sustainability Education and urban sustainability.

Part II
Case Studies: Sustainable Cities

A Diversity of Eco-Developments: An Overview and Comparison of Sustainability in Six Eco-Developments

Dexter Villanueva and Edmund Horan

Abstract The notion that our future will be determined by how sustainable we make urban regions has gained momentum over the past few decades with countries all over the world now aspiring towards sustainable urban development. While several sustainability assessment tools are available as means of accreditation for these developments, the subjective nature of weighting and scoring, lack of representation of the social and economic dimensions and overall interpretation of indicators represented in these tools has garnered increasing attention and discussion in academia. This paper attempts to formulate a multi-dimensional pathway to guide the formation of a substantial comparative framework for a wide diversity of eco-developments. This method may be helpful in shedding light on eco-development trends from which sustainability assessment tools can base their frameworks to address shortcomings.

Keywords Ecocities · Sustainability assessment tools · Eco-developments

1 Introduction

Over the past few decades, concepts of sustainable development have manifested in a diversity of ecocities, towns, districts and villages, with countries all over the world aspiring towards sustainable urban development. However, it is found that certain claims of sustainability are met with scepticism; terms such as 'carbon zero' and 'climate positive' for example, used to describe some successful

D. Villanueva (✉)
Environmental Engineering, RMIT University, 124 La Trobe Street, Melbourne, VIC 3000, Australia
e-mail: dextervill28@gmail.com

E. Horan
Program Director Master of Sustainable Practice, RMIT University, 124 La Trobe Street, Melbourne, VIC 3000, Australia
e-mail: Edmund.horan@rmit.edu.au

W. Leal Filho et al. (eds.), *Sustainable Development Research in the Asia-Pacific Region*, World Sustainability Series, https://doi.org/10.1007/978-3-319-73293-0_14

eco-developments are not always clearly defined (Rauland and Newman 2015). To address the challenges of urban sustainability, several assessment tools have been developed to aid with decision-making processes allowing for more verifiable means of achieving sustainability. Under increasing attention and scrutiny, it has been found that ambiguities in weighting and scoring, and deficits in the coverage of social, economic and institutional dimensions of sustainability in several sustainability assessment tools (Sharifi and Murayama 2013) all raise concerns regarding the integrity of these tools.

This paper represents exploratory work towards the process of crystallising meaningful holistic criteria in order to gain a more objective analysis of sustainability performance and the possibility of developing sustainability analysis criteria, which can be applied across a wide variety of eco-developments.

2 Critical Analysis of the Current Scenario of Sustainability Assessment Tools (SATs)

Many Sustainability Assessment Tools have been developed over the past few decades and are under constant review and revision as they continue to mature. Although the diversity of assessment tools testifies to its complex and contextual nature, literature has presented common trends amongst these tools. In order to gain some perspective in their evolution, the themes explored in this review will begin with the core principles of sustainability relating to the trends in how these are included in SAT frameworks. Then, sustainability criteria will be explored in terms of how they are weighted and scored. Lastly, the topic of sustainability indicators used in sustainability tools will be discussed in terms of their efficacy in monitoring progress. These three themes were chosen with the intention of providing a setting with which past influences, current analysis and possible future directives may be observed.

3 The Pillars of Sustainability in Sustainability Assessment Tools

While interpretations of sustainability and sustainable development have varied, the core of sustainability is defined as the balance between environmental, social and economic aspects concerning our basic needs and wants. In the Johannesburg World Summit for Sustainable Development (2002), the inclusion of an institutional dimension emerged in order to address the driving forces behind environmental and development problems, which led to the proposal and categorisation of institutional indicators (Spangenberg, Pfahl and Deller 2002). These indicators however, were criticised for not adequately addressing the institutional dimension due to minimal data availability, lack of clarity and non-relevancy (Spangenberg et al. 2002). Relating this pathway of sustainability pillars in SATs, Sharifi and Murayama (2013), in their critical analysis of seven neighbourhood sustainability assessment tools have found several lack the coverage of the institutional dimension. In this analysis, it is made explicit that sustainability includes institutional aspects as its fourth dimension. However, there does not seem to be an explicit mention of the kinds of specific indicators and criteria sought out to make this claim. Furthermore, the analysis does not seem to explain possible reasons as to why institutional dimensions were excluded in SATs. Dawodu, Akinwolemiwa and Cheshmehzangi (2017) make point of this, and make the argument in relating this to the initial criticisms of early proposals of the institutional indicators.

When considering the conventional three pillars of sustainability adopted in SATs, there seems to be a growing consensus towards the emphasis of environmental aspects of sustainability over social and economic aspects. Though this trend has been suggested and/or acknowledged in studies from Ameen, Mourshed and Li (2015), Berardi (2013), Dawodu et al. (2017), Haapio (2012) and Medved (2016), literature has revealed varying perspectives on how indicators have been assigned under the pillars, revealing some interesting disparities. One instance can be highlighted in a comparison of the economic coverage analysis for the sustainability tool, Comprehensive Assessment System for Built Environment Efficiency – Urban Development (CASBEE-UD). In Sharifi and Murayama's analysis, one of the frameworks was developed consolidating criteria into themes; Resources and environment, Transport, Social, Economic, Location/Site selection, Pattern and design and Innovation. Indicators from the seven case studies are categorised accordingly under these themes, and compared based on the percentage of coverage across these themes. This method revealed CASBEE-UD as having no indicators relating to the economic theme. On the other hand, Dawodu et al. based their study on the multi-dimensional nature of indicators, characterising the Point Aspect (one-dimensional consideration of indicator), Linear Aspect (two-dimensional) and Planar Aspect (three-dimensional) in terms of Tractability, Relationality, Adaptability and Contextuality. This study revealed that when considering indicators in a multi-dimensional perspective, CASBEE-UD incorporates economic considerations, albeit as offshoots of social and environmental considerations. Sharifi and Murayama do not seem to clarify whether multi-dimensional considerations were made when categorising indicators. Haapio (2012) and Ameen et al. (2015) have acknowledged the multi-categorical nature of indicators; Haapio gives an example of pedestrian and bicycle ways as possibly being categorised in both Transport and Well-being themes, while Ameen et al. have recognised the possibility of indicators affecting all dimensions of sustainability.

3.1 Sustainability Criteria—Weighting and Scoring

Many sustainability tools acknowledge the complexity of applying sustainability in developments, and this has led to the practice of weighting and scoring to address this in SATs. The idea of weighting and scoring is to place emphasis on the significance of different criteria and aid in gauging priorities. However, the objectivity of weighting and scoring does not seem to be well established according to literature. Ambiguities in weighting and scoring of sustainability criteria have been indicated to be recurring issues with sustainability assessment tools (Sharifi and Murayama 2013), and this results from the largely subjective nature of weighting and scoring different criteria (Garde 2009; Vakili-Ardebili and Boussabaine 2007). Retzlaff (2009) further adds to this by claiming "any assignment of weights is essentially a subjective exercise".

Nonetheless, Sharifi and Murayama's analysis provides some insight on how some SATs have attempted to form a more objective basis in weighting and scoring. Similar weighting strategies were undertaken in the tools, LEED-ND, Ecocity (ECC), and VicUrban's Sustainable Community Rating (SCR), where the impact of each criterion were considered in assigning weights. The criteria are benchmarked based on standards from various institutes and organisations and points are awarded for meeting and exceeding these. Though the aspect of weighting here is where the practice becomes prone to subjectivity, Sharifi and Murayama recognise the benchmarking process of verification with several institutes and organisations as a good effort towards a more objective analysis.

3.2 Performance Monitoring

Performance monitoring becomes a crucial theme to consider in SATs as this provides future directives for developments to maintain the sustainability standards achieved. Consistent monitoring will help to identify problems as they arise, which can facilitate decision making towards targeting these areas. However, it has been noted in SAT analysis' that many SATs, including the major and internationally well-known ones (LEED-ND, CASBEE-UD and BREEAM Communities) are primarily static. Sharifi and Murayama demonstrate this in the Presentation of Results section of their analysis. They claim that out of seven tools, only one (Haute Qualité Environnementale et Economique Réhabilitation/High Quality Environment and Economic Regeneration [HQE^2R]) accounted for temporal changes in sustainability indicators. Berardi (2013), providing an analysis of the major SATs previously mentioned, also recognises the static nature of current SATs, stating that these systems are barely influential in the day-to-day lives of communities as these are primarily prescribed by developers only, and that there is need to incentivise continuous monitoring while increasing interactivity. This follows suit with claims that public participation (King 2016) and diverse stakeholder input (Ameen et al. 2015) are important in the monitoring of sustainability in developments. Still, with findings in the Participation section of Sharifi and Murayama's study, it is revealed that only HQE^{2}R and Ecocity tools set a framework for public participation of the choice of core criteria.

4 Methodology and Approach

With the aim of exploring the diversity of eco-developments, sustainability initiatives of selected eco-developments is investigated in two phases. First, initiatives are mapped out on the basis of multi-dimensional aspects to provide a simple yet balanced overview of how developments present sustainability. For simplicity, only the three main pillars are included in this overview, hence eco-developments is analysed in terms of Environment (E), Social (S), Economic (Ec), Enviro-Social (E-S), Enviro-Economic (E-Ec), Socio-Economic (S-Ec) and Enviro-Socio-Economic (E-S-Ec) perspectives. In exploring different eco-development scales, the selected eco-developments consist of both neighbourhood (Hammarby Sjöstad, Kronsberg, Vauban) and 'village' scale developments (BedZED, Findhorn, Crystal Waters). 'Village' scale refers to eco-developments with populations in the hundreds.

In phase 2, an in-depth analysis of eco-developments is then formulated based on the gaps revealed in phase 1for which focus-point (qualitative) indicators are considered in attempting to flesh out these dimensions in the comparative framework. Due to their prevalence and tangibility compared with social and economic indicators, temporal indicators for environmental aspects will be prioritised. Indicators are organised based on themes comprising the elements of a sustainable neighbourhood. These are:

- Natural Resources and Waste Conservation
- Sustainable Transport
- Sustainable Urban Planning and Ecology
- Socio-Enviro-centric economy
- Institutional initiatives

Only themes most representative of the focus-point and temporal indicators are applied in demonstrating how eco-developments compare within these frameworks. The provision of both these types of indicators will provide a more balanced framework when comparing developments as focus points help to determine what needs to be targeted, while temporal indicators measure the performance of those once they're implemented.

5 Phase 1—Multidimensional Overview of Eco-Developments

Sustainability initiatives featured are consolidated to give a simple overview of how sustainability is addressed in the various eco-developments. Tables 1 and 2 use Findhorn as an example to demonstrate how sustainability initiatives can be interpreted with respect to multi-dimensional considerations; the same approach is applied to the other eco-developments (Fig. 1).

All eco-developments seem to present environmental initiatives most prominently, while economic and E-S-Ec are the least featured. Economic sustainability initiatives appear to be addressed mainly in terms of environmental and social aspects of sustainability such as mix-use development (Socio-Economic), community-owned renewable energy systems (Enviro-Economic) and hosting permaculture workshops

Table 1 Environmental, social and economic initiatives of Findhorn

Environment	Social	Economic
– Ecologically benign building design – Wildlife corridors – Wind energy supply – Biological wastewater treatment – Biomass boiler – Comprehensive recycling system – Rainwater harvesting – Microgeneration hydropower	– Shared facilities – Local food production – Internationally linked communities – Local activities—visitation for local school – Findhorn applied ecovillage living and ecovillage design education forum	– Findhorn foundation supports 400 jobs (locally?)

Table 2 Multi-dimensional initiatives in Findhorn

E-S	E-Ec	S-Ec	E-S-Ec
– Open recreational green spaces – Bike & pedestrian friendly	– Car sharing club – Community owned wind turbines – Findhorn consultancy	– Findhorn caravan park tourism enhancement – Local exchange trade system – Organic food cooperative – Outreach workshops – Community forum	Ecovillage workshops—permaculture

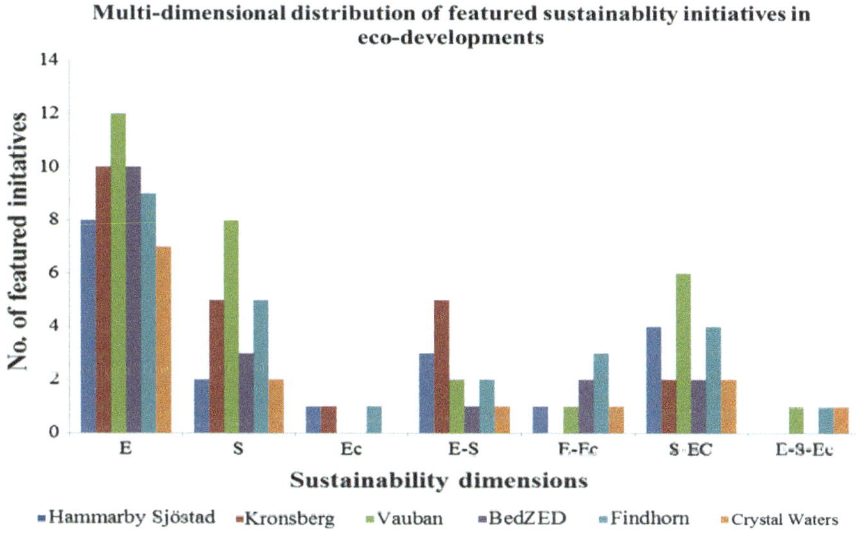

Fig. 1 Sustainability initiative distribution

(Enviro-Socio-Economic). While social initiatives appear most prominently for Vauban, social aspects in general appear spread across the multi-dimensional aspects. Hence, the importance of social and economic aspects seems most relevant when considered in relation to each other dimensions and the environmental dimension.

6 Framework Formulation and Analysis

It is revealed that the general trend in the way sustainability is featured in eco-developments is centred in environmental aspects, while social and economic dimensions do indeed fall behind. Therefore, in an attempt to 'flesh-out' the social and economic dimensions, focus-point indicators will be targeted for the formulation of a comparative framework. With the environmental aspect already well established, temporal indicators will be prioritised to gain an idea of their performance. To demonstrate the application of temporal and focus point frameworks, "Natural Resources and Waste Conservation" and "Sustainable Transport" are used to represent the former, while "Socio-Enviro-centric economy" will demonstrate the latter (Table 3).

Table 3 Resource consumption, conservation and passive housing

	Resource consumption and conservation							Passive housing			
	Building Energy consumption (electricity and heat) (kWh/m^2/y)	Energy consumption (kWh/p/y)	Renewable energy Supply (%)	Water consumption (kL/p/y)	Recycled water supply (%)	Household waste produced (kg/p/y)	Waste recovered/ recycled (%)	Provision	Energy rating target (kWh/ m^2/y)	No. of houses	Proportion (%)
Hammarby Sjöstad	157 113	11,540	25 50	55	0	450	50 (energy conversion) 33 (material recycling) 16 (biogas)	✓	60	1	N/A
Kronsberg	125	4444	72	51	0	67	30 (household pre-sorting)	✓	15	32	1
Vauban	50	2493	65	29	–	–	–	✓	15	200 + 59	13
BedZED	82.4	3139	20	26	18	104	50	✓	90	–	–
Findhorn	–	–	28	–	–	–	–	✓	–	61	59
Crystal waters	–	–	–	–	–	–	–	✓	–	–	–

6.1 Natural Resource and Waste Conservation

Consumption of resources were categorised in three main sources; energy, water and waste and were measured on a per capita basis. Renewable energy supply was based on the percentage of energy used as renewable sources which include solar, wind, biogas, CHP and wastewater energy. Recycled water was based on the combined percentage of non-mains water usage which can include stormwater, water treatment effluent, and rainwater capture. Waste recycling includes efforts to recycle as well as energy and material gain from waste. Hammarby Sjöstad, Kronsberg, Vauban and BedZED data were documented most widely, while Findhorn and Crystal Water figures did not seem to be established in literature.

Hammarby Sjöstad was revealed to have the highest energy consumption, a figure more than doubling when compared to the second highest, Kronsberg. However, this figure, along with BedZED's, accounts for both heat and electricity consumption (Jernberg, Hedenskog and Huang 2015); this type of consumption could not be verified in the case of Kronsberg and Vauban as they were converted from gigajoules using the conversion factor of 277 kwh per gigajoule. Building energy consumption was also highest in Hammarby Sjöstad when considering the figure from Fraker (2013a). This, combined with the relatively low enthusiasm towards sustainability when compared to developments such as Vauban and Kronsberg, could possibly be a factor in higher consumption rates. Nonetheless, some temporal changes can be observed here suggesting this is improving according to the latest figures.

Water consumption of BedZED and Vauban was approximately half that of Hammarby Sjöstad and Kronsberg. When accounting for the rate of recycled water used however, it is possible that the larger developments of Kronsberg and Hammarby Sjöstad may indeed have the largest consumption of mains water as these developments do not include means of rainwater capture for reuse.

Household waste was significantly lower in Kronsberg than in other developments, owing to the encouragement of a low waste lifestyle and the inclusion of several repair shops in the district. Hammarby Sjöstad virtually has all waste reused as either for energy production and recycling, with 1% being hazardous waste which is disposed (Iveroth, Johansson and Brandt 2013).

Crystal Waters, BedZED and Findhorn were counted to provide some potential for passive housing, though these were not explicitly mentioned, instead only featuring elements of passive housing such as solar orientation, or in Findhorn's case, referring these to as 'Ecological housing'. The most ambitious targets came from Vauban and Kronsberg, though Vauban provides a significantly higher proportion of not only passive housing but 'Energy-Plus' housing as well.

6.2 Public Transport Sustainability

Sustainable transport here is measured as the distribution of transport preferences, and the provision, distance and frequency of public transport options. In all cases documented, private car use was the least chosen mode of transport compared to pedestrian/cycling and public transport options; a good indicator of car-use reduction initiatives being a success. The most preferred transport mode in all documented cases except Vauban was public transit. Vauban's figures however, were taken prior to the connection of light rail to the district (Field 2011). No official survey was conducted for Kronsberg, though anecdotally it is claimed to have achieved these proportions more or less (Fraker 2013b). Though no data was found for Crystal Waters, its location on steep ridges makes car use a necessity to travel between sites (Table 4).

Hammarby Sjöstad, Kronsberg, Vauban and BedZED appeared to be most well connected in public transport provision, locality and frequency. Hammarby Sjöstad was shown to lead in this manner, with major transport nodes being no more than 350 m from the furthest residential areas and a frequency of services of no more than 10 min intervals. As a rural eco-village, Crystal Waters' isolation means private vehicle is the only means of reliable access.

Though Findhorn was documented to have the least car travel distances per resident, these figures are presented in terms of ecological footprint (Tinsley and George 2006); whether this figure measures actual distance travelled was unclear. While Findhorn is less isolated than Crystal Waters, the only means of public transport connection is a bus service with a frequency of every hour.

6.3 Socio-Enviro-Centric Economy

It can be said that economic sustainability should be considered in reference to the social and environmental implications which may be tied in economic activities in order to enhance the liveability of an eco-development and foster well-being in citizens. Here, several focus-points are consolidated with the aim of mapping out how each development essentially adopts these elements of sustainability in their respective economic and social frameworks, whilst still accounting for environmental needs within them (Table 5).

Vauban sets itself up here as one of the most exemplary examples of community building. Roots in community building development have allowed for strong decision making power at the community level, fostering a real sense of ownership. This bottom up approach set the foundation for a strong community, featuring a number of community associations and local press for advertising local events. Social diversity is high with a population consisting of young families to the elderly, and accommodates students, researchers, eco-tourists and professionals alike. Findhorn adheres to similar sets of community-based principles, featuring a

Table 4 Resident transport habits and public transport accessibility

	Proportion of resident transportation method (%)			Private vehicle travel (km/person/y)	Public transport accessibility and frequency (max. distance from residents and frequency of services)									
	Public transit	Pedestrian/bike	Private vehicle		Train		Light rail/tram		Bus		Car share		Other	
					Distance from residential area (m)	Frequency (min)	Distance from residential area (m)	Frequency (min)	Distance from residential area (m)	Frequency (min)	Distance from residential area (m)	Frequency (min)	Distance from residential area (m)	Frequency (min)
Hammarby Sjöstad	52	27	21	1927	1500	N/A	300	8	300	8	350	N/A	800	10
Kronsberg	40	40	20	4854	3600	N/A	600	10	600	30	–	–	–	–
Vauban	19	65	16	–	2800	N/A	300	15	–	–	–	–	–	–
BedZED	61	22	17	2318	700	N/A	1500	5	50	14	Onsite	N/A	–	–
Findhorn	–	–	–	539	6000	N/A	–	–	600	60	Onsite	N/A	–	–
Crystal Waters	0	–	–	–	–	–	–	–	–	–	–	–	–	–

Table 5 Socio-economic and enviro-economic sustainability

	Sense of community and identity							Accessibility to jobs, skills and training			
	Public input in plan decision making	Social diversity	Community-based social framework	Community activities and associations	Local network/ forum	Community centre	Local community food garden	Local economy (jobs per resident)	Direct access to major economic zones	Internet access	Local skills and training
Hammarby Sjöstad	✗	✗	✗	✓	✗	✗	✓	0.31	✓	✓	–
Kronsberg	✓	✓	–	✓	✗	✓	✓	0.20	✓	✓	✓
Vauban	✓	✓	✓	✓	✓	✓	✓	0.12	✓	✓	✓
BedZED	✓	✓	–	–		✗	✓	–	✓	✓	✓
Findhorn	✓	✗	✓	✓	✓	✓	✓	Up to 0.97	✗	✓	✓
Crystal waters	✓	✗	–	✓	✗	✓	✓	–	✗	✗	✓

	Affordable housing		Socio/enviro-centric economic activity					
	Average house price relative to surrounding area	Proportion of low-income housing (%)	Local Exchange trading system (LETS)	Local markets	Local renewable energy cooperative	Organic food cooperative	Car share initiative	Tourism prospects
Hammarby Sjöstad	High	–	✗	✗	✗	✗	✓	–
Kronsberg	–	37	✗	✗	✗	✓	✗	–
Vauban	–	10	✓	✓	✗	✓	✓	✓
BedZED	5–10% more	25	✗	✓	✗	✗	✓	✓
Findhorn	60–80% more	–	✓	✓	✓	✓	✓	✓
Crystal waters	–	–	✓	✓	✗	✓	✗	✓

well-connected community reaching locals and internationally. Findhorn Foundation forms the hub of community interaction in the community, hosting forums to provide advice and education to those interested in eco-village development. While social diversity would be diverse with the interconnected community set by Findhorn, local communities appear to be united by common philosophical agendas, and so it is possible this attracts mostly those with similar agendas.

Both Crystal Waters and Kronsberg also score fairly well in this analysis, both hosting a number of community activities, local community gardens and implement public input in plan decision making. Crystal Waters falls somewhat short in terms of social networks and diversity, mainly due to its isolation and demographic of mostly young families (Barton 2013). In Kronsberg's case, local networks in engaging communities also appears lacking, while there does not seem to be any special social frameworks in particular.

Hammarby Sjöstad appears to score the least in these criteria having much to do with the homogenous population of mostly wealthy residents. This resulted from a change in government selling of the development to the private sector which saw an increase in housing prices. While this has made it one of the most financially successful in its area, this is not reflected well in this framework in addressing social and environmental aspects of economic enhancement. Nonetheless, there are efforts to increase community engages via the HS2020 community initiative and the provision of small urban food boxes (Medved 2016)

Findhorn however, appears to do well in regard to socio-enviro-centric economics, a diversity of activities from permaculture workshops to local markets and a community energy cooperative. Vauban also hosts many of the same economic initiatives, only missing out on a renewable energy cooperative among locals. As a major contender in increasing the tourism profile of the area, Findhorn house prices were found to be higher than the surrounding area. It is unknown if Findhorn has subsidised housing of any sort.

7 Discussion of Comparisons and Limitations

7.1 *Village Versus Neighbourhood Scales*

With the intention of gaining some perspective between village and neighbourhood scales, some discernible aspects can be drawn in this framework; isolation in typical rural Eco developments appears to undermine overall sustainability in terms of transport, and rural areas adopt LETS to facilitate economic activity. However, lack of data remains a major drawback in deriving any real comparisons and insights, as several quantitative figures were simply not available in literature for these scales of development.

7.2 Temporal Data

There were very limited instances of temporal changes observed, occurring only in the 'Building Energy Consumption' and 'Renewable Energy Provision' indicators for Hammarby Sjöstad. Even so, these figures were derived from separate and different sources. Consistent monitoring is important in helping gauge the performance of sustainability initiatives.

7.3 Data Verification

The methods used to derive figures are widely lacking in sources and assumptions and/or estimates may have to be made when converting data. Figures for Kronsberg and Vauban 'Energy Consumption' per capita converted from GJ to Kwh were assumed to refer to electric energy only. Hence, consistency is lost and comparisons can become redundant.

8 Additional Frameworks and Further Adjustments

The following are conceptual frameworks for further progress in the determination of meaningful criteria in sustainable planning and ecological planning, and an experimental mapping framework of institutional partnerships associated with eco-developments (Tables 6 and 7).

With respect to further developing the focus-points constructed in the "Socio/Enviro-centric economy" theme, temporal measures can be considered to track the dynamics of community initiatives and activities, and monitor the successes of new initiatives which aim to facilitate a stronger engagement from the local population. Quantitative indicators of interest would include:

- No. of workshops for skills and training
- Proportion of population involved in local renewable energy cooperative
- Proportion of people involved in organic food cooperative

Table 6 Sustainable urban planning and ecology

Urban planning	Urban ecology
• Mix-use	• Proportion of open green space (%)
• Central amenities	• Ecological construction materials
• Access to diverse amenities	• Natural systems technologies and infrastructure
• Heterogeneous housing designs	• Passive energy harnessing (orientation)
• Bike/Pedestrian friendly design	• Biodiversity enhancement
• Maximum distance to local park (m)	• Wildlife corridors

Table 7 Institutional partnerships and associations

Decision making	No. of associated institutions and organisations	Institutional programs and partners
• Institutional • Private		• Local council • Stakeholder engagement • Environmental management • Community participation • Community networks • Investors • Sustainability education and awareness • Research engagement

- Proportion of people involved in LETS
- Proportion of people using local networks/forums
- Distribution of local population demographics (elderly, children, adults, teens, ethnicity, income, etc.)
- Number of local community food gardens
- No. of community associations
- Proportion of people using community centre and frequency of use

9 Conclusions

In summary the following conclusions can be made

- A multi-dimensional perspective of sustainability can aid in calibrating aspects addressed in sustainability initiatives, and provide some direction in determining gaps in sustainability
- Quantitative performance data for rural eco-developments appears absent in literature
- Temporal trends are not monitored consistently, and figures sometimes have to be derived from separate sources
- Methodologies aren't always given for figures presented

It is paramount that we are able to make meaningful comparisons in order to learn from and acquire the best possible practices eco-developments have to offer. The major challenge in achieving this is providing a balance between simplicity and breadth, while considering target focus points to aid sustainability planning and temporal indicators to provide future directives and guide sustainability action.

References

Ameen, R. F. M., Mourshed, M., & Li, H. (2015). A critical review of environmental assessment tools for sustainable urban design. *Environmental Impact Assessment Review, 55,* 110–125.

Barton, H. (2013). *Sustainable communities: The potential for eco-neighbourhoods*. Taylor & Francis.

Berardi, U. (2013). Sustainability assessment of urban communities through rating systems. *Environment, Development and Sustainability, 15,* 1573–1591.

Dawodu, A., Akinwolemiwa, B., & Cheshmehzangi, A. (2017). A conceptual re-visualization of the adoption and utilization of the pillars of sustainability in the development of neighbourhood sustainability assessment tools. *Sustainable Cities and Society, 28,* 398–410.

Field, S. (2011). Vauban. In S. Field, & N. Foletta (Eds.), *Europe's vibrant new low car (bon) communities* (pp. 96–106).

Fraker, H. (2013a). Hammarby Sjöstad, Stockholm, Sweden. In The Hidden (Ed.), *Potential of sustainable neighborhoods: Lessons from low-carbon communities* (pp. 43–67). Washington, DC: Island Press/Center for Resource Economics.

Fraker, H. (2013b). Kronsberg, Hannover, Germany. In *The hidden potential of sustainable neighborhoods: lessons from low-carbon communities* (pp. 69–95). Washington, DC: Island Press/Center for Resource Economics.

Garde, A. (2009). Sustainable by design?: Insights from U.S. LEED-ND pilot projects. *Journal of the American Planning Association, 75,* 424–440.

Haapio, A. (2012). Towards sustainable urban communities. *Environmental Impact Assessment Review, 32,* 165–169.

Iveroth, S. P., Johansson, S., & Brandt, N. (2013). The potential of the infrastructural system of Hammarby Sjöstad in Stockholm, Sweden. *Energy Policy, 59,* 716–726.

Jernberg, J., Hedenskog, S., & Huang, C. (2015). Hammarby Sjöstad an urabn development case study of Hammarby Sjöstad in Sweden, Stockholm.

King, L. O. (2016). Functional sustainability indicators. *Ecological Indicators, 66,* 121–131.

Medved, P. (2016). A contribution to the structural model of autonomous sustainable neighbourhoods: new socio-economical basis for sustainable urban planning. *Journal of Cleaner Production, 120,* 21–30.

Rauland, V., & Newman, P. (2015). Eco-precincts. In *Decarbonising cities: Mainstreaming low carbon urban development* (pp. 65–94). Cham: Springer International Publishing.

Retzlaff, R. C. (2009). Green buildings and building assessment systems a new area of interest for planners. *Journal of Planning Literature, 24,* 3–21.

Sharifi, A., & Murayama, A. (2013). A critical review of seven selected neighborhood sustainability assessment tools. *Environmental Impact Assessment Review, 38,* 73–87.

Spangenberg, J. H., Pfahl, S., & Deller, K. (2002). Towards indicators for institutional sustainability: lessons from an analysis of Agenda 21. *Ecological Indicators, 2,* 61–77.

Tinsley, S., & George, H. (2006). *Ecological footprint of the Findhorn foundation and community*. Forres, Moray: Sustainable Development Research Centre.

Vakili-Ardebili, A., & Boussabaine, A. H. (2007). Application of fuzzy techniques to develop an assessment framework for building design eco-drivers. *Building and Environment, 42,* 3785–3800.

Evolving a Locally Appropriate Indicator System for Benchmarking Sustainable Smart Cities in India

Sarbeswar Praharaj, Jung Hoon Han and Scott Hawken

Abstract Urban development initiatives in India are largely shaped by a 'patchwork of programmers' launched periodically by the Government of India. Recently, the Indian government has launched multiple urban transformation programmes aiming at smart city development, urban renewal, and heritage development. But, India's urban policy interventions lacks effective mechanisms for measuring the impact of such programmes over time. Indicator systems can provide insight into local urban challenges and issues and offer a basis for measuring urban progress as a result of public investments. Although there are multiple global indicator facilities available for adoption, considerable challenges remain in selecting an appropriate set of indicators for local conditions. With Indian urbanisation intensifying and an accompanying proliferation of urban development programmes, there is an increasing need for a set of indicators suited to local context. A local approach can objectively measure the existing performance of Indian cities vis-a-vis quality of life and sustainability, and can evaluate the outcome and benefits of new local urban development programmes as they are implemented. To address this emerging need, this research aims to provide a bridge between local urban performance and the capacity to meet local and global standards. By the application of factor analysis (Principal Component Analysis) on 59 indicators chosen from various urban sectors, this study attempts to derive the most significant factors to measure the performance of 98 Indian cities and assess their potential to become sustainable smart cities. The study enhances the selection and development of policies by city makers. It also establishes an approach to develop urban typologies for different Indian cities. Such a categorisation makes the bewildering range of Indian cities and their challenges easier to comprehend and address.

S. Praharaj (✉) · J. H. Han · S. Hawken
Faculty of Built Environment, The University of New South Wales,
UNSW Sydney, Sydney, NSW 2052, Australia
e-mail: s.praharaj@unsw.edu.au

J. H. Han
e-mail: h.han@unsw.edu.au

S. Hawken
e-mail: s.hawken@unsw.edu.au

W. Leal Filho et al. (eds.), *Sustainable Development Research in the Asia-Pacific Region*, World Sustainability Series, https://doi.org/10.1007/978-3-319-73293-0_15

Keywords Urban sustainability indicators · Performance assessment of cities Sustainable smart cities · Factor analysis · 100 smart cities mission in India

1 Introduction

Resources for urban infrastructure development in the global south are scarce. Urban investments therefore require a strategic approach that relies on objective identification of advantage and disadvantages of cities before projects are developed and substantial financial commitments are made. Assessment of sustainability outcomes from investment is critical too vis-a-vis social, economic and environmental benefits (Giddings et al. 2002). Measuring urban progress provide cities the scope to adjust and tailor projects, address the urban deficiencies and lead to just utilisation of scarce financial resources. There is a strong realisation among the academic community that monitoring and review of outcomes from urban initiatives in the global south are not sufficiently robust to shape policies and implementation strategies over time (Kamath and Zachariah 2015; Shen et al. 2011). The new batch of urban reforms therefore require "strong mechanisms for benchmarking, review and support processes to ensure adequate resource commitment and adjustments to implementation strategies over time" (Cities Alliance 2015).

The notion of urban competitiveness which has gained ground over last decade rests heavily on the measures of sustainability. Often the outcomes from cross-city comparisons on the measures of sustainability is used to highlight the competitive advantage of cities on a regional scale. This is particularly important for cities as they are not only competing for investment and jobs, but also for attracting generation Y and generation Z people (Harrison and Donnelly 2011). The people Richard Florida termed as 'Creative Class' (Florida 2005), whose decision to move into a city is influenced by its brand, image and sustained performance. Robust assessment of cities across indicators of sustainability is therefore key to promoting competitiveness and has great implications for local economy. Beyond cities and regions, these assessments help countries to report against the Sustainable Development Goals (SDG) and advance understanding of global progress on sustainability targets.

Although benefits from sustainability assessment is well documented in public policy, the methodology for such assessment and the measures have been inconsistent. There are diverse approaches towards ranking of cities based on parameters related to sustainability, liveability, quality of life and of late smartness. These thematic assessments use parameters that are often seemingly different from each other, sometimes overlapped, leading to divergent outcomes from the assessments. Moreover, the academic community has not given enough attention to take account of complex links between indicators and variables (Jiang and Shen 2010) that can potentially skew the ranking of cities. These sustainability rankings although are important pointers of civic performance, often criticised due to the indicators of assessment not tailored for specific contexts (Miller et al. 2013). Especially for global south, where cities are facing rapid urbanisation and dealing with unique challenges, a different approach is essential for benchmarking cities with unique indicators which can support informed policy targeting.

This paper offers a novel approach to urban sustainability assessment at the city scale across 98 cities in India. Rather than merely adopting indicator sets from established literature, we have identified unique measures suited to Indian context to assess and compare performance of Indian cities. The indicators are defined in this paper with due consideration that performance is not only about how sustainable or smart a city is, but a more holistic and complex understanding about measuring the advance of a city towards its capacity to deliver a better quality of life and undertake transformative practices (Dameri and Rosenthal-Sabroux 2014). This paper adopts robust statistical analysis to identify the most relevant factors that influence the sustainability outcomes in cities and are non-correlated. This data intensive, statistically sound investigation offers a methodology to transform urban monitoring and shape policy in a profound way. Such approaches can help develop emerging urban typologies, identify potential areas for optimization and establish reference points for cities in designing policies to become sustainable and smart cities.

2 Why Measuring Performance Is the Key to India's Urban Prosperity

The urban landscape in the world's largest democracy and fastest growing economy—India, is changing dramatically than ever before. From 17.96% urbanisation level in the year 1961, the share of urban population rose to 31.16% in the year 2011; and the country is expected to house half of its population in urban areas by the year 2050 (Census of India 2011). By 2030, 590 million people will live in cities in India, with a net increase of 270 million working age population from 2011. There will be 68 million plus cities in the country by 2030, whereas the entire Europe has only 35 such urban agglomerations today (McKinsey and Company 2010). But, increasing pace of urbanisation in India has not been matched by adequate planning, sector policies, governance and infrastructure development. In fact, the Government of India is yet to develop a country wide strategic policy for urban development. On the contrary, the central governments approach has been sectarian and event based, rolling out schemes and programmes as a firefighting measure without any objective assessment of the outcomes. In this background, the Strategic Plan of Ministry of Urban Development, Government of India for 2011–2016 commented that "to ensure competitiveness of cities while offering basic services to citizens, urgent steps are required to harness the opportunity that the scale of urbanisation presents and to avoid urban decay" (Government of India 2011).

Seizing the existing challenges and despairing context of urbanisation, in 2015, the Government of India has initiated a courageous mission to transform 100 urban centres across the country into smart cities. The central objective of the initiative as per the smart cities mission statement and guideline is to improve the quality of life in cities with more investment in core infrastructure that leads to sustainable development (Government of India 2015). The Government of India has also launched several other high-visibility urban development programmes such as Atal Mission for Rejuvenation and Urban Transformation (AMRUT), Housing for all, Heritage City Development and Augmentation Yojana (HRIDAY) etc. While the intent of these initiatives is promising, there is little emphasis on building monitoring frameworks and indicator systems that can offer quantification and objective identification of issues and create opportunities for dialogue on local conditions (Bhattacharya et al. 2016). There seems to be no lessons learnt from just concluded Jawaharlal Nehru Urban Renewal Mission (JnNURM), the largest urban reform and infrastructure development programme in India, that has often been criticized for not achieving its intended objectives in infrastructure and housing development due to weak monitoring and evaluation mechanisms (Grant Thornton 2011). These programmes as it stands have more emphasis on physical outputs, in the form of built infrastructure, and lack focus on measuring outcomes to assess the impacts of the projects on the baseline conditions (Planning Commission 2012a, b). Therefore, on the one hand, India's present approach to policy making and project development lacks frameworks for objective identification of targeted issues for urban development; and on the other, it fails to measure how project implementation impacts the society and the overall sustainability agenda.

We found that different agencies of the Government of India have made discrete attempts in developing urban performance assessment framework. The Ministry of Urban Development (MoUD) back in February 2009 launched a handbook on service-level benchmarking (SLBs) to collate data on urban service levels and facilitate comparison between cities and track changes in performance over time (Ministry of Urban Development 2009). These SLB's were extensively referred by cities while preparing City Development Plans (CDPs) under JnNURM. But, this initiative essentially remained as a tool that was used to make a case for seeking investment for specific infrastructure and failed to emerge as a comprehensive urban performance assessment model. The second major step towards developing performance assessment of cities came under the National Mission for Sustainable Habitat in 2011 (Town and Country Planning Organisation 2011). It attempted to design parameters for measuring the sustainability of habitats within the constitutionally provided framework for urban planning in India. However, the recommendations remain in paper only and has not been meaningfully adopted in the actual planning processes. In purview of the smart cities mission, the Bureau of Indian Standards (BIS) has published Smart Cities—Indicators ICS 13.020.20 on October 2016. The Indicators defined by BIS are accumulated from 17 different sectors with 46 core indicators and 47 supporting indicators (Bureau of Indian Standards 2016). However, we found these indicators were primarily picked up from the ISO 37120:2014 Sustainable Development of Communities: Indicators for city services and quality of life. Many of the indicators identified by BIS lack local context awareness and does not suit Indian cities present state of development.

What India needs is the measure of readiness of cities to become smart and not assessment of how smart they are at present, which the current indictors are trying to assess. Moreover, the data on high number of indicators are presently unavailable, practically making impossible for cities to use the framework.

The Ministry of Urban Development has recently published a set of 'Liveability Standards in Cities' to develop 'Liveability Index' and rate cities. The publication titled 'Methodology for Collection and Computation of Liveability Standards in Cities' (Ministry of Urban Development 2017) lists a total of 79 indicators (57 Core Indicators and 22 Supporting Indicators) grouped under 15 thematic categories. These categories are designed as part of four broad pillars of sustainable development—institutional, social, economic and physical. The document proposes to develop various 'Category Sub-Indexes' and a composite 'City Liveability Index' based on the performance of cities against the various core and supporting indicators. For calculating the index, a differential weightage is assigned against each category. The physical infrastructure category is assigned with the highest weightage of 45%, whereas the category under economy is assigned 5% weightage. Governance and social categories are equally weighted at 25% each. In First of its kind, each of the indicators identified are matched with its reference in the SDG targets.

Undoubtedly, this is the most significant step by Government of India towards developing a comprehensive ranking of cities. However, various concerns have emerged from a critical analysis of the prescribed methodology. The proposed framework appears to be biased towards large cities as the designated benchmarks are set very high. The smaller and medium sized towns and cities in India are set to lose out in this ranking as the level of capacity and basic infrastructure in these cities are much lower than their larger counterparts. It was essential for this methodology, to devise differential benchmark standards according to the size of cities. It would have been then possible to develop indexes for each category of cities, providing a level playing field among similarly capacitated cities. Also, the methodology fails to devise a mechanism to recognise and eliminate the causalities among the indicators. For example, percentage of budget allocated towards cultural/sports activities and the number of cultural/sports events hosted by city, which are independently weighted in the given framework, does appear to have casual relation. A bi-variate correlation analysis was essential to determine and eliminate such indicators which can potentially distort the outcomes from the index.

It is evident from the above analysis that the Government of India has shown persistent interest in developing a sound methodology for ranking of cities based on their performance. However, there is lack of continuity and linkage among the different initiatives undertaken at different period by various central government agencies. Also, an obvious lack of consensus is observed whether the ranking should be devised on the level of sustainability, liveability or smartness; concepts which are seemingly different from each other. With India's urbanisation on the verge of a massive leap, the time is now to converge the different initiatives into a uniform concept and develop a statistically robust framework for performance assessment of cities of different scale.

3 Selecting Locally Appropriate Urban Sustainability Indicators for India

Many researchers have attempted to document the level of sustainability in cities using indicators to reveal the challenges that cities faces. In fact, the use of social and economic indicators to inform policy decisions began as late as 1960s in the west (Taylor 1981). Unfortunately, the initial boom in the 'social indicators movement' has suffered setback in the late 20th century due to lack of sound methodological basis of the early studies (Carley 1981) and missing links of those research with policy concerns of the city makers (Knox 1978). Increasingly a need was realised to change the selection process of indicators by focussing more on the ones that are likely to produce the most accurate understanding about the status of practice, rather than generating general information. The United Nations Statistical Institute for Asia and Pacific in the year 2007 has contributed to this debate by proposing a rule that indicators must be SMART—Specific, Measurable, Achievable, Relevant and Time-related (UNSIAP 2007).

The research approach presented in this chapter here follows five basic principles for indicator selection as outlined by Coombes and Wong: (1) availability of data; (2) geographical unit consistency; (3) time-series prospects; (4) implementability; and (5) interpretability (Coombes and Wong 1994). Data availability is one of the most fundamental problem that has high impact on indicator based sustainability assessment. This issue is more prominent in the developing world where data is highly fragmented due to multiplicity of governance actors and lack of institutional linkages. Moreover, a considerable share of data relevant to urban and regional development policy rests with the private sector which often becomes inaccessible. Therefore, both availability and access to data are critical for determining indicators for measuring urban prosperity. The emphasis of this paper has been to identify small number of basic yet critical indicators on which reliable data is available from public data sources. We have taken specific care that lack of data availability does not influence selection of proxy measures which is often found in academic literature.

The challenges of scale mismatch which is also known as "Areal Unit Problem" (Fotheringham and Wong 1991) is the second major concern as established in the literature concerning to urban indicator system that heavily relies on spatial data. To address this challenge, we have used data at the level of City Municipal Corporation on a consistent basis. This helps in drawing meaningful comparisons at the city scale. By using data from Census of India publications, we have ensured that the data is highly reliable and that the time series analysis can be undertaken. Datasets available from various Government of India publications and municipal websites have been referred for collating data on urban governance and physical infrastructure delivery.

The core focus of this paper is to identify indicators that are locally appropriate and have implications for public policy. As part of this research we have conducted questionnaire survey of 179 urban development professionals across India, cutting across government, industry and academia. We have picked the indicators around the most significant challenges and opportunities identified from those responses. Basic infrastructure level, urban governance, education and health, community awareness and behaviour were identified by respondents as key measures to India's urban prosperity. After identifying these categories, we have picked indicators relating to each theme from established literature and Government of India published indexes. We have also identified indicators that are never cited in literature yet we believe significant for India's present urban context. For example, percentage of non-slum households, female share of workers, percentage of households not defecating in open, percentage of households using bicycle etc. as measures for urban performance is chosen in this study for the very first time in such assessment. These indicators may seem to be quite elementary, but holds high relevance in view of present state of Indian cities.

A total of 59 indicators were selected at the beginning of this study. Those were grouped under 7 themes. After an initial analysis 12 measures were dropped as they were highly correlated with other indicators. The final list of 47 indicators shown in Table 1 is used as part of this analysis. Seven indicators were grouped under

Table 1 List of indicators used in the analysis

Demographic and social	Population density in km/km^2
	Decadal population growth rate in percentage
	Sex ratio
	Female share of workers
	Literacy rate
	Creative and agile population (15–44 age group)
	Percentage of non-slum households
Economic	Per capita GDP
	Work participation rate
	Working age population
	Share of main workers
	Share of secondary and service sector workers
Education and health	Gross enrolment ratio in higher education
	Number of colleges/lakh population
	Number of universities/lakh population
	Number of college graduates each year
	Hospital beds/1000 population
	Number of physicians/1000 population
Physical infrastructure and services	Road Density in km/km^2
	Mode Share of Public Transport
	Percentage of households having access to treated tap water
	Percentage of households with access to drinking water within premises
	Percentage of households having electricity connection
	Percentage of households having access to latrine within premises
	Waste water connected to closed drainage (percentage of households)

(continued)

Table 1 (continued)

	Household level coverage of municipal solid waste (MSW)
	Efficiency of collection of MSW (in percentage)
	Extent of MSW recycled and recovered (in percentage)
	Percentage of households having enclosed bathroom
	Percentage of households availing banking services
	Percentage of households having computer/laptop with Internet
	Percentage of households using mobile phones
Shelter	Percentage of households with condition of house as good
	Percentage of households having permanent house structures
	Percentage of households residing in own houses
Behavior and lifestyle	Percentage of households using solar energy
	Percentage of households using cleaner energy for cooking
	Percentage of households having bicycle
	Percentage of households not defecating in open
Governance	Per capita municipal capital expenditure
	Availability of Master plans
	Availability of dedicated urban planning cadre within local bodies (ULB)
	Availability of dedicated local government website
	Status of publication of service level benchmarks
	Status of setting up of single window civic service centres in ULB

demographic and social theme including, population density, population growth rate, sex ratio, female share of workers, literacy rate, creative and agile age group population and the share of non-slum population. Social development is often seen as a weak link in India's urban policy dynamics. The issues of women equality (Chant 2013; Phadke 2007) and increasing concentration of poor communities in the slums has not found its desired importance in the conventional master planned development practice (Ghertner 2011). If India needs to move forward with its urban development in a harmonious and sustainable way, these soft issues must be brought into light so that adequate policy and resources are channelled to address them.

The second group of 5 indicators were placed under economic potential of cities. These indicators include, per capita GDP, work participation rate, working age population, share of main workers and share of secondary and service sector workers. Indian cities are seeing a trend of massive young working age population emerging (James 2011). In keeping its economic momentum alive, India must harness the benefits of demographic dividend and should create more job opportunities in the service sector (Chandrasekhar et al. 2006). These group of indicators can substantially help in defining the level of economic performance of cities and prospects of new policy interventions.

Six indicators were selected in the analysis under the education and health category. Functions related to education and health does not come under the ambit of municipal governance in India. Rather it is increasingly being privatised, which highly impacts the affordability of these services (Baru 2003). We as part of this research are keen to investigate if education and health issues impact the overall sustainability outcomes in cities.

A maximum of 14 indicators were selected under physical infrastructure and municipal service delivery. These indicators were primarily concentrated around level of water supply and waste water, sanitation and access to digital infrastructure. Municipal service delivery is the weakest link in achieving India's sustainable urban development goals (McKinsey and Company 2010; Government of India 2011). Rapid urban growth in Indian cities and demand for civic services outpaces the existing supply mechanism. These set of indicators will allow us to measure the varying level of service delivery in cities and lead towards the identification of advantage and disadvantages across the infrastructure sectors.

The remaining 13 indicators were grouped under three different themes; shelter, behavior and lifestyle and governance. Housing has remained a critical sector in India's policy agenda reflected through various schemes laid out by Government of India such as "Housing for all mission", "Basic services for urban poor" and "Rajiv Awas Yojana". The measures grouped under 'shelter' category will indicate how these initiatives has impacted the access to shelter in different cities across India. In order to analyse the smart community potential, 4 different indicators were selected that can measure sustainable behavioural practice across Indian cities. Six different indicators are outlined to assess the governance efficiency in cities and to recognize the potential for optimisation in urban management practice.

4 Methodology

To identify the most significant indicators that influence the sustainability outcomes of cities in India, we have undertaken factor analysis using the dimension reduction tool in SPSS statistical computer package. Factor analysis is particularly useful to identify a relatively small number of factors which can represent the relationships among large set of variables (Coombes and Wong 1994). Through this statistical procedure, all the raw data are automatically standardised as Z-scores before the most critical factors are extracted. The aim of this technique is to summarise as much as possible the variance in the given dataset with the minimum number of factors. Although the single-factor solution is the most popular choice (Thompson 2004), a multi-factor solution is expected in this case due to substantial number of indicators used in the analysis and their heterogeneity. This technique has two unique advantages; on the one hand, the method counters with the intercorrelation within the dataset by producing a correlation matrix, and on the other, it provides an automatic statistical weighting of each variable on each factor which can be used for indexing or ranking.

In this research, we have used Principal Components Analysis (PCA) as the extraction method to emphasize the variation and bring out strong patterns in the dataset (Wold et al. 1987). We began the analysis by running a correlation coefficient model to understand the pattern of relationships among the variables. In this stage, we have scanned the significance values for variables that are greater than 0.05 and dropped 12 indicators from the initial list because of singularity in the data. The final list of 47 indicators that are used for further analysis in this paper are generally uncorrelated. In the next stage, we have conducted the Kaiser-Meyer-Olkin measure of sampling adequacy and the Bartlett's test of sphericity test. This is to ensure that the dataset comply with the recommendation of Kaiser (1974), that any values greater that 0.5 are acceptable (Field 2005). To find out if the factor analysis as a technique was appropriate for the data set, we have confirmed that the desirable Bartlett's test significance value ($p < 0.001$) is achieved.

Communalities and total variance are analysed as per Principal Components method that can indicate the amount of variance in each variable that is accounted for. From the communalities analysis output, we inspected for any values after extraction that are below 0.5, which is generally considered unacceptable for inclusion in further modelling stages. We then analysed the eigen values from the total variances explained associated with each linear component (factor) after extraction and after rotation. As SPSS is used for analysis, by default all the factors with eigen values greater than 1 are extracted. This analysis has helped in determining the number of factors whose sum is equal to number of items which are subjected to factor analysis. Lastly, rotated component matrix is presented in this paper through application of Varimax with Kaiser Normalisation method. This method is particularly suitable for this analysis as we have ensured that the variables were non-correlated. The rotated component matrix provides the loading on each factor and helps make interpretation of the analysis easier by flagging significantly higher loadings (Kaiser 1958). In this analysis, we have designated the value of 0.4 and above as significant loadings for the ease of recognizing patterns in the dataset.

5 Analysis and Results

Table 2 highlights the SPSS outputs for the Kaiser-Meyer-Olkin measure of sampling adequacy and the Bartlett's test of sphericity. KMO values of 0.653 is achieved, which is considered acceptable (>0.5) for such analysis. Bartlett's Test of Sphericity results shows a significance value of 0.000, which confirms that factor analysis as a technique was suitable for the data set. The table also shows the

Table 2 KMO and Bartlett's test

Kaiser-Meyer-Olkin measure of sampling adequacy		0.653
Bartlett's test of sphericity	Approx. Chi-Square	3484.119
	df	1081
	Sig.	0.000

approximate Chi-Square value and difference value as default outputs from the SPSS statistical computer package.

Table 3 represents the Communalities before and after extraction. Principal Component Analysis works on an initial assumption that all variance is common; therefore, before extraction the communalities are all given the value of 1. Extraction communalities are estimates of the variance in each variable accounted for by the components in the factor solution. For interpretation, we can say from the Table 3 that 79.9% of the variance associated with variable 1 is common, or shared, variance. More importantly, none of the variables selected for this study show a value below 0.5 which we have identified as baseline value for communalities after extraction. Therefore, there is no need for dropping any variable from the model as they all seem to fit well with the factor solution.

Table 3 Communalities

	Initial	Extraction
Log population density	1.000	0.799
Population growth rate	1.000	0.887
Sex ratio	1.000	0.875
Female share of workers	1.000	0.830
Literacy rate	1.000	0.739
Creative and agile population	1.000	0.874
Percentage of non-slum households	1.000	0.689
Log per capita GDP	1.000	0.738
Work participation rate	1.000	0.790
Working age population	1.000	0.840
Main workers share	1.000	0.607
Secondary and service sector workers share	1.000	0.722
Gross enrolment ratio higher education	1.000	0.841
Colleges per lakh population	1.000	0.820
Universities lakh population	1.000	0.822
Log college pass outs each year	1.000	0.877
Secondary schools per 1000 children	1.000	0.655
Hospital beds per 1000 population	1.000	0.766
Physician per 1000 population	1.000	0.723
Road density in km per km^2	1.000	0.729
Mode share of public transport	1.000	0.839
Average trip length in km	1.000	0.723
Access to treated tap water	1.000	0.777
Access to drinking water within premises	1.000	0.805
Households having electricity connection	1.000	0.781
Access to latrine within premises	1.000	0.721

(continued)

Table 3 (continued)

	Initial	Extraction
% of households connected to closed drainage	1.000	0.808
Household level coverage of MSW	1.000	0.740
Efficiency of collection of MSW	1.000	0.689
Extent of MSW recycled and recovered	1.000	0.756
Households availing banking services	1.000	0.865
Households having enclosed bathroom	1.000	0.890
Households having Computer/laptop with internet	1.000	0.865
Households using mobile phones	1.000	0.801
Households with condition of house as good	1.000	0.769
Households having permanent house structures	1.000	0.837
Families residing in own houses	1.000	0.775
Households using solar energy	1.000	0.822
Households using cleaner energy for cooking	1.000	0.720
Households having bicycle	1.000	0.727
Percentage of households not defecating in open	1.000	0.778
Log per capita ULB capital expenditure	1.000	0.653
Availability of master plans	1.000	0.621
Dedicated urban planning cadre within ULB/DA	1.000	0.704
Dedicated local government website	1.000	0.725
Publication of service level benchmarks	1.000	0.822
Single window civic service centres in ULB	1.000	0.513

Extraction Method Principal Component Analysis

Table 4 highlights the total variance explained by each of the underlying factor. SPSS by default, divides the table into three sub-sections, i.e. initial eigen values, extraction sums of squared loadings and rotation sums of squared loadings. For analysis and the purpose of interpretation we are mostly concerned with the extraction sums of squared loadings. It can be observed that the first factor accounts for 20.520% of the variance, the second 12.522% and the third 7.378%. A total of 13 factors were extracted by SPSS that can be labelled as significant. These 13 factors have an eigen value above 1 which was identified as the baseline value for this analysis. The 13 factors that were extracted define a cumulative 76.910% of the variance in the dataset.

For ease of interpretation, in this paper we have presented the rotated component matrix output instead of the simple component matrix output table before rotation. Table 5 shows the loadings of the 47 variables on the 13 factors extracted. The higher the absolute value of the loading, the more the factor contributes to the variable. But, it is possible that in large datasets such as the one used for this analysis, to see items with large loadings on several of the factors. The idea of rotation is to reduce the number of factors on which the variables under investigation have high loadings for the ease of interpretation. We have used the Varimax

Table 4 Total variance explained

Component	Initial eigen values			Extraction sums of squared loadings			Rotation sums of squared loadings		
	Total	% of variance	Cumulative (%)	Total	% of variance	Cumulative (%)	Total	% of variance	Cumulative (%)
1	9.644	20.520	20.520	9.644	20.520	20.520	5.713	12.156	12.156
2	5.885	12.522	33.042	5.885	12.522	33.042	4.969	10.572	22.728
3	3.466	7.375	40.418	3.466	7.375	40.418	3.682	7.835	30.563
4	2.936	6.246	46.664	2.936	6.246	46.664	3.127	6.653	37.216
5	2.377	5.058	51.722	2.377	5.058	51.722	2.779	5.914	43.130
6	2.142	4.558	56.280	2.142	4.558	56.280	2.736	5.820	48.950
7	1.868	3.974	60.254	1.868	3.974	60.254	2.313	4.920	53.871
8	1.716	3.651	63.905	1.716	3.651	63.905	2.271	4.833	58.703
9	1.524	3.243	67.149	1.524	3.243	67.149	1.981	4.215	62.918
10	1.306	2.778	69.927	1.306	2.778	69.927	1.899	4.040	66.958
11	1.177	2.503	72.430	1.177	2.503	72.430	1.812	3.856	70.814
12	1.095	2.329	74.759	1.095	2.329	74.759	1.613	3.432	74.246
13	1.011	2.150	76.910	1.011	2.150	76.910	1.252	2.664	76.910
14	0.914	1.945	78.855						
15	0.882	1.876	80.731						
16	0.813	1.729	82.450						
17	0.771	1.641	84.102						
18	0.690	1.469	85.570						
19	0.626	1.331	86.901						
20	0.610	1.297	88.199						
21	0.542	1.153	89.352						
22	0.515	1.096	90.448						
23	0.458	0.975	91.423						
24	0.419	0.892	92.315						

(continued)

Table 4 (continued)

Component	Initial eigen values			Extraction sums of squared loadings			Rotation sums of squared loadings		
	Total	% of variance	Cumulative (%)	Total	% of variance	Cumulative (%)	Total	% of variance	Cumulative (%)
25	0.388	0.826	93.141						
26	0.346	0.735	93.876						
27	0.328	0.698	94.574						
28	0.277	0.589	95.163						
29	0.267	0.568	95.731						
30	0.230	0.489	96.220						
31	0.217	0.461	96.681						
32	0.201	0.428	97.109						
33	0.194	0.413	97.522						
34	0.160	0.340	97.862						
35	0.146	0.311	98.173						
36	0.134	0.285	98.458						
37	0.120	0.255	98.713						
38	0.106	0.225	98.938						
39	0.091	0.194	99.132						
40	0.080	0.170	99.302						
41	0.078	0.166	99.468						
42	0.057	0.121	99.590						
43	0.056	0.118	99.708						
44	0.047	0.101	99.808						
45	0.040	0.084	99.893						
46	0.028	0.059	99.952						
47	0.022	0.048	100.000						

Extraction Method Principal Component Analysis

with Kaiser Normalization method and suppressed smaller values below 0.4 to achieve a simpler pattern. We can clearly see in Table 5 that variables associated with economic potential loads highly on Factor (component) 1, while indicators on basic service levels loads highly on Factor 2, and, education and health related parameters are substantially loaded on Factor 3. These factors with high loadings and clear patterns can be used as variables for further analysis.

6 Discussion and Way Forward

Classification of urban areas based on indicators is, in essence, a simplification by which large volumes of data for a set of measures gets processed to achieve a manageable and meaningful set of carefully measured groups (Mikelbank 2004). This paper attempts to derive the most critical factors that can measure the readiness of 98 Indian cities towards becoming sustainable smart cities. Using factor analysis on the data derived from different sectors, i.e. demographic, social, economic, education and health, basic infrastructure services, shelter, behavior and lifestyle and governance, we have developed a structure among a large number of selected variables that can generate insights into the state of cities. Rather than following an arbitrary approach (Coombes and Wong 1994), we have derived indicators in a systematic manner that can reliably lead to creation of sustainable city index for Indian cities. From the very beginning, this study was shaped by seeking the answers to basic questions, such as, 'what is the connection of this study with specific policy or programme objectives?'; and 'what policy instruments can use these findings? Our aim was to provide an assessment framework that can capture how different central government schemes (JnNURM, Smart Cities Mission, AMRUT etc.) have been impacting sustainability outcomes of Indian cities.

The results from the Principal Component Analysis suggest that the discussion on urban performance assessment in Indian cities should concentrate around three major topical issues: (1) social and economic opportunities, (2) access to education and health services and (3) availability of core infrastructure. Social and economic opportunities have emerged as the most significant measure. Social indicators such as female share of workers, creative and agile populations, and economic indicators including work participation rate, working age population, and share of service sector workers has shown significant loadings on Factor 1 explaining one fifth of the variance in the dataset. We have also observed highly significant loadings on factors related to education and health that include, number of universities per lakh population, number of college pass outs each year, hospital Beds per 1000 Population and Physician per 1000 Population. It is quite interesting to see that while the Government of India's Smart Cities Mission and other high profile urban development programmes are concentrating on building digital applications and high-tech urban precincts (Brookings 2016), the real need seems to rest with soft issues around socially equitable development and knowledge and health capital.

Table 5 Rotated component matrix[a]

	Component												
	1	2	3	4	5	6	7	8	9	10	11	12	13
Log population density					0.568								
Population growth rate			0.875										
Sex ratio												0.533	
Female share of workers	0.426							0.473					
Literacy rate	0.794												
Creative and agile population												−0.744	
Percentage of non-slum households											−0.553		
Log per capita GDP				0.402									
Work participation rate	0.674							0.410					
Working age population	0.800												
Main workers share	0.437												
Secondary and service sector workers share	0.670												
Gross enrolment ratio higher education								0.773					
Colleges per lakh population							0.866						
Universities lakh population			0.848										
Log college pass outs each year			−0.532										
Secondary schools per 1000 children							0.607						
Hospital beds per 1000 population			0.642										
Physician per 1000 population			0.767										
Road density in km per km^2					0.734								
Mode share of public transport				0.442								0.500	

(continued)

Table 5 (continued)

	Component												
	1	2	3	4	5	6	7	8	9	10	11	12	13
Access to treated tap water		0.602											
Access to drinking water within premises		0.615			0.401								
Households having electricity connection	0.439	0.733											
Access to latrine within Premises					0.719								
% of households with Waste water outlet connected to closed drainage		0.523											
Households with no wastewater drainage connection		−0.406								−0.432			
Household level coverage of MSW									0.627				
Efficiency of collection of MSW											0.755		
Extent of MSW recycled and recovered									0.824				
Households availing banking services						0.871							
Households having enclosed bathroom		0.844											
Households having Computer/laptop with internet						0.712							
Households using mobile phones	0.591					0.501							
Households with condition of house as good	0.508	0.650											
Households having permanent house structures		0.732											

(continued)

Table 5 (continued)

	Component												
	1	2	3	4	5	6	7	8	9	10	11	12	13
Families residing in own houses	−0.663												
Households using solar energy		−0.586									0.447		
Households using cleaner energy for cooking		0.461											
Households having bicycle										−0.434			
Percentage of households not defecating in open					0.677								
Log per capita ULB capital expenditure				0.728									
Availability of master plans													0.524
Dedicated urban planning cadre within ULB/DA	0.412												
Dedicated local government website										0.820			
Publication of service level benchmarks	−0.566												
Single window civic service centres in ULB				0.679									

[a]Rotation converged in 30 iterations
Extraction Method Principal Component Analysis
Rotation Method Varimax with Kaiser Normalization

The significance of India's demographic dividend, surging young and working age population and its impact on urban landscape is being debated for long now. A decade ago assessing the impacts of demographic dividend Chandrasekhar et al. found that the absorption of Indian youth into the labour force is not as high as one would expect (Chandrasekhar et al. 2006). This is primarily due to the poor employability of the workforce, which is severely affected by a deficit in educational attainment and health as identified by Ernst & Young and the Federation of Indian Chambers of Commerce and Industry in a report published in 2013 (Ernst & Young and FICCI 2013). Clearly there is circular linkage between demographic dividend, the need for jobs, skills and employability, access to education and health infrastructure. In fact, Reddy and Tiwari while measuring urban sustainability in India weighted education and health parameters more than any other measures (Reddy and Tiwari 2016). These discussions indicate that Indian cities need to emphasise on developing social infrastructure to promote sustainability, more than developing digital services, which the current regime is aggressively promoting. The significance of the social, economic, education and health factors in measuring sustainability outcomes of cities emerged from this study is therefore well substantiated with previous findings in scientific literature.

The results from this study also indicate that access to basic services (treated drinking water, waste water drainage, electricity and enclosed toilet-bathrooms) and shelter have considerable loadings on Factor 2. This implies that any assessment on urban sustainability performance of Indian cities must take into account the level of household access to water and sanitation related services and access to housing. This study outcome is strongly complimented by the observations of the High Powered Expert Committee for Water Supply and Sewage Sector formed by the Government of India in 2010 (Ministry of Urban Development 2010). The group found that nearly 40% of the households in urban India does not have access to piped water supply and 4861 out of the 5161 cities/towns in India do not have sewerage network (National Mission for Sustainable Habitat 2011). The committee recommended periodic benchmarking of water and sanitation services to assess the level of sustainability of Indian cities vis-à-vis municipal service delivery. The challenges around access to good quality housing is also echoed by many researchers and agencies. The National Institute of Urban affairs in its report highlighted that around 30% of the urban population in India live in poor quality, overcrowded accommodation with inadequate or no provision for basic infrastructure and services (National Institute of Urban Affairs 2011). It remarked, although public initiatives such as 'Basic Services for Urban Poor' conceived under JNNURM and 'Rajiv Awas Yojana' has to some extent been able to address issues around access to housing; persistent efforts are needed to improve housing delivery mechanism for enhancing overall sustainability standards in Indian cities. It is quite apparent from these discussions that housing and access to core services remains the key to India's urban sustainability. Therefore, indicator based assessments must take account of these domains to ideally reflect the present state of Indian cities and their future need.

Although the results from this study are promising, it has limitations in regard to sample size in the adopted statistical analysis. We have used factor analysis with a sample size of 98, whereas a sample size of 200 or more is considered as standard for such analysis. The deviation is justified as this paper primarily focusses on the 98 smart cities that are selected under the Governments of India's Smart Cities Mission. The focus of this paper is therefore more on targeting policy linkages of such analysis rather than merely conforming to the sampling adequacy. In future, we will explore the scope for increasing the sample size by including cities selected under other complimentary initiatives sponsored by Government of India. Furthermore, we will conduct cluster analysis and discriminant analysis to build typology of advantage and disadvantages in Indian cities that will help in identification of patterns of growth and progress vis-a-vis sustainability. Analysis such as these helps in improved targeting of policies and contribute to coordination of different agencies through unification of the current 'patchwork of programmers' related to urban development. Outcomes from these studies also help in linking the local urban performance with global development goals and sustainability targets.

References

Baru, R. (2003). Privatisation of health services: A South Asian perspective. *Economic and Political Weekly, 38*(42), 4433–4437.

Bhattacharya, S., Patro, S. A., & Rathi, S. (2016). Creating Inclusive cities: A review of indicators for measuring sustainability for urban infrastructure in India. *Environment and Urbanization ASIA, 7*(2), 214–233.

Brookings. (2016). *Smart cities mission: Installing digital technologies alone will not deliver results*. Retrieved 04 15, 2017, from https://www.brookings.edu/opinions/smart-cities-mission-installing-digital-technologies-alone-will-not-deliver-results/.

Bureau of Indian Standards. (2016). *Smart cities - indicators ICS 13.020.20*. New Delhi: Bureau of Indian Standards, Smart Cities Sectional Committee.

Carley, M. (1981). *Social Measurement and Social Indicators*. London: Allen & Unwin.

Census of India. (2011). *Provisional population totals urban agglomerations and cities*. New Delhi: Government of India. Retrieved 01 10, 2011, from http://censusindia.gov.in/2011-prov-results/paper2/data_files/India2/1. Data Highlight.pdf.

Chandrasekhar, C., Ghosh, J., & Roychowdhury, A. (2006). The 'Demographic Dividend' and Young India's Economic Future. *Economic and Political Weekly*, 5055–5064.

Chant, S. (2013). Cities through a "gender lens": a golden "urban age" for women in the global South? *Environment and Urbanization, 25*(1), 9–29.

Cities Alliance. (2015). Sustainable development goals and habitat III: Opportunities for a successful New Urban Agenda. Brussels: Cities Alliance Discussion Paper.

Commission, Planning. (2012a). *Report of the committee on JnNURM-2*. New Delhi: Ministry of Urban Development.

Commission, Planning. (2012b). *Report of the steering committee on urbanisation*. New Delhi: Planning Commission, Government of India.

Coombes, M., & Wong, C. (1994). Methodological steps in the development of multivariate indexes for urban and regional policy analysis. *Environment and Planning, 26*, 1297–1316.

Dameri, R. P., & Rosenthal-Sabroux, C. (2014). *Smart city: How to create public and economic value with high technology in urban space*. Verlag Berlin Heidelberg: Springer.

Ernst & Young, & FICCI. (2013). *Reaping India's promised demographic dividend - industry in driving seat.* Kolkata: Ernst & Young LLP.

Field, A. (2005, 10 12). *Factor analysis using SPSS.* Retrieved from http://users.sussex.ac.uk/~andyf/factor.pdf.

Florida, R. (2005). *Cities and the creative class*. New York: Routledge.

Fotheringham, A., & Wong, D. (1991). The modifiable areal unit problem in multivariate statistical analysis. *Environment and Planning, 23,* 1025–1044.

Ghertner, D. (2011). Rule by aesthetics: World-class city making in Delhi. In A. Roy, & A. Ong (Eds.), *Worlding cities: Asian experiments and the art of being global* (pp. 279–306). Blackwell Publishing Ltd.

Giddings, B., Hopwood, B., & O'Brien, G. (2002). Environment, economy and society: Fitting them together into sustainable development. *Sustain Dev, 10,* 187–196.

Government of India. (2011). *Strategic plan of ministry of urban development for 2011–16.* New Delhi: Ministry of Urban Development, Government of India.

Government of India. (2015). *Smart cities mission statement and guidelines.* New Delhi.

Grant Thornton. (2011). *Final report: Appraisal of Jawaharlal Nehru National Urban Renewal Mission (JnNURM).* New Delhi: Grant Thornton India. Retrieved from http://www.cmamp.com/CP/FDocument/JnNURMvolumeII.pdf.

Harrison, C., & Donnelly, A. (2011). A theory of smart cities. In *Proceedings of the 55th Annual Meeting of the ISSS,* Hull UK.

ISO 37120. (2014). *Sustainable development of communities—Indicators for city services and quality of life*. Retrieved 4 18, 2016, from https://www.iso.org/obp/ui/#iso:std:iso:37120:ed-1:v1:en.

James, K. (2011). India's demographic change: Opportunities and challenges. *Science, 576,* 576–580.

Jiang, Y., & Shen, J. (2010). Measuring the urban competitiveness of Chinese cities in 2000. *Cities, 27*(5), 307–314.

Kaiser, H. F. (1958). The varimax criterion for analytic rotation in factor analysis. *Psychometrika, 23*(3), 187–200.

Kaiser, H. F. (1974). An index of factorial simplicity. *Psychometrika, 39*(1), 31–36.

Kamath, L., & Zachariah, Y. (2015). *Impact of JNNURM and UIDSSMT/IHSDP Programmes on Infrastructure and Governance Outcomes in Cities/Towns in India.* Tata Institute of Social Sciences.

Knox, P. (1978). Territorial social indicators and area profiles. *Town Planning Review, 49,* 75–83.

McKinsey & Company. (2010). *India's urban awakening: Building inclusive cities, sustaining economic growth*. New York: McKinsey & Company.

Mikelbank, B. A. (2004). A typology of U.S. suburban places. *Housing Policy Debate, 15*(4), 935–964.

Miller, H., Frank, H., & Tribby, C. (2013). Developing context-sensitive livability indicators for transportation planning: A measurement framework. *Journal of Transport Geography, 26,* 51–64.

Ministry of Urban Development. (2009). *Handbook of service level benchmarking*. New Delhi: Government of India.

Ministry of Urban Development. (2010). *National mission on sustainable habitat*. New Delhi: Government of India.

Ministry of Urban Development. (2017). *Methodology for collection and computation of liveability standards in cities*. New Delhi: Government of India.

Ministry of Urban Development, G. o. (2016, 6 20). *Smart cities mission.* Retrieved from Smart Cities Mission: http://smartcities.gov.in/.

MoUD Govt. of India. (2011). *Report on indian urban infrastructure and services.* New Delhi: Ministry of Urban Development, Government of India.

National Institute of Urban Affairs. (2011). *Report on indian urban infrastructure and services.* New Delhi: Ministry of Urban Development, Government of India.

National Mission for Sustainable Habitat. (2011). *Report of the committee set up to frame national sustainable habitat standards for the urban water supply and sewerage sector*. New Delhi: Government of India.

Phadke, S. (2007). Dangerous liaisons women and men: Risk and reputation in Mumbai. *Economic and Political Weekly*, 1510–1518.

Reddy, B., & Tiwari, A. (2016). *Picking the winner: Measuring urban sustainability in India*. Mumbai: Indira Gandhi Institute of Development Research.

Shen, L., Ochoa, J., Shah, M., & Zhang, X. (2011). The application of urban sustainability indicators e A comparison between various practices. *Habitat International, 35,* 17–29.

Taylor, C. (1981). *Indicator systems for political, economic, and social analysis*. Cambridge, MA: Gunn & Hain.

Thompson, B. (2004). *Exploratory and confirmatory factor analysis: Understanding concepts and applications*. Washington, DC: American Psychological Association.

Town and Country Planning Organisation. (2011). *Report of the sub committee on development of sustainable habitat parameters in the field of urban planning*. New Delhi: Government of India.

UNSIAP. (2007). Building administrative data systems for statistical purposes addressing training issues and needs of countries. In *Inception/regional Workshop on reta6536: Improving Administrative Data Sources for the Monitoring of the mdg indicators.* Bangkok, Thailand: United Nations Statistical Institute for Asia and Pacific.

Wold, S., Esbensen, K., & Geladi, P. (1987). Principal Component Analysis. *Chemometrics and Intelligent Laboratory Systems, 2,* 37–52.

Analysing the Role of India's Smart Cities Mission in Achieving Sustainable Development Goal 11 and the New Urban Agenda

Anurita Bhatnagar, Triveni Prasad Nanda, Shilpi Singh, Kruti Upadhyay, Anil Sawhney and D. T. V. Raghu Rama Swamy

Abstract India is in a state of significant urban transition and transformation, which has resulted in the share of the urban population increasing from 18% in 1960 to 31% in 2011. It is expected that this growth would further accelerate, centred around cities, which are the existing centres of economic, social and urban importance. Recognizing the significance of cities in its future, the Government of India (GoI), in June 2015, launched the Smart Cities Mission (SCM), which focuses on the sustainable and inclusive development of Indian cities. This research aims to assess and analyse the synchronisation achieved between the two global treaties namely Sustainable Development Goal 11 (SDG 11) and New Urban Agenda (NUA) and SCM, twenty months post the commencement of SCM. The research analyses secondary data from existing SCM statements and guidelines and assesses them with respect to SDG 11 and NUA. Subsequently, it focuses on top nineteen cities under SCM. The research examines synergies between SDG 11, NUA and SCM, benchmarks the achievements and lacunae of the mission in pursuit of developing safe, resilient and sustainable cities.

Keywords Sustainable development · Smart cities · Urbanisation India · Sustainable development goals · New urban agenda

A. Bhatnagar · T. P. Nanda · S. Singh (✉) · K. Upadhyay · D. T. V. R. R. Swamy
RICS School of Built Environment, Amity University, 5th Floor, Block F-2, Sector 125 Noida, Noida, India
e-mail: shilpimago@gmail.com

A. Bhatnagar
e-mail: anurita.bhatnagar@gmail.com

A. Sawhney
Department of the Built Environment, Liverpool John Moores University, Liverpool, UK

W. Leal Filho et al. (eds.), *Sustainable Development Research in the Asia-Pacific Region*, World Sustainability Series, https://doi.org/10.1007/978-3-319-73293-0_16

1 Introduction and Background

Every civilization that has come into existence since the last eight millennia has gravitated towards a city-centric society (Pardo and Echavarren 2002). Be it, Jericho, in ancient Israel or the later centres of urban excellence such as Harappa (Indus Valley) or Samarkand (present-day Iran), cities grew from being mere trade centres to political and academic seats of authority, exemplified by the Greek city-states of Athens and Sparta. From ancient Rome to Constantinople, cities have served as religious loci and eventual points of confluence of wealth and power (Kostof 1993). Each city of importance, both in the past or the present, has its own significance and has been identified by a key characteristic such as fashion, finance, architecture or entertainment (Cohen 2006). The state of cities often signifies the maturity or lack of it for any civilization. Eminent urban journalist Jane Jacobs in her seminal work (Jacobs 1961), wrote, that "cities have the capability of providing something for everybody, only because, and only when, they are created by everybody". Some of the greatest cities in history have been magnets for human beings; attracting dwellers from near and far, with the promise of better quality of life attained through meaningful employment and enterprise. But post the Second World War, the world has seen an unprecedented rise in the number of cities and their inhabitants leading to adverse impact on the living environment. This has pushed the academic and bureaucratic discourse towards synthesizing a harmonious solution to the current predicament.

2 Literature Review

Large scale human settlements, when not well conceived are often accused of indulging in the indiscriminate use of natural resources, generating a large amount of waste and impacting the environment negatively (Hiremath et al. 2013). There is a growing consensus among nations, both developing and developed, to inhibit impending global environmental crises and introduce a collaborative statute which could be adopted and implemented by all. Being the foremost platform for international collaboration, the United Nations (UN) shifted its policy imperative from the Millennium Development Goals (MDGs) to the SDGs (UN 2015a, b), which establish the framework for development for the next fifteen years. The UN identified cities as the de facto centres of economic growth and has oriented the SDGs for greater sensitization towards the environment and sustainable development for all. Along with the SDGs, a framework for transformative commitments for sustainable urban development was introduced into the discourses in the form of the NUA (UN-Habitat 2016). Ratified by 193 member states of the UN, NUA is a voluntary, non-binding statute to guide urban development in a holistic manner, in accordance with the principles of SDGs.

The 17 SDGs represent diverse areas of action but are interlinked such that fulfilment of one goal may contribute to other goals too. Goal 11, as outlined in SDGs, is of particular interest for urban development. The goal is dedicated to '*Make cities inclusive, safe, resilient and sustainable*' (UN 2015a, b). According to Cities Alliance 2015 "The SDGs have the most direct relevance for Habitat III due to the inclusion of a stand-alone urban goal SDG 11, and the consideration of urban issues in several targets of other goals". The list of targets provided under SDG 11 has been included in Appendix 1.

The United Nations Conference on Housing and Sustainable Urban Development (**Habitat III**) was aimed at 'reinvigorating' the global political commitment to the sustainable development of towns, cities and other human settlements, both rural and urban. UN-Habitat deliberated on arriving at a consensus on an urban agenda that would commit countries to sustainable urbanization "which is now more critical than ever as populations, social interactions, economic activities, and environmental impacts are increasingly concentrated in cities" (UN-Habitat 2016). The product of that reinvigoration, along with pledges and new obligations, is the ***New Urban Agenda***. This agenda has set a new global strategy around urbanization for the next two decades. The Agenda includes four key transformative commitments for sustainable urban development:

- *Sustainable urban development for social inclusion and ending poverty*
- *Sustainable and inclusive urban prosperity and opportunities for all*
- *Environmentally sustainable and resilient urban development*
- *Effective implementation: Building the urban governance structure and Planning and managing urban spatial development* (UN-Habitat 2016).

According to Cities Alliance 2015 "The SDGs reveal a substantial but implicit urban dimension; the NUA could add value by formulating a clear aspirational vision, which can mobilize relevant urban stakeholders and guide local implementation." This thought has been one of the guiding elements of the current research.

The India Habitat 3: National Report details out the ways in which these international developments have had an overbearing effect on the long-term economic development paradigm for India, as the cities here are no different from their global counterparts (MoHUPA 2016). Coming from a long history of exceptional cities amidst a largely agriculture-based hinterland (MGI 2013), India is now facing an unprecedented rush of economic migrants to the existing pool of cities (Ramaswamy and Madakam 2013), that are overstrained due to excessive population and inadequate infrastructure (Hoelscher 2016). Today, urban India's population constitutes approximately 32% of the national population and the resulting urban activities, having accrued nearly 60% of the country's gross domestic product, have resulted in a noticeable shift in the structure of the Indian economy (MGI 2013). The potential of the urban economy, however, is crucially dependent on the levels of employment, socio-economic and demographic profile, spatial planning, inflation, investments (Kumar and Dahiya 2017). This rapid transition to a highly urbanized population has created several challenges for the planning, development, and operation of cities that are stimulating new thinking among architects, urban planners and designers, transportation engineers, utilities engineers, social scientists, environmental scientists, public finance and policy experts, municipal government officials, and, most recently, information technologists (Harrison and Donnelly 2011). As a measure to restrict further spread of decadence (MGI 2013) in the cities, a series of structured federally funded programs have been introduced over the years by the Indian government. Ambitious, mega projects such as the Jawaharlal Nehru National Urban Renewal Mission (JNNURM) and Rajiv Awas Yojna (RAY) were prime examples of many such schemes. Started with much flourish, these programs failed to deliver on the expected infrastructure augmentation and an integrated approach to urban development (World Economic Forum 2016). In 2014, the Indian government brought with it an array of urban development initiatives that were designed to address the challenges and tap the opportunities of urbanization (MoHUPA 2016). *SCM*, the flagship initiative of the Indian government's Ministry of Urban Development (MoUD), aims at streamlining urban growth and is one such mission which has been conceptualised to set the course for developing the blueprint for future Indian cities and human settlements.

Smart Cities provide a new form of instrumentation for observing the way people use a built space and enable an insight in developing new approaches to spatial development. Through new sources of information, cities hope to create insight, innovation, opportunity and real jobs that will increase prosperity and quality of life (Harrison and Donnelly 2011).

The objective of SCM is to promote cities that provide core infrastructure and give a decent quality of life to its citizens, a clean and sustainable environment and application of 'Smart' solutions. The focus of the mission is on sustainable and inclusive development. In the mission statement issued by the GoI, the mission "would set examples which may be replicated within and outside the cities, catalysing the creation of similar smart cities in the various parts of the country" (MoUD 2015). The development under SCM has been strategized such that local area development is prioritised with the use of technological solutions that lead to smart outcomes. Under the mission development is sought under two broad categories:

The Area-based development (ABD), which concentrates on retrofitting and redevelopment of existing areas within the city, to transform existing areas, including slums, into better-planned ones, thereby improving the livability of the whole city. There are three strategies adopted for ABD viz. retrofitting, redevelopment and Greenfield development.

Pan-city development concentrates on the application of selected smart solutions to the existing city-wide infrastructure. These solutions focus on the use of technology and relevant information to improvise on existing infrastructure and services. These include measure such as intelligent transportation systems, smart metering, integrated waste management, etc.

The selected cities under the mission were expected to develop proposals and subsequent implementation plans for a compact area which should include at least one of the ABD strategies as well as a pan-city solution. Usage of ICT-enabled 'smart' solutions has been indicated as a priority feature.

Much like their European counterparts, the Indian smart cities would be delivering incremental benefits primarily through urban retrofitting that integrates technological systems and solutions into existing economically developed cities (Hoelscher 2016).

Prima facie, SCM and the other urban development missions are a strong step ahead in imbibing these global sustainable development visions into India's urban policy mandate, supported by an implementable action plan.

But, with the shifting global focus towards inclusiveness, liveability, and sustainability, will the Indian cities fulfil the expectations of its citizens? Are the Indian cities capable of handling the current pressure as well as be resilient enough to ensure a sustainable future for the generations to come?

The current study, aimed at assessing the hypothesis that '*SCM translates the vision of global treaties like SDGs and NUA into implementable action plans at national and regional levels*' is part of an ongoing research project funded by the School of Construction, RICS School of Built Environment (SBE).

3 Research Aim and Objectives

This research aims to assess and analyse the synchronisation achieved between SCM, SDG 11 and NUA in the Indian context; twenty months post the commencement of SCM.

The research has been designed with the following objectives:

- Study the principles of SDG 11, transformative commitments of NUA and guidelines of SCM and identifying synergies between SDG 11, NUA and SCM
- Develop an assessment matrix for evaluating incorporation of sustainability principles in SCM implementation tool—the smart city proposals
- Assess the smart city proposals of nineteen smart cities identified during the first round of SCM in accordance with the assessment matrix
- Benchmark the achievements and lacunae of the mission implementation tools in pursuit of developing safe, resilient and sustainable cities
- Suggest measures for SCM course-correction, if any, for addressing the principles of sustainability.

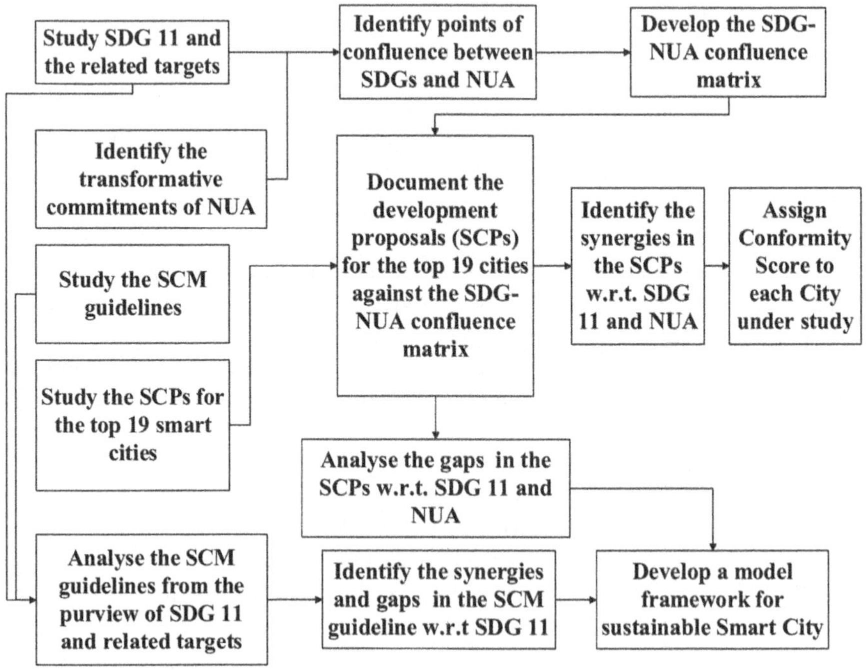

Fig. 1 Research methodology adopted

4 Research Methodology

Before identifying the synergies and the gaps between SDGs, NUA, and SCM, it is important to understand each of them separately. Further, synergies between SDG 11 and NUA need to be identified and analysed. Once the similarities were established, and anomalies identified, proposals of the nineteen[1] identified cities were assessed with respect to this matrix. Each city was scored for conformance to inclusiveness and sustainability principles identified under the SDGs and NUA. To do this, the following methodology was adopted (see Fig. 1).

To understand the **linkage** between an agenda of global importance and an urban development program of national importance, the research team concentrated on understanding the ethos and principles of SDG 11, NUA and SCM. To begin with, the ten targets under SDG 11 and the transformative commitments for sustainable urban development under the NUA were studied to establish the synergies between the SDG 11 and commitments for NUA. These synergies were translated into a matrix-based analytical framework. This matrix was further used to assess the Smart City Proposals (SCPs)—the tool for implementation of SCM, for nineteen

[1]Out of the twenty smart cities selected by GoI, New Delhi Municipal Corporation (NDMC) has been omitted from the analysis due to the unique governance characteristic of the ULB.

Table 1 Comparative analysis of SDG 11 and SCM guideline

SDG 11 targets[a]	Provision in SCM guideline
11.1	Yes
11.2	Yes
11.3	Yes
11.4	Yes
11.5	Yes
11.6	Yes
11.7	Yes
11.8	**No**
11.9	Yes
11.10	**No**

[a]Refer Appendix 1 for SDG 11 targets

cities with respect to addressing the principles of sustainability. Given the elementary stage of SCPs, the analysis was conducted by applying simple text analysis technique.

A preliminary analysis of the SCM statement and guidelines with respect to the SDGs reveal that the SCM guidelines, which define its broad framework for development and implementation, conforms to eight out of ten SDG 11 targets ***(80% conformance)***. The SCM guidelines do not take into cognizance Targets 11.8 "*Support positive economic, social and environmental links between urban, peri-urban and rural areas by strengthening national and regional development planning*" and 11.10 "*Support least developed countries, including through financial and technical assistance, in building sustainable and resilient buildings utilizing local materials*". The comparative analysis of SDG 11 targets and SCM guideline is presented in Table 1.

5 Expert Consultation

To better understand the SCM from a broader perspective, the research team approached several urban development experts including field practitioners, consultants, and representative of multi-lateral agencies like UN-Habitat (UNH), India. Based on these consultations it could be concluded that India's flagship mission is an excellent example of proactive approach with a matured thought process that considers the world vision. Unlike preceding urban renewal schemes, the mission has been strategized in a two-tiered format in which each state has been recognised as a decision maker, and the respective Urban Local Body (ULB) of each city has been recognized as the key implementing agency, whereas the central government's role has been relegated to an advisory one. SCM is perceived as the stepping stone for introducing integrated planning in India. The mission is being heralded as a remarkable step in contextualizing the global visions and providing clarity in the envisaged course of urban development in India.

6 Analysis of Synergies Between the Global Agenda and the Local Mission

To understand the extent to which India's key urban development mission addresses the principles of sustainability, it was important to first understand the relationship between the two globally important agenda of sustainable development and urban development. Figure 2 represents the methodology adopted for this assessment and development of matrix.

The targets of SDG 11 were compared with the commitments of NUA to establish the conformance and identify the outliers. A sample of this analysis has been included in Fig. 3.

Similarly, the remaining targets under SDG 11 were compared with NUA commitments.

It was found that forty-six transformative commitments of NUA conformed to the ten targets of SDG 11 as depicted in Fig. 4.

However, few of the commitments like equal opportunities, contribution of informal sector sustaining urban economies, etc. were found to align with other SDGs, and three commitments including tenure security, harnessing ageing factor and enabling policy framework for participatory planning and management of urban development have been identified as stand-alone commitments based on the current urban development patterns. These commitments have been highlighted in Appendices 2 and 3.

The nineteen identified smart cities' proposals were analysed with respect to the matrix thus developed. The research primarily focused on analysing each city's

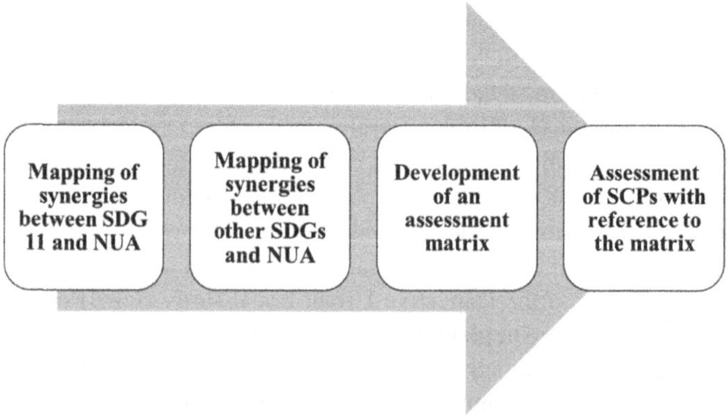

Fig. 2 Methodology for development of assessment matrix

SDG	SDG Description	NUA Sr. No.	NUA Description
11.1	Ensure access for all to adequate, safe and affordable housing and basic services and upgrade slums	1.1	eradicating poverty in all its forms and dimensions
		1.6	strengthening the coordination role of national, subnational and local governments; promoting adequate services, accommodation and opportunities for decent and productive work for crisis-affected persons in urban settings
		1.8	promoting national, subnational and local housing policies that support the progressive realization of the right to adequate housing for all
		1.9	development of integrated and age- and gender-responsive housing policies and approaches across all sectors
		1.10	stimulating the supply of a variety of adequate housing options
		2.4	promoting the role of affordable and sustainable housing and housing finance, including social habitat production
		3.15	strengthening the resilience of cities and human settlements, including through the development of quality infrastructure and spatial planning, by adopting and implementing integrated, age- and gender-responsive policies and plans and ecosystem-based approach

Fig. 3 SDG 11 and transformative commitments of new urban agenda (partial matrix)

SDG 11 Targets	Transformative Commitments of NUA									
11.1	1.1	1.6	1.8	1.9	1.10	2.4	3.15			
11.2	1.13	2.9	2.13							
11.3	1.16	1.17	1.18	1.19	2.7	2.10	3.4	3.11		
11.4	1.3	1.15	2.5							
11.5	2.14	3.5								
11.6	1.7	3.6	3.12							
11.7	1.14	2.12	3.5							
11.8	1.2	2.8	2.9	2.11						
11.9	2.2	2.3	2.13	3.1	3.3	3.5	3.9	3.16	3.17	3.18
11.10	3.8	3.13	3.14							

NUA – 1 "Sustainable urban development for social inclusion and ending poverty"
NUA – 2 "Sustainable and inclusive urban prosperity and opportunities for all"
NUA – 3 "Environmentally sustainable and resilient urban development"
NUA – 4 "Effective implementation: Building the urban governance structure and planning and managing urban spatial development"

Fig. 4 NUA commitments with respect to SDG 11 targets

SDG 11 Target & Description	Smart Cities: Vizag	Guwahati	Coimbatore	Kochi	Bhubaneswar	Jabalpur	Surat	Belagavi	Chennai	Pune	Indore	Ahmedabad	Solapur	Kakinada	Ludhiana	Davangere	Bhopal	Jaipur	Udaipur	NUA Transformative Commitment and description	
11.1 Ensure access for all to adequate, safe and affordable housing and basic services and upgrade slums	1	0	0	1	0	0	0	0	0	0	0	1	0	1	0	0	0	0	0	1.1	Eradicating poverty in all its forms and dimensions
	1	1	1	1	1	0	0	0	0	1	0	1	0	0	0	1	0	0	0	1.6	Strengthening the coordination role of national, subnational and local governments; promoting adequate services, accommodation and opportunities for decent and productive work for crisis-affected persons in urban settings
	1	0	1	1	1	1	1	1	1	1	1	1	1	1	0	1	1	0	1	1.8	promoting national, subnational and local housing policies that support the progressive realization of the right to adequate housing for all
	0	0	0	0	0	0	0	0	0	1	0	0	0	1	0	0	0	0	0	1.9	development of integrated and age- and gender-responsive housing policies and approaches across all sectors
	1	0	1	1	1	1	1	1	0	0	0	1	0	1	0	0	0	0	1	1.10	stimulating the supply of a variety of adequate housing options
	1	0	1	1	1	1	1	1	0	1	1	1	1	0	0	0	0	0	1	2.4	promoting the role of affordable and sustainable housing and housing finance, including social habitat production
	1	1	0	1	0	1	0	0	1	0	0	1	0	0	0	0	0	0	0	3.15	strengthening the resilience of cities and human settlements, including through the development of quality infrastructure and spatial planning, by adopting and implementing integrated, age- and gender-responsive policies and plans and ecosystem-based approach

Fig. 5 Partial analytical matrix

vision, strategic goals and focus and the implementation plan detailed in SCPs with respect to SDG-NUA analytical matrix. Points of conformity between the SCPs, targets of SDG 11 and transformative commitments of NUA were awarded a “1” and non-conformities were allocated “0”. Partial representation of the analytical matrix may be found in Fig. 5.

Research Findings

Following emerged as essential elements of most SCPs while conducting the analysis:

Affordable housing and provision of basic infrastructure services: 85% of the cities analysed have provisions for affordable housing as well as slum up-gradation in their SCPs. The focus has been on stimulating the supply of adequate, suitable housing options for the impoverished sections of society. 50% of the cities have the provisions to stimulate the supply of a variety of housing options. All cities have mentioned implementation measures to improvise on existing basic infrastructure facilities like water supply, sanitation as well as solid waste management. Service level up-gradation and implementation of smart solutions have also been included to ensure standardized service delivery and monitoring facilities.

Sustainable and intelligent transportation systems: 90% of the cities have identified sustainable, energy efficient and intelligent transportation systems as one of the significant intervention areas. As per the needs of each city, improvised transportation solutions have been identified. This feature of the SCPs emphasizes the use of non-motorised transportation, providing access to pedestrian-friendly walkways and identification of non-vehicle streets or zones, provision of public transportation facilities, seamless multi-modal mobility and intelligent transport management systems.

Cleaner cities and sustainable human settlements: As an effort to ensure visible improvement in the areas of intervention, all cities have identified measures to ‘clean-up’. This includes the provision of solid waste management facilities,

24 × 7 water supply and sanitation facilities, digitalization of citizen services and smart metering options for ease of monitoring and control. The introduction of planned development approach is seen as an integral part of the SCPs (Bhattacharya et al. 2015). Measures like transit-oriented development (TOD), mixed land-use, and urban renewal have been suggested on a need basis to streamline the urban development of the cities under consideration. Each city that has been analysed has identified one or the other urban development strategy for implementation in the identified area of intervention.

40% of the cities have identified implementation measures to include spatial development strategies for urban extension and renewal. 40% of the cities have identified long-term urban and territorial planning processes for water resource management, an essential feature for sustainability of human settlements.

90% of the cities have committed themselves to adopt a smart-city approach that makes use of opportunities from digitalization, clean energy, and technologies, as well as innovative transport technologies.

Resource conservation and efficiency: As a signatory to the Paris Agreement (UNFCC 2015), India needs to reduce its carbon footprint by 30–35% from its 2005 levels. In addition to this, India will have to diversify its power generation sources and shift them significantly towards renewable energy sources to reduce volumes of emissions per unit of GDP. In numbers, by 2025, India will need a 175-gigawatt power production capacity from non-fossil fuel sources (NIUA 2015). As an endeavour to encourage alternative sources of vehicular fuel, street lighting, sourcing of electricity and to give impetus to energy efficiency and water conservation, the city plans have provisions to encourage use of electric vehicles, LED-powered street lighting, tapping of solar and wind energy for power generation, encouraging energy-efficient buildings and including provisions for rainwater harvesting. 90% of the cities have identified measures to stimulate generation and use of renewable energy, waste reduction, and water conservation.

Integrated rural-urban development approach: One of the targets of SDG 11 is aimed at supporting rural-urban continuum. NUA too has dedicated transformative commitments to address the need for an 'integrated approach to support territorial systems that address rural-urban continuum' (UN-Habitat 2016). Even though regional and urban planning approaches have been suggested to streamline the existing urban sprawl, a consolidated/integrated urban development policy approach is missing. None of the cities under analysis reported an intention to address this transformative commitment of NUA. Consequently, there was no visible emphasis for knowledge exchange with international and national organizations for enhanced coordination in urban-rural development strategies.

Leveraging the proximity of resources to address sustainability challenges: One of the key features of ensuring sustainability is to ensure that the use of local material should be given precedence over distant sourcing (UNEP 2011). None of the cities make provisions for bringing this aspect into practice. For example, a significant proportion of construction work is envisaged as part of retrofitting and redevelopment while none emphasize the need to source this development via local means.

Human rights of refugees and land tenure security: All SCPs are silent on this aspect.

Inclusiveness, social equality and environmental sustainability: Inclusiveness is one of the building blocks of SDGs and NUA. Even though the smart city proposals have been developed after citizen consultations, adequate provisions are not visible for ensuring age and gender responsive development. Only one city has addressed this aspect. Similarly, only one city has identified the potential of harnessing experienced ageing population for its advantage. Even though 65% of the cities express commitment to developing vibrant, sustainable and inclusive urban economies, only 10% identify provisions to ensure business environments based on the principles of sustainability.

Recognition of informal sector and eradicating poverty: Emphasis has been placed on capitalising on the economic development of the city, but the role of informal sector has not been duly recognised. Only two cities (10% of the sample size) acknowledge the informal sector and address this aspect. Few provisions have been suggested to empower and streamline the informal sector which forms the backbone of our economy (Bhattacharya et al. 2015). SDGs identify eradication of poverty as a salient feature of an ideal urban development, but only 20% of the smart cities analysed have suggested measures incorporating/targeting this feature.

Climate Change Mitigation and Adaptation: India experiences climatic extremities which pose a serious hazard to human health, safety, and wellbeing (NDMA GoI 2016). Despite the risk of exposure, very few cities have identified this feature as a priority action area in the city proposals. Only 20% of the cities cite measures for addressing climate change impact.

Sustainable and resource efficient construction: Despite the scale of anticipated construction activity in India, only 20% of the cities identify sustainability and resource efficiency in construction as an area of intervention.

Urban economies, productivity, and employment: As stated earlier, 65% of the cities express commitment to the development of vibrant, sustainable and inclusive urban economies, only 20% cities seek to promote full and productive employment through identified measures and 25% cities seek to increase economic productivity.

Disaster risk reduction and resilience: Many of the cities that are being analysed are prone to natural disasters like earthquakes and cyclones and all urban settlements are at a risk of human-made disasters (UNEP 2007). Despite this fact, only 30% of the cities identify disaster risk reduction and resilience as a city plan feature.

Considering a consolidated view of the cities' performance, the research team found that the SCPs of five cities exhibited more than 50% conformance to the SDG11 targets and transformative commitments of NUA. Three cities' plans exhibited less than 25% conformance to SDG 11 targets as well as transformative commitments of NUA. The conformance score for each city is presented in Table 2.

Table 2 Smart city conformance score

Government of India ranking	City	Score	Conformance percentage (%)
8	Vizag	35	58.3
17	Guwahati	34	56.0
13	Coimbatore	33	55.0
5	Kochi	30	50.0
1	Bhubaneswar	30	50.0
7	Jabalpur	27	45.0
4	Surat	26	43.3
15	Belagavi	26	43.3
18	Chennai	25	41.7
2	Pune	22	36.7
11	Indore	22	36.7
6	Ahmedabad	21	35.0
9	Solapur	21	35.0
14	Kakinada	20	33.3
19	Ludhiana	18	30.0
10	Davangere	17	28.3
20	Bhopal	14	23.3
3	Jaipur	11	18.3
16	Udaipur	7	11.7

Source For GoI rankings: MoUD, GoI (2015)

7 Conclusions

It may be concluded that SCM is a commendable endeavour by the Indian government. The mission provides the basic framework to strengthen the core infrastructure and essential services. SCM has been successful in introducing layer of the sustainable thought process in the overall urban development fabric.

Considering the fact that the selected smart cities would become 'beacons of excellence' for the next tranche of seventy-eight smart cities (MoUD 2015), the mission needs greater rigor of incorporating inclusiveness and sustainability as guiding principles. Most of the cities have focused on adopting ICT based solutions for monitoring and improving the existing civic services but their SCPs have not outlines how they will address the 'quadruple bottom-line' (Deng et al. 2017) in its entirety.

In most of the cities' development plans, key sustainability characteristics such as disaster resilience, climate change adaptation, rural-urban connect, use of local material for construction, etc. have not been included.

Another conspicuous aspect is the fact that, the intervention areas in each of the SCPs are relatively small as compared to the respective city sprawl. The suggested interventions and action plans may work in a compact setting or isolated area but the mission lacks a definitive framework to replicate the same on an absolute urban scale.

In addition, the mission is seen to be focus on a 'part-to-whole' approach where smaller interventions come together to make a larger impact. However, these 'parts' also have their own urban interactions with impact towards sustainable development goals. The research team believes that if these 'urban interactions' are not duly leveraged by the cities, the development could be highly fragmented and fail to achieve the overarching sustainable goal. There is an imminent need to establish a framework to "shift scales from the parcel, block and neighbourhood to district, place, and region" (Chapple 2015).

As is evident, a long road remains ahead for Indian cities to take forward the sustainability mandate. It is pertinent to put measures in place today to ensure that urban development progresses in a "sustainable manner" and Indian cities emerge to be at par with their global counterparts.

8 Limitations and Scope for Further Research

Having said that, the research team recognises the difference in timelines of SDGs, NUA, and SCM. It also acknowledges the nascence of SCM as a flagship urban development mission and thus the lapse in integration with the world vision. The team also acknowledges the fact that the existing state of cities has not been analysed with respect to the sustainability principles.

Given the preliminary stage of SCM implementation, the team has not commented on the status of projects proposed in SCPs. This presents itself as a topic for further research. In addition to this, once the project implementation begins, the team would be assessing/evaluating the cities' performance against global benchmark such as ISO 37120 in the future.

Appendix 1: SDG 11 and Its Targets

SDG goal 11 targets	
11.1	Ensure access for all to adequate, safe and affordable housing and basic services and upgrade slums
11.2	Provide access to safe, affordable, accessible and sustainable transport systems for all, improving 11.2 road safety, notably by expanding public transport, with special attention to the needs of those in vulnerable situations, women, children, persons with disabilities and older persons
11.3	Enhance inclusive and sustainable urbanization and capacity for participatory, integrated and sustainable human settlement planning and management in all countries
11.4	Strengthen efforts to protect and safeguard the world's cultural and natural heritage
11.5	Significantly reduce the number of deaths and the number of people affected and substantially decrease the direct economic losses relative to global gross domestic product caused by disasters, including water-related disasters, with a focus on protecting the poor and people in vulnerable situations
11.6	By 2030, reduce the adverse per capita environmental impact of cities, including by paying special attention to air quality and municipal and other waste management
11.7	By 2030, provide universal access to safe, inclusive and accessible, green and public spaces, in particular for women and children, older persons and persons with disabilities
11.8	Support positive economic, social and environmental links between urban, peri-urban and rural areas by strengthening national and regional development planning
11.9	By 2020, substantially increase the number of cities and human settlements adopting and implementing integrated policies and plans towards inclusion, resource efficiency, mitigation and 11.9 adaptation to climate change, resilience to disasters, and develop and implement, in line with the Sendai Framework for Disaster Risk Reduction 2015–2030, holistic disaster risk management at all levels
11.10	Support least developed countries, including through financial and technical assistance, in building sustainable and resilient buildings utilizing local materials

Source UN (2015a, b). SDGs

Appendix 2: Other SDGs and Transformative Commitments of NUA

SDG	SDG description	NUA Sr. No.	NUA description
SDG 10	Reduced inequalities	1.4	Promoting equally shared opportunities and benefits that urbanization can offer; achieve their full human potential
SDG 10	Reduced inequalities	1.5	

(continued)

(continued)

SDG	SDG description	NUA Sr. No.	NUA description
			Ensuring full respect for the human rights of refugees; strengthening synergies between international migration and development at the global, regional, national, subnational and local levels by ensuring safe, orderly and regular migration through planned and well-managed migration policies, and to supporting local authorities in establishing frameworks that enable the positive contribution of migrants to cities and strengthened urban-rural linkages
SDG 8	Decent work and economic growth	2.1	Sustained, inclusive and sustainable economic growth, with full and productive employment and decent work for all
SDG 8	Decent work and economic growth	2.15	Increasing economic productivity
	Decent work and economic growth	2.16	Promoting, as appropriate, full and productive employment
SDG 12	Responsible consumption and production	2.17	Commit ourselves to promoting an enabling, fair and responsible business environment based on the principles of environmental sustainability
SDG 8	Decent work and economic growth	2.18	Recognizing the contribution of the working poor in the informal economy, particularly women, including unpaid, domestic and migrant workers, to the urban economies, taking into account national circumstances
SDG 9	Industry, innovation and infrastructure	2.19	Sustaining and supporting urban economies to transition progressively to higher productivity through high-value —added sectors, by promoting diversification, technological upgrading, research and innovation
SDG 4	Quality education	2.20	Harnessing the urban demographic dividend, where applicable, and to promoting access for youth to education, skills development and employment

Source UN-Habitat (2016). New Urban Agenda

Appendix 3: Standalone Commitments of NUA

Stand alone commitments	
1.12	At the appropriate level of government, including subnational and local government, increased security of tenure for all, recognizing the plurality of tenure types, and to 1.12 developing fit-for-purpose and age-, gender- and environment-responsive solutions within the continuum of land and property rights, with particular attention to security of land tenure for women as key to their empowerment, including through effective administrative systems
2.21	Addressing the social, economic and spatial implications of ageing populations, where applicable, and harnessing the ageing factor as an opportunity
4.1	Require enabling policy frameworks at the national, subnational and local levels, integrated by participatory planning and management of urban spatial development

Source UN-Habitat (2016). New Urban Agenda

References

Bhattacharya, S., Rathi, S., Patro, S. A., & Tepa, N. (2015). *Reconceptualizing smart cities: A reference framework for India*. Bangalore: Center for Study of Science, Technology and Policy.

Chapple, K. (2015). *Planning sustainable cities and regions: Towards more equitable development* (1st ed.). New York: Routledge.

Cities Alliance. (2015). *Sustainable development goals and habitat III: Opportunities for a successful New Urban Agenda*. Brussels: Cities Alliance.

Cohen, B. (2006). Urbanization in developing countries: Current trends, future projections, and key challenges for sustainability. *Technology in Society, 28* (1–2), 63–80.

Deng, D., Liu, S., Wallis, L., Duncan, E., & McManus, P. (2017, February 8). Urban sustainability indicators: How do Australian city decision makers perceive and use global reporting standards? *Australian Geographer*, 1–16.

Harrison, C., & Donnelly, I. (2011). A theory of smart cities. In Proceedings of the 55th annual meeting of the ISSS. Hull: ISSS.

Hiremath, R. B., Balachandra, P., Kumar, B., Bansode, S. S., & J.Murali (2013). Indicator based urban sustainability—A review. *Energy for sustainable development, 17*, 555–563.

Hoelscher, K. (2016). The evolution of the smart cities agenda in India. *International Area Studies Review, 19*(1), 28–44.

Jacobs, J. (1961). *The death and life of great American cities*. New York: Random House.

Kostof, S. (1993). *The city shaped: Urban patterns nad meanings through history* (2nd ed.). New York: Bulfinch.

Kumar, T. V., & Dahiya, B. (2017). Smart economy in smart cities. In T. V. Kumar (Ed.), *Advances in 21st century human settlements* (pp. 3–76). Singapore: Springer.

MGI. (2013). *India's urban awakening*. Mckinsey Global Institute.

MoHUPA. (2016). *India habitat 3: National report*. New Delhi: Ministry of Housing and Urban Poverty Alleviation (MoHUPA), GoI.

MoUD. (2015). *Smart city: Mission statement and guidelines*. New Delhi: Ministry of Urban Development, GoI.

NDMA GoI. (2016). *National disaster management authority*. Retrieved March 28, 2017, from http://ndma.gov.in/en/vulnerability-profile.html.

NIUA. (2015). *Exploratory research on smart cities: Theory policy and practice*. New Delhi: NIUA.

Pardo, M., & Echavarren, J. M. (2002). Urban development and its forms: Origins and new challenges for the twenty-first century. In *Encyclopaedia of life support systems*. UNESCO.

Ramaswamy, R., & Madakam, S. (2013). The state of art: Smart cities in India: A literature review report. *International Journal of Innovative Research and Development, II*(12), 115–119.

UN. (2015a). SDGs. Retrieved March 10, 2017, from https://sustainabledevelopment.un.org/post2015/transformingourworld.

UN. (2015b). UN. Retrieved March 10, 2017, from http://www.un.org/sustainabledevelopment/page/2/.

UNEP. (2007). *Cities and urban vulnerability in the context of urban environmental management*. UNEP.

UNEP. (2011). *Decoupling natural resource use and environmental impacts from economic growth, A report of the working group on decoupling to the International Resource Panel*. Geneva: UNEP.

UNFCC. (2015). UN: Paris Agreement. Paris: UN.

UN-Habitat. (2016, October 22). *New Urban Agenda*. Retrieved January 5, 2017, from https://unhabitat.org/new-urban-agenda-adopted-at-habitat-iii/.

World Economic Forum. (2016). *Reforms to accelerate the development of India's smart cities shaping the future of urban development & services*. Geneva: World Economic Forum.

The Housing Gap—Sydney, Australia

Janet Chappell and Nicole Campbell

Abstract The median house price in Sydney is $1.12 million (Domain,). For many households, the gap between incomes and median house prices is wide. Those ranging from pensioners to teachers earn less than the amount needed to service a loan to buy a median priced apartment—even if they had the necessary $71,000 deposit (UrbanGrowth NSW 2016). Sydney recently leap-frogged Vancouver becoming the second most expensive city in the world, behind Hong Kong (Day et al. 2016). With population growth continuing to drive demand for housing, the gap between incomes and house prices shows no sign of shrinking. Landcom (formerly Urban Growth NSW), the NSW Government's land and housing development agency, is required to respond, as stipulated in the NSW Government's strategy *'The government will...provide affordable housing in government-led urban renewal projects and on government-owned sites to meet the shortfall in affordable housing'* (NSW Government 2014). There is an abundance of academic research on housing and housing affordability This paper proposes a very simple one-page diagram which details an analysis of housing price-points compared to householder's ability to pay. The purpose of keeping to just one page is to respond to busy lives, short attention spans and the need to convey meaning in that environment. Award wages are charted for example households and compared with incomes needed to attain housing on a spectrum from social to private market housing. There is need for diverse housing for rent or purchase for those on moderate incomes and above *and* more subsidised Affordable Housing to rent for those on very low to moderate incomes and more social housing safety net and the private market can continue to take care of the high income households. Households working in many occupations are affected, not solely those on very low incomes. This supports looking to innovate in the private housing market with excellent design, innovation and ingenuity in construction and financing and

---

J. Chappell (✉) · N. Campbell
MLC Center, Level 12, 19 Martin Place, Sydney, NSW, Australia
e-mail: jchappell@landcom.nsw.gov.au

N. Campbell
e-mail: ncampbell@landcom.nsw.gov.au

W. Leal Filho et al. (eds.), *Sustainable Development Research in the Asia-Pacific Region*, World Sustainability Series, https://doi.org/10.1007/978-3-319-73293-0_17

planning/regulatory efficiencies to increase the supply of both more rent-controlled Affordable Housing and better overall housing affordability.

Keywords Housing · Affordability · City · Sustainability · Sydney

1 Introduction

Articulating the mismatch between incomes and household need, and available housing, is vital to address failure to meet what the New Urban Agenda (Habitat 11 2016) refers to as the basic social function that a city should serve being the right to adequate housing as a component of the right to an adequate standard of living (along with drinking water, sanitation, food security, education etc.). Linked to this provision of housing shelter, is the inter-related social system in our cities that reduces inequality within countries—as is the focus of Goal 10 of the United Nations Sustainable Development Goals, but also a key component of achieving Goal 11 to make cities inclusive, safe, resilient and sustainable.

There is a broad and growing preoccupation on housing affordability in Australia and particularly Sydney. In terms of research and opinion there is an enormous amount of material. It would be impossible in a paper this size to chronicle comprehensively the situation and that is not the intention. Rather, this paper outlines the process taken and response to this complexity and the diverse range of information, viewpoints and confusing terminology that resulted eventually in one simple diagram. The diagram summarises the critical thinking that took place leading to identification of a broad section of the community that have very limited housing choices within their budgets. The diagram encourages focus on the spectrum of areas that require action in relation to housing affordability and diversity.

This paper has three specific elements:

Part One contextualises the issue of housing affordability in metropolitan Sydney and describes why the approach taken is to provide a holistic view of housing need that takes in all income levels, all types of housing and tenures and allays some widely held myths about who needs more affordable housing.

Part Two explains the critical thinking behind the diagram and explains all the parts of the diagram itself—unpacking this simple visual tool with definitions of key terms such as affordable housing and the housing gap, which gives this paper its title and which uses the term, ‘diverse housing’ to explain the diagram and the policy opportunities.

Part Three considers what the diagram identifies as the actions needed to address housing affordability and diversity in metropolitan Sydney to support the delivery of a broad range of housing provided by a broad mix of players and involving collaboration across all levels of government and industry.

2 Part 1 Context and Why This Approach

Key findings from a recent NSW Parliament review (Angus 2017) indicated that:

- Greater Sydney's dwelling prices have risen by an average of 6.6% each year since 1991
  - Median price, Sept 1991: $158,000
  - Median price, Sept 2016: $776,000.
- Median dwelling prices in Sydney are up to 69% higher than the NSW median ($618,000)
  - Inner Sydney: $1.042 m (+69% NSW median)
  - Middle Sydney: $840,000 (+36%)
  - Outer Sydney: $685,000 (+11%).
- Sydney house prices have outpaced wage growth in NSW over the past 25 years
  - Cumulative house price growth: 180%
  - Cumulative wage growth: 99.6% (Angus 2017).

The main causes and challenges are well explained by the work of 100 Resilient cities organisation (2017) which says:

> Housing affordability is a significant problem in both the rental (Rental Affordability Index 2017) and owner occupied housing markets (Eslake 2017) across metropolitan Sydney. Market prices for property are being driven up by population growth and immigration, desirability of inner city living, demand for investment properties and low interest rates. The fact that income growth has not kept up with purchase and rental prices compounds the problem. The most affordable housing for buyers and renters alike is found in the west and southwest of Sydney. These areas are located far from jobs forcing residents to commute long distances to work. Poor public transport in the west and southwest mean this commute is largely by car. This is particularly an issue as population growth is expected to be strongest in the West Central and Southwest Districts, nearly doubling by 2031 to 2.7 million, about 45% of Metropolitan Sydney's total population. There is also a lack of diversity of housing catering to Sydney's growing, aging and diverse population, with a need for user centred and people driven responses to housing affordability issues, including housing type and financing structures.
>
> The median housing price in Sydney is now 12.2 times the median household income. Rental affordability in metropolitan Sydney is in a critical position with the average household required to spend 28% of income to access a rental dwelling, very close to the 30% housing stress threshold, and more than 30% within the inner city.

However, in addition to the key statistics, the main causes and challenges broadly summarised above, there are many more issues and aspects, ideas and contributing factors that stem from it.

For example about the need for private investment capital to fund housing such as purpose build-to-rent apartments that are a common form of housing in many other countries and attract good levels of private investment but not yet in Australia (Fizzy Living). Other funding measures such as providing an aggregated bond model to improve lending conditions for the Community Housing sector as described by Lawson (2014). Increasing levels of homelessness such as is measured by the City of Sydney local government (2017).

There is also debate on the appropriate level of foreign investment in housing which has lead to intervention by the Commonwealth Treasurer (Morrison 2017). Regulation and governance discussion relating to housing affordability has included debate about federal tax settings including negative gearing (Daley et al. 2016) and capital gains tax and family home exclusion from asset testing for pensions.

State government based taxation such as land tax and stamp duty discouraging relocation to appropriate accommodation is also an important area of policy discussion (NSW Business Chamber and NCOSS 2016).

Another related area of concern is the level of Australian household debt, which is currently the second highest in the world after Switzerland (Angus 2017).

Also concerning are the wasted housing opportunities with ‘cling-film’ empty apartments that investors are leaving uninhabited as reported by McKenny et al. (2016)—once again recently responded to by the most recent Federal Budget (2017).

Another global tourism impact affecting housing affordability is the rise of holiday letting making housing out of reach for locals (Sansom 2016).

Another important element playing a large or small role in limiting housing supply depending on who you ask, is anti-development sentiment by communities in local areas, as mentioned by Fielding (2014).

Ageing in place and facilitating downsizing for empty nesters to free-up larger homes for larger households is also topical (Saner et al 2017).

The rapid increases in house prices in Sydney has led to much hypothesising of whether there is a real estate bubble or correction coming that may lead to better affordability but devalue properties with large mortgages (Ryan 2017) which in-turn leads to attention on whether bank lending practices may be too lax, leaving many households exposed if such a price drop should occur or even if interest rates rise a small amount (Shapiro 2017).

Meanwhile there are many ideas being floated to try to help a generation of Sydney-siders to purchase their first home—for example by allowing them to use their superannuation funds to put a deposit down on a home (Janda 2017) and the consternation of many who see the inflationary effect this could create and other unintended consequences.

All of the above issues are research-worthy, but there is one particularly fundamental conflict in Sydney’s populous—being the fact that one household’s un-affordable housing nightmare is another household’s well-earned retirement nest egg.

The net effect of the multitude of housing affordability related factors and opposing forces is often polarising or paralysing action to address housing affordability, and is certainly not a recipe for co-ordinated response.

The use of a diagram helps to overcome barriers that exist between different cultures and expertise silos that occur for example between state and local government, between neighbours and prospective residents, between the social services sector and the private market housing development sector, between generations who saw bricks 'n mortar ownership as a must and millennials and everyone in between.

A common language helps describe the way all these perspectives form part of a picture.

In researching housing policy and through conversations with colleagues, stakeholders, friends and family around the dinner table it becomes readily apparent that:

- not everyone knows that social housing, housing affordability and Affordable Housing are not the same thing
- depending on the socio-economic group you're speaking to, affordability in one person's view is about how hard it is to buy a home to live in or invest in, while others are referring to the basic need to get a roof overhead and avoid homelessness
- local opposition to increased density and change is not always understood to be linked to a lack of housing options in your area for adult children, downsizing or ageing relatives
- the reason you may not have childcare places that allow parents to do paid work might also be related to the cost of housing in the area
- people who qualify for Affordable Housing are not a bunch of undesirables but were 60% of all households in 2011 i.e. most people
- the increasing need for housing supply is a factor of more births than deaths in our city, divorce and ageing—immigration is not the main cause of population growth in Sydney.

Drawing on the comprehensive academic research from the collective Australian Housing and Urban Research Institute (AHURI 2017) and other operational sources, the approach taken in this paper has been to interpret the information diagrammatically.

3 Part 2 The Diagram and Definitions

Consolidating key information on housing affordability into a simple diagram is the approach that has been successfully applied by Landcom, the New South Wales state government developer and state owned corporation.

The use of localised housing affordability data, coupled with need-based indicators, has facilitated greater clarity by policy makers on the key housing affordability issues. This approach has informed the development of Landcom's Housing Affordability and Diversity Policy (released in 2017). The diagram has also facilitated collaboration between Landcom and other government agencies as well as the private development sector in terms of working with others toward a shared and broadly understood purpose.

Conveying meaning in a one-page diagram helps explain key terms and housing types. It is obviously important to balance sophisticated information relating to a truly complex topic, with the need to inform but not overwhelm the audience. This may be an approach that is useful to other aspects of sustainability to better enable action.

This particular diagram highlights the fundamental problem that incomes, without inheritance contributions, do not match the cost of housing in Sydney. The diagram also highlights the significant discrepancy between minimum Award wages and actual housing costs.

The top section of the diagram shows household income levels—with lowest on the left $0 income, through to highest on the right $200,000 per annum and beyond.

Below that (represented by coloured blocks) are the broad types of housing that align in terms of ability to pay for affordably within the household budget, for that household income range. These types of housing use terms like Affordable Housing etc. are each defined below. These terms are used carefully since widely differing definitions distracts attention from the matter at hand and a common language allows discussions and hopefully solution-finding. From left:

Social Housing—the pale blue coloured arrow represents the households in the community who qualify for Social Housing. That is, they earn less than 50% of the median income level. It is important to understand that there is not a matching

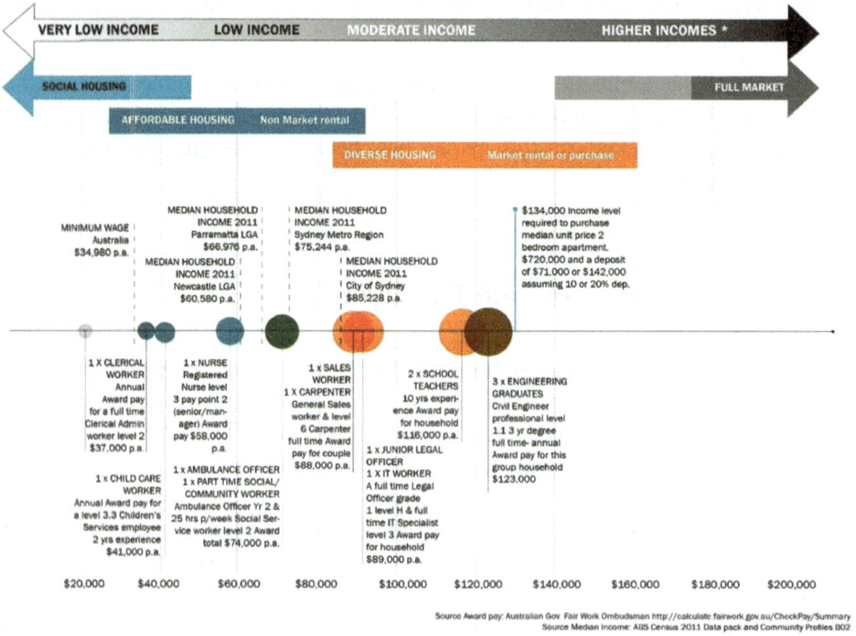

Fig. 1 Diagram of incomes and housing

amount of homes for this group—many people struggle by in homelessness, or paying more than they can afford in the private rental market, or in possibly overcrowded housing shared, in their cars etc. while the waiting list for a 2 or 4 bedroom home is over 10 years in Sydney according to the Centre for Affordable Housing. This could be described as 'safety net housing' paid for by broad based taxation for those in the community that are most in need.

Affordable Housing—(the darker blue colour) is housing with controlled rents, managed by Community Housing Providers who are not for profit organisations. Households who qualify for Affordable Housing earn up to 120% of the median income, however they can also earn less than 50% of the median income—therefore overlapping with the Social Housing definition. At time of writing this is approximately a maximum of just over $90,000 gross per annum, but varies according to the size of the household. As a rule of thumb, housing is usually considered affordable if it costs less than 30% of gross household income (Centre for Affordable Housing 2017a, b, c).

Full Market housing—(shown in two shades of grey arrow), this is where newly constructed housing is typically priced. The market will naturally provide new housing to high income households. The lighter grey indicates the need to move the grey arrow to the left to improve affordability, to meet the needs of more than just the wealthiest households and investors. The price of land in Sydney along with other demand forces such as low interest rates for bank loans mean that this is now standard pricing of most existing housing stock as well as newly built homes.

Diverse Housing—This is a new term that is needed to span the very wide gap shown in orange, between the affordable housing and full private market housing. It can refer simply to the range of price points for purchase or rental and also refer to having a range of tenure options such as long term rental, shared or community ownership, and the suitability of housing types in matching household sizes, and ways of living. In other words, the variety of housing should match the variety we have in our community.

The bottom section of diagram shows a series of households as coloured dots.

This is a practical and real type of analysis which compares housing price-points to householder's ability to pay. Award wages (Fair Work Ombudsman 2017a, b) are charted for example households and compared with incomes needed to attain housing on along the housing spectrum.

The award wage data, is used to create fictitious examples of households with a mix of occupations and part-time, full-time and home duties. What is immediately apparent is that most households are on the left of the diagram. In other words, many combinations of occupation, that are reasonably well remunerated and are important to the ability for the city to function, are still a long way from the income level needed to purchase housing in Sydney.

The mid-section of diagram also includes a number of key statistical markers. These provide an indication of where in the housing spectrum various benchmark incomes align. These include:

- Minimum wage in Australia $34,980 p.a. (Fair Work Ombudsman 2017a, b)
- Newcastle median household income
- Parramatta LGA median household income
- Sydney Metro region median household income. When the 2011 census was conducted, this was $75,244 p.a (ABS 2011). This is, it seems by regional and national standards, a high income level. It may therefore come as a surprise to many, that this level of income qualifies at the upper end of the range for Affordable Housing. This income level would not be able to service a loan to purchase a property and may also struggle to pay private market rent without being in housing stress. That is, more than 30% of total income required to pay for housing, leaving insufficient funds to cover the costs of food, utilities, transport etc.
- City of Sydney local government area median household income
- The blue line shows that an income level of $134,000 (ANZ 2017) is required to purchase the median 2 bedroom apartment which is at-time-of-writing $720,000 in metropolitan Sydney. This would require a deposit of $72,000 based on a 10% deposit for a principal and interest loan over 30 years for an owner occupier—however some lenders may require 20% deposit. This is based on a scenario where the household pays exactly 30% of their gross income to mortgage payments however is conservative given it does not include the cost of strata fees or maintenance costs or stamp duty payment on the transaction (UrbanGrowth NSW 2016).

4 Part 3 What The Diagram Tell Us

The diagram indicates our housing is not suited to the Sydney metropolitan population in a number of ways:

- there is not enough Social Housing
- there is not enough Affordable Housing
- the market is focusing on providing for the wealthiest households
- there is a big gap between incomes and market housing costs.

There is broad consensus that we need more new homes in Sydney, for example as put by Pradolin in the Australian Financial Review (2017), or at least, as put by Pawson et al. (2015) that *action to maximise supply is unquestionably part of the required strategy* but that alone will not resolve affordability issues. We need to work on *what sort of housing* (author's emphasis) we are supplying. Apart from the need to reduce overheated demand for housing as Visentin quoted Phibbs (2016) there are at least four key housing supply aspects to address:

1. Providing social housing to those most in need
2. Providing affordable housing to rent for those on very low to moderate incomes
3. Creating housing diversity to match the budgets and circumstances of a broader range of renters and potential home owners on moderate incomes

4. Maximising delivery of housing by community housing providers and industry through innovation and leveraging a range of planning, design and financial support.

If not addressed, the urban renewal projects, and probably all other new housing developments occurring within and around metropolitan Sydney, will not feature any housing that is 'affordable'. Meaning housing that rents between $236 and $567 per week, or for sale for between $224,000 and $538,000—considered affordable for very low to moderate income households according to the Centre for Affordable Housing (2017a, b, c).

There is need for diverse housing for rent or purchase for those on median incomes and above *and* more subsidised Affordable Housing.

Households working in many occupations are affected, not solely those on very low incomes. This supports looking to innovate in the private housing market with excellent design, construction and planning/regulatory efficiencies to increase the supply of more Affordable Housing and overall housing affordability.

The diagram also indicates the interconnectedness of the housing spectrum. If no purchase options are available, especially for those on reasonably high incomes, these households will necessarily rent and therefore compete for limited rental properties with those for whom renting will be the only option for most if not all of their lives. As described by Yates (2014) in her submission to the Senate Economic References Committee on Affordable Housing—this increase in rental demand from thwarted buyers *in combination with a decline in social housing stock has created additional demand pressures on the private rental market, felt most severely at the lower end of the market.*

Although homelessness is not shown on the diagram it could be considered that by not meeting the needs of those in the middle to right hand upper incomes, the result is to knock everyone to the left of the diagram and the extreme left is falling off altogether into homelessness. A future expansion of this diagram would include crisis accommodation and shelters on the left.

There is a need for the private sector/non-subsidised housing to contribute to addressing affordability. The amount of housing needed for lower income households is too great to be met solely by the Community Housing Provider sector—a combination of partly subsidised and responsive market housing is important.

The diagram obviously doesn't show everything, but sets a framework to consider related unsustainable outcomes of unaffordable housing, for example the spatial effect. In Sydney, and other parts of the world, expensive housing clusters in areas convenient to employment opportunity and high amenity. This has both social and economic impact in compounding disadvantages such as income-earning potential, requiring long distance commuting and the flow on effects on family and leisure and environmental damage of sprawl, high levels of private vehicle travel and an unhealthy monoculture in the wealthy exclusive areas and concentrations of disadvantage many kilometres away. These differences and inequities are being passed to future generations since some have benefited from enormous capital gain windfalls on their homes allowing them to accumulate multiple properties, while others have been potentially homeless or couch surfing or in unstable rental accommodation with the reduced educational/employment opportunities that can stem from that for both adults and children in that situation.

This diagram provides a tool to speak to existing communities and the development industry to help put forward the collective benefit of providing a range of housing for a range of households to live in every part of Sydney.

5 Conclusion

A simple one-page diagram has proven to be a highly useful tool for Landcom in focussing attention across a diversity of stakeholders on the housing affordability challenges of metropolitan Sydney.

It has been successfully used as a point of reference for discourse between different levels of government, across state agencies, academia, financiers, developers, not-for-profit organisations, and the general public.

One diagram combined with localised needs-based indicators, enables the use of market data and other evidence to drive action and cooperation toward a sustainable city with housing that is affordable, suitable, diverse and available for everyone.

Of course, there is no single step to resolve the housing affordability challenges. A multi-pronged approach which includes public sector provided, not-for-profit subsidised and private sector innovation are all needed across the spectrum of social, affordable and diverse market housing—which are all important and inextricably linked—because a failure at one point affects all.

There is a need for government to act and particularly on government owned land (NSW Department of Planning 2014). How then, do we collectively respond? Landcom is coalescing efforts toward collective government and private sector delivery of more social, affordable and diverse private market housing, while also encouraging industry to innovate. Landcom also has a key role across government in ensuring optimal overall housing supply—which appropriately responds to the diverse needs of our local communities.

References

ABS Australian Bureau of Statistics. (2011). Census 2011, Source Median Income: ABS Census 2011 Data pack and Community Profiles B02.

ANZ Home Loan Calculator. (2017). Based on a loan amount of $648,000 and interest rate of 4.65% and monthly repayments of $3,342 as calculated by the ANZ bank on 13 April 2017 see https://www.anz.com.au/personal/home-loans/calculators-tools/calculate-repayments/.

Angus, C. (2017). Demand, deposits, debt: Housing affordability in Sydney, Briefing Paper No 1/2017, NSW Parliamentary Research service, March 2017.

AHURI (Australian Housing Urban Research Institute). (2017). http://www.ahuri.edu.au/?gclid=CK_24ci0oNMCFdAKKgodZ8sHHw.

Centre for Affordable Housing. (2017a). www.housing.nsw.gov.au/centre-for-affordable-housing.

Centre for Affordable Housing. (2017b). Housing pathways website accessed 11 April 2017 http://www.housingpathways.nsw.gov.au/how-to-apply/expected-waiting-times/sydney,-south-eastern-sydney-and-northern-sydney.

Centre for Affordable Housing. (2017c). What is affordable housing? Accessed 11 April 2017 http://www.housing.nsw.gov.au/centre-for-affordable-housing/about-affordable-housing.

Centre for Independent Studies The Grattan Institute. https://grattan.edu.au/wp-content/uploads/2016/03/593_transcript_neg_gearing_CGT.pdf.

City of Sydney. (2017). Street Count website access December 2017 http://www.cityofsydney.nsw.gov.au/community/community-support/homelessness/street-count.

Day, B., Cox, W., & Pavletich, H. (2016) 12th Annual Demographia international housing affordability survey: 2016, 2015: 3rd Quarter, Table 9, p. 15.

Daley J., & Cowan S. (2016) The policy pitch—Negative gearing and capital gains tax reform: What's in the public interest?—Melbourne 22 March 2016.

Domain House Price Report. (2016). http://static.domain.com.au/domainblog/uploads/2016/10/25164342/DM15498_Domain House_Price_Report_September_2016_v2_FINAL-1.pdf.

Eslake, S. (2017). No place like home, The impact of declining home ownership on retirement, prepared by Saul Eslake for AIST March 2017, Australian Institute of Superannuation Trustees.

Fair Work Ombudsman. (2017a). Award pay: Australian Government http://calculate.fairwork.gov.au/CheckPay/Summary.

Fair work Ombudsman. (2017b). Minimum wage fact sheet, weekly figure multiplied by 52 weeks as of 11 April 2017 https://www.fairwork.gov.au/how-we-will-help/templates-and-guides/fact-sheets/minimum-workplace-entitlements/minimum-wages#current-national-minimum-wage.

Federal Budget. (2017). Reducing Pressure on Housing Affordability Fact Sheet 1.6 - Stronger rules for foreign investors owning Australian housing. See http://www.budget.gov.au/2017-18/content/glossies/factsheets/html/HA_16.htm.

Fielding Z. (2014). Property Observer, 11 August 2014, Numerous Sydney areas rezoned to allow for higher density developments. http://www.propertyobserver.com.au/forward-planning/adding-value/subdivision-zoning/34334-numerous-sydney-areas-rezoned-to-allow-for-higher-density-development.html.

Habitat 11. (2016). New Urban Agenda, Draft outcome document for adoption in Quito, October 2016.

Housing NSW. (2015). www.housing.nsw.gov.au/centre-for-affordable-housing/about-affordable-housing.

Janda M. (2017). ABC news, Superannuation for housing deposits would facilitate intergenerational theft, Michael Janda, 16 Mar 2017. See http://www.abc.net.au/news/2017-03-16/super-for-housing-deposits-intergenerational-theft/8360890.

Lawson, J., Berry, M., Hamilton, C., & Pawson, H. (2014) Enhancing affordable rental housing investment via an intermediary and guarantee, AHURI Final Report No. 220.

McKenny L., & Ting I. (2016). Sydney Morning Herald, March 28 2016, Thousands of empty homes adding to Sydney's housing crisis, experts say. See http://www.smh.com.au/nsw/thousands-of-empty-homes-adding-to-sydneys-housing-crisis-experts-say-20160323-gnpc52.html.

Morrison S. (2017). Media release, The Hon Scott Morrison MP, Treasurer of the Commonwealth of Australia. http://sjm.ministers.treasury.gov.au/media-release/006-2017/.

NSW Business Chamber and NSW Council of Social Service. (2016). February 2016, Taking on tax: Reforming NSW property taxes. http://www.nswbusinesschamber.com.au/NSWBC Website/media/Policy/Thinking%20Business%20Reports/FINAL-NSWBC-NCOSS-Taking-on-Tax-Report.pdf.

NSW Council of Social Service (NCOSS). (2016). Switch stamp duty for land tax to boost housing affordability and jobs: NCOSS, media release. https://www.ncoss.org.au/news-and-events/media-releases/switch-stamp-duty-for-land-tax-to-boost-housing-affordability.

NSW Department of Planning, A Plan for Growing Sydney. (2014). Action 2.3.3, Sydney. http://www.planning.nsw.gov.au/Plans-for-your-area/Sydney/A-Plan-for-Growing-Sydney.

Pawson H., Randolph B., Yates J., Darcy M., Gurran N., Phibbs P., & Milligan V. (2015). Tackling housing unaffordability: A 10-point national plan. June 24, 2015. https://theconversation.com/tackling-housing-unaffordability-a-10-point-national-plan-43628.

Pradolin R. (2017). Australian Financial Review, Feb 12 2017, If Australia has a housing crisis, supply is the answer.http://www.afr.com/real-estate/if-australia-has-a-housing-crisis-supply-is-the-answer-20170210-guaf1p#ixzz4e5k4iQrw.

Rental Affordability Index. (2017). https://nested.com/research/rental/2017/global.

Rockefeller Foundation. (2016). 100 resilient cities' in discovery area 4, 100 resilient cities—Pioneered by the Rockefeller Foundation (100RC) is dedicated to helping cities around the world become more resilient to the physical, social and economic challenges that are a growing part of the 21st century. http://www.100resilientcities.org/cities/entry/sydney#/-_/.

Ryan, P. (2017). ABC news, Reserve Bank concerned on highly indebted households, real estate bubble, The World Today, 21 Mar 2017. http://www.abc.net.au/news/2017-03-21/reserve-bank-warns-housing-bubble-in-sydney-and-melbourne/8372856.

Saner E., Usborne S., Benedictus L., Khaleeli H., & Pidd H. (2017). The guardian, housing crisis: The empty nester's guide to downsizing, Thursday 9 February 2017. See https://www.theguardian.com/society/2017/feb/08/housing-crisis-the-empty-nesters-guide-to-downsizing.

Sansom M. (2016). Government news, Airbnb: What NSW councils need to know, October 21, 2016. See http://www.governmentnews.com.au/2016/10/airbnb-local-councils-need-know/.

Shapiro J. (2017). Australian Financial Review, ASIC eyes interest-only bank lending in mortgage crackdown, Apr 3 2017. See http://www.afr.com/business/banking-and-finance/financial-services/asic-eyes-interestonly-bank-lending-in-mortgage-crackdown-20170403-gvc5v4#ixzz4dttNiBTB.

United Nations. (2015). United nations sustainable development goals: 17 Goals to transform our world, Goal 10 & Goal 11. http://www.un.org/sustainable development/cities/. Last accessed May 24, 2017.

UrbanGrowth NSW. (2016). Calculation using the ANZ mortgage calculator, 30-year loan, 10% deposit, variable interest rate and maximum 30% of gross income as affordable cost for housing. https://www.anz.com.au/personal/home-loans/calculators-tools/calculate-repayments/.

Visentin L. (2016). Sydney Morning Herald, December 22 2016 Record number of homes built in Sydney, but it's still unaffordable. See http://www.smh.com.au/nsw/record-number-of-homes-built-in-sydney-but-its-still-unaffordable-20161221-gtfnau.html.

Yates J. (2014). Affordable housing submission 53 issues and policy options submission to the senate economic references committee on affordable housing, Judith Yates, Honorary Associate Professor in Economics, University of Sydney, p. 3, 2014.

Creating Public Value Through Collaborative Governance—Case Study: The Strategic Development of the Bays Precinct, Sydney Transformation Plan

Alexandra Vella and Campbell Nicole

Abstract The paper presents an approach developed by UrbanGrowth NSW known as the *City Transformation Life Cycle*™ (UrbanGrowth NSW 2014a) and its application to developing a 25-year urban renewal strategy (a transformation plan) for an area known as The Bays Precinct, Sydney. This case study provides a powerful demonstration of collaborative governance and public value creation (Moore in Creating public value: strategic Management in government, Cambridge, MA, Harvard University Press, 1995; Stoker in Public value management: a new narrative for networked governance?, American Review of Public Administration, 2006). UrbanGrowth NSW, as the lead government agency, had the leadership forcsight to genuinely share its role as the 'place-maker' with other diverse actors to establish the principles that ultimately underpin *The Bays Precinct, Sydney Transformation Plan* (UrbanGrowth NSW 2014b, c).

Keywords Place-making · Urban transformation · Public value
Collaborative governance · Stakeholder participation

This paper is divided into four parts.

Part 1 identifies three significant challenges for associated with complex urban transformation projects: managing broader public expectations; recognising cultural impacts and appreciating the multitude of interdependencies associated with these complex place making projects. Other key theoretical concepts are also briefly outlined: place-based leadership; governance, stakeholder engagement and public value creation. UrbanGrowth NSW's *City Transformation Life Cycle*™ (UrbanGrowth NSW 2014a) is described as a new approach to iterative business planning for urban transformation, a process that embodies a collaborative and organic approach to place making.

A. Vella · C. Nicole (✉)
Landcom, Level 12 MLC Centre, 19 Martin Place, Sydney, NSW, Australia
e-mail: ncampbell@landcom.nsw.gov.au

A. Vella
e-mail: ap.vella@iinet.net.au

W. Leal Filho et al. (eds.), *Sustainable Development Research in the Asia-Pacific Region*, World Sustainability Series, https://doi.org/10.1007/978-3-319-73293-0_18

Part 2 of the paper provides a detailed case study of how the *City Transformation Life Cycle*™ (UrbanGrowth NSW 2014a) was applied to the stakeholder engagement associated with the strategic development of a transformation plan for the Bays Precinct. The case study details the engagement approach and the outputs that emerged from that engagement. It also relates the objectives to deliver The Bays Precinct, Sydney Urban Transformation Program that are underpinned by the twenty (20) high-level Principles of *The Bays Precinct, Sydney Transformation Plan* (UrbanGrowth NSW 2015c) to the United Nations Sustainable Development Goals (Australian Government 2015).

Part 3 of the paper assesses the degree to which the engagement process associated with the development of *The Bays Precinct, Sydney Transformation Plan* (UrbanGrowth NSW 2015c) demonstrated collaborative governance (see Ansell and Gash 2007) as well as public value creation by applying Moore's Public Value Chain (Moore 2007 in Grant et al. 2014).

Part 4 concludes the paper and summarises key learnings arising from the public engagement process.

1 Introduction

Urban transformation projects are by nature highly complex due to the need to respond to broader public expectations, cultural impacts and the many interdependencies associated with planning and delivering the project.

2 Broader Public Benefit Expectations

It is a public expectation that urban transformation projects are expected to be part of the solution for broader complex issues like climate change and societal disadvantage (see Jones 1998).

Learnings from waterfront regeneration projects (Jones 1998) in the United Kingdom and United States of America in late last century suggest these projects were generating wider economic, social and environmental benefits at the expense of local communities' benefits. The later European experience in waterfront regeneration like projects in Barcelona and Copenhagen had taken a more place-based approach to "*emphasise small-scale and publicly-orientated, as well as innovative regeneration schemes*" (Jones 1998, p. 440).

A land use framework and other inter-related plans endeavour to resolve value conflicts through illustrating what balancing broader public benefit and local place making means (Godschalk 2004). They change the urban environment and, as such, it is important that an over-arching strategy (transformation plan) which underpins them captures shared aspirations on what a place might be.

2.1 Cultural Impacts

Because urban transformation projects change the urban environment, they impact on how local people exist within it. The importance of place in these projects cannot be underestimated; it is here where sustainable place shaping occurs, motivated by values engrained in culture (Horlings 2015).

It is at the pedestrian scale where spaces are cherished by people and where social practices occur (Friedmann 2010). These are the spaces in between buildings (Hyslop 2014) that expand the public realm for social practices to express culture.

What is culture in the context of transformative projects? Landry's discussion on cultural literacy notes, "*Culture is who we are, the sum of our beliefs, attitudes and habits. It is seen in customary ways of behaving—making a living, eating, expressing affection, getting ahead or, in the urban context, behaving in public places.*" (Landry 2006, pp. 245–249). Landry emphasises the criticality of appreciating culture in times of "*dramatic transformation*" "*because it is then that the culture needs to absorb, digest and adjust.* Acknowledging culture fosters change through "*creativity, innovation and renewal*" which are critical ingredients in sustainable change. However, if culture "*feel threatened or weak*" openness to change may diminish.

Horlings (2015) muses that "*Culture plays a mediating role between people or society and the environment, influencing people's intentions, way of life, sense of place, practices, norms and rules. (Dessein, Battaglini, & Horlings, in press; Horlings, in press). In its variety, culture—including tangible as well as intangible aspects—is one of the sources as well as an outcome of distinctiveness between places.*" (Horlings 2015, p. 259). It could be argued that when urban transformation is done in a sustainable way and with a focus on the common good, then people are willing to accommodate change.

In examining why people would accommodate change and participate in place-shaping processes O'Brien notes that "*Transformation to sustainability is not only driven by practices and political structures, but also by individual and shared beliefs, values, worldwide views and paradigms that influence attitudes and actions*" (O'Brien 2012, 2013 in Horlings 2015, p. 259). It is the latter that motivates stakeholders to find common ground.

Balancing broader public benefit and local place making necessitates a holistic approach where introduction of new or renewal of built-form (housing, employment and infrastructure uses) acknowledges culture, and culture confidently plays that mediating role for the common good.

2.1.1 Many Interdependencies

Urban transformation projects comprise many different and interconnected parts that are contemplated simultaneously—these interdependencies can be difficult to grasp and take time to unravel.

Landry (2006) discusses complexity and human mindset construction and suggests it is not equipped with unravelling complex issues in the emerging world we live. This is based on the evolution of cities from their initial industrial origins to service-based cities. Landry (2006) proposes a conceptual framework that considers the 20-year horizon and beyond—"*through which it may be easier to focus on the significant and strategic, to unravel the trivial from the profound, and to understand timelines and connections*" (Landry 2006, p. 191).

The success of achieving broader public benefit and authentic place making are mutually dependent on each other. It begs the question of where does the stakeholder thinking need to be to deal with complexity, interdependencies and those broader issues?

3 Part 1: Theoretical Overview of Key Concepts

Successful urban transformation projects require consideration and application of a broad array of theoretical concepts in both the planning and delivery of the project. Some key concepts are outlined in this section.

3.1 Place-Based Leadership

The role of place-based leadership is core to informing the approach to stakeholder engagement and, ultimately the decision-making process. Hambleton (2015) identifies political, public managerial/professional, community, business and trade union as the five roles of place-based leadership and notes that overlapping these roles creates potential innovation zones. In this innovation zones lies the opportunity where "*place-based leadership can shape the quality of the exchanges*" (Hambleton 2015, p. 127). While the innovation zones can be where conflicts could arise, they can also represent areas of 'common ground'.

Crosby et al. (2016), explored "*orchestrated collaborative work*" as a mechanism for moving away from the traditional top-down business-as-usual approach to stakeholder engagement and postulating an alternative approach where "*leaders must act as sponsors, champions, catalysts and implementers*" (Crosby et al. 2016, pp. 5–6). Employing multi-leadership roles helps break down barriers, legitimise and energise collaboration, bring diverse skills and knowledge and deliberately and appropriately disrupt the collaboration process to encourage stakeholders "*to think outside the box*" (Crosby et al. 2016, pp. 5–6).

Other scholars (Crosby et al. 2016; Hambleton 2015; Horlings 2015; Stoker 2006), suggest place-based approaches are suited to complex-problem solving through stakeholder engagement where the emphasis is placed on collaboration, inclusiveness, taking a holistic view, adaptive, open, and interactive (learning together).

3.2 Stakeholder Engagement

Insights by Zivkovic (2015) in complex adaptive systems theory suggest that communities are "*complex adaptive systems* and *focus areas for building the adaptive capacity of communities* are: *create a disequilibrium state, amplify action, encourage self-organisation, stabilise feedback and enable information flows*" (Zivkovic 2015, pp. 2, 4–6).

A stakeholder engagement framework that includes an iterative process is necessary in strategy formulation when dealing with complex problems. Complex problems change character (Zivkovic 2015), and while a transformation plan for a place will have a vision, principles and objectives, it is important as society evolves, that the vision, principles and objectives are checked for relevance.

In the context of a complex urban transformation project, stakeholder engagement is an iterative and cyclical process with a defined scope that can change over the life of the transformation project. Urban transformation projects are 20–30-years in the making and stakeholder engagement must be sustained over this period, recognising that within the lifetime of the project new stakeholders will emerge with new generations, each with new ideas.

3.3 Public Value

Public Value has been described in terms of citizens being 'shareholders' in their community. "*The value may be created through economic prosperity, social cohesion or cultural development. Ultimately the value—such as better services, enhanced trust, or social capital… is decided by the citizen… not just through the ballot box, but through taking part in consultations and surveys, for example*" (Horner and Hazel 2005 in O'Flynn 2007).

> …governments across Australia are continually engaged in Public Value Creation – defined *as the process of adding value to the public sector through the exercising of managerial authority* – all the time. However, managers cannot decide for themselves what Public Value is. Rather, they should *seek the views of all in a complex authorising environment* comprised of elected officials and a range of stakeholders…significant community engagement is required to assess and measure Public Value, as opposed to private value (Grant et al. 2014, p. 1).

Moore (1995) postulated the concept of Public Value by describing the relationships between public sector managers, elected representatives and other stakeholders as 'The Strategic Triangle' (see Fig. 1), where the authorising environment is aligned with the operational environment and administrative capabilities to create public value (Moore 1995 in O'Flynn 2007).

Moore expanded this concept further to describe the "Public Value Chain" (see Fig. 2)—which enables measurement of the public value created and critically recognises that 'outputs' and the ultimate 'outcomes' are not the same. Assessing

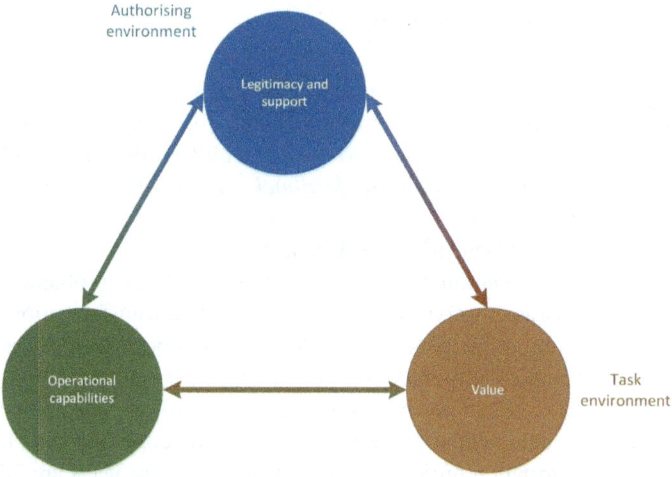

Fig. 1 Moore's public value: the strategic triangle Grant et al. (2014). Adapted from Alford and O'Flynn (2009, p. 173)

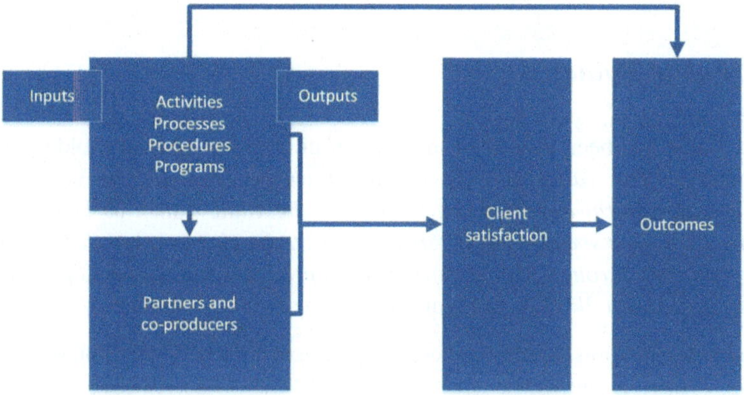

Fig. 2 Public value chain (Moore 2007 in Grant et al. 2014)

the ultimate 'outcomes' of a process is more indicative of the true public value created (Moore 2007 in Grant et al. 2014).

3.4 Collaborative Governance

Closely connected to public value creation is the concept of collaborative or networked governance. Stoker (2006) describes networked governance as "*a particular*

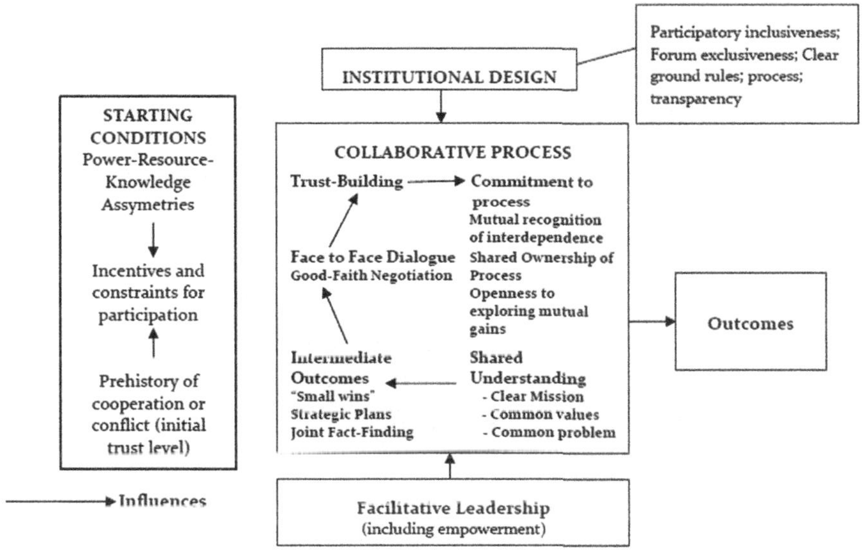

Fig. 3 Collaborative governance (Ansell and Gash 2007)

framing of collective decision-making that is characterised by a trend for a wider range of participants to be seen as legitimate members of the decision-making process in the context of considerable uncertainty and complexity" (Stoker 2006, p. 41). Building relationships based on values of mutual respect and shared learnings resonate with this approach. Ansell and Gash's model for collaborative governance (see Fig. 3) highlights the importance of trust as a critical factor in the success of collaborative governance mechanisms (Ansell and Gash 2007).

The rationale for the above model is that a citizenry actively engaged in shared decision-making processes and operating based on mutual respect and trust is less likely to disengage—which can strengthen governance structures because of increased public scrutiny of the decisions and behaviours of the government.

4 Part 2: Case Study—Strategic Development of the Transformation Plan for the Bays Precinct, Sydney

In 2014, UrbanGrowth NSW (State-Owned Corporation) was tasked with leading the transformation of The Bays Precinct, Sydney (The Bays Precinct) (UrbanGrowth NSW 2014c, d).

The Bays Precinct is currently a place that sleeves the iconic Sydney Harbour surrounded by well-established urban villages (see Fig. 4). The Bays Precinct. Its waterways (94 ha of Sydney Harbour) and most of its land (95 ha) is Government

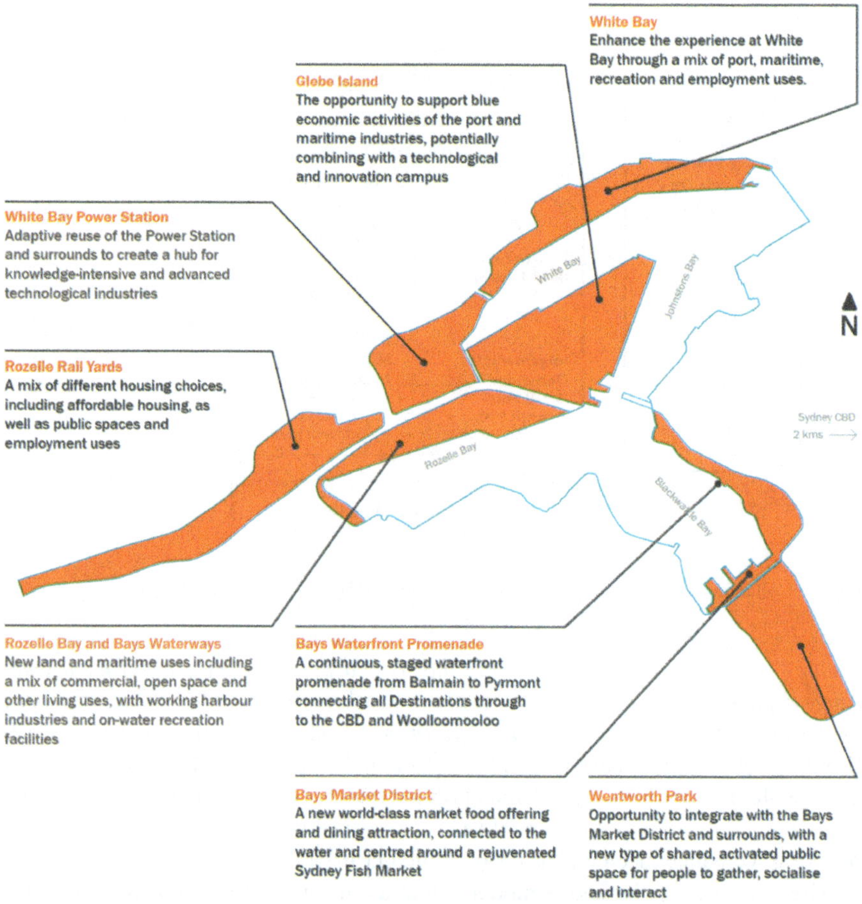

Fig. 4 The Bays Precinct, Sydney (UrbanGrowth NSW 2015a, b, c, d)

owned and predominately used for port, maritime and commercial uses. However, a significant part of its 5.5 km of foreshore is not publicly accessible. The Bays Precinct contains several significant heritage items like White Bay Power Station and Glebe Island Bridge.

The Bays Precinct has a rich indigenous, multicultural and industrial history of transformative functions for Sydney and Australia. It was a place of trade between Aboriginal clans to essential maritime commerce in the new colony to industrial and recreation uses. In the early 1900s White Bay Power Station was built on the shores of The Bays Precinct to power Sydney's train network. The Bays Precinct was used for Second World War purposes. Its current occupiers are Sydney Fish Market, White Bay International Passenger Terminal and other maritime uses (The People for Places and Spaces 2015).

Recent history suggests that the place has importance not only to local community but also the wider population and visitors: "*The area is of special interest because it fulfils unique operational and recreational needs and has the potential to meet a wider range of operational, open space and development needs. The Precinct is where adjacent new and long standing residential, waterfront industrial, waterfront commercial, recreational and cultural interests of different scales come together... The future of the Bays Precinct is important to both the local residential and business community, to the wider population of Sydney and New South Wales and to visitors to the State.*" (NSW Government 2012, pp. 9–10).

4.1 UrbanGrowth NSW's City Transformation Life Cycle™

UrbanGrowth NSW's operations must have regard "*to compliance with the principles of Ecologically Sustainable Development contained in s6(2) of the Protection of the Environment Administration Act 1991*" (Landcom Act 2001). This Act states that "*ecologically sustainable development requires the effective integration of economic and environmental considerations in decision-making processes*" (Protection of the Environment Administration Act 1991).

UrbanGrowth NSW's *City Transformation Life Cycle*™ (UrbanGrowth NSW 2014a) recognises the importance of the above legislative requirements and customises its stakeholder engagement processes that are based on practical experience and consistent with the International Association for Public Participation (IAP2) principles: collaborative; purposeful; proactive; accountable and; inclusive (IAP2 2014).

The *City Transformation Life Cycle*™ (UrbanGrowth NSW 2014a) (see Fig. 5), endeavours to challenge the 'business-as-usual approach' to urban transformation projects. Why challenge the 'business-as-usual' approach? We now function in a highly connected globalised world with access to learnings, ideas and many communication options. Our cities are evolving from industrial base to service base (see Landry 2006). The world faces environmental and social challenges that drives cities to be more sustainable.

Simply, it is expected that city projects individually and collectively contribute to a city's sustainability. In this example, managing complexity like cultural impacts, many interdependencies and broader public benefit expectations required leadership centred on place and involving citizens (Hambleton 2015; Stoker 2006) to create public value.

The City Transformation Life Cycle™ (UrbanGrowth NSW 2014a) postulates that urban transformation projects should be contemplated in four elements: Thinking Cities, Funding Cities, Building Cities, and Living Cities suited to the longer time-frames of these projects' type and where change is a continuum.

Fig. 5 UrbanGrowth NSW: City Transformation Life Cycle™ (UrbanGrowth NSW 2014a)

- Thinking Cities—Establishing a single ambition and objectives that focus on the strategic, economic and cultural significance of a place.
- Funding Cities—Determining funding and finance model(s) that supports investment certainty for the ambition and objectives.
- Building Cities—Integrating new and existing uses within a place and its surrounds for example, land, water, infrastructure and services through transparent processes and staged delivery.
- Living Cities—Creating sustainable places, spaces and opportunity centred on resilience, happiness and prosperity and; through managing and monitoring, review the strategy.

The City Transformation Lifecycle™ recognises that implementing urban transformation projects is not linear and involves unravelling complex issues and as such, the elements should be re-visited periodically and to act in response to effects such as changes to the funding and finance setting, innovation and disruptive events. Taking a non-linear approach is echoed in the recent World Economic Forum Roadmap for Urban Transformation (World Economic Forum 2016).

Gradual change can lead to significant outcomes in the physical sense of a place but also in people's well-being and in the health of the environment in which they live. Engagement with stakeholders about this type of change requires a mindset motivated by vision (Landry 2006) as primacy and less on the built-form. The City Transformation Life Cycle™ (UrbanGrowth NSW 2014a) approach was an attempt to equip stakeholders for a non-traditional mindset in developing transformation projects through broadening public participation in decision-making as a public value creation opportunity.

4.2 The Stakeholder Engagement Process for the Development of the Transformation Plan

Transformation projects on public lands requires approaches that identify public value, in the context of the potential economic, cultural and social benefits for the local and broader community of Sydney.

For The Bays Precinct, these were challenging questions because over the previous 15 years, The Bays Precinct had undergone extensive consultation on land use issues, a strategic land-use framework and strategic planning principles (NSW Government 2012). As such, some stakeholders had deep content knowledge, and others, very little and there were often clear competing agendas. Public trust in the process had also been eroded over this period.

The City Transformation Life Cycle™ (UrbanGrowth NSW 2014a) was applied in the stakeholder engagement program in developing The Bays Precinct's strategy known as *The Bays Precinct, Sydney Transformation Plan* (The Transformation Plan) (UrbanGrowth NSW 2015c). The stakeholder engagement program's objectives were:

- *Ensure broad participation for Sydneysiders to be informed, consulted and involved,*
- *Build relationships to increase opportunities for involvement and collaboration,*
- *Engage future users that may live, work or visit The Bays Precinct in the future,*
- *Listen and be responsive by acting on public participation feedback where possible and,*
- *Demonstrate how feedback was considered clearly to all participants* (UrbanGrowth NSW 2016b, p. 2).

In the case of The Bays Precinct, treating community as a complex adaptive system (see Zivkovic 2015), could enable a better understanding of the key challenge: its complexity and interdependencies, determine what the behaviour needs to be and problem-solve through an interactive platform.

As a starting point for this engagement process, UrbanGrowth NSW utilised *The Bays Precinct Strategic Framework Report* to the NSW Government (NSW Government 2012) to establish an agreed baseline. While this Report did not include a program to prioritise and resolve identified issues, it did offer valuable insight into issues and opportunities that built on previous consultations—which identified clear value conflicts between local and broader public benefits.

> A key challenge in developing a land use framework for the Precinct is to balance the economic and regional needs of broader Sydney with protection of local residential amenity, enhancement of recreational and open space and foreshore access and urban renewal. It is important that careful decisions are made for the best possible use of these valuable lands. (NSW Government 2012, p. 4).

Addressing these value conflicts was essential for The Bays Precinct because its transformation ambition is one that shifts it from an industrial place to a service-based place and in doing so requires a leadership type that shifts the mindset appropriate for industrial city ambition to one that fits a service based city. This type of leadership mindset is based on being visionary and courageous, (Landry 2006) where many people have roles to play and where there is recognition that complex urban regeneration projects have a timeframe that extends beyond political electoral cycles.

At the time of the development of The Transformation Plan the goal for the Bays Precinct in Government's 20-year plan for Metropolitan Sydney Area (*A Plan for Growing Sydney*) is to transform its "*currently underused areas for the economic, cultural and social benefit of Sydney and the State*" (NSW Government 2014, p. 26).

Subsequently, Government's ambition for The Bays Precinct is "*[T]o drive an internationally competitive economy, through the creation of great destinations on Sydney Harbour that will transform Sydney, New South Wales and Australia.*" (UrbanGrowth NSW 2015a, p. iii).

4.3 Anchor Events

Hambleton (2015) and Stoker (2006) emphasise the importance of creating opportunities to exchange international learnings and share local knowledge as a critical element in driving place-based change and realising the related public benefit outcomes. Hambleton (2015) outlines a framework for understanding international learnings based on formal and informal learning that can lead to technical advancements, changes in policy, practice and governance.

Hambleton notes this application in examples of waterfront renewal projects in the United Kingdom—Bristol Floating Harbour and Cardiff Bay, and United States of America—Baltimore Inner Harbour. Equally important is the local knowledge because this describes peoples' experience in place—its identity, its beliefs, its issues, its strengths and its environment. (Hambleton 2015). Stoker makes the point of the "*need to give more recognition to the legitimacy of a wide range of stakeholders*" this includes "*neighbourhood leaders*" who have the knowledge as "*users*" (Stoker 2006, p. 47).

One of the key principles to drive the success of 10-Step Action Plan in the World Economic Forum Roadmap for Urban Transformation is "*learning from other cities*" (World Economic Forum 2016, p. 51). It also identifies the importance of city identity and citizen-centric collaborative approaches in evolving in the global context.

The Transformation Plan was built through utilising a staged approach through creating platforms for participation and feedback. This approach was considered innovative because it "*helped to overcome the challenges that had stalled its progress previously*" (UrbanGrowth NSW 2016b, p. 8).

UrbanGrowth NSW established key anchor events that also generated publicity that captured a wider audience. The anchor events were supported by smaller targeted activities with stakeholders. Information was gathered and shared through a variety of ways reflecting different cognitive preferences (UrbanGrowth NSW 2016a, b).

These anchor events included: The Bays Precinct, Sydney International Summit 2014 and the Finance and Investment International Summit. Collectively, these engagement activities brought together a broad range of stakeholders (350 participants) from community, government, business, industry groups and academia with local, national and international practitioners to primarily discuss learnings from other cities and collaborate to develop a shared ambition.

These anchor events enabled formal and informal knowledge sharing through one on one discussions to forums (see UrbanGrowth NSW 2016a, b). Knowledge capture from the 1200 participants occurred via traditional and modern technology tools and included 2700 feedback notes and 146,548 YouTube views of video content during the consultation (see UrbanGrowth NSW 2016a, b).

Learnings arising from the Summits encouraged UrbanGrowth NSW to open up the largely publicly inaccessible site to the public. "Discovery Day" (12 April 2015) attracted 25,000 participants who experienced the place in its current state, learned about its past and had the opportunity to contemplate its future (see UrbanGrowth NSW 2016a, b).

The Sydneysiders Summit and Leadership Forums were held in May 2015 and included opportunities for school students, industry, local residents and the broader Sydney community to contribute their thoughts on the future for The Bays Precinct. A discussion paper was developed to present the case for change, the rationale and evidence-base that had informed the transformation scope, the project's key challenges and possible mix of uses in The Bays Precinct (UrbanGrowth NSW 2015a) as part of a broad public engagement program.

The Call for Great Ideas (UrbanGrowth NSW 2015b) was a further platform to draw out innovative opportunities for The Bays Precinct. Two hundred and thirteen submissions were made by local residents, interstate and overseas countries (UrbanGrowth NSW 2016a, b). These ideas were assessed by an Independent Assessment Panel supported by a Technical Advisory Panel and Community Advisory Panel. The Independent Assessment Panel recommendations also informed the development of The Transformation Plan.

4.4 *Outputs—Principles for the Transformation Plan*

Sessional Papers for The Bays Precinct, Sydney International Summit (2014c, d) and the Finance and Investment International Summit were developed to set the context and pose questions and ideas from local, national and international experience. The *City Transformation Life Cycle*™ (UrbanGrowth NSW 2014a) was applied at these summits to ensure ongoing and interactive engagement that was based on the following objectives for the strategy to develop The Transformation Plan for The Bays Precinct:

- *develops a set of principles and ambitions for the Precinct that will contribute to the long-term needs of an international, regional and local community*
- *ensures these ambitions can be delivered through innovative and resilient funding models*
- *sees the design response as an essential outcome but not a driving part of the process*
- *builds a platform for active and ongoing collaborations with the community, stakeholders and partners*
- *installs transparent governance arrangements that enable more efficient decision making* (UrbanGrowth NSW 2014a, p. 7).

These objectives reflected expectations associated with the strategy to develop The Transformation Plan: long-term vision, evidence-based, collaborative and transparent. Through an interactive process, UrbanGrowth NSW captured the knowledge from these summits and co-created with stakeholders twenty (20) high-level Principles (listed below) with the themes of: public benefit, recognition of culture, transparency and long-term sustainable responses.

1. Build on the unique history of The Bays Precinct.
2. Establish a powerful and enduring governance model based on whole of-government collaboration that fearlessly pursues public benefit.
3. Be transparent and communicate the issues and challenges we face and the investments needed to realise the Precinct's potential.
4. Allow the time to invest in genuine and early engagement with, and broad acceptance of our plans from, all categories of the public, government and industry.
5. Unlock public access to the Harbour's edge and waterways along the entire coastline.
6. Develop an overall Bays Precinct Transformation Program to prioritise major projects and define the staging for integrated development and land use.
7. Establish a whole-of-precinct transport infrastructure plan early, based on connectivity, accessibility and active transport
8. Prioritise planning for public spaces, White Bay Power Station and Sydney Fish Market.
9. Generate optimal housing supply outcomes based on a model of diverse housing options, the highest design principles and activated public spaces.
10. Ensure the land use and associated development is diverse, beautifully designed and creates 'great places and great spaces'.
11. Build the capacity for The Bays Precinct to be a place that contributes to healthy, prosperous and resilient lifestyles.
12. Support economic development and growth that can drive a strong, digitally connected, innovative and diverse knowledge economy.
13. Plan for future generations by being open to new ideas and embracing emerging trends.

14. Adopt world-class energy generation systems that maximise efficiency and establish The Bays Precinct as the exemplar for 'big city' energy provision.
15. Introduce environmental and ecological systems to improve water quality, address ongoing sources of water pollution and encourage public recreation.
16. Support the economic activities of maritime industries and celebrate the authenticity of the working harbour.
17. Provide the platform for investment from Australia and abroad, and from public and private sectors.
18. Incorporate a strong funding and financial strategy to enable innovative, leading-edge and productive investment vehicles that promote investor appetite.
19. Seek broad sources of funding for urban transformation across a range of investors, including superannuation and pension funds, and philanthropy.
20. Employ an ethical procurement process that optimises value for government and taxpayers while being attractive to investors.

In September 2015, the Australian Federal Government endorsed the United Nations Sustainable Development seventeen (17) Goals (see Australian Government 2015). A number of these goals are pertinent to the objectives to deliver The Bays Precinct, Sydney Urban Transformation Program that are underpinned by the twenty (20) high-level Principles (see Table 1).

4.5 Outcomes

An important outcome from The Bays Precinct, Sydney International Summit 2014 and the Finance and Investment International Summit was the highly interactive engagement process, which facilitated visionary thinking. Participants had the benefit of shared knowledge and learnings, and started from an agreed baseline to develop the strategy for The Transformation Plan for The Bays Precinct.

The final Transformation Plan's objectives, mix of uses and inclusion of large adjoining public space was significantly influenced by the feedback garnered through the engagement process.

To demonstrate transparency and clarity on how the feedback influenced The Transformation Plan and seek Government's approval a companion document (GHD 2015) was prepared. This companion document detailed "*the changes from the Discussion Paper to the [Transformation] Plan, and how the Discussion Paper feedback and the Call for Great Ideas influenced changes*". It also highlighted that, "*It is clear in the 'Informing the Transformation Plan' document that the public feedback and ideas had a strong influence on the final Plan*" (UrbanGrowth NSW 2016a, b, p. 6).

The Transformation Plan translates the ambition for The Bays Precinct into a policy framework (UrbanGrowth NSW 2015a, b, c, d, pp. 11–12) to guide UrbanGrowth NSW in its collaboration with State agencies and other entities.

Table 1 Comparison of sustainable development goals with the objectives for The Bays Precinct, Sydney Urban Transformation Program

Sustainable development goals	Sub goal	The objectives for The Bays Precinct, Sydney Urban Transformation Program
8 Decent work and economic growth	8.2, 8.3, 8.5, 8.9	1. To deliver a hub of export-oriented knowledge-intensive jobs that can increase Sydney's global competitiveness
10 Reduced inequalities 11 Sustainable cities and communities	10.2 11.7	2. To deliver enduring, socially inclusive and great places to benefit Sydney siders and national and international communities
11 Sustainable cities and communities	11.1	3. To deliver housing choices, including affordable housing options, through design, finance and construction excellence
11 Sustainable cities and communities	11.2	4. To deliver a world-class mass and active transit solution that unlocks the economic and human potential of The Bays Precinct and demonstrates a model of environmental excellence
11 Sustainable cities and communities	11.7	5. To achieve building design excellence and quality urban design in all destinations
3 Good health and well being 6 Clean water and sanitation	3.9 6.3	6. To provide ecological and marine water quality improvements to enable abundant biodiversity
7 Affordable and clean energy 9 Industry, innovation and infrastructure 11 Sustainable cities and communities	7.1 9.1 11.6	7. To deliver integrated utilities solutions that enable advanced energy generation and technologies
4 Quality education 11 Sustainable cities and communities 13 Climate action 14 Life below water 15 Life on land 16 Peace, justice and strong institutions	4.7 11.3, 11.6, 11b 13.1 14.2 15.8, 15.9 16.7	8. To apply integrated planning within a land and water context that considers strategic policy decisions and the interrelationships between biophysical, social and economic aspects
11 Sustainable cities and communities	11.4	9. To celebrate heritage and culture by creating new experiences throughout The Bays Precinct

This collaboration endeavours to bridge the divide between public policy and market reality to maximise public value on public lands.

Through a cabinet process The Transformation Plan was adopted by the NSW Government (UrbanGrowth 2015c).

5 Part 4: Demonstration of Collaborative Governance and Public Value Creation

In the context of The Transformation Plan for The Bays Precinct, if Moore's model for Public Value Creation is applied (see Moore 2007 in Grant et al. 2014), then clearly the engagement process by UrbanGrowth NSW in the development of The Transformation Plan for The Bays Precinct provides an applied demonstration of public value creation.

The innovative approach taken by UrbanGrowth NSW to 'share power'—and genuinely open the decision-making to include other diverse points of view was a clear departure from traditional bureaucratic 'consultation'. While challenging to undertake, the outputs of the engagement process have resulted in tangible outcomes that reflect the high degree of stakeholder participation—in particular, the underlying principles for The Transformation Plan that will guide the future development of The Bays Precinct.

Similarly, in terms of collaborative governance, assessing the engagement approach for the development of The Transformation Plan reflects the criteria established by Ansell and Gash (2007) as demonstrated in Table 2:

Legacy learnings arising from developing a strategy for a complex urban transformation project anchored by a commitment to broad public participation include:

- Future proofing: The process enables future adaptation of the strategy so that the Transformation Plan will remain relevant over the lifetime of the project. The Transformation Plan is a living document (action-orientated and priorities) and has capacity to respond to input by current generations as well as enable contributions by future generations. The Transformation Plan "*acts as a reference and guiding document for practitioners over the life of the Transformation Program*" that outlines high-level spatial planning framework focused on evidence-based analysis, integration and holistic thinking (UrbanGrowth 2015c, pp. 57–70).
- Public value creation through collaborative governance: Enabling broader public participation in decision-making has delivered a strategy with strong support from stakeholders and established a strong foundation of public trust in the process. The Transformation Plan embeds a commitment to ongoing public engagement that includes a reference group, established through an open expression of interest process.
- Embedding life-long learning opportunities, innovation and creativity into urban transformation projects. The development of the strategy for The Transformation Plan and the application of the City Transformation Lifecycle™ as the mechanism for broader stakeholder participation catalysed UrbanGrowth NSW's subsequent commitment to embedding learning and research programs into all of its urban transformation projects (schools, vocational educational providers and universities).

Table 2 Assessment of collaborative governance principles applied to the transformation plan for The Bays Precinct (adapted from Ansell and Gash 2007)

Criteria for collaborative governance	Assessment of engagement process for the transformation plan for The Bays Precinct
Was the *City Transformation Life Cycle*™ (UrbanGrowth NSW 2014a) and the subsequent engagement strategy for The Bays Precinct, Sydney transformation plan initiated by public agencies or institutions?	Yes. Urbangrowth NSW (a state-owned corporation) initiated the *City Transformation Life Cycle*™ (UrbanGrowth NSW 2014a) which established an iterative process for engagement, starting from an agreed baseline and working collaboratively to share knowledge and learnings
Are participants in the forum non-state actors?	Yes; international representatives with expertise in urban transformation, local residents, businesses and community organisations actively participated in the initial, The Bays Precinct, Sydney International Summit and Finance and Investment International Summit Broader public participation occurred through the Sydneysider Summit, leadership forums, discovery day and public involvement in the call for great ideas
Are participants engaged in decision-making (not just consultation)?	Yes, the twenty (20) high-level principles that underpin the transformation plan were co-created by the participants at the initial summits and UrbanGrowth NSW Feedback was genuinely considered in the development of the transformation plan and this was reflected in a public document (GHD 2015), detailing how the feedback had been incorporated into the plan In addition, community and state agency representatives were included on the assessment panel for considering submissions to the call for great ideas
Is the forum formally organized and does it meet collectively?	The engagement activities were formally organised—and an ongoing Bays Precinct Reference Group continues to meet with UrbanGrowth NSW on a regular basis
The forum aims to make decisions by consensus (noting that may not always occur)	The engagement was not predicated on achieving consensus but rather providing an opportunity for everyone to be heard. The twenty (20) high-level principles that underpin the transformation plan were developed by participants at the initial summits
The focus of the collaboration is on the public policy or public management	Yes, the collaboration was focussed on determining the future for an iconic waterfront location in Sydney and the transformation plan represents a key public policy document for the NSW Government

The success of the future implementation of The Transformation Plan is dependent on effective decision-making by the implementers and stakeholders in the areas of leadership, whole-of-government partnership, ongoing public engagement, working with industry, assuring excellence through a design directorate, holistic placemaking, efficient transport and mobility, precinct wide funding and finance model and, open and transparent procurement at major project stages (see UrbanGrowth NSW 2015c).

The Transformation Plan's policy framework includes accountability through its statement of commitments under the themes: diverse and socially inclusive; globally competitive; connected; heritage and culture; built environment and natural environment and in the planning and urban design strategic framework process (see UrbanGrowth NSW 2015c). It affords the opportunity through the organisation's sustainability policy and urban design framework that the delivery of these commitments is monitored overtime.

6 Conclusion

This paper presented UrbanGrowth NSW: *City Transformation Life Cycle™* (UrbanGrowth NSW 2014a) and its application to the developing an urban renewal strategy for The Bays Precinct. It also reflected on the public engagement process through understanding current practices in leadership suited to transformation projects pursuing public value. These leadership practices are centred around place-based leadership and the importance of place to citizens (Crosby et al. 2016, Hambleton 2015; Horlings 2015; Stoker 2006; Zivkovic 2015).

The Bays Precinct key challenge evokes value conflicts between local and broader public benefits. The high-level spatial planning framework in The Transformation Plan aims to guide the transformation. This framework is underpinned by twenty (20) high-level Principles and feedback from stakeholders throughout the strategy formulation phase. This was done by involving public from the onset through an innovative engagement program and utilising UrbanGrowth NSW: City Transformation Life Cycle™ (UrbanGrowth NSW 2014a).

In the past, progressing The Bays Precinct transformation was challenged by complexity, concerns and mixed expectations. Using this model equipped stakeholders through preparing the mindset that motivated by vision (Landry 2006). Here preparedness met opportunity through the diverse activity base (be it large-scale events or the many intimate encounters) that created the opportunity for sharing ideas and knowledge that was captured, digested and utilised and shared with the public.

By applying the City Transformation Life Cycle™ (UrbanGrowth NSW 2014a) and embedding an iterative and deep approach to stakeholder engagement, The Transformation Plan for the Bays Precinct, Sydney was delivered as a genuinely co-created product. This demonstrated UrbanGrowth NSW's strong commitment place-based leadership, collaborative governance and public value creation (Moore 1995; Stoker 2006).

Shortly after the adoption by the NSW Government of The Transformation Plan for The Bays Precinct, the commitment to its core principles were tested by the market. UrbanGrowth NSW released a Request For Proposals (February 2016) for the development of one of the destinations, the White Bay Power Station. Submissions closed in June 2016 and were assessed. All submissions were rejected by UrbanGrowth NSW and the NSW Government because they were inconsistent with The Transformation Plan (UrbanGrowth NSW 2016a, b).

This demonstrates that the rigour of the engagement processes used to develop The Transformation Plan were highly valued by the NSW Government.

As the development of The Bays Precinct evolves over the next 20–30-years, the public value created through the unique and iterative engagement processes enabled by the City Transformation Life Cycle™ (UrbanGrowth NSW 2014a) will become a lasting legacy and provides strong learnings for other complex urban transformation projects.

References

Alford, J., & O'Flynn, J. (2009). Making sense of public value: Concepts, critiques and emergent meanings. *International Journal of Public Administration, 32*(3&4), 171–191.

Ansell, C., & Gash, A. (2007). Collaborative governance in theory and practice. *Journal of Public Administration Research and Theory, 18,* 543–571.

Australian Government. (2015). *Global development.* http://dfat.gov.au/aid/topics/development-issues/global-development-agenda/Pages/2030-agenda-for-sustainable-development.aspx. Accessed May 23, 2017.

Crosby, B., t' Hart, P., & Torfing, J. (2016). Public value creation through collaborative innovation. *Public Management Review.* https://doi.org/10.1080/14719037.2016.1192165.

Friedmann, J. (2010). Place and place-making in cities: A Global perspective. *Planning Theory & Practice, 11*(2), 149–165. https://doi.org/10.1080/14649351003759573.

GHD. (2015). UrbanGrowth NSW the bays precinct urban transformation program consultation report on transforming city living. *The Bays Precinct Discussion Paper.*

Godschalk, D. (2004). Land use planning challenges. *Journal of the American Planning Association, 70*(1), 5–13. (ISSN: 0194-4363, Winter 2004, 9p.3 Diagrams, 1 Chart).

Grant, B., Tan, S. F., Ryan, R., & Nesbitt, R. (2014). *Public value summary background paper—Prepared for the local government business excellence network (LGBEN).* Australian Centre of Excellence in Local Government (ACELG): University of Technology Sydney.

Hambleton, R. (2015). *Leading the inclusive city: Place-based innovation for a bounded planet.* Policy Press.

Horlings, L. (2015). Values in place; A value-oriented approach toward sustainable place-shaping. *Regional Studies, Regional Science, 2*(1), 257–274. https://doi.org/10.1080/21681376.2015.1014062.

Hyslop, D. (2014). *The bays precinct, Sydney international summit keynote address: White bay session.*

International Association for Public Participation Australasia. (2014). IAP2 public participation spectrum. http://www.iap2.org/. Accessed March 10, 2016.

Jones, A. (1998). Issues in waterfront regeneration: More sobering thoughts —A UK perspective.

Landry, C. (2006). *The art of city-making*. Earthscan.
Moore, M. (1995). *Creating public value: Strategic management in government*. Cambridge, MA: Harvard University Press.
New South Wales Government. (1991). *Protection of the Environment Administration Act 1991*. Sydney: NSW Government.
New South Wales Government. (2001). *Landcom Corporation Act 2001*. Sydney: NSW Government.
New South Wales Government. (2012). *Strategic framework report to the NSW government bays precinct taskforce* (Vol. 1). Sydney: NSW Government.
New South Wales Government. (2014). *A plan for growing sydney*. Sydney: NSW Government.
O'Flynn, J. (2007). From new public management to public value: Paradigmatic change and managerial implications. *The Australian Journal of Public Administration, 66*(3), 36–353.
Stoker, G. (2006). Public value management: A new narrative for networked governance? *American Review of Public Administration*.
The People for Places and Spaces. (2015). *Critical considerations for 21st century living at the bays precinct, Sydney*.
UrbanGrowth NSW. (2014a). *City transformation life cycle*. http://www.urbangrowth.nsw.gov.au/our-approach/city-transformation-lifecycle/. Accessed March 10, 2016.
UrbanGrowth NSW. (2014b). *About urbangrowth NSW*. http://www.urbangrowth.nsw.gov.au/about-us/about-urbangrowth-nsw/. Accessed March 10, 2016.
UrbanGrowth NSW. (2014c). *The Bays Precinct Sydney international summit: Make new sessional paper*. Sydney: NSW Government.
UrbanGrowth NSW. (2014d). *The Bays Precinct Sydney international summit november 2014: Statement of principles* (3rd ed.). Sydney: NSW Government.
UrbanGrowth NSW. (2015a). *A discussion paper, transforming city living: The Bays Precinct—Ideas to drive Sydney's global success by transforming The Bays Precinct*. Sydney: NSW Government.
UrbanGrowth NSW. (2015b). *The Bays Precinct Sydney: Call for great ideas*. Sydney: NSW Government.
UrbanGrowth NSW. (2015c). *The Bays Precinct Sydney: Transformation plan*. Sydney: NSW Government.
UrbanGrowth NSW. (2015d). *Making it happen in The Bays*. https://thebayssydney.com.au/making-it-happen-in-the-bays-2/. Media Release Accessed April 18, 2017.
UrbanGrowth NSW. (2016a). *UrbanGrowth NSW to become master developer of new tech hub at white bay power station*. https://thebayssydney.com.au/urbangrowth-nsw-to-become-master-developer-of-new-tech-hub-at-white-bay-power-station-2/. Media Release Accessed April 18, 2017.
UrbanGrowth NSW. (2016b). *International association for public participation: 2016 core values awards submission: Development of the transformation plan: The Bays Precinct, Sydney*. Sydney: NSW Government.
World Economic Forum. (2016). *Inspiring future cities & urban services: Shaping the future of urban development & services initiative*. World Economic Forum.
Zivkovic, S. (2015). *A complexity based diagnostic tool for tackling wicked problems*. https://journal.emergentpublications.com/.

Greening Regional Cities: The Role of Government in Sustainability Transitions

Simon Wright, Samantha Sharpe and Damien Giurco

Abstract The scale and complexity of the 'wicked problems' posed by sustainability are forcing collaborations between unlikely partners. In many instances, government is choosing to play a critical role in the sustainable innovation process. Yet much of the innovation literature pushes to the fore the ingenuity of the firm, leaving government to play a secondary supporting role. Drawing on ideas from innovations and transitions theory on the role of government in supporting green niches, this paper analyses the example of an evolving biomass project in regional NSW. In particular, the paper focuses upon the role that regional government plays in supporting this community-led collaboration. Based on circular economy principles, the project aims to achieve energy independence whilst simultaneously generating bio-products for the agricultural sector. Utilising a case study methodology including in-depth semi-structured interviews with more than 20 key stakeholders from government, business and the local community, analysis of the data suggests that government is a critical actor in the innovation process; plays a multiplicity of roles across the network; and that these roles vary to a greater degree than previously suggested in the literature. A number of factors are also identified that shape these roles at different stages of the innovation process. This paper sheds new light on the critical role played by government in facilitating and leading sustainability transitions and contributes to our knowledge of sustainable innovations more broadly. It also highlights a need for more research to improve our understanding of appropriate actors at different stages of sustainability transitions.

Keywords Innovation · Role of government · Sustainability transitions
Green cities

S. Wright (✉) · S. Sharpe · D. Giurco
Institute for Sustainable Futures, University of Technology Sydney,
235 Jones Street, Ultimo, NSW 2007, Australia
e-mail: simon.d.wright@student.uts.edu.au

W. Leal Filho et al. (eds.), *Sustainable Development Research in the Asia-Pacific Region*, World Sustainability Series, https://doi.org/10.1007/978-3-319-73293-0_19

1 Introduction: Third Party Actors and Innovation for Sustainability

The role of third party actors in the innovation process recurs constantly in the literatures, particularly in the formation and development of networks and within innovation systems themselves. This cast of third party actors is considerable and varied. Freeman (1991), for example, emphasises the need for diversity in the set of actors within the national system of innovation including industry, government and research institutes.

A broad range of terminology has been used to describe these invariably external actors ranging from 'third parties' (Mantel and Rosegger 1987) and 'intermediaries' (Stankiewicz 1995) to 'brokers' (Hargadon and Sutton 1997; Provan and Human 1999) and 'bridgers' (Bessant and Rush 1995; Zaheer and McEvily 1999). Unsurprisingly the roles of these actors can also be multifaceted. Social network theory identifies other specific roles for third parties such as plugging holes and bridging knowledge flows (Burt 1992). A host of other functions drawn from the networks and innovation literatures can be identified that include acting as catalysts for innovation, information brokers and intermediaries (Popp 2000), information 'bridgers' (Zaheer and McEvily 1999) and facilitating knowledge and technology transfers 'across people, organisations and industries' (Hargadon and Sutton 1997). In some instances, third parties can play a critical role, bringing credibility to a network and enhancing connectedness; in other instances, the role of the third party may be quite minor. For example, case study research of 22 innovation intermediaries in the UK identified a number of differing roles for third parties (Howells 2006). Studies on cross industry innovation conducted by Gassmann et al. (2011) also define three high-level groups of innovation intermediaries, again playing varying roles.

The sustainability literature carries a substantial body of research on third parties in various forms and with broadly similar themes (Blacconiere and Northcut 1997; Bush et al. 2015; Sbragia 2000; Van Kleef and Roome 2007). However, in the context of sustainable development it becomes evident that the theoretical and empirical research on the role of intermediaries in the innovation process itself is relatively scarce (Patala et al. 2014). Hargreaves et al. (2013) reinforce this, arguing that 'very little work has examined the role of intermediaries in sustainability niches and still less has examined the nature and extent of the roles they may play in helping grassroots innovations to develop and grow.'

Indeed it is the role of government as an intermediary in the sustainable innovation process that is most pertinent to this research. Drawing from the industrial ecology literature of direct relevance to the CLEAN case study discussed later in this chapter, there are contrasting and conflicting views about this role. Although industrial ecology appears to promote collaborations between firms along with a more holistic and sustainable approach to business rather than old style 'command and control' by government (Ehrenfeld 2000; Lifset and Graedel 2002), there is limited empirical evidence of this actually occurring. In fact, most research seems to advocate for government intervention, largely by capacity building through public policy (Wallner 1999), legislation that supports project development (Carr 1998) and the provision of economic and regulatory incentives for collaborative networking across regional platforms (Brand and de Bruijn 1999). Public sector agencies also have a role to play in facilitating and supporting collaboration (Benz and Fürst 2002).

Wading deeper into the broader sustainability literature, we see a varied landscape with a myriad of often contrary findings. Our introduction noted the complexity and scale of many of the challenges emerging under the banner of sustainability and the need for partnerships or multi-stakeholder collaborations (Roberts 2000; Selman 1998). Government is not exempt from this and has started to link with 'unlikely partners' through public/private partnerships to foster collaboration, access information and resources and reduce the risk associated with innovation (Bocken et al. 2014). Drawing on transitions theory and with clear implications for government as a creator of public policy, Van den Bergh et al. (2011) conclude that there is no definitive ranking of the effectiveness of policy instruments on environmental innovations.

With much in common with transitions theory and of direct relevance to the theme of green infrastructure more broadly, the literature on public private partnerships or 'PPPs' unsurprisingly touches on a number of topics related to the role of government. In their wide-ranging performance review of international PPPs, Hodge and Greve (2007) remind us of the confusion surrounding these partnerships and the contradictory results to date relating to roles and their effectiveness. Importantly, a need for improved evaluation of programs is identified, particularly in the social (and environmental) spheres.

Finally and of pertinence to this paper, again from the field of transitions theory, is the potentially critical role played by government in supporting innovation in green niches (Hargreaves et al. 2013; Smith et al. 2016), a theme that will be explored in greater detail in the following paragraphs.

2 Greening Regional Cities—The Case of CLEAN Cowra

Cowra Low Emissions Action Network or 'CLEAN' was established in 2007 as a community-led group originally to coordinate a bulk buy of solar. Since 2011 CLEAN has been collaborating with Local and State Governments, Industry and Community to develop a local biomass to energy project. The goals of the project are threefold—to empower the local agricultural community to generate their own energy from organic waste streams using a model of decentralised energy generation; to distribute the energy through a localised network; and finally, to develop a business model that allows for community involvement and investment. Ultimately CLEAN is hoping to forge a template or framework for regional, decentralised energy generation that is both scalable and replicable (Fig. 1).

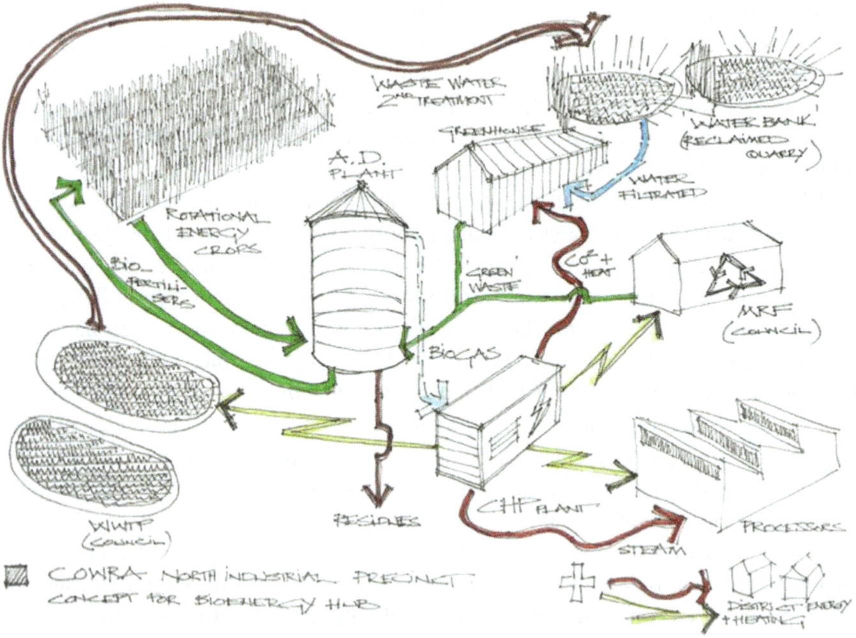

Fig. 1 The CLEAN process (CLEAN Cowra Inc. 2017)

In many ways, the evolution of the CLEAN network has been typical of many community collaborations. Led by a dynamic and committed local architect, the network has relied on the goodwill and support of a range of third parties who have been drawn to the project, primarily because of its highly innovative and unique nature or by the passion and persistence of the lead entrepreneur with a strong community focus. A range of technical and subject matter experts have given their time to share knowledge and introduce other third parties to assist the CLEAN team to take the project to the next stage. As a result of funding and in-kind support primarily from state and local governments, the project achieved a recent critical milestone, completing a week-long trial to assess the suitability of local biomass for gas and energy generation. The project team is now engaged in capital raising for the next stage, namely the much anticipated construction of a 'proof of concept' energy plant in the next 12 months. Ultimately CLEAN hopes to generate 12 MW of renewable energy from local waste, sufficient to power every home in Cowra.

One major differentiator for CLEAN has been the high degree of support the project has received from Government, both local and state. For example, Cowra Council is a key stakeholder on a number of levels. Not only has Council provided financial and in-kind support to the project, it will also be a significant potential supplier of organic matter to CLEAN through its waste water treatment plant and as a primary source of municipal solid waste for processing. To some degree Council is also viewed as a surrogate and representative for local community (Fig. 2).

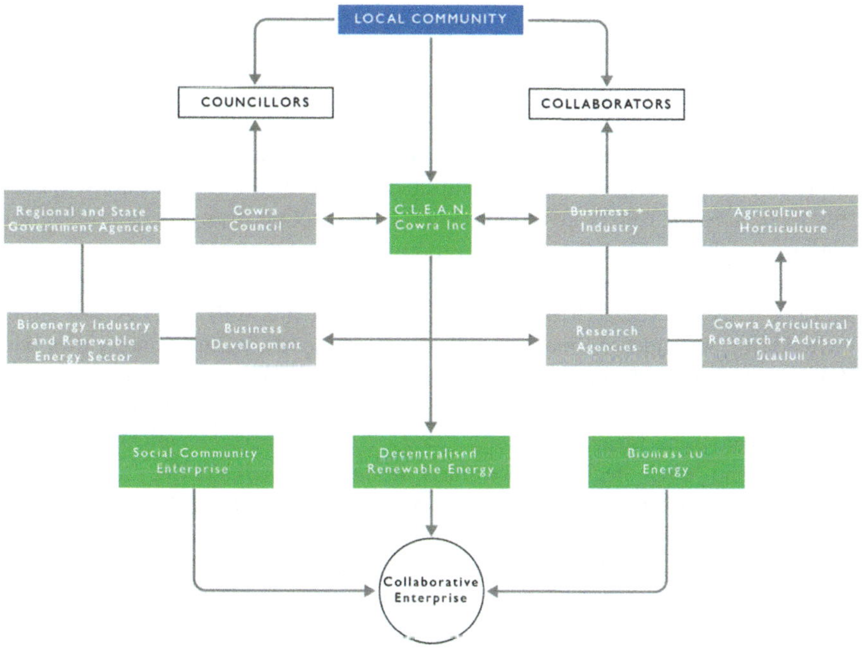

Fig. 2 An overview of CLEAN's stakeholders (CLEAN Cowra Inc. 2017)

Another critical source of momentum has been the support received from the New South Wales State Government's Office of Environment and Heritage (OEH) in the form of the much-feted Sustainability Advantage ('SA') Program supported by the Regional CLEAN Energy Program (RCEP). The overarching objective of SA is to increase the competitiveness and enhance the profitability of member organisations striving to demonstrate environmental leadership in their industry sector. This is achieved by assisting organisations to identify resource efficiencies in the areas of waste, water and energy and to exploit these opportunities through the support of a network of subsidised OEH-affiliated consultants. As a finalist in the 2015 Global Circulars Accenture Award for Circular Economy Pioneers, SA has been acknowledged as an effective and strategic broker in the emerging circular economy, building bridges between unlikely collaborators and traditional competitors. CLEAN epitomises this collaborative approach, indeed it is the role of OEH within the CLEAN network that forms the focus of this paper.

3 Research Approach

Based on a review of the extant literature on innovations, networks and transition studies, a number of potential research areas were identified that would significantly enhance our knowledge of the role played by government in the sustainable innovation process. These were concentrated into core research questions, specifically to explore the role of government in the CLEAN network and determine how it influences the process of innovation more broadly.

In terms of research approach, qualitative enquiry was deemed most appropriate determined by the research question and purpose (Maxwell 2012) A case study methodology was adopted incorporating in-depth, semi-structured interviews with 21 key stakeholders from regional and state government agencies (3), local government (3), CLEAN Cowra Inc. (3), business and industry (3) agriculture and horticulture (3), the local community (3), specialist bio-energy industry representatives (3) and 'other' technical and OEH affiliated consultants (3).

Using open ended questions, research participants were typically interviewed at their place of work and interviews lasted around 40 min on average. The questions posed in the course of the interview and hence the thematic headings in the discussion were strongly influenced by the academic literature and designed to address the research question. Responses were recorded and subsequently transcribed before being aggregated and analysed using established QDA software (NVivo). Focusing on a sample set of interviews, themes or codes were attributed to portions of the text and grouped under appropriate headers. Codes were then rationalised and reorganised before being applied to the broader data set. A further process of rationalisation and regrouping of codes was performed before a final analysis of the data was undertaken.

Drawing on the work of Fontana and Frey (2000), ethical considerations of most relevance included security of the participants during the interview, obtaining informed consent from the respondents and ensuring privacy and confidentiality of the data and the recordings.

4 The Role of Government—A Discussion of Research Findings

In contrast to the management literature suggesting a restricted number of roles played by intermediaries in the innovation process and extending the work undertaken by Howells (2006), interview findings from CLEAN suggest that government plays a wide variety of roles and that these roles vary across the innovation cycle.

4.1 Government as Banker and Funding Advisor

In an atmosphere of resource and capacity constraints that is typical of community projects, it is perhaps unsurprising that one of the two most important and oft-cited roles played by OEH should be as a funder or provider of funding advice. Respondents acknowledged the critical role played by State Government in committing funding to the project by way of SA.

> 'Was the funding important?' 'Very, we wouldn't be where we are without it'.

> They bring some money to the table which at the end of the day's fairly crucial.

Funding typically created capacity for the project team, most notably enabling the lead entrepreneur to dedicate time to project planning, engaging third party assistance and building the CLEAN network.

Introductions to potential funders, sometimes in other State Government departments, were seen as equally important to the successful continuation of the project, as was the case with RCEP.

> Well they brought money ok, the community grant was a very important milestone for the project….

Equally, employing consultants through SA whose primary job was to work with the project team to chase funding was both viewed positively by stakeholders and proved successful, often enabling the project to move to the next stage of evolution.

> so really my involvement for Cowra has been helping them go for grants…. and to some degree that's been supported by the state government, by the treasurer and her team.

Even without conscious introductions, the very presence of OEH as a key supporter was often sufficient in itself to convince others to support and sometimes even fund the project. Council in particular viewed the involvement of OEH as enhancing the credibility and longevity of the project and an important reason to lend their weight to the initiative.

> … when you start getting the long term guys that are prepared to stand behind and put their money where their mouth is and send representatives out, it's a different game.

Whilst the funding in itself was undeniably important and created capacity in the immediate term, the very act of supporting the project financially indicated that government had belief in the project over the longer term. This delivered other tangible and intangible benefits, most notably serving to energise the core network and bring credibility and sustained interest to the project in the eyes of other stakeholders, most importantly potential supporters and funders.

> … key turning point was the initial Community Energy Grant 18 months into the project which served to keep people involved and focused and generated lots of conversations.

4.2 Government as a Relationship Bridger

Government as a 'bridger' or facilitator of relationships was identified by interviewees as being equally as important as government's role as a funder. For a resource constrained community project in search of support and strategic advice, identifying potential trusted collaborators willing to invest time in the project to share expertise and progress the discussion is clearly critical. Moreover, consultants and subject matter experts recommended by government avoided wasting valuable time looking for appropriate advisors and ensured they were both credible and came equipped with the prerequisite knowledge. Government's commitment and interest in successful outcomes for the project added a further driver to ensure the most qualified individuals were singled out.

> They are the facilitator, I guess, their role is to put you in touch with the right people to give you access to the right people and right analysis and advice without you tripping over and going down the wrong oath etc. etc. So OEH has provided good guidance in that respect.

The very fact that potential collaborators were requested by Government to assist also ensured a degree of commitment on their part, not least because invariably they sit on an approved panel of service providers and generate significant income from the broader SA program.

Of significance to key stakeholders was government's ability to link the project to specialists from the private sector but also to contacts in other government departments and of course to potential funding. Once finalised, this funding in turn enabled CLEAN to employ some of these consultants on a short term basis to progress the project to the next milestone and to extend the network.

> Once we got the money from OEH… that gave us real cash to then spend on other consultants so that sort of network started.

Whilst of less overt significance than perhaps much of the networks literature suggests and although not cited explicitly by many interviewees, it is apparent that the issue of trust sits beneath the surface of many of these conversations and remains of particular relevance to relationships with potential collaborators. As noted above, in connecting CLEAN to consultants, OEH lowers the risk inherent in these engagements and instils credibility in the minds of the CLEAN network.

> Yep, so sometimes we help them by weeding out the cowboys I guess, so helping them with finding consultants who've got some wings under the belt in areas of expertise.

4.3 Bring Credibility, Provide Stability

The relative anonymity of the CLEAN project, at least in the early years, combined with the volatility and uncertainty of the renewable energy landscape posed particular challenges in terms of attracting support and investment. Along with funding and facilitating relationships, bringing credibility to the project and providing stability were viewed by participants as a critical role and benefit delivered by government involvement. OEH's willingness to invest both time and money in the project both placated anxious stakeholders and attracted other key players who would ordinarily struggle to engage in a project that might be perceived as higher risk when compared to their day-to-day business.

> Once one government entity is involved in it, it provides a lot of credibility.

This was particularly true of Local Government. Whilst hesitant to take a leadership position, Local Government has recognised the enormous potential benefits the project can deliver, both to local rate payers as well as to Council's bottom line through reduced costs for waste processing, landfill and energy.

> They've been very supportive obviously in funding as well as attending meetings and giving their input and their expertise… and that's one of the reasons why… council has joined….

Indeed, OEH acknowledges that through the SA program, delivering credibility to the project to entice others to join is a key strategic objective.

> The fact that I listen… and continue to come to meetings tells them that well maybe they do have something to offer.

Credibility breeds interest and commitment which results in a degree of stability for the project, a critical factor especially for community-led ventures scrambling from one round of funding to the next.

> They (OEH) bring support references and I think that allows it to gain momentum and I think that was instrumental like I said in pulling some more money out of council so it begins to snowball.

4.4 Support Specific Outcomes

Largely due to the strategic objectives of the SA program and the policy goals of State Government, OEH has worked with CLEAN to provide funding or in-kind support for specific outcomes. These have included resource audits, technical reviews and assistance with funding applications. In some ways this has forced structure and focus on what can definitely be described as a predominantly informal, loose and sometimes chaotic network, as well as instilling process and discipline to the project.

> It's great that they did come on board but I guess they could only really support specific outcomes.

A more defined process and structure have certainly resulted in attracting unlikely partners to the project who have proven critical to it success. Again Council would probably be the most significant of these.

> … support for the council to know, they want a robust process that's gonna be ongoing, not some sort of pie in the sky, they actually want to know that there's a process behind it, I guess that's it.

Whilst enforced structure and process have enabled the project to progress and to entice other parties to collaborate, it could be argued that it has obliged CLEAN to take a different path towards its end goal. It is possible that Government intervention has actually complicated the innovation process and delivered outcomes that may be of limited relevance to the project but provide a means to an end, in this instance further rounds of funding. This will be revisited later in the chapter.

4.5 Support Early Stage, Replicable Innovation

Much of the innovations literature typically portrays government as a minor actor in the system. By contrast, empirical evidence from early stage innovations suggests that government along with industry can play a key role in building momentum for new products and services. Similarly, transitions theory advocates a critical role for government in supporting grassroots innovation. CLEAN certainly seems to support this based upon the findings from a broad range of stakeholders, indeed support for early stage innovation was cited as one of the most critical aspects of government's involvement in the project.

> I don't think it would've got off the ground without OEH. OEH opened the door to money, to reinforce to A2 that if they put a good case up there was funding to push themselves forward with turning ideas and vision into something more than that….

Stakeholders internal to the core network and external stakeholders alike recognise the importance of government support in developing the project to proof of concept. This is particularly important in the context of a community led project which by definition tends to evolve more slowly than its industry-led counterpart, largely due to resource constraints. Innovation in this instance tends to be characterised by short bursts of activity and progress fuelled by an injection of capital or expertise followed by periods of consolidation as results are reviewed and plans are hatched to attract the next tranche of funding or in-kind support.

> They were really important catalyst in getting the project started and that's the hardest thing.

Ideology and personal relationships are two factors that played a critical role in Government's decision to lend support to CLEAN's early stage innovation. It is apparent that the innovation in this instance stems less from product or technology and more from the collective packaging of three distinct yet highly complementary initiatives—decentralised energy, circular economy principles and a community led project respectively. Many stakeholders cite the potential for this template to be replicated across regional towns and cities and OEH in particular identify this as a major motivator for their commitment to the project.

> Cowra support was due to replicable innovation, a look at a way to solve regional energy and waste problems with a promising project, which needed support for proof of concept.

Government engagement was strengthened by the personal commitment of the individual tasked by OEH to support the project. Across the network they were seen by many stakeholders as a major asset and enabler for the project, both in terms of finding ways to accommodate the project and its key stakeholders within a very tightly defined sustainability program (SA) as well as her dedication and commitment to ensure the project's longevity and success.

> My only experience with OEH has been A6[1] so it's a very positive bonus for the network, I'd sorta say A6's almost part of that network as much as I am as, if not from a professional role from a personal commitment to it.

Equally there is an acknowledgement by many stakeholders that beyond this phase of early stage innovation, it is highly likely that Government will play a much lesser role or indeed exit the network entirely at some point, particularly when significant funding is needed closer to full commercialisation and implementation.

4.6 Accelerate the Project to the Next Stage

The early focus of government in the project was to support the network and take the project to the next stage. Key network members identified this as an early contribution of the SA program and helped determine the *modus operandi* for the broader collaboration with OEH. Indeed, OEH themselves were conscious of this and saw it as a means to deliver tangible successes, thus energising the network to focus on next steps. This was achieved typically through the involvement of a number of third parties subject matter experts, other personnel from within OEH (e.g. RCEP) or funding or other in-kind support.

> There was very early involvement and support around developing the project to the next level.

The open, loose and agile nature of the network recognised by internal and external members alike allows for the dynamic entry and exit of change agents through the innovation process to inject specialist knowledge and momentum and take the project to the next stage. OEH has facilitated much of this dynamic process ideally suited for this kind of community network. In the words of OEH, they view their role as saving valuable time or providing 'shortcuts' for the project or provide resources when outside help is needed.

4.7 Broker Knowledge, Plugging Gaps

Related to OEH's ability and willingness to broker relationship has been their ability to plug knowledge gaps particularly in terms of the development process. An extensive network of more than 500 businesses has enabled OEH to build experience and knowledge across a broad range of sustainability projects and to share this across the member network. Evidently this has yielded benefits for stakeholders within CLEAN.

> Their role is to put you in touch with the right people, to give you access to the right people and right analysis and advice… OEH has provided good guidance in that respect.

> They have a great understanding of the process and have ID'd I believe gaps in knowledge and expertise.

Given OEH's desire to replicate the model and share the template with similar potential projects, a free flow of knowledge between parties has not only plugged key gaps but has arguably accelerated the process of innovation, again facilitated by an open and dynamic network structure.

4.8 Other Contingent Roles

A number of other roles played by State Government were identified by stakeholders as being important to the network at varying times. Implicit in some of the other roles mentioned is the ongoing stakeholder engagement undertaken by OEH beyond the involvement of third party consultants. Perhaps the most critical of these for both the community and the larger project has been the relationship with local government which has emerged as a critical stakeholder. In a similar vein, the commitment of OEH to a community-led venture has ensured that the 'keep it local' approach has continued past OEH's involvement, something that would have altogether more challenging with the entry of a private investor. On occasions, OEH has also been viewed as a project manager and mediator, able to give technical expertise and resolve issues hitherto unaddressed by the group. At other times, OEH was seen as a critical yet trusted friend of the project, enabling them to obtain unique perspectives on the working of the project and engage in conversations with core members that might ordinarily have been improbable including offering advice and guidance to the CLEAN team.

Other roles that emerged from interviewees included government as a risk taker and accommodator respectively. Risk taking is perhaps more apparent, implicit as it is in supporting early stage innovation in an emerging field such as renewable energy. The adaptability and flexible approach necessary to accommodate this type of undefined and emergent project in a program as well defined and established as SA further demonstrates the extent of OEH's belief and commitment to CLEAN as a whole.

Finally, OEH was seen to play a lead role in resource selection, allocation and direction. OEH itself identifies resource stewardship as one of its three core functions within the project, along with stakeholder engagement and investor management.

5 Government Support of Innovation—A Double Edged Sword?

The interview data from across the stakeholder group affirms a uniform set of largely positive perceptions about the role of state government in the CLEAN project. Yet all relationships experience peaks and troughs and stakeholders certainly saw the relationship with OEH as no exception.

Whilst undoubtedly a network enabler in many facets, stakeholders of all types identified a number of challenges in a relationship of this nature. Funding and program constraints, a not uncommon feature of programs reliant upon the public purse, was the most cited impediment for network members both within and outside of OEH. Breaking down the vision into fundable 'chunks' was a challenging and sometimes frustrating exercise for all parties, as was the process of repositioning these component parts such that they complied with the funding guidelines without diluting the very essence of the project.

> … in the course of 'dumbing it down' to meet government requirements for the grants, they may be losing what they saw as innovative, the real thing that makes a real community project.

Another feature of working with government is the vagaries of political cycles. Whilst not a feature of this particular collaboration, a cloud of concern nevertheless hung over the project with stakeholders worried about how long OEH would remain a facilitating partner in the face of ministerial changes and funding pressures. Similarly, given the relatively high risk nature of the project, many stakeholders working with or for other government departments harboured concerns about State Government's longer term appetite for engagement in the project, particularly in the face of project delays or failures. The relatively slow pace of government decision making combined with the constraints attached to funding served to prompt questions in the minds of the core network about whether a slower, more circuitous route could have been avoided. Mention was made by a few of other minor concerns about the relationship including the vacillating strategic priorities of environmental programs, loss of key personnel and knowledge in head office and the language barrier that existed between bureaucrats and industry.

In summary, only one in four stakeholders harboured any concerns relating to government involvement in the project despite direct prompts in the course of the interview. This was overshadowed by the overwhelming focus by respondents on the positive roles played by OEH to support the CLEAN project.

6 Conclusions: The Changing Role of Government

CLEAN is but one example of an highly innovative community waste to energy project still at a relatively early stage of development. The most striking initial observation from our data is that the CLEAN network has survived and prospered due significantly to the intervention of government, in this instance OEH. That is not to say that another suitor would not have materialised in due course and had a greater or lesser impact.

Government has taken different approaches and played a number of different and largely constructive roles in supporting this network. These have ranged from the provision of small amounts of capital to pay for people and expertise at the early stages of innovation to ongoing project advice and guidance; brokering relationships with subject matter experts to bringing much needed credibility to the network to enable stakeholders to engage directly with other parties. The impacts have been many and varied, resulting in the acceleration of specific phases of the project and continued if intermittent progression to the end goal.

Similar to all collaborations, issues arise that need to be addressed and differences emerge between parties. In the context of CLEAN, stakeholder concerns have related to funding requirements and constraints, political cycles and government's usually low appetite for risk, the speed of government decision making and a host of other minor concerns that whilst apparent but do not appear to have had a significantly detrimental impact on the project.

This paper builds on broader research into third party intermediaries and extends more recent research on the role of the government in supporting sustainability transitions. The findings from CLEAN suggest that in this instance, government is indeed a critical actor in these sustainability focused innovation networks, plays a multiplicity of roles in these projects and that these roles vary to a greater degree than previously suggested in the literature. Whilst the necessity of omnipresence is far from proven, this case study provide some clear examples of where government can and is able to make a considerable impact on innovation networks and provide clear, acknowledged leadership in the area of sustainability in terms of tangible, on ground projects, as well as education and the co-creation of a marketplace for industrial waste.

This in itself is an important finding, not least since much of the existing literature sees government as a supporting rather than lead actor, focusing on the role of the firm as a primary source of innovation. Similarly and at an international level, much of the focus on the role of government in advancing sustainability has been at a federal level with only limited coverage of the state, regional and arguably local levels. Looking specifically at the literature on multi-stakeholder partnerships for sustainability in the context of regional development, the role that local government plays in the creation of sustainable development is largely ignored. Whilst not explored in particular detail, this research touches on the critical role played by local government as a funder, supporter and potential beneficiary of the CLEAN project. Whilst a single case study is bounded by obvious limitations, it is hoped that this work will stimulate broader investigation into the potentially significant role government might play in the sustainable innovation process.

References

Benz, A., & Fürst, D. (2002). Policy learning in regional networks. *European Urban and Regional Studies, 9*(1), 21–35.

Bessant, J., & Rush, H. (1995). Building bridges for innovation: the role of consultants in technology transfer. *Research Policy, 24*(1), 97–114.

Blacconiere, W. G., & Northcut, W. D. (1997). Environmental information and market reactions to environmental legislation. *Journal of Accounting, Auditing & Finance, 12*(2), 149–178.

Bocken, N. M. P., Short, S. W., Rana, P., & Evans, S. (2014). A literature and practice review to develop sustainable business model archetypes. *Journal of cleaner production, 65*(0), 42–56. https://doi.org/10.1016/j.jclepro.2013.11.039.

Brand, E., & de Bruijn, T. (1999). Shared responsibility at the regional level: The building of sustainable industrial estates. *European Environment, 9*(6), 221–231. https://doi.org/10.1002/(SICI)1099-0976(199911/12)9:6<221:AID-EET209>3.0.CO;2-Z.

Burt, R. S. (1992). *Structural holes: The social structure of competition*. Cambridge, MA: Harvard University Press.

Bush, S. R., Oosterveer, P., Bailey, M., & Mol, A. P. J. (2015). Sustainability governance of chains and networks: A review and future outlook. *Journal of cleaner production, 107*, 8–19. https://doi.org/10.1016/j.jclepro.2014.10.019.

Carr, A. J. P. (1998). Choctaw Eco-Industrial Park: An ecological approach to industrial land-use planning and design. *Landscape and Urban Planning, 42*(2), 239–257.

Ehrenfeld, J. R. (2000). Industrial ecology paradigm shift or normal science? *American Behavioral Scientist, 44*(2), 229–244.

Fontana, A., & Frey, J. H. (2000). The interview. From structured questions to negotiated text. *Handbook of Qualitative Research, 2*(6), 645–672.

Freeman, C. (1991). Networks of innovators: A synthesis of research issues. *Research policy, 20* (5), 499–514. https://doi.org/10.1016/0048-7333(91)90072-X.

Gassmann, O., Daiber, M., & Enkel, E. (2011). The role of intermediaries in cross-industry innovation processes. *R&D Management, 41*(5), 457–469.

Hargadon, A., & Sutton, R. I. (1997). Technology brokering and innovation in a product development firm. *Administrative Science Quarterly*, pp. 716–749.

Hargreaves, T., Hielscher, S., Seyfang, G., & Smith, A. (2013). Grassroots innovations in community energy: The role of intermediaries in niche development. *Global Environmental Change, 23*(5), 868–880. https://doi.org/10.1016/j.gloenvcha.2013.02.008.

Hodge, G. A., & Greve, C. (2007). Public–private partnerships: An international performance review. *Public Administration Review, 67*(3), 545–558.

Howells, J. (2006). Intermediation and the role of intermediaries in innovation. *Research policy, 35*(5), 715–728. https://doi.org/10.1016/j.respol.2006.03.005.

Lifset, R., & Graedel, T. E. (2002). Industrial ecology: Goals and definitions. In *A handbook of industrial ecology* (pp. 3–15).

Mantel, S., & Rosegger, G. (1987). *The role of third-parties in the diffusion of innovations: A survey* (pp. 123–134). Innovation: Adaptation and Growth.

Maxwell, J. A. (2012). *Qualitative research design: An interactive approach* (Vol. 41). Thousand Oaks, CA: Sage Publications.

Patala, S., Hämäläinen, S., Jalkala, A., & Pesonen, H.-L. (2014). Towards a broader perspective on the forms of eco-industrial networks. *Journal of cleaner production, 82*(0), 166–178. https://doi.org/10.1016/j.jclepro.2014.06.059.

Popp, A. (2000). "Swamped in information but starved of data": Information and intermediaries in clothing supply chains. *Supply Chain Management: An International Journal, 5*(3), 151–161.

Provan, K. G., & Human, S. E. (1999). Organizational learning and the role of the network broker in small-firm manufacturing networks. *Interfirm networks: Organization and industrial competitiveness* (pp. 185–207).

Roberts, N. (2000). Wicked problems and network approaches to resolution. *International Public Management Review, 1*(1), 1–19.

Sbragia, A. (2000). Environmental policy. In *Policy-making in the European Union* (pp. 293–316).

Selman, P. (1998). Local Agenda 21: Substance or spin? *Journal of Environmental Planning and Management, 41*(5), 533–553.

Smith, A., Hargreaves, T., Hielscher, S., Martiskainen, M., & Seyfang, G. (2016). Making the most of community energies: Three perspectives on grassroots innovation. *Environment and Planning A, 48*(2), 407–432.

Stankiewicz, R. (1995). The role of the science and technology infrastructure in the development and diffusion of industrial automation in Sweden. In *Technological systems and economic performance: The case of factory automation* (pp. 165–210). Berlin: Springer.

Van den Bergh, J. C., Truffer, B., & Kallis, G. (2011). Environmental innovation and societal transitions: Introduction and overview. *Environmental Innovation and Societal Transitions, 1* (1), 1–23.

Van Kleef, J., & Roome, N. (2007). Developing capabilities and competence for sustainable business management as innovation: A research agenda. *Journal of Cleaner Production, 15*(1), 38–51.

Wallner, H. P. (1999). Towards sustainable development of industry: Networking, complexity and eco-clusters. *Journal of Cleaner Production, 7*(1), 49–58. https://doi.org/10.1016/S0959-6526(98)00036-5.

Zaheer, A., & McEvily, B. (1999). Bridging ties: A source of firm heterogeneity in competitive capabilities. *Strategic Management Journal, 20*(12), 1133.

Author Biographies

Simon Wright is a Ph.D. candidate at the Institute for Sustainable Futures investigating networks and the role of third parties in the sustainable innovation process as part of and supported by the CSIRO's Wealth from Waste Cluster.

Samantha Sharpe is a research and policy professional at the Institute for Sustainable Futures with more than 10 years' experience in public policy research and economic analysis. Her research focuses on regional economic development and innovation within firms and associated public policy of each. Samantha has been lead investigator on research projects funded by UK EPSRC, European Union and OECD. Outcomes of this research are policy development and industry advice around the support of innovative activity in places, the incubation of new technology, and the role public policy can play in establishing emerging markets for environment and 'green' technologies particularly in energy efficiency and renewable energy.

Professor Damien Giurco is Director (Innovation) at the Institute for Sustainable Futures. Concurrently, he is Professor of Resource Futures, team leader for the research areas of Resources and Energy. With a focus on strategies for responsible prosperity, Damien has worked collaboratively with government and industry clients spanning the minerals, water, waste and energy sectors to create change towards sustainable futures in a digital age. On behalf of ISF, Damien is also leading the CSIRO Wealth from Waste Cluster to identify pathways for creating wealth from waste containing metals, including e-waste. The $9 m three year collaboration partners with researchers at The University of Queensland, Monash, Swinburne and Yale.

A Circular Economic Model for a Sustainable City in South Asia

K. K. K. Sylva

Abstract In the process of development, beginning with the initial conceptualizations that involve any disturbance or alteration to the natural environment, it is mandatory to consider sustainability, a concept which includes a set of considerations that immerged to reduce global environmental threats. Since detrimental effects of rapid development are not confined to the immediate context of application, this consideration takes on ever greater urgency. Although rapid growths of economies have been achieved in the South Asian region, the development patterns that are observable are not conducive to either human society or to the natural environment. This paper proposes a sustainable circular economic model for a South Asian city by extending a case study done for a selected demographic and ecological boundary in Sri Lanka. The model is based on the contemporary illustration of 'strong sustainability' as three nested circles, which is different to the traditional illustration of sustainability that uses the Venn diagram: the confluence of the three pillars indicated by three circles. This concept paper, which elaborates the afore-mentioned model, provides a basic guideline for further development of cities in contexts rich in natural resources and habitats in South Asia, allowing for the preservation or refinement of natural settings. The output of the research, the Circular Economic Model applied to a Sri Lankan city, will serve as a model to support the reduction of indiscreet development that is taking place, and allow the user to envision appropriate routes towards sustainable development in the South Asian region.

Keywords Sustainable cities · Circular economy · South Asia

K. K. K. Sylva (✉)
Department of Engineering Management, Faculty of Engineering,
University of Peradeniya, Peradeniya, Sri Lanka
e-mail: kamanisylva@pdn.ac.lk

W. Leal Filho et al. (eds.), *Sustainable Development Research in the Asia-Pacific Region*, World Sustainability Series, https://doi.org/10.1007/978-3-319-73293-0_20

1 Introduction: Sustainable Development Through Economic Models

Detrimental effects of human activities have reached a state that cannot be rectified by either natural or artificial means. The most visible environmental problems in the twentieth century have been due to emissions and toxic waste (Ayres 2008), and these local problems in the West, which began with industrialization, were not limited to their geographical borders, but gave birth to global problems. Sharp growth attempts of the East in the twenty first century added largely to this, making environmental degradation the most critical problem for the sustainability of living organisms. These unprecedented growth attempts have become endless and the problems associated are not limited to environmental pollution, but have led to a concurrent critical condition regarding the sustainability of human beings—a resource shortage for our survival. In open economies, consumption was regarded as a good sign and the success of the economy was measured by the amount of throughput from the factors of production, a phenomenon identified as 'cowboy economy' by Boulding (1966). Extraction from natural reservoirs counted as positive while no consideration was given to the output as pollution until it reached a stage at which it was not controlled naturally. Although many remedial measures have been taken to rectify the resultant situation as end-of-pipe solutions—treatment of waste and pollutants in air and water—most of these attempts do not produce the expected outcomes because human activities that cause these conditions have not been mitigated with these methods.

With the failure of attempts to maintain expected standards of the environment with end-of-pipe solutions, the next attempt was to formulate regulations to either reduce the pollutants to accepted levels or eliminate the usage of such sources that contribute to pollution as attempts of source reduction. The concept of sustainable, green economies was developed within this context as a more appropriate survival strategy for human beings to protect the environment for both the present and the future while achieving economic growth. This contemporary concept 'circular economy' was proposed by Boulding (1966 cited in Wang et al. 2014) as a more appropriate solution to both critical factors: pollution and resource scarcity. According to Wang et al. (2014), a circular economy can transform a traditional linear economy which depends on raw resource consumption for growth, into a circular economy which relies on the circulation of available resources. The basic principles used in this attempt are 'reduce, reuse, and recycle', which would lead to lesser consumption of raw resources and lesser emission related to growth (Ghisellini et al. 2015).

This concept paper is based on a theoretical study carried out to apply sustainable economic approaches to a Sri Lankan City, Gampaha, where environmentally damaging development practices were observed (Department of Census and Statistics 2014; CEA 2014; Atukorala 2011; Bambaradeniya 2008). The study was then extended to investigate and verify the patterns of growth of several other major cities—Kandy, Colombo, Galle, Badulla and Kurunegala—that share similar conditions. The scope of the study was to build a generalized conceptual model of a sustainable city that would reduce the detrimental effects of economic growth and development processors on the natural environment and on society. The context, common to the selected cities in Sri Lanka, is generally rich in natural resources and habitats and the practice until recent times was to secure these spaces as nature reserves. However, recent unprecedented growth attempts have disregarded such positive practices. Although many growth activities are taking place in these cities, no collaboration takes place between the public and private sector investments (Ministry of Power and Energy 2015; Atukorala 2011) in order to ensure sustainability. Many ad hoc practices have been initiated by both the public and private sectors that pay little or no attention to the system as a whole. Flooding and landslides have been reported as common natural disasters in these cities mostly due to human activities.

The proposed circular economic model focuses on upgrading indigenous industries as the main economic source for the city as a pledge to environmental friendliness. Native industries would allow higher adapting and adopting possibilities for the inhabitants. Any extraneous additions to the economic system are to be prevented by policies and indicators for sustainability. Considerations such as green energy generation plans; waste management strategies; proper land use and management of the built environment; watershed and water resources management strategies were incorporated into the model to meet the targets of sustainability. The goals of the model were to propose minimum disturbance to the living organisms due to any development activity, to give priority to source reduction than end-of-pipe solutions for sustainability, and to not compromise the non-use-valued items in the process of development. Poverty reduction through education and employment and a healthier society through indigenous approaches of healthcare too were incorporated into the model for social wellbeing.

2 Methodology

First, a conceptual model for a sustainable city using the circular economic approach was developed based on a study done on a selected demographic and ecological boundary in Sri Lanka, the city of Gampaha. This model was then verified, in terms of similar patterns pertaining to the existing conditions and growth trends, to be applied to five other cities in Sri Lanka, namely Kandy, Colombo, Galle, Badulla and Kurunegala. This conceptual application led to a generalized model.

The conceptual model for the city of Gampaha was developed on the basis of the contemporary illustration of 'strong sustainability,' as three nested circles: the economy exists within society, and both the economy and society exist within the environment (Wu and Wu 2012). This conceptualization is different to the illustration of sustainability presented by the Venn diagram that shows a confluence of the three pillars indicated by three circles. The nested circles were selected assuming that society has some control on the economy but it has less control on the environment since the environment is shared with other societies who could either support or oppose in protecting the environment.

The proposed generalized model was conceptualized for contexts rich in natural resources and habitats allowing for the preservation or refinement of natural settings. Such contexts also allowed the model to maintain the wellbeing of society within the boundary while development was taking place. All aspects of human welfare proposed by Hueting (2008), such as the package of goods and services produced, scarce environmental functions, leisure time, the distribution of scarce goods, income distribution, the conditions under which scarce goods are acquired, employment, and the future security of society were considered when building the model.

3 Indicators for Sustainability

The measuring indicators were proposed to be more inclined towards source controlling, also called 'feed-in' indicators in this paper, rather than end-of-pipe indicators. But if the system is to add any pollutants to the environment, such as waste water or emissions, indicators were proposed to allow for measuring and rectifying any detrimental effects to the system.

Indicators were generated to represent the three pillars of sustainability: environmental, social, and economic, through a comprehensive literature review (UNDP—HDP 2015; Scott et al. 2014; Barbier and Markandya 2013; Puppim de Oliveir 2012; Wu and Wu 2012; Arendse and Godfrey 2010; Ayres 2008; Geng and Doberstein 2008; United Nations 2007; Zhijun and Nailing 2007; Milani 2005; Nanayakkara 2005; Mederly et al. 2003; Victor 1991). Since society would not have total control over the environment, the factors that could be controlled at the level of society to prevent affluent entering the environment were given priority. The environmental indicators were developed especially as feed-in indicators rather than allowing the system to depend on end-of-pipe controls such as air quality, water quality, etc. The proposed indicators are presented in Table 1.

In selecting social indicators, it was assumed that people who are employed and who have a substantial income could afford to live a comfortable life (Hueting 2008). Since the city is planned in a way to offer many 'green jobs', it was assumed that the unemployment factor would reduce with these green jobs. Although people are employed, if the national grid cannot supply electricity to that group of people, this may lead to a deprived society since access to many other facilities including information and education depends on the availability of electricity. Therefore access to electricity was included. Access to clean water was selected as a basic need to prevent waterborne diseases and lead to a healthy society. Many natural water sources are presently contaminated due to industrial activities taking place in these regions. Therefore, it was expected that purified water should reach every citizen. Although 'happiness' as a concept could be different from person to person, this indicator was selected as an end-of-pipe indicator for societal wellbeing due to its intersection with different facets of development.

Table 1 Proposed indicators for sustainability

Dimension of sustainability	*Feed-in* indicators (source controlling indicators)	*End-of-pipe* indicators
Environment	% Natural forest cover	% of waste reused
	% Plantation cover	% of waste recycled
	Water quality in natural sources	Natural disaster resilience measures
	Air quality at periphery of the city	Air quality within the city
	Soil quality at periphery of the city	Soil quality within the city
	% Green jobs	% Non-green industries
	% Green material inflow	% Green product outflow
Society	% Employed	Happiness indicator
	% of households with electricity	% Access to education
	% of households with access to clean water	% Access to health facilities
Economy	% Contribution to infrastructure development	Contribution to GDP of the country
	% Contribution to indigenous industrial development	% Contribution to GDP from small (green) industries

4 Common Zoning of the Cities

The six cities studied revealed that a common zoning pattern can be developed for modeling purposes. Every city has a highly populated area towards the city center, which has been defined as the 'high residential zone'. Towards the suburbs of the city there is a reduction in population and these areas are named 'medium-low residential zones.' It was predicted that these areas would grow rapidly in population. A common positive characteristic identified in the cities was the reservations, especially for water sources such as rivers and lakes. Many types of sanctuaries marked off for the purpose of maintaining environmental balance were also identified in these major cities. Some of these are: Udawatta Kele Sanctuary in the city of Kandy, Viharamahadevi Park in the City of Colombo, Henarathgoda Botanical Gardens in the City of Gampaha, Bickay Nature Reserve in the city of Galle, Botanical Gardens in the city of Badulla, and the Badagamuwa conservation forest in Kurunagala. Many possible areas to be developed as parks were also identified and all these areas were commonly named as 'park/reservation zone.' Areas that can be developed as indigenous industries in the region were termed 'industrial zones.' Zoning of the city was carried out in order to create minimum disturbance to the existing lifestyle of the residents and any other living organisms. The zoning for Gampaha City is presented in Fig. 1.

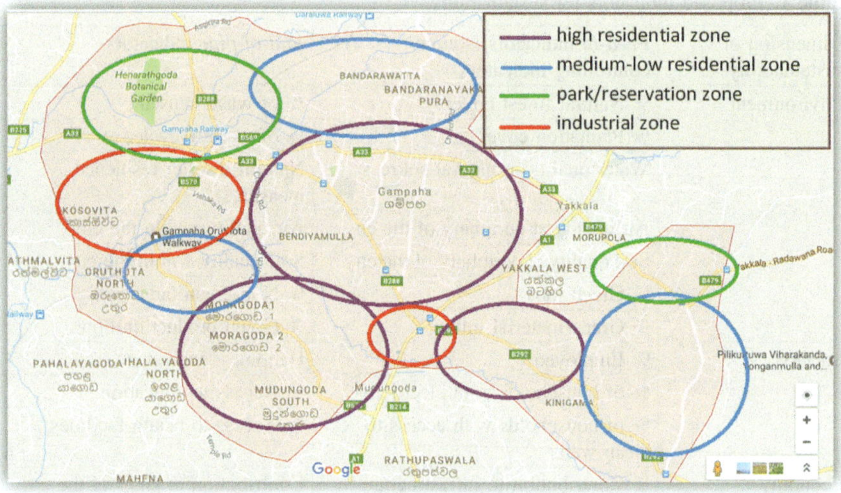

Fig. 1 Zoning of Gampaha city

5 Development Plans for a Circular Economic Model

The model was divided into three main areas as economy, society, and environment to comply with the sustainability model that illustrates 'strong sustainability' as three nested circles. Planning of the economy was the first step in the proposed development process and it was based on indigenous industries of the area. The welfare of the society was then considered in relation to the proposed industries while considering the possibility that some groups may seek employment in other cities as well. Source reduction of pollution and watershed management were incorporated in the planning activities. Consequently, it was expected that environmental pollution would be at minimum levels and the wellbeing of society would reach a high level through the new economic model.

5.1 The Economy of the Circular Economic Model for the City

The economic plan was proposed based on the concept of green industries as a source reduction mechanism to avoid possible pollution. As the cities selected still have their original indigenous industries at a small scale, it was proposed to improve such industries as the main measure of economic development. Many cities have their traditional treatment facilities that provide *ayurvedic* treatment (indigenous medicine) which is environmentally friendly. For example, Gampaha is well known for *ayurvedic* treatment. For the development of the economy, *ayurvedic* centers and educational facilities were proposed around the Gampaha Wichramarachchi *Ayurvedic* Hospital and higher education institute. The *Ayurveda* Teaching Hospital, Borella (Colombo) and the Institute of Indigenous Medicine too are similar institutions that of indigenous industries that can be developed. In almost all selected cities, the *ayurvedic* industry is growing in the form of spa treatment centers at a small scale linked to the tourism industry. Many green industries in relation to food, textile, furniture, livestock, and pottery, which are operating at a small scale, are also proposed to be developed as supplementary industries for production and generation of wealth of the system. All these income-generating sources are selected based on the concept of source reduction of pollution. It is proposed that these economic activities be located in the industrial park so that minimum relocation of people is necessary, or be introduced as small industries in the medium-to-low residential zones without disturbing the lifestyle of residents. Tourism related industries are proposed to be integrated with the business opportunities in the green reservation stretchers identified. In addition, such industries can also be developed around the *ayurvedic* spa centers in the city model to enable income generation for city dwellers. The infrastructure and services related to all these industries is an integral part of this development.

5.2 *Society in the Circular Economic Model for the City*

The wellbeing of the community is proposed to superimpose the economic plan. Highly residential areas and medium-to-low residential areas are to be kept intact without disturbance to the lifestyle of the residents. In the medium-to-low residential areas, business centers are proposed to be developed in different locations for local as well as foreign communities to purchase the production of the cities. Park lands are proposed to provide the community with recreational facilities. *Ayurvedic* treatment centers with green treatment approaches, which cater to tourists at present, are proposed to be developed in order to provide general herbal healthcare facilities to the community. To complete the circulation process of the model, it is also proposed that rain water be collected in the residential zone, that the residential zone contribute to the green cover, and it contribute towards the waste recycling process wherever possible. Access to electricity will be provided either through the national grid or through the energy generated from the waste circulation.

5.3 *The Environment of the Circular Economic Model for the City*

The environment preserved as sanctuaries and reservations are proposed to be protected or enhanced through the model. It is also proposed that the green cover in community residential zones be improved by helping the community to grow indigenous plants and air and water cleaning plants. The secondary aims of the proposal are to supply raw material for the main *ayurvedic* industry and to use any waste for power and fertilizer production.

Three zones for watershed management are identified as the Retention Zone: lowest elevations in the city; Detention Zone: shallow depth to bedrock (therefore cannot infiltrate the rain water); and Infiltration Zone: greater depth of bedrock and soil that accept infiltrating rainwater. In the Retention Zone, rain water retention is facilitated with the use of basins or ponds, constructed wetlands, and subsurface storage that are recommended in these areas. Medicinal plants that absorb harmful chemicals and purify water are proposed to be planted to maintain the standards of water quality in the area since the continent is abundant in such resources. In the Detention Zone, rainwater harvesting, green rooftops, and vertical gardening are recommended for detaining rainwater above ground. This zone is located within residential areas of the model and it is recommended that the residents grow medicinal plants on their rooftops or in gardens in the available areas as a water detention mechanism as well as a secondary income-generation source. The Infiltration Zone provides the best opportunity for rain water to infiltrate spaces using vegetated swales, infiltration plants, rainwater trees, rainwater gardens, and permeable pavements. Through watershed management it is expected to reduce the detrimental effects of human triggered natural disasters such as landslides and flooding. Green jobs and green raw materials will be a natural by-product of the development of the proposed economic activities.

5.4 The Circulation of Resources in the Circular Economic Model for the City

Green sources such as biomass, solar, and wind are proposed for energy and power supply as a source reduction mechanism and to reduce the use of non-renewable energy sources in the national energy supply scheme. A polygeneration plant is proposed in a less residential area to supply the energy demands of the industries and the community as direct supply or through the national grid which enables energy storage facilities when there is excess production. In a polygeneration system, a centralized gasification plant, a solar PV plant, and a wind turbine can be used in an intergraded manner. Coconut residue, paddy residue, and Municipal Solid Waste (MSW), which is in abundance, can be used as inputs for the gasification units.

In the development plan of the city, botanical gardens have been considered wherever possible as a major contributor towards the natural beauty and the enhancement of the tourist industry. As the bio-diversity of the Asian continent is high, it is proposed that existing botanical gardens be upgraded or that such areas be introduced as eco-tourism zones to add to the economic and social development of the city. The surrounding area is proposed as public parks with walkways for jogging and cycling tracks for the wellbeing of the community. Floral plants that cleanse the air are proposed to be planted in these areas so that they overlap with the

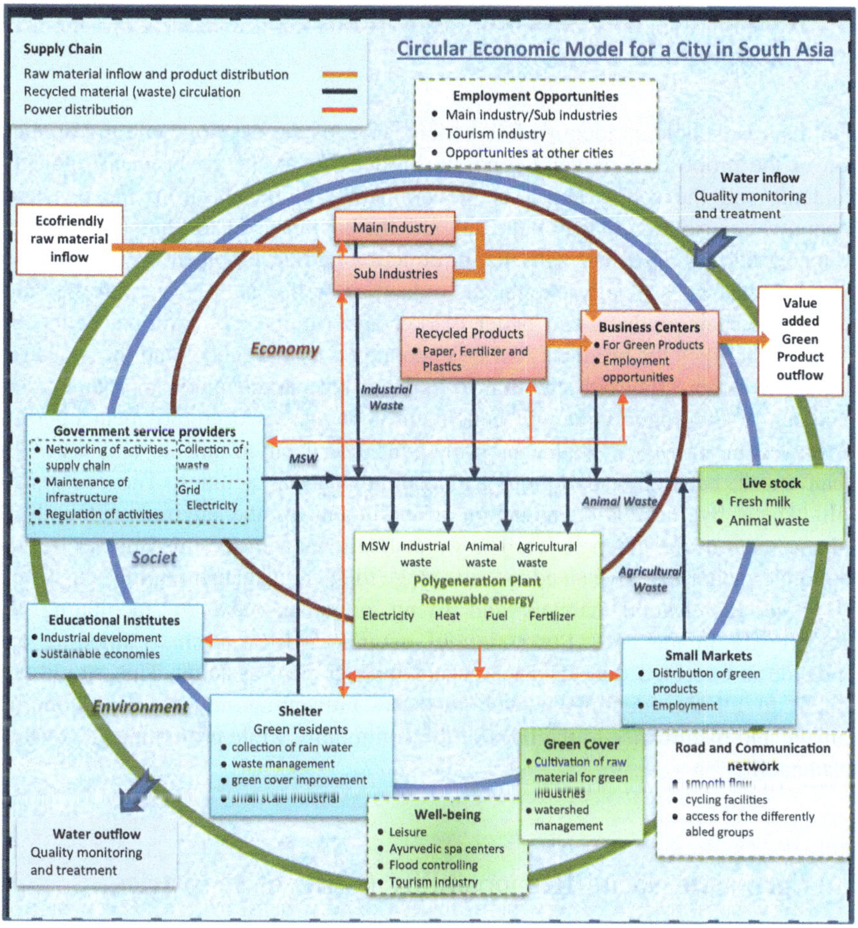

Fig. 2 Circular economic model city for South Asia

residential areas and improve the wellbeing of the dwellers while maintaining ecological balance. Construction of tourist villas and cottages in the surrounding area with *ayurvedic* spas which provide herbal and aroma therapy treatment is proposed as a means of income generation. The area surrounded by the water ponds proposed for water retention can also be used as public parks and recreational facilities for city dwellers and for tourists. Well-planned floating markets and floating restaurants could enhance the socioeconomic value of the area.

The model recommends maintaining the existing road and communication networks by developing them to a better standard. Sub-routes and walkways in the areas are proposed to be developed with provision for cycling facilities that do not disturb pedestrians. The main items of the circular economic model are presented in Fig. 2.

6 Stakeholder Analysis and Stakeholder Collaboration Plan

The main stakeholders identified as beneficiaries of the development and application of the proposed Circular Economic Model City is the government, industry, residents and the eco-system. The government will benefit from the enhanced economic contribution made by the city and the happiness factor that is expected to be increased in this society with the introduction and development of the proposed model. Industry, especially indigenous industries, will benefit by way of the value addition to their products and better market opportunities. In addition, better distribution facilities for products will be encouraged by the model and the industries will receive more focused attention to expand the market for their products. The residents or the community will benefit through increased employment opportunities; leisure facilities; health support; educational environments and access to clean water; better transportation; and communication facilities. The eco-system will benefit by the source reduction of pollutants to the environment and the enhancement of the green cover. The watershed management superimposed on the economic model will give an added advantage to the management of the eco-system by preserving natural habitats and living organisms and by facilitating the bio-diversity in the area. Prevention of natural disasters such as flooding and landslide is expected due to proper water management systems. This would add benefit to all the identified stakeholders by preventing unnecessary economic damage and disturbance to the lives of the community while promoting eco-system sustainability.

7 Corporate Social Responsibility (CSR) of Industries

The model expects the industries to be driven by a high level of corporate responsiveness to adhere to the sustainability concepts by using a proactive strategy to embed CSR into its activities. This would be done by taking a different approach to answer the basic economic questions, for whom to produce, what to produce, and how to produce. They should be able to first produce for the community within the boundary established, allowing lesser preservation, storage, and transportation of products, complying with the goals of a circular economy. They should also plan to produce nature-friendly products either with new raw material supplied by the community or using recycled inputs. Industry should comply with the requirement of using environmentally-friendly methods of production while allowing sufficient green employment opportunities for the community to fulfill CSR goals.

8 Work Breakdown Structure and Scheduling

The application of the proposed circular economic model to the city should be initiated by planning the economy. Simultaneously, the supply chain should be strengthened by introducing regulations and responsibilities for government agencies and the community as stakeholders. This move would ensure an uninterrupted supply of raw materials because production has to be ensured by either government agencies or identified public-private partnerships. The waste of the system should then be directed for recycling or power generation through proper networking. The work breakdown structure proposed is presented in Fig. 3.

9 Risk Analysis

Non-use values such as that of living beings, habitats, and endangered species will not be at risk due to the application of the model because the very reason for the proposed development process is to either not disturb these habitats or to enhance favorable conditions for such habitats.

There would be a moderate risk of natural disasters until the system is established with a superimposed watershed management plan which may take a few years to come into effect. Community acceptance of the model is assumed to be low and the impact of this factor would be high. With educational programs to make people aware of the social wellbeing resulting from the model, this risk factor can be minimized. After educating the community, it is expected that they would comply with simple techniques that are to be introduced for source reduction of

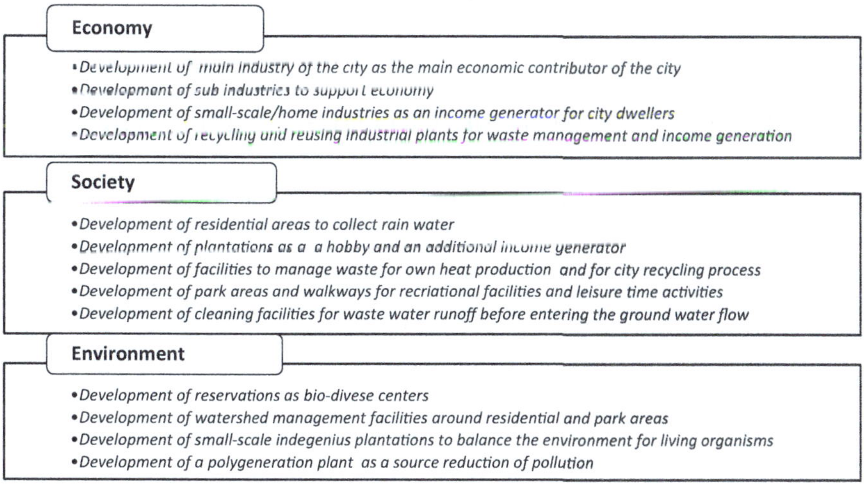

Fig. 3 Work breakdown structure

Impact / Occurrence	Low	Medium	High
Low	Sabotage by the community	Emissions from industries	Non-acceptance by the community
Medium	Improper waste water treatment by community	Natural Disasters	Supply Chain breakdowns
High	Migration of new dwellers	Difficulties in formulating policies	Difficult Public-Private Partnerships

Fig. 4 Risk matrix of the proposed circular economic model

environmental pollution. In order to initiate new immigrants into the model city, frequent inspection and educational programs are proposed.

The highest risk factor identified was the integration and collaboration of industry and public organizations in the region. The general expectation is that the public sector will be less flexible in adjusting to new regulations. Applications of the 'Triple P' concept—the Public-Private Partnership programs—were identified as a preventive measure to circumvent this risk factor when implementing the model. Through this approach, the private sector or industry would collaborate by financing projects with substantial resources from the public sector to initiate activities. This integration would assure sustainability of the supply chain for production as well as for community cohesion building. The risk analysis matrix is presented in Fig. 4. The other high risk areas such as supply chain management and formulating and executing appropriate policies will be minimized by assigning responsibilities to the bodies developed through public-private partnerships with proper guidance regarding the smooth application of the model.

10 Intended Outcomes of the Circular Economic Model

The proposed Circular Economic Model will create little or no disturbance to existing activities in the city and will consider the following areas:

- development of indigenous industries to upgrade the economy of the city
- development of residential, recreational, and green healthcare facilities for the wellbeing of the society
- promotion of development activities based on 'green concepts' while adhering to environmental policies and regulations for maintaining and upgrading the environment as 'feed-in' indicators while providing end-of-pipe solutions to any detrimental effects of development

The expected outcomes of the circular economic model are: increase in green production, better employment opportunities, healthier community, increase in 'happiness,' better waste management, enhancement of green cover, reduction of the use of non-renewable resources for power generation, and reduction of flooding and natural disasters through the superimposed watershed management in the selected area. No major structural changes to the existing condition of the city are proposed through the Circular Economic Model other than an enhancement of traditional industries to upgrade the economy, which does not involve an addition of high-tech solutions. Green technology and environmentally-friendly living are promoted through the model, aiming for source controlling for environmental protection rather than planning for end-of-pipe treatment of pollution. It is recommended that the context of application be further studied before applying this model in order to identify any specific resources as raw material and related industries.

References

Arendse, L., & Godfrey, L. (2010). Waste management indicators for national state of environment reporting. South Africa: Environmentek, CSIR. http://www.unep.or.jp/ietc/kms/data/2010.pdf. Last accessed December 12, 2016.

Atukorala, R. N. A. (2011). Gampaha as a prototype district in Sri Lanka. Available at: http://www.ft.lk/2011/12/13/gampaha-as-a-prototype-district-in-sri-lanka. Last accessed October 15, 2016.

Ayres, R. U. (2008). Sustainability economics: Where do we stand? *Ecological Economics, 67,* 281–310. www.sciencedirect.com. Last accessed December 12, 2016.

Bambaradeniya, C. (2008). *Western Province biodiversity and conservation action plan.* [pdf] Sri Lanka: Biodiversity Secretariat.

Barbier, E. B., & Markandya, A. (2013). *A blueprint for a green economy* (pp. 94–100). United Kingdom: Routledge.

Boulding, K. E. (1966). The economics of the coming spaceship earth. www.ub.edu/prometheus21/articulos/obsprometheus/BOULDING.pdf. Last accessed October 01, 2017.

Central Environmental Authority (CEA). (2014). *District profile of Gampaha.* Available at: http://www.cea.lk/web/index.php/en/publication?id=101. Last accessed October 15, 2016.

Department of Census and Statistics. (2014). *District statistical handbook.* Sri Lanka: Gampaha. [pdf] Available at: http://www.statistics.gov.lk/DistrictStatHBook.asp?District=Gampaha&Year=2014. Last accessed October 15, 2016.

Geng, Y., & Doberstein, B. (2008). Developing the circular economy in China: Challenges and opportunities for achieving leapfrog development. *International Journal of Sustainable Development and World Ecology, 15,* 231–239.

Ghisellini, P., Cialani, B. C., & Ulgiati, C. S. (2015). A review on circular economy: The expected transition to a balanced interplay of environmental and economic systems. *Journal of Cleaner Production, 114,* 11–32.

Hueting, R. (2008). Why environmental sustainability can most probably not be attained with growing production. In *Proceedings of the First International Conference on Economic De-Growth for Ecological Sustainability and Social Equity.* Paris, 18–19 April 2008. Available at: http://www.events.it-sudparis.eu/degrowthconference/en/appel/Degrowth%20Conference%20-%20Proceedings.pdf. Last accessed October 01, 2017.

Mederly, P., Novacek, P., & Topercer, J. (2003). Sustainable development assessment: Quality and sustainability of life indicators at global, national and regional level. *Foresight, 5*(5), 42–49.

Milani, B. (2005). *Building materials in a green economy: Community-based strategies for dematerialization*. Ph.D. dissertation, Department of Adult Education, Community Development and Counselling Psychology, Institute for Environmental Studies, University of Toronto.

Ministry of Power and Energy. (2015). *Sri Lanka energy sector development plan for a knowledge-based economy 2015–2015*. [pdf] Ministry of Power and Energy. Available at: http://powermin.gov.lk/sinhala/wp-content/uploads/2015/03/ENERGY_EMPOWERED_NATION_2015_2025.pdf. Last accessed June 15, 2016.

Nanayakkara, A. G. W. (2005). *Selected millennium development goals (MDG) indicators*. [pdf] Colombo: Department of Census and Statistics. Available at: http://www.statistics.gov.lk/mdg/mdg.pdf.

Puppim de Oliveir, J. A. (2012). *Green economy and good governance for sustainable development: Opportunities, promises and concerns*. Japan: United Nations University Press.

Scott, E. M., Cocchi, D., & Gemmell, J. C. (2014). Defining a fit for purpose statistically reliable sustainability indicator. *Sustainability Accounting, Management and Policy Journal, 5*(3), 262–267.

United Nations. (2007). *Indicators of sustainable development: Guidelines and methodologies*. New York: United Nations.

United Nations Development Programme 2015, Human Development Report. (2015). *Work for human development, Briefing note for countries on the 2015 Human Development Report Sri Lanka*.

Victor, P. A. (1991). Indicators of sustainable development: Some lessons from capital theory. *Ecological Economics, 4,* 191–213.

Wang, P., Che, F., Fan, S., & Gu, C. (2014). Ownership governance, institutional pressures and circular economy accounting information disclosure—An institutional theory and corporate governance theory perspective. *Chinese Management Studies, 8*(3), 487–501.

Wu, J., & Wu, T. (2012). Sustainability indicators and indices: An overview. https://www.researchgate.net/publication/228456338_Sustainability_indicators_and_indices_an_overview. Last accessed October 01, 2017.

Zhijun, F., & Nailing, Y. (2007). Putting a circular economy into practice in China. *Sustainability Science, 2,* 95–101.

Author Biography

K. K. K. Sylva has spent 17 years teaching, researching and consulting in engineering management, sustainable development and related areas. She received her Bachelor's in Civil Engineering from the University of Peradeniya, Sri Lanka (1992) and went on to receive two Master's, one from AIT, Bangkok in Geotechnical Engineering (1995) and an MBA from the Post Graduate Institute (PIM) of the University of Sri Jayawardenapura, Sri Lanka (2013). She is also in the final stage of completing a Master's in Sustainable Energy Engineering offered by KTH and University of Gävle, Sweden. She first taught engineering management and sustainable development at the University of Ruhuna, Faculty of Engineering (from 2000 to 2011), and has been teaching the same areas at the University of Peradeniya since 2014. As an accredited professional, she contributes to the GBCSL, (the Green Building Council of Sri Lanka) and to other projects pertaining to sustainable development. She would like to thank KTH, Sweden and the Open University of Sri Lanka for envisioning this area of study through the summer program (2016) where initial work was laid out for the present study.

Urban Regeneration Process: The Legacy Village in the Urban City of Hong Kong

Esther H. K. Yung and Maria Yu

Abstract It is increasingly recognized that embracing heritage conservation in urban regeneration plan can enhance sustainable development. However, the complex juxtaposition of conservation and urban regeneration has to be clearly addressed in cities relying heavily in property development and economic growth. The paper investigates the Nga Tsin Wai old village in the urban city of Hong Kong. It aims to explore the contested roles of the key stakeholders in the debate of the conservation and redevelopment of the village over the past twenty years. Focus group meetings of the original inhabitants and in-depth interviews of different stakeholders groups were conducted. The case vividly demonstrates the key challenges of integrating heritage conservation in the redevelopment of the old village including incompatible interests of different stakeholders groups, fragmented institutional arrangements, and lack of supportive government policies. This paper recommends a stronger collaboration between urban planners, conservationists and policy makers for better integration of the different regimes of economic development, real estate development, land use planning, and housing and urban policy. Most importantly, supportive government policy, especially in land use planning throughout the whole process is fundamental to resolve the conflicts arise from different interest groups.

Keywords Urban regeneration · Heritage conservation · Old village
Challenges · Hong Kong

E. H. K. Yung (✉) · M. Yu
Department of Building and Real Estate, The Hong Kong Polytechnic University, Hung Hom, Hong Kong SAR, China
e-mail: esther.yung@polyu.edu.hk

W. Leal Filho et al. (eds.), *Sustainable Development Research in the Asia-Pacific Region*, World Sustainability Series, https://doi.org/10.1007/978-3-319-73293-0_21

1 Introduction

Literature on urban regeneration and urban renewal has grown tremendously in the last two decades. However, there is still no comprehensive definition of it. It can be seen that the mode of urban redevelopment has evolved from urban reconstruction to urban renewal around 1960s–1970s, to urban regeneration around 1980s–1990s in the UK (Roberts 2000).

It has been stressed that regeneration should consider how physical, social and economic initiatives can come together to deliver the best outcomes through a well articulated strategy (Rhodes et al. 2007). Leary and McCarthy (2013) provides a more useful definition which states that regeneration is "area-based intervention which is public sector initiated, funded, supported, or inspired, aimed at producing significant sustainable improvements in the conditions of local people, community and places suffering from aspects of deprivation, often multiple in nature" (Leary and McCarthy 2013, p. 9). It furthers supplement that private sector partners are also vital for successful regeneration, and the role of community enterprises is also emerging.

There also exists a contested relationship between heritage conservation and regeneration and the tension are particularly difficult to address in the urban cities undergoing rapid dynamic changes. Conservation has increasingly put emphasis on preserving local culture and buildings with local significance rather than the World Heritage sites. Redevelopment in old districts or towns inevitably has taken place (Shin 2010; Ryberg-Webster and Kinahan 2014). Old villages in urban cities usually has long history which dates back to at least a few thousands years ago. These Villages contain rich cultural values and traditional customs for the future generation to treasure. However, due to lack of maintenance, these villages usually have deteriorated physical conditions which do not fit the modern standard of living. In addition, with the fast changing political, economic, social and cultural context, these historical villages fits uneasily with the surrounding urban setting and poses challenges on the future role in which they can play in the regenerated city. The case study of Nga Tsin Wai, the last village still surviving in the urban area in Hong Kong is a case in point. It is a Walled villages which are now mainly remained in the New Territories and this is what made the case of Nga Tsin Wai Village so unique.

This study aims to explore the complex juxtapositions between heritage conservation and urban regeneration, using the case of an old village in Hong Kong. A focus group with the previous and very few remaining inhabitants and in-depth interviews with different stakeholders groups were conducted to gather the roles of different stakeholders play in the debate on conservation and redevelopment which has lasted for more than two decades. The case illustrates the many facets of challenges of integration heritage conservation into the urban regeneration planning process.

2 Conservation Under Urban Regeneration

Urban regeneration has been predominantly driving by themes of property-led, economic profit driven, improving living conditions, and meeting housing needs. Scholars criticize the demolition in legacy cities as twenty-first century urban renewal (Gratz 2010; Florida 2012).

Only until recently, historic preservation is increasingly recognized as an urban renewal strategy (Ryberg 2012). However, it is by no means a straight forward issue.

Most historic areas in the urban cities suffer deterioration and decay. In addition, the mismatch between the historic area in regard to functions, location, image and legal aspects and the contemporary needs is another major issue Doratli et al. (2004). At the same time, the historic urban areas also face tremendous intensities of redevelopment pressure. Doratli et al. (2004) states that successful revitalization of historic urban quarters shows effective organizational, legal, financial mechanisms and correct identification of the values which make the areas worthy to be preserved and revitalized are all important considerations.

Under the trend of global urbanization and dynamic changing context in the cities, the challenges in protecting the integrity in heritage conservation have been even more severe. Although there has been a paradigm shift towards the conservation of local cultures of local significance rather than preservation of monuments, even with the unprecedented emphasis on heritage being promoted as economic assets around the world, conservation of villages in urban cities is definitely not without enormous opposing forces.

3 Urban Regeneration in Hong Kong

It was not until the mid-1980s, urban redevelopment of dilapidated old areas has become one of the major concern of the Hong Kong government. A prominent mode of delivering urban regeneration in Hong Kong was the Land Development Corporation (LDC) established under the British administration as a statutory body in 1988. The LDC is empowered to carry out and facilitate redevelopment project within the older urban areas, either on its own or in joint-venture partnerships with developers or owners (Ng 2002). The government established LDC with the view that the private sector should be the dominant party in a redevelopment project and the speed of the urban renewal could be accelerated by contracting out the redevelopment project to a specialist organization with strong development interests (Adams and Hastings 2001). In the first phase of urban renewal, LCD adopted a partnership approach working very closely with key private property developers and this has aroused much criticisms (Adams and Hastings 2001). In the later stage, it started to develop a different operational model for redevelopment which could pursued on its own or in partnership with the existing owners in the early 1990s.

However, this model has been a failure as illustrated in the pilot case of the K11 redevelopment project in Tsim Sha Tsui. Nga Tsin Wai Village was also one of the twenty-five identified sites at this stage. Similar to the K11 project, the residents/ land owners rather sold their property or land to the private developer to get the money instead of waiting for about twenty or thirty years for the uncertain profit.

Urban Renewal authority (URA) is a statutory quasi-government body established in 2001 to replace the LDC to speed up renewal of old areas. A major difficulty for URA to undertake redevelopment is the resumption of land. Under the Urban Renewal Authority Ordinance (URAO) enacted in 2003, the URA may apply to the Secretary for Planning and Lands (SPL) requesting him to recommend to the Chief Executive in Council the resumption of land required for urban renewal. Private interests of land can be resumed under the Land Resumption Ordinance. Under the compensation policy, eligible owners of domestic properties will be offered the statutory compensation and an ex-gratia home purchase allowance or supplementary allowance as appropriate (Development Bureau 2011). Although the URA may request resumption of land for redevelopment under the Ordinance, it should consider acquiring land by agreement before making such a request to SPL (Development Bureau 2011). However, the process of discussions with existing owners was complex and lengthy (Adams and Hastings 2001) and complains about unfair compensation and rehousing have often been heard from the affected residents.

In contrary to LDC's piece meal approach, URA recognized the need for a wider district-based approach to urban regeneration. Heritage conservation's and urban renewal's discourse did not clearly intersect until the URA adopted the 4R business strategy, i.e., redevelopment, rehabilitation, preservation and revitalization, to rejuvenate older urban areas. In addition, the awareness and request on heritage conservation had been relatively low until the debate of the Star Ferry Pier and the Queen's pier conservation (Yung and Chan 2011). However, it is not difficult to find criticism on the projects completed by the URA. The Lee Tung Street in Wanchai which used to accommodate small businesses involved in printing of traditional Chinese wedding invitation cards is one of the projects. The project has been criticized for its lost of old atmosphere and local character in the wake of redevelopment into luxury shopping street and residential towers.

4 Methodology

In the research study, qualitative methods were adopted. A detailed case study was chosen and focus group meeting and in-depth interviews were carried out. Eleven inhabitants living in the Nga Tsin Wai Village were invited to participate in the focus group to understand their own experience and feelings on the place and the ongoing physical changes over the last two to three decades. The focus group meeting took around one and a half hour and was held at the Village office. After that, twelve representative of the different stakeholders including Legislative

Council members, scholars, concern groups members, former and current senior staffs of Urban Renewal Authority (URA) and Antiquities and Monument Broad (AAB) members, professional bodies' members, Ng's Clan residents and shop owner were participated in the in-depth interviews. The interviews intend to obtain the different stakeholders' views on their concerns and the issue of conservation and redevelopment of Nga Tsin Wai Village.

5 The Case of Nga Tsin Wai Village

- *Historical background*

  The Nga Tsin Wai is located in the Kowloon City, at east Kowloon, Hong Kong which was established in the form of an "unwalled" village and had existed since Yuen dynasty (1206–1370). The village was established by the Ng, Chan and Lee clans in the mid-14th century. During 1644–1661, Qing government ordered those who lived along the coastal province, should remove inwards 50 km from the boundaries for defending the pirates. The Village was cleared. In 1668, this policy ended and the inhabitants went back to the Village. However, the pirates who came from Taiwan always plundered the villages. The Ng's Clan daily life was affected. In order to defend the village against the bandit and pirate attack, Ng's clan accomplished with Chan'sand Lee's Clan, whose lived in nearby buildings and villages to build a walled village in 1724 (Cheung 2013; Hase 1999). In the period of Japanese Occupations (*1941–1945*), the villages nearby the Nga Tsin Wai were demolished by the Japanese government due to the construction of runway of Kai Tak Airport. Immigrates from China came to occupy those villages in Kowloon after the World War II in 1949. The government regarded the village houses as squatters and the inhabitants were not allowed to rebuild. Thus, the government tried to construct public housing in nearby areas and intended to clear the Nga Tsin Wai Village in 1965 (Public Record Office 1965). There had been about 110 households and one thousand people lived in the Village (Focus group meeting, 2016). The village covers an area of approximately 0.4 ha. The village houses are separated by three streets and six narrow alleys in a rectangular grid layout (Fig. 1).

  Apart from the architecture or relics, the location, setting, pattern and spatial layout of a walled village are also worth preserving. The village can be entered from the central axis of the village and the parallel rolls of village houses, are very unique. In regard to the broader setting, the village is located in front of the Kowloon Walled City, has its own river (now named Kai Tak nullah) for protection which is set beside the Kowloon Bay, and there are all significantly different from the other remaining Walled Villages in Hong Kong. The village is also rich in intangible values which should be retained. These values include strong social network, sense of place and belonging within the village.

Fig. 1 Nga Tsin Wai Village. *Source* URA website (2017)

- *Redevelopment until Now 1970–2016*
  The Evolution of the redevelopment-conservation debate of Nga Tsin Wai has lasted for more than four decades. Starting in the late 1970s, the developer, Cheung Kong Properties Limited had acquired and demolished two-third of the village houses in the Village. The redevelopment project of Nga Tsin Wai Village, with an area of about 6000 square meters, was one of the 25 urban renewal projects announced by the former Land Development Corporation in 1998. The Antiquities Advisory Board (AAB) discussed this project in December, 1994 and raised no objection to the redevelopment. They re-confirmed their decisions subsequently in 1999 and 2000. During the negotiation with LDC, the residents were not satisfied with the compensation offered and the acquisition of land failed. The residents started to negotiate with the private developer for the sale of their houses. However, they were also not satisfied with the price offered by the developer which they thought the price would not allow them to buy another flat elsewhere.
  In fact, the inhabitants were eager to start the redevelopment program for improving their living conditions since 2005. In 2006, the developer had already obtained about 70% of land in the Village. The developer has returned all the acquired land to the government and let the government and URA to acquire all the remaining land because URA is entitled to acquire land under the Urban Renewal Authority Ordinance and the Land Resumption Ordinance. During the

interviews, the villagers' representative said that the developer offered about HK $1.8 million to buy the house before the URA took over the project, whereas the URA offered HK$4 million for the compensation after several years later. Thus, more residents, especially the indigenous inhabitants were willing to leave. After acquisition, URA and the private developer joined together in partnership to develop the site. The negotiation between the developer and the government, definitely involves a lot of uncovered deals and business considerations. In response to the growing community concerns on heritage conservation in Hong Kong, URA engaged a conservation consultant team in 2006. The URA announced its plans to redevelop Nga Tsin Wai Village and partner with the developer in 2007 and proposed the "Nga Tsin Wai Village Project" Hong Kong Legislative Council (2007). The proposal (Fig. 2) proposed to construct a conservation park and develop two 40 meters high residential blocks of 750 residential flats on the two sides which claimed to balance the old and the new setting Anonymous (2015). Eight village houses will be kept for reconstruction as retails and the three historic relics designated as Grade 3 historic building by AAB on 4th March, 2014 will be conserved (Antiquities and Monument Advisory Broad 2016). However, unlike declared monuments, graded historic buildings do not have statutory protection.

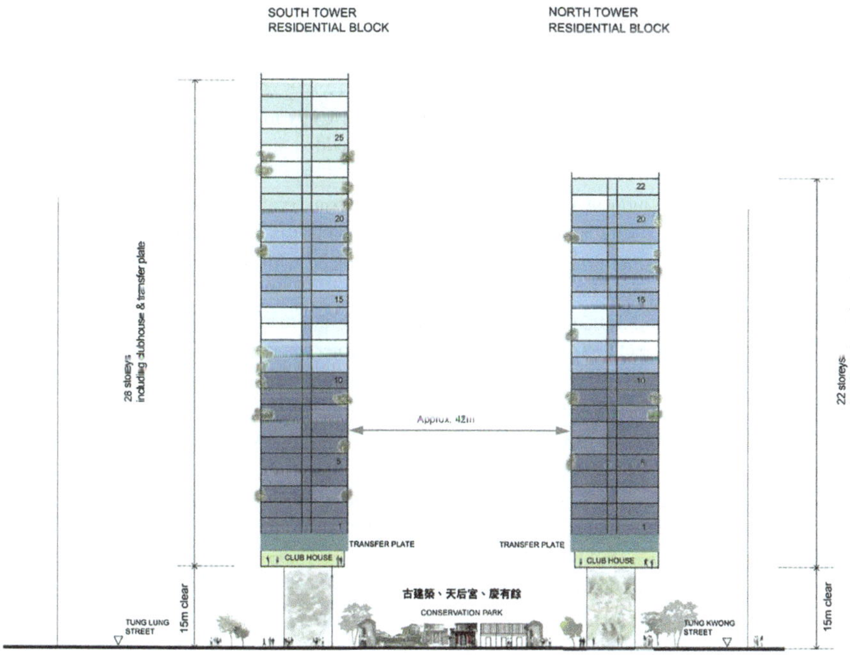

Fig. 2 Redevelopment scheme proposed by URA. *Source* Legislative Council Panel on Home Affairs Subcommittee on Heritage Conservation (2007)

The URA submitted an application requesting the resumption of land for implementation of the redevelopment project under the URAO in July 2009. Twenty private land ownership have been resumed under the Lands Resumption Ordinance in Oct. 2011. URA proposed compensation to the tenants, however, some tenants were dissatisfied with the compensation offered by the URA and set up a concern group for petition to government. The URA proposes to relocate the affected shop tenants to the Nga Tsin Wai Village after the redevelopment is completed and offer a monthly rent of HK$600 for the first three years and $3000 for the fourth year and $6000 for the fifth year. However, this proposal was rejected by the shop owners and tenants. In 24th January, 2016, URA completed the clearance of Nga Tsin Wai Village Anonymous (2016). Table 1 is the summary of the key events happened in the redevelopment and conservation debate of Nga Tsin Wai Village.

Table 1 Major Chronological events and the stakeholders involved in the redevelopment and conservation of the Nga Tsin Wai Village

1970s	Chun Kong Property Development started to buy the village houses and demolished them
1994	The Antiquities Advisory Board raised no objection to the redevelopment of NTW Village. AAB also subsequently reconfirmed it in 1999 and 2000
1998	Land Development Corporation (LDC) announced Nga Tsin Wai Village as one of the 25 redevelopment projects
2001	URA was established and take over the 25 redevelopment projects from LDC
2005	Wong Tai Sin District Council requested URA to expedite the redevelopment and to preserved the gatehouse, the embedded stone tablet, and the Tin Hau Temple
2006	URA engaged internationally renowned conservation expert to carry out conservation study and proposed the concept of "Conservation by Design"
2007	URA announced the project was commenced with partnership with the private developer, Chun Kong Property Ltd.
2009	URA submitted an application requesting the resumption of land for implementation of the redevelopment project under the URAO
July 2011	The Lands Department announced the resumption of land for the implementation of the Nga Tsin Wai Village redevelopment project
Oct. 2011	The affected private land and property interests reverted to the Government under the Lands Resumption Ordinance
2011	Concern group was set up for petition to government to express their dissatisfaction towards the compensation and relocation arrangements
2014	The Ancestral Hall, Tin Hau Temple and the Entrance Gate are graded as Grade 3 historic building by AAB
Dec. 2015	The Lands Department issued an official notice in accordance with the Lands Resumption Ordinance ordering the remaining illegal occupants to leave by 25 Jan. 2016
25th Jan. 2016	Clearance took place, the last two remaining occupants accepted the compensation and left the Village

Table 1 summarizes the different roles and interests of the different stakeholder groups in the debate of redevelopment and conservation of Nga Tsin Wai Village.

6 Challenges of Integrating Heritage Conservation in the Redevelopment of Nga Tsin Wai Village

The following issues are identified by intensive analysis of various kinds of sources including government records, newspapers, and Legislative Council Hansards etc. In addition, the issues were also raised by different stakeholders in the in-depth interviews.

7 Difficulties in Fulfilling Different Stakeholders Groups' Needs

- ***Incompatible interests of different stakeholders groups***
  There are many different stakeholder groups involved in the redevelopment and conservation of the Nga Tsin Wai Village. They include the immediately affected groups, such as residents group comprises of indigenous inhabitants, inhabitants, tenants, shops owners and tenants. The general public and community including the Wong Tai Sin District council, Nga Tsin Wai Concern group, professionals and academics. The public sector includes the LDC and URA, the relevant government departments, and the commercial section which is the private developer. Inevitably, there exists many incompatible interests of the different stakeholder groups.
  An official of the URA clearly expressed that the government has never stated that it is a conservation project from the very first day. He claimed that URA has been embracing its social responsibility in preserving the three historic relics and the eight houses. From the developer's view, the main concern is the selling price of the completed residential units and it is very rare if they have thought about the feasibility of self-financing for other alternative uses of the site if the village was to be preserved.
  Even within those groups which advocate for conservation of the village, whether the integrity of the whole village has been kept is still controversial. The legislators, concern group's members and scholars prefer preserving the whole or larger part of the village while others are satisfy with the latest proposal of the URA to keep only the three historic buildings.
- ***Poor living condition of the Nga Tsin Wai Village***
  Some of the inhabitants indicated that prefer redevelopment. The inhabitants sent a letter to Wong Tai Sin District Council, Lands Department, Home Affairs

respectively for expressing their urgent needs for redevelopment (Wong Tai Sin District Council 2005).

> We request the URA to redevelop the Village. We only want to improve our living condition ......and have standard facilities inside our house. If the village is preserved as heritage, URA could not undertake redevelopment. Our living condition cannot be improved.'.......Nowadays, somebody or legislators who oppose the redevelopment of Nga Tsin Wai Village, do not have any properties rights or they are not the indigenous of the village. Neither redevelopment nor preservation of the village does affect them much. They do not realize the villagers' utmost urgent housing needs.

In a focus group interview, an inhabitant told that

> My grandfather kept this house for my father and my father kept this house for me. However, in my son generation, he can't live in the house because it is no deteriorated.

A member of AAB doubts if anyone would consider the wills of those inhabitants.

> Nobody want to move in, the condition of the place is poor. I will move out if I have had money and live outside......This is a social status problem. ....I will be labeled as "poor guy" by others if I stilllive there.

– ***Authenticity of the Nga Tsin Wai Village after regeneration***
  It is also questioned that whether the authenticity of the village can be preserved. Since many of the original village houses have been demolished or seriously deteriorated, it is at the most a reconstruction of the place for people to recall the memories and history of the Nga Tsin Wai Village, or at worst it will become a 'Disney land' type of imitation.

A former official of URA emphasized that it is only a redevelopment projects and add some conservation elements inside while some stakeholders groups complained that the proposal could not keep the authenticity and its value of the village. The Urban Renewal official claimed that

> If nobody lives inside, what do we preserve?

An AAB member raised that after the regeneration, Nga Tsin Wai Village will become a park, the use has been changed, it is not a place for residence.

> I don't think it is a conservation project... however, to make it to become a park is better than doing nothing, at least there are a few historic structures there.

In his opinion, the intangibles can hardly be preserved as people have all left. It is hope that the preservation of the temple and provision of the park can allow some traditional customs to be held there after regeneration. However, the URA official expressed that

> there is no intangible value there since there is no one remained in the village, how can there be any traditional custom. Thus, there is hardly any intangible to be preserved.

On the other hand, a member of a professional body thought that it is not the physical buildings which can be conserved, it is the culture and the traditional customs which can create social cohesion and maintain the social networks.

For instance, if there is a place for gather the villagers who had moved out to join the special festival and traditional events. In regard to URA's redevelopment proposal, he concerned the continuity of its cultural value after the completion of the residential buildings. He worried that if the developer is the one to manage the conservation park and he said

> Would the inhabitants still be allowed to hold their own traditional cultural events in the park, particularly during the festival?

8 Knowledge About Property Rights—Inhabitants, Occupants and Developers

The awareness of respecting property rights has increased in the society, particularly after 1970s. "*Properties right above all things*" tends to be in the mindset of the public and the developers. Therefore, it is a complex issue when the private owners have the property rights of the heritage buildings and use it as means to object conservation.

In the case of Nga Tsin Wai Village, the knowledge of property rights varies among the inhabitants in the different time period. In the old days, the inhabitants in Nga Tsin Wai had weak knowledge about property and land rights. This is very different from those indigenous villagers of the big clans in the rural areas who have a strong sense of having the privileges of the small house policy.

On the other hand, Nga Tsin Wai Village is located in the south of the Boundary Street which all land are included in the cession of Kowloon to the Britain. Unlike rented land in the New Territories, the British government has the right to develop the surrendered land. Thus, the government did not see Nga Tsin Wai as a village in the urban areas as those traditional village in the New Territories nor see the villages as indigenous inhabitants who have the legal land ownership in the colonial period. The government has recorded these village houses as squatter houses. As a result, *the village houses have been confiscated by government easily in the old days*. When their ancestors were under the ruling of the British government, most of their cultivated lands were confiscated by the government with little compensation because they have no legal proof of their ownership. The inhabitants said that "*Our ancestors with little knowledge about this, even the confiscated land lots were published in the Gazettes*." In government's view point, Nga Tsin Wai and all the other villages in the New Kowloon should be demolished to make way for public housing, this was agreed by the Legislative Council in the Colonial period.

In contrary, during the redevelopment undertaken by the URA, the inhabitants have stronger awareness of their property rights. They thought that they have been living there for the previous several generations without any deed, they do not accept that they cannot continue to live there. With the implementation of the compensation mechanism, they have been negotiated with URA for the amount they expected. For the non-residents, stores' owners, they claimed that they brought their stores with the village chief in the early years, but did not have a reasonable relocation nor compensation for their loss of property rights. In 2016, the remaining shop owners have attempted to ask for a rent of HK$600 per month forever. The URA official sees this is unreasonable compensation and said that "*they have only occupied the land without having the legal right and refused to leave*."

In regard to the developer's concerns, as it has already obtained 70% of the land at the end of 1990s and this has made the government very difficult to undertake conservation of the entire village. A concern group's member stressed that the developer has much more power than the URA in the redevelopment of Nga Tsin Wai Village. Although the government stipulated that URA has the authority to acquired land for urban renewal under the Urban Renewal Authority Ordinance and the Land Resumption Ordinance, URA may need to go through the litigation process if the developer is dissatisfied with the compensation for his deprivation of their property rights. One possible way the government could do is to utilize the Town Planning control to restrict the developers' development rights with reasonable planning conditions. The URA official claimed that the major obstacle is to make the developers to compromise with their redevelopment with a sub-theme of conservation as they would reduce the full development potential of the site. In fact, this is very difficult to be implemented, or even not feasible under the current town planning system. The ex-URA official and also a current AAB member said that

> the other more possible control mechanisms may be to impose some restrictions to the development and/or land use through the lease conditions or the building control mechanism.

9 Unsupportive and Inflexible Land Use Planning Policy

- *The issue of private property rights*

  In the case of Nga Tsin Wai Village, the compensation for properties rights to the land owners was the major issue of preservation. The developer owns majority of the land interests of the village. If the government wants to acquire the land for heritage conservation, it involves a huge amount of public money in order to purchase the land. In the last few decades, the private developer had bought the village houses and demolished them one by one, except those which are not owned by a single land owner such as the temple, the ancestral hall, the gatehouse (Interview with ex-AAB member, 2016).

  An architect who is a member of the concern group, had suggested to use the transfer of development rights to compensate the developer in exchange of the preservation of a large part of the village in 2003. He suggested keeping the same plot ratio and exchanging the land for carpark near the adjacent public housing estate which was owned by the Housing Authority. However, the Housing Authority refused the exchange of land proposal and claimed that the authority is not responsible for preservation. The legislator, scholar and concern groups all criticized the lack of cooperation between the different government departments.

- *Land use zoning*
  Another land use planning policy which could possibly be better utilized to enhance preservation of an area is the zoning mechanism. However, the Town planning had done the opposite. An Antiquities Advisory Board (AAB) member indicated that Nga Tsin Wai Village has been zoned "R/C" (residential/ commercial) since the 70s, thus, it has become more likely to attract developers for demolition of the village houses than for preservation. On contrary, if it is zoned "V" zone as those located in the New Territories, it would be much more difficult to rezone it to "R/C" and they would not be easily redeveloped Building and Land Department (1959). However, when you look back to history, the cession of Kowloon made all land had a 999 years ownership term whereas the land in New Territories is only leased until Hong Kong was returned to China in 1997 and now the lease is granted for another 50 years. In addition, since the government has recorded the village houses as squatter area, it would be very unlikely to designate Nga Tsin Wai Village as a 'V' zone as those traditional villages in the New Territories.
- *Maintenance and improvement of village houses*
  In general, the government provided no financial support for the maintenance of the village houses in the previous years. The government had also not provided any subsidize for the improvement works for the living quality such as problem of water leakage, sanitary provisions, and air conditionings, etc.
  There has also been some saying of the inhabitants that since the villages houses were recorded as squatter houses by the government after the World War II, no alteration nor additional works to these illegal structures were permitted. However, there has been no clear evidence on the extent of maintenance works that were allowed. Some residents did spent their money to do minor repair works such as rectifying water leakage or adding insulation materials on the roof. In general, lack of maintenance has accelerated the pace of deterioration of the village houses.

10 Conclusion and Policy Implications

The case of Nga Tsin Wai Village portrays the many facets of issues in regeneration of old urban areas, especially in historical villages which contain valuable history and culture which is worth preserving. On one hand, land in urban area is always scared and land value is extremely high in cities like Hong Kong, thus, redevelopment is inevitable. The challenge is how to make a balance or a tradeoff between redevelopment and conservation. It also poses the question whether the two cannot be co-existed.

This paper analyzes the redevelopment story of the old urban village in the past few decades. It suggests the government decision makers and the relevant departments to explore the opportunities and feasibility of imposing restrictions on buying and selling of the historic properties, and the development potential through town planning control and land administration mechanism. Property rights and development rights may not necessarily be always complementary. There is room to explore whether the land owner can remain the property rights and have the restricted development rights.

Public policy making is dominant by quantitative approach and in contrast, the concepts related to heritage conservation are in general difficult to measure, such as character, collective and individual memory, place identity, etc.) The contribution of heritage preservation in urban revitalization is long-term, thus, a more comprehensive assessment of the heritage values and what is worth preserving can enhance incorporating the conservation element into regeneration and redevelopment projects.

References

Adams, D., & Hastings, E. M. (2001). Urban renewal in Hong Kong: Transition from development corporation to renewal authority. *Land Use Policy, 18,* 245–258.

Anonymous. (2015). The new resettlement proposal in Nga Tsin Wai Village, low rent for shop tenants,«衙前圍村新安置方案原址平租舊商戶», *Sing Tao Daily*, September 21, 2015 (in Chinese).

Anonymous. (2016). Resumption of land ownership in nine years, Nga Tsin Wai Village was finally cleared. *Sing Tao Daily News*, January 26, 2016 (in Chinese).

Antiquities and Monument Advisory Broad. (2016), *List of new items and new categories with assessment results*. Updated at April 18, 2016. http://www.aab.gov.hk/form/list_new_items_assessed.pdf.

Building and Land Department. (1959). *K.P.A. No.11- Tsz Wan Shan, Diamond Hill and San Po Kong, Tai Hom K.P.A. No.8—Kowloon Tsai, Wan Tau Hom and Tung Tau, (1978).* P.W.D.8773/59. Public Record Office, HKRS Record No. 896-1-112

Building and Land Department. (1965). *Nga Tsin Wai Village, Clearance of …*". 5/4802/65. Public Record Office, HKRS Record No. 337-4-1915.

Cheung, S. W. (2013). *Demolish village: The lost of Kowloon Village «拆村:消逝的九龍村落»*. Hong Kong: Joint Publishing. (in Chinese).

Development Bureau. (2011). *Urban renewal strategy, urban renewal authority*. https://www.devb.gov.hk/filemanager/en/Content_3/URS_eng_2011.pdf.

Doratli, N., Hoskara, S. O., & Fasli, M. (2004). An analytical methodology for revitalization strategies in historic urban quarters: A case study of the Walled City of Nicosia. North Cyprus, *Cities*, *21*(4), 329–348.

Florida, R. (2012). *The rise of the creative class revisited*. New York: Basic Books.

Gratz, R. B. (2010). *The battle for Gotham: New York in the shadow of Robert Moses and Jane Jacobs*. New York: Nations Book.

Hase, P. H. (1999). Beside the Yamen: Nga Tsin Wai Village. *Journal of the Royal Asiatic Society Hong Kong Branch, 39,* 1–82.

Hong Kong Legislative Council. (2007). Presentation materials provided by the Administration on "NgaTsinWai Village Project". Minutes of meeting of the Subcommittee on Heritage Conservation, 2nd October 2007, Hong Kong: The Authority, 2007. http://library.legco.gov.hk:1080/record=b1065904.

Leary, M., & McCarthy, J. (2013). The Routledge companion to urban regeneration. In M. E. Leary & J. McCarthy (Eds.), *2013*. London and New York: Routledge.

Legislative Council Panel on Home Affairs Subcommittee on Heritage Conservation. (2007). Nga Tsin Wai Village project, Legislative Council paper LC paper no. (2) 2782/06-07.

Ng, M. K. (2002). Property-led urban renewal in Hong Kong: Any place for the community? *Sustainable Development, 10*(3), 140–146.

Rhodes, J., Tyler, P., & Brennam, A. (2007). The single generation budget: Final evaluation. Cambridge: Department of Land Economy University of Cambridge.

Roberts, P. (2000). The evolution, definition and purpose of urban regeneration. In P. Roberts & H. Sykes (Eds.), *Urban regeneration: A handbook* (pp. 9–36). London: Sage.

Ryberg, S. R. (2012). Historic preservation's urban renewal roots: Preservation and planning in midcentury Philadelphia, *Journal of Urban History, 39*(2), 193–213.

Ryberg-Webster, S., & Kinahan, K. L. (2014). Historic preservation and urban revitalization in the twenty-first century. *Journal of Planning Literature, 29*(2), 119–139.

Shin, H. B. (2010). Urban conservation and revalorisation of dilapidated historic quarters: the case of Nanluoguxiang in Beijing. *Cities, 27*, S43–S54.

Wong Tai Sin District Council. (2005). Nga Tsin Wai village inhabitants land owners, land owners, business shop owners requested the urban renewal authority starting the redevelopment project as soon as possible. Annex B. In "*Future Development of NgaTsinWai Village*", Wong Tai Sin District Council Paper No. 79/2005 (in Chinese). http://www.districtcouncils.gov.hk/archive/wts_d/pdf/2005/WTS_2005_079_TC.pdf. Last accessed April 15, 2017.

Yung, E. H. K., & Chan, E. H. W. (2011). Problem issues of public participation in Built heritage conservation: Two controversial cases in Hong Kong. *Habitat International, 35*, 457–466.

Characteristics of Ridesharing as a Sustainable Transport Tool in Metro Manila

Jose Regin F. Regidor and Ma Sheilah G. Napalang

Abstract The increasing requirement for mobility in urban centers and the declining level of service of public transportation has led to worsening congestion. The introduction of smartphone application-based modes of transportation has been described as an innovative strategy to address the problem. This paper presents the outcomes of studies on ridesharing in Metro Manila where ridesharing services are classified under Transport Network Services (TNCs). The main objective is to determine the reasons behind the popularity of ridesharing services, specifically Uber and GrabCar, in Metro Manila, and the corresponding decline of preference and patronage of conventional taxis. A comparative analysis of TNCs and conventional taxis is presented using key indicators. These include travel speed, reliability, passenger expense, and quality of service. For completeness, the study included GrabTaxi services, which represented a basic upgrade of conventional taxis through an app-based system designed to facilitate getting rides. Travel diaries from regular users of ridesharing services were collected and analyzed. Other surveys performed included perception surveys among commuters using TNCs and conventional taxis. Based from the results of the analysis, ridesharing services, specifically Uber and GrabCar are preferred over conventional taxis. TNCs perform better in terms of reliability and safety. Uber and GrabCar were found to have better quality of service compared to conventional taxis. Meanwhile, it was also found that the advantages and attractiveness of ridesharing are very similar to app-enhanced taxis (i.e., GrabTaxi). The study presents recommendations on how both ridesharing and conventional taxis may be improved in the context of sustainable transport as ridesharing is perceived to reduce dependence on private cars but at the same time attracts commuters away from other public transport modes in

J. R. F. Regidor (✉)
Institute of Civil Engineering, College of Engineering, University of the Philippines, Diliman, 1101 Quezon City, Philippines
e-mail: jfregidor@up.edu.ph

M. S. G. Napalang
School of Urban and Regional Planning, University of the Philippines, Diliman, 1101 Quezon City, Philippines
e-mail: mgnapalang@up.edu.ph

W. Leal Filho et al. (eds.), *Sustainable Development Research in the Asia-Pacific Region*, World Sustainability Series, https://doi.org/10.1007/978-3-319-73293-0_22

Metro Manila, including the necessity of government to regulate certain aspects of the TNC operations.

Keywords Ridesharing · Uber · Grab · Conventional taxis · Quality of service Reliability

1 Introduction

The United Nations (2012) estimates that 50% of the world's population now reside in urban centers and projects that this will reach 80% by 2050. The Philippines is an example of this phenomenon where urban population increased from 29% in 1955 to 45% in 2015. By 2050, it is projected that the urban population of the Philippines will reach 60%. This increase in urban population has consequently increased the need for mobility in urban centers. In a study conducted by the Japan International Cooperation Agency (JICA 2014), the traffic demand in Metro Manila was estimated at 12.8 million trips per day and 69% of these total trips use public transportation. However, transportation infrastructure development is slow and many projects including much-delayed mass transit lines are only beginning to be constructed, in the final stages of design or scheduled for bidding out for the private sector to implement. Due to this, innovative application-based modes of transportation have become popular in key urban centers in the Philippines, particularly in Metro Manila where the urban transportation problem is perceived to be most severe.

This paper seeks to describe the characteristics of application-based modes of public transportation currently operating in the Philippines. Issues and concerns pertaining to their operations are identified and discussed. The paper also attempts to evaluate how and if these modes contribute to the promotion of sustainable transportation. Included, too, are discussions on the role and responses of the government in as far as regulation is concerned.

2 Sustainable Transportation

Although there are several definitions of sustainable transportation found in literature, this paper adopts that of the Centre for Sustainable Transportation (CST 2005, as quoted by Littman 2016, p. 8), which reads:

A sustainable transportation system is one that:

- *Allows the basic access needs of individuals and societies to be met safely and in a manner consistent with human and ecosystem health, and with equity within and between generations;*

- *Is affordable, operates efficiently, offers choice of transport mode, and supports a vibrant economy;*
- *Limits emissions and waste within the planet's ability to absorb them, minimizes consumption of non-renewable resources, limits consumption of renewable resources to the sustainable yield level, reuses and recycles its components, and minimizes the use of land and the production of noise.*

The above quoted characterization of sustainable transportation is accepted by many experts including the Transportation Research Board's (TRB) Sustainable Transportation Indicators Subcommittee and the European Council of Ministers of Transport (Litman 2016). This paper will focus on four aspects of sustainable transportation in its analysis of emerging modes of public transportation: (1) efficiency, as manifested in reliability of service, (2) safety, (3) affordability, and (4) minimization of consumption of non-renewable resources such as fuel by reducing congestion.

3 Ridesharing and Ridesourcing

Ridesharing refers to a mode of transportation in which individual travelers share a vehicle for a trip and split travel costs such as gas, toll, and parking fees with others that have similar itineraries and time schedules. Conceptually, ridesharing combines the flexibility and speed of private cars with the reduced cost of fixed line systems, at the expense of convenience (Furuhata et al. 2013). Ridesharing is a powerful strategy to address traffic congestion, fuel emissions and gasoline dependency. A simple concept is followed: fill up empty seats to maximize the vehicle's occupancy potential and reduce vehicles on the roadway. Unlike taxis and Transportation Network Companies (TNCs), drivers are not motivated by profit but by reduced travel cost. Passengers also have a common origin and destination with the driver (Rayle et al. 2014; Shaheen 2014).

On the other hand, ridesourcing companies utilize technology to provide on-demand transport services with the promise of higher reliability and reduced wait times (Rayle et al. 2014). These are generally referred to Transportation Network Companies (TNCs) and defined as 'an organization that provides pre-arranged transportation services for compensation using an internet-based technology application or a digital platform technology to connect passengers with drivers using their own personal vehicles' (DOTC 2015).

4 App-Based Public Transportation in Metro Manila

Based on the definition given in the previous sub-section, the more prominent innovation in the transportation system operating in the Philippines is ridesourcing as described in the following sections.

4.1 Grab

Grab started with GrabTaxi, a subsidiary of Malaysia's MyTeksi, which is a smartphone-based hailing and booking service first introduced in Metro Manila in 2013 (Grab Philippines). The service was later opened in three other major urban centers, Cebu City, Davao City, and Iloilo City, and expanded its operations to other cities as well. The company has expanded its services to include GrabCar (private cars that can be hired through the application on demand and is much like Uber) and Grab Express (express delivery service). At one time, the company attempted to provide motorcycle taxi service on demand via GrabBike. This was disallowed by the LTFRB as motorcycle taxis are officially illegal in the Philippines.

4.2 Uber

Uber was introduced in the Philippines in 2013. Unlike GrabTaxi, Uber provides pre-arranged transportation services for compensation, using an online-enabled application or platform technology to connect passengers with drivers using their own personal, non-commercial vehicles (Dela Paz 2015).

In 2016, Uber introduced its ride-sharing services, UberPool and UberHop. UberPool matches riders 'coming from the same area heading at the same direction at the same time' for more efficient energy use and reduction of fuel emissions (UberPool 2016a). On the other hand, UberHop 'enables riders heading in the same direction to share a ride during rush hour for a flat fare' (UberHop 2016b). As of April 2016, there were 9 pick-up and drop-off points in Quezon City, Mandaluyong City, and Makati City during the morning service and 9 "pick-up" and "drop-off" points in the three cities, with Bonifacio Global City (BGC), a mixed use development, as an additional area for the evening service.

4.3 Other Ridesharing and Ridesourcing Services

Another service available in Metro Manila is Wunder, a carpooling service, which was introduced in 2016. This service was established in Germany and is probably operating closest to the original concept of carpooling in that it mainly depends on available vehicles for the typical commutes in the morning (e.g., home to workplace) and afternoon (e.g., workplace to home). That is, vehicles generally have only two trips per day and do not roam the streets for passengers like Grab and Uber vehicles. It claims the following benefits (Wunder 2016):

- Sharing costs and saving money
- Meeting new people
- Reducing the number of cars on the road
- Avoid crowded public transportation
- Helping the environment

A new entrant into the transportation market in Metro Manila is the ridesourcing motorcycle taxi service called "Angkas" (backride in the local language). It is unclear who developed or manages the smartphone application for this service. However, it is enterprising enough to take advantage of the worsening traffic situation in Metro Manila coupled with the surge in the number of motorcycles. Motorcycle taxis are generally illegal in the Philippines and particularly in urban areas mainly due to safety concerns but are unregulated yet popular in rural areas where there is a lack of public transport services and poor road infrastructure. Grab, for example, ceased operations of its GrabBike when the LTFRB issued a memo reminding about the prohibition against such services for public transportation.

5 Government Regulations of TNC/TNVS

In the Philippines, the Land Transportation Franchising and Regulatory Board (LTFRB), a line agency under the Department of Transportation (DOTr) is tasked with economic regulatory functions for road public transport services. It has the authority over public land transport services in terms of:

a. Route/area of operation prescription and regulation in terms of viable route capacities;
b. Issuance of the Certificate of Public Convenience (CPC), otherwise called as franchise, to entities worthy to be public transport operators with corresponding franchising terms and conditions;
c. Prescription of fares/charges on public transport services;
d. Promulgation and enforcement of rules and regulations pertaining to public transport service operations.

Table 1 TNC/TNVS requirements

For submission upon application	For submission upon hearing
• Formal offer of documentary evidence • Motion for Application of Provision Authority • Filled out application form • Publication in major newspapers • Proof of Citizenship • Proof of Good Standing (TNC) • Proof of Accreditation of the Vehicle (TNC)	• Photocopy of Passenger Insurance Policy (LTFRB Board Accredited Insurance Provider) • List of TNVS drivers and vehicles • 2 copies of operator data sheet with recent 2″ × 2″ picture • Statement of Financial Capability Form and Proof of Entries • ITR or BIR Certificate of Registration • Certificate of Business Name (DTI) • Location Map of Garage or Address of Operator
Requirements for the applicants driver/s	Requirements for the applicant's vehicle
• Proof of accreditation of the Driver by TNC • Professional Drivers' License • NBI Clearance • PNP Clearance	• OR/CR with Year Model or Delivery Receipt/Sales Invoice

The Agency is tasked to ensure the safety of passengers as well as safeguard them against sudden increase in fares, beyond their paying capacity. Towards the fulfillment of these tasks, the LTFRB monitors the entry of new public transportation providers.

Hence, when e-hailing transport services where introduced in the Philippines, the LTFRB issued guidelines in the form of four Memorandum Orders (LTFRB 2015a, b, c, d) to regulate the operation of the two key players: Transportation Network Companies (TNCs) and its contractor/partner, the Transportation Network Vehicle Service (TNVS). The requirements are summarized in Table 1.

It must be highlighted that one of the key requirements is the requirement for TNCs to obtain permit to operate is the passenger insurance coverage, which is not covered by the comprehensive vehicle insurance. Another safeguard for passenger well-being, apart from the NBI and PNP clearances, is the requirement for drivers to be accredited by the TNC, hence improving accountability. Even with the stringent regulation for entry, the number of applications for transport network vehicle service (TNVS) has ballooned to almost 30,000 in 2016 from only 3000 in 2015 (Tan 2016).

Another aspect of ridesourcing services that have been subject to deliberation for regulation is the matter of surge pricing, or the spiking of fares during peak periods due to increased demand. It was reported that during the Christmas rush in 2016, surge prices in Metro Manila were reported to vary between ₱2000 and ₱28,000 per ride (CNN Philippines 2016). This has prompted the LTFRB to issue a memorandum to the two prominent apps-based services in Metro Manila to put a cap on their fares. Specifically, Uber is directed "that the maximum allowable price surge on the fare shall be twice the rates for time covered and distance travelled excluding the base fare". It likewise directed Grab Philippines to "lower its fare per kilometer from ₱12.00–₱16.00 to ₱10.00–₱14.00 depending on the type of vehicle used."

The LTFRB stopped granting franchises to Uber in July 2016 due to complaints from taxi companies. It also ordered Angkas and Wunder to cease operations due to safety concerns (i.e., absence of passenger insurance) without a proper franchise.

6 Basic Characteristics of Ridesourcing in Metro Manila

The number of TNVS vehicles has quickly overtaken the number of conventional taxis in Metro Manila. Data from the LTFRB has shown that as of 2015, for example, there were 3628 ordinary taxis registered in the National Capital Region (NCR) while TNVS units were at 9735 or almost triple the number of taxis. Almost all the TNVS units are either Uber (76.64%) or Grab (23.36%) vehicles with the exception of ordinary taxis that are registered also as GrabTaxis. It is reasonable to assume that since then, there has been a significant increase in the number of TNCs due in part to their popularity that has translated into an attractive option for those looking for income through transportation service provision.

Since it was first introduced in late 2014, the number of Uber users, as denoted by the number of sign-ups has grown to over 433,000 (Uber 2016a, b), most of whom are from within Metro Manila but with more recent registrations from surrounding provinces.

6.1 *Trip Purpose*

Based on the 2016 Uber Manila survey covering 1450 respondents out of the 15,360 users who received the survey instrument, the most popular trip purposes for use are commuting to work/home (67%), social activities (49%), going shopping (30%), and travelling to/from business meetings (29%). Figure 1 shows the outcomes of the survey conducted by Uber where respondents could provide multiple answers for their trip purposes. This can be compared with the results of the study by Nistal and Regidor (2016) that covered both Uber and Grab Car. Figure 1 shows the main trip purposes of Uber and Grab Car users.

6.2 *Passenger Profiles, Preferences and Perceptions*

Out of the total number of respondents of the Uber Manila Survey (2016), 57% do not own a car and 43% own one. Of those owning a car, 52% indicated that they are driving less and 15% stated that they have not changed frequency of driving due to Uber.

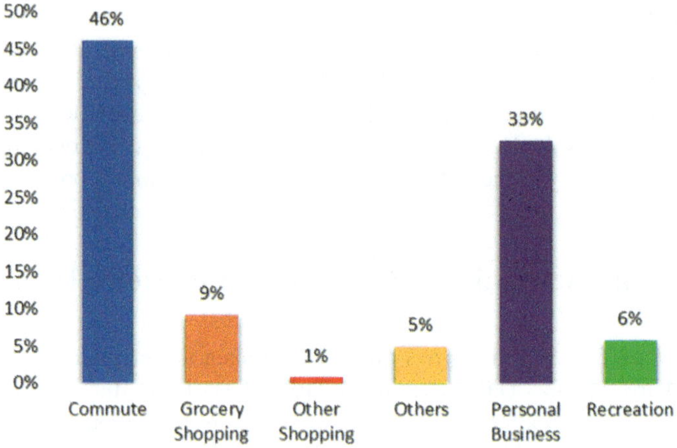

Fig. 1 Primary trip purposes of Uber and Grab Car users (Nistal and Regidor 2016, with permission)

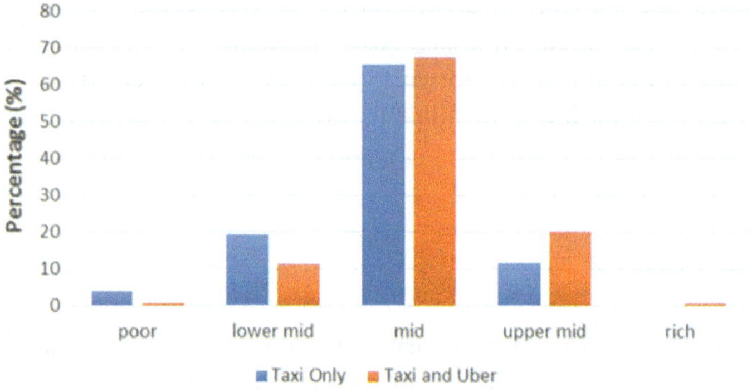

Fig. 2 Income Class Distribution of Respondents (Nistal and Regidor 2016, with permission)

The respondents' profiles in the study by Nistal and Regidor (2016) indicated that users of both ordinary taxis and Uber belong to the middle income class. This is shown in Fig. 2.

In a study by Dela Peña and Dizon (2016), the reasons for passengers preferring GrabTaxi to conventional taxis are mainly according to convenience, reliability, and safety. These are shown in Fig. 3, where responses are categorized accordingly.

Convenience appears to be the top reason for preference of app-enhanced taxis for a variety of reasons, including easy access of service through smartphone application, reducing the necessity to wait at curbside as when hailing regular taxi and the fare has already been set prior to the ride, eliminating negotiations with the driver. Incidentally, convenience was also found to be the top reason for preferring

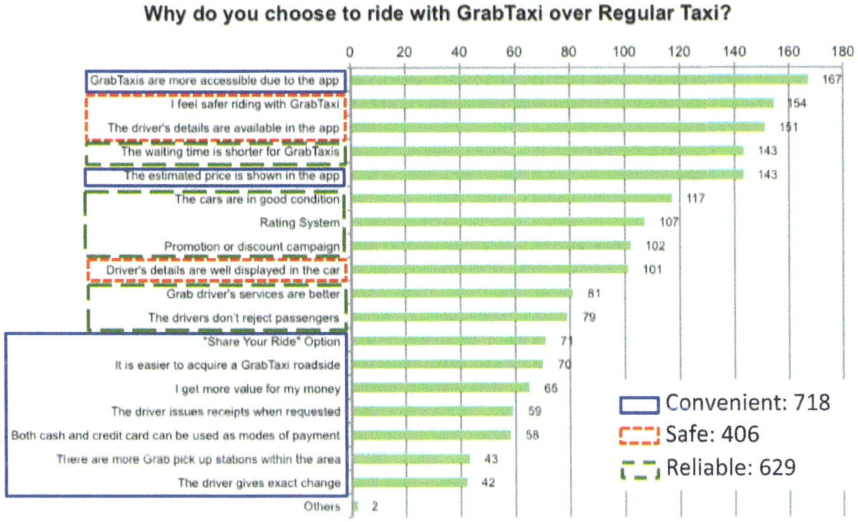

Fig. 3 Reasons for preference of GrabTaxi over regular taxi (Dela Peña and Dizon 2016, with permission)

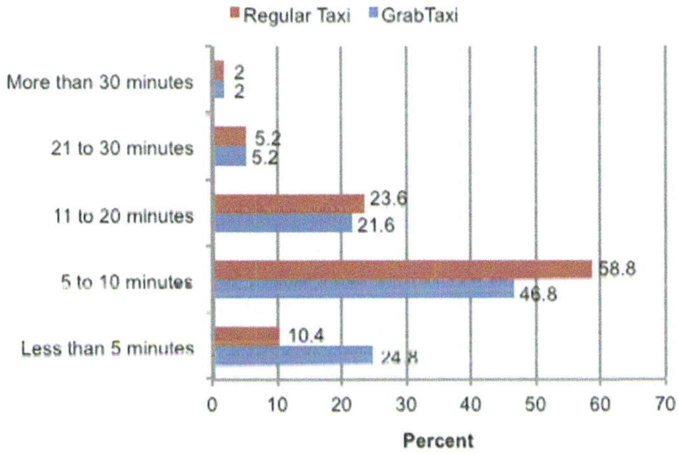

Fig. 4 Perceived waiting time for Regular Taxi and GrabTaxi (Dela Peña and Dizon 2016, with permission)

ridesourcing services like Uber and Grab in the study by Paronda, Regidor and Napalang (2016); with safety coming in second and reliability third.

One key indicator for reliability is waiting time, which in the study of Dela Peña and Dizon (2016) is perceived to be longer for regular taxis, with an average weighted mean of 10.11 min for taxis and 9.27 min for GrabTaxi as shown in Fig. 4.

Transportation Network Vehicle Services (TNVS) are perceived by commuters to have the potential to contribute to the overall transportation system efficiency by (1) improving private and public transportation connectivity, (2) make ride-sharing more convenient, and (3) reduce need for self-driving according to the survey by Uber Philippines (2016).

7 Evaluation of Ridesourcing Services

The ridesourcing services in Metro Manila can be evaluated from the perspective of sustainable transport. In this case, sustainability is discussed in terms of reliability, safety, affordability, and traffic congestion reduction. These are discussed in the following sections.

7.1 Reliability

Recent researches on e-hailing services like Uber and GrabTaxi and their variants in the Philippines have revealed that these have delivered on their promise of improved reliability and shorter waiting time for the most part. However, there have been reported instances when drivers of these services refuse to book passengers during inclement weather or holiday rush, particularly when the known destination is congested.

7.2 Safety

To date, there have only been a few reported cases of threats to the safety and security of passengers using TNVS in Metro Manila with a bigger concern expressed for drivers taking advantage of passengers during times of high demand (Lozada 2015). In fact, safety is the second highest rated reason for commuters to choose TNVS over regular taxis. However, the bigger issue is the lack of insurance for passengers using TNVS, which are largely non-commercial private vehicles. Comprehensive vehicle insurance only covers damage to own property as well as that of the third party in case of road crashes and does not include passengers.

7.3 Affordability

Ride hailing services like Grab used to allow passengers to offer tips to TNVS drivers to book a ride. This was perceived as a form of bidding practice where the driver will have the option to choose the passenger with the highest tip. In addition, as mentioned in the sub-section on regulation of TNCs/TNVS, fares can spike drastically during peak periods and inclement weather due to surge pricing. These pricing mechanisms puts the commuters at the disadvantage and will potentially hinder those that are not willing to pay more from availing of the ridesourcing services. The 2016 Forbes survey conducted by Ford Motor Company confirmed that "*60% of the Filipino respondents said their commutes are getting more expensive, mainly because of higher fares, higher fuel costs, and choosing taxis and ride-hailing services over cheaper options*" (Tordesilla 2016). Grab removed this option for tipping and drivers can no longer use this as basis for choosing passengers.

Uber has long incorporated surge pricing in their fares. Surge pricing is basically additional charges levied on passengers for times when traffic congestion is severe and/or there are a limited number of Uber vehicles available for engagement. Surge pricing is typically used as an incentive to Uber drivers to go into an area where there is a demand for their services. This, of course, originally was premised on the assumption that most Uber drivers were part-timers and not operating like taxis.

7.4 Reduction of Congestion

Based on the data from Uber, the availability of TNVS has reduced the need for self-driving. Due to this, demand for parking spaces may be reduced. In fact, TNVS has been used in New Jersey as a parking solution (Hawkins 2016). However, it still does not address the issue of extensive utilization of low occupancy vehicles like private cars and taxis/TNVS which has been identified as one of the root causes of congestion in urban centers. It is, however, commendable that there are recent initiatives from the TNCs to move towards ridesharing (i.e., UberPool and UberHop) and improved access to mass transportation (i.e., Grab to MRT promo launched in the Metro Manila for a limited period where passengers booking a trip to any of the MRT stations will be given 50 peso discounts on their fares).

8 Conclusion

Although it has been established in the studies that have been conducted and cited in this paper that apps-based transportation services provide attractive alternatives to the riding public, it is also clear that most users are those who mainly use taxis or other public transport modes rather than commuters shifting from private vehicles (Dela Peña and Dizon 2016; Nistal 2016; Paronda et al. 2016). Add to this the industry report that TNVS have boosted car sales in the Philippines (Lorenciana 2017), with small sedan model cars being the most popular among people who have the intention of using these for Uber or Grab. As such, the experience in Metro

Manila is far from what ridesharing/ridesourcing companies claim as their contribution towards achieving sustainable transport. In fact, it can be argued that TNVS popularity has led to more vehicles roaming Metro Manila roads as their operations are practically like that of taxis, with most Uber and Grab vehicles being driven full-time rather than part-time.

It is also clear that government regulations are necessary to protect the rights and well-being of the commuters. Moreover, it is also the government's task to ensure that ridesharing and ridesourcing should be part of rational, integrated transport system, complementing other more efficient and higher-capacity modes.

To ensure that the potential benefits of TNVS are realized for a sustainable transportation system, the following issues must be addressed:

- Mode shift—it must be evaluated whether passengers shifted from lower or higher capacity vehicles. If it is the latter, then this could translate to increase in car use, thereby exacerbating congestion. On the other hand, when passengers shifted from private transport to Uber/Grab, then this does not necessarily mean less car use because the Uber/Grab car simply replaced the private vehicle in terms of road space usage.
- Trip chain—there is also a need to determine the typical trip chains for multi-mode use by commuter, whether Uber or Grab served the main mode or used for the initial or last miles.

Acknowledgements The authors would like to acknowledge the Engineering Research and Development for Technology (ERDT) Program for supporting this research. The authors express their gratitude to the Land Transportation Franchising and Regulatory Board and Uber Philippines for providing data for this study. Acknowledgements are also due to research assistants Jasper Dela Peña, Micah Angela Dizon, Patrick Nistal and Arden Paronda.

References

CNN Philippines. (2016). *LTFRB orders Grab, Uber surge cap*. CNN Philippines. October 12, 2016. http://cnnphilippines.com/news/2016/12/27/LTFRB-orders-Grab-Uber-surge-cap.html. Last accessed 6/14/2017.

Dela Paz, C. (2015). *Taxis beware: Government introduces 4 new transport categories*. Rappler Philippines. May 11, 2015. http://www.rappler.com/business/industries/infrastructure/92857-new-transport-categories. Last accessed 6/14/2017.

Dela Peña, J. A., & Dizon, M. A. P. (2016). *Comparative study of GrabTaxi and regular taxis within Metro Manila*. Unpublished research report. Institute of Civil Engineering, University of the Philippines, Philippines (53 pp).

DOTC. (2015). Department order no. 2015-011-further amending department order no. 97-1097 to promote mobility. Department of Transportation and Communications. Philippines. http://dotr.gov.ph/images/issuances/DO/2015/DO2015-11.pdf. Last accessed 12/18/2017.

Furuhata, M., Dessouky, M., Ordónez, F., Brunet, M., Wang, X., & Koenig, S. (2013). Ridesharing: The state-of-the-art and future decisions. *Transportation Research Part B, 57,* 28–46.

Grab. (2013). A Southeast Asian Journey, https://www.grab.com/ph/about/. Last accessed 6/12/2017.

Hawkins, A. J. (2016). New Jersey town decides to pay Uber instead of building a parking lot. The Verge. October 3, 2016. https://www.theverge.com/2016/10/3/13147680/uber-new-jersey-free-ride-parking-lot-train-commute. Last accessed 6/14/2017.

JICA. (2014). *Roadmap for Transport Infrastructure Development for Metro Manila and Its Surrounding Areas (Region III and Region IV-A)*. Philippines: Final Report. Japan International Cooperation Agency.

LTFRB. (2015a). Memorandum Circular No. 2015-015—Rules and Regulations to Govern the Accreditation of Transportation Network Companies (TNCs). Land Transportation Franchising and Regulatory Board. Philippines. http://ltfrb.gov.ph/media/downloadable/MC_NO._2015-015_.pdf. Last accessed 6/16/2017.

LTFRB. (2015b). Memorandum Circular No. 2015-016—Terms and Conditions for Transport Network Companies (TNCs) Land Transportation Franchising and Regulatory Board. Philippines. http://ltfrb.gov.ph/media/downloadable/MC_NO._2015-016_.pdf. Last accessed 6/16/2017.

LTFRB. (2015c). Memorandum Circular No. 2015-017—Implementing Guidelines for Transportation Network Vehicle Service (TNVS). Land Transportation Franchising and Regulatory Board. Philippines. http://ltfrb.gov.ph/media/downloadable/MC_NO._2015-017_.pdf. Last accessed 6/16/2017.

LTFRB. (2015d). Memorandum Circular No. 2015-018-Terms and Conditions of a Certificate of Public Convenience to Operate a Transportation Network Vehicle Service (TNVS). Land Transportation Franchising and Regulatory Board. Philippines. http://ltfrb.gov.ph/media/downloadable/MC_NO._2015-018_.pdf. Last accessed 6/16/2017.

Litman, T. (2016). *Well measured: Developing Indicators for sustainable and livable transport planning*. Victoria Transport Policy Institute. http://www.vtpi.org/wellmeas.pdf. Last accessed 6/14/2017.

Lorenciana, C. S. (2017). *Uber and Grab boost car sales*. The Freeman. March 3, 2017. http://www.philstar.com:8080/cebu-business/2017/03/16/1681568/uber-and-grab-boost-car-sales. Last accessed 6/13/2017.

Lozada, D. (2015). *Are Grab drivers abusing commuters' rights?* Rappler Philippines. September 9, 2015. http://www.rappler.com/move-ph/105269-grab-taxi-car-abuse-commuters-rights. Last accessed 6/13/2017.

Nistal, P. D. (2016). *Comparative study of Uber and regular taxi service characteristics*. Unpublished research report. Institute of Civil Engineering, University of the Philippines, Philippines (120 pp).

Nistal, P. D., & Regidor, J. R. F. (2016). Comparative study of Uber and regular taxi service characteristics. In *Proceedings of the 23rd Annual Conference of the Transportation Science Society of the Philippines*, Quezon City, Philippines, August 8. http://ncts.upd.edu.ph/tssp/wp-content/uploads/2016/08/Nistal-Regidor.pdf. Last accessed 6/15/2017.

Paronda, A. G. A., Regidor, J. R. F., & Napalang, M. S. G. (2016). Comparative analysis of transportation network companies (TNCs) and conventional taxi services in Metro Manila. In *Proceedings of the 23rd Annual Conference of the Transportation Science Society of the Philippines*, Quezon City, Philippines, August 8. http://ncts.upd.edu.ph/tssp/wp-content/uploads/2016/08/Paronda-et-al.pdf. Last accessed 6/15/2017.

Rayle, L., Shaheen, S., Chan, N., Dal, D., & Cervero, R. (2014). *Ap-based, on-demand Francisco*. Working Paper. University of California Transportation Center, University of California, Berkeley, United States. https://www.its.dot.gov/itspac/dec2014/ridesourcingwhitepaper_nov2014.pdf. Last accessed 6/15/2017.

Shaheen, S. (2014). Introduction to ridesharing: Overview of definitions and setting the stage. In *ACT International Conference*, University of California, Berkeley, August 5, 2014. http://actweb.org/wp-content/uploads/2014/12/ACT_Powerpoint_Ridesharing_Shaheen.pdf. Last accessed 6/15/2017.

Tan, K. J. (2016). Uber to offer fixed fare in Metro Manila starting Wednesday. ABS CBN News. October 8, 2016. http://news.abs-cbn.com/business/10/08/16/uber-to-offer-fixed-fare-in-metro-manila-starting-Wednesday. Last accessed 6/13/2017.

Tordesilla, K. (2016). 3 in 10 Filipinos dread their commute, says survey. CNN Philippines. May 12, 2016. http://cnnphilippines.com/metro/2016/05/12/filipinos-commute-worst-part-of-day-forbes-survey.html. Last accessed 6/13/2017.

Uber. (2016a). UberPool Mnl FAQs. https://newsroom.uber.com/philippines/uberpool-mnl-faqs/. Last accessed 6/13/2017.

Uber. (2016b). UberHop Pickup and Dropoff Info and FAQ. https://newsroom.uber.com/philippines/uberhop-info-and-faqs-2/. Last accessed 6/13/2017.

UN. (2012) World Urbanization Prospects, The 2011 Revision, Department of Economic and Social Affairs. United Nations. http://www.un.org/en/development/desa/population/publications/pdf/urbanization/WUP2011_Report.pdf. Last accessed 6/15/2017.

Wunder. (2016). https://www.wunder.org. Last accessed 6/12/2017.

Author Biographies

Prof. Regidor obtained his doctorate in Civil Engineering at the Yokohama National University in Japan and his master's degree at the University of the Philippines Diliman. He is currently with the Institute of Civil Engineering of the University of the Philippines Diliman. He is also a Research Fellow at the National Center for Transportation Studies of which he was Director from 2006 to 2012, and engaged in research mainly on environmentally sustainable transport, leading the study team that formulated the "National EST Strategy for the Philippines" from 2008 to 2011 with support from the UNCRD. He has also worked for the World Bank, contributing to the formulation of a "Transport Infrastructure Framework Plan and Roadmap for the Philippines" in 2013–2014, and crafting working papers on the "Politics of Transport Reform in Metro Manila" and "Promoting Partnerships for Transport Planning." He was part of a study team of experts that undertook "A Study of Long-Term Transport Action Plan for ASEAN" in a project supported by the Clean Air Asia and the Institution for Transport Policy Studies of Japan.

Dr. Napalang obtained her doctorate in Civil Engineering at the Tokyo Institute of Technology in Japan and her master's degree at the Virginia Polytechnic Institute and State University. She is member of the faculty of the School of Urban and Regional Planning and serves as the current Director of the National Center for Transportation Studies of the University of the Philippines Diliman Her research focus is on social impacts of transportation reforms and equitable transportation. She has been engaged in projects for the implementation of the Bus Rapid Transit in the Philippines and led the study to evaluate the 'Impacts of Urban Transport Reforms on the Jeepney Operations in Cebu City'. She is one of the core members of the recently established *Women in Transport Leadership (WiTL)*, a knowledge sharing network aimed at creating knowledge, fostering innovation and empowering female-to-female collaboration to address diverse transport challenges, particularly issues related to gender and transport. She was Chair of the Department of Civil Engineering at Xavier University in Cagayan De Oro City in the southern Philippines.

Part III
Case Studies: Sustainable Infrastructure

Analysis of Rooftop Solar Photovoltaic System Across the Indian States: Learnings for Sustainable Infrastructure

Brijesh Bhatt and Anjula Negi

Abstract Facing twin challenges of providing energy access to its citizens and complying with international protocols on climate change mitigation, India has set ambitious targets for solar power generation. A major policy initiative was undertaken with the launch of Jawaharlal Nehru National Solar Mission (JNNSM) in January 2010. JNNSM aimed to increase solar power capacity from 0.0178 GW (Gigawatt) in 2010 to 22 GW by the year 2022. With the new government formation in 2014, these targets were increased fivefold to 100 GW by the year 2022. However, most solar power projects are large utility scale (LUS) in comparison to rooftop solar photovoltaic (RSPV). LUS projects have long gestation period and require huge tracts of land; a scarce resource given India's high population density. Consequently, adequate attention needs to be given to RSPV. RSPV also increases overall system efficiency as transmission and distribution losses are minimized. Targets for RSPV projects in JNNSM have been enhanced to 40% of the total solar targets. This target does appear formidable as it has to be implemented across the Indian states, which have varying RSPV policy outlook. Thus, whether state wide policy measures would support this scaling up may need to be understood. While there is sufficient literature focusing on the success of LUS solar projects, limited literature exists on challenges and successes of RSPV in India. Based on literature review and semi-structured interviews with key stakeholders, the study identifies few scaling up challenges (or barriers) for the adoption of RSPV. Thereafter, the study maps if policies of states having high RSPV installed capacity, have addressed the identified barriers. Given multiple challenges and an evolving sector, inter alia, major issues collated in the study include wide spectrum of regulations across states, varying consumer tariffs, market readiness for adopting RSPV, lack of awareness amongst consumers, anchors and skilled persons missing to support the initiative and limited institutional capacities, low usage of subsidies offered and

B. Bhatt (✉) · A. Negi
School of Infrastructure, RICS School of Built Environment,
Amity University Noida, Noida, UP, India
e-mail: bbhatt@rics.org

A. Negi
e-mail: anegi@rics.org

W. Leal Filho et al. (eds.), *Sustainable Development Research in the Asia-Pacific Region*, World Sustainability Series, https://doi.org/10.1007/978-3-319-73293-0_23

high installation costs. In conclusion, from the comparative analysis of RSPV policies of top three states having high installed capacity, the study examines how these challenges have been addressed therein, and further, the paper identifies policy features that can lead to the creation of sustainable infrastructure for RSPV.

Keywords Roof solar photovoltaic · Renewables · Electricity · Barriers Sustainable infrastructure

1 Introduction

Sustainable infrastructure development and universal access to renewable energy are complementary in nature. At 21st meeting of Conference of Parties (COP 21), a global summit of climate change convention held at Paris, India presented one of the goals to increase non-fossil based electricity to 40% by the year 2030 (UNNFCC 2017). Presently, the renewable generation capacity of India is mere 17% and comprises of four sources; wind, solar, small hydro, biomass cogeneration and waste to energy[1]. Solar, followed by the wind has the second largest installed capacity among renewables. Although the proportion of solar is low, India has high solar energy potential, capable of producing 5000 trillion kW of clean energy, with around 300 sunny days in a year and national average solar insolation of 4–7 kWh per m^2 per day (MNRE 2012). Harnessing this efficiently, can reduce energy deficit of the entire country and with near nil carbon emissions (MNRE 2012). Solar energy can be harnessed using either solar photovoltaic (PV) or concentrated solar thermal technology. International Energy Agency (IEA) forecasts that by the year 2050, solar photovoltaic and concentrated solar thermal would contribute about 16 and 11% of the total electricity consumption respectively in the world (IEA 2014). In India, the majority of installed solar capacity is based on solar photovoltaic technology (Hairat and Gosh 2017).

The solar PV plant of larger capacity (in MW[2]) with PV modules mounted on the ground are known as large utility scale (LUS) whereas, plants of smaller capacity (in kW[3]) with PV modules mounted on consumers rooftop are known as rooftop solar photovoltaic (RSPV). LUS are not best suited with regards to grid stability and efficient resource utilization. RSPV has following advantages over LUS (Gosh et al. 2015; CARE 2015):

[1]As on 31st May 2017, the total generation capacity of India is 330 GW. Thermal comprising of coal, gas, and diesel has the largest generation capacity of 67%. Nuclear and hydro forms 2 and 13% of total generation capacity respectively.

[2]MW = Megawatts.

[3]KW = Kilowatts.

- Avoids additional land requirement dedicated to energy generation as RSPV uses rooftops that otherwise may remain unused
- Reduced transmission and distribution losses as power is produced near point of consumption
- Lower gestation period
- Reduction in system congestion and reduced requirement of transmission networks
- Development towards smarter grid with more decentralized generation
- Social advantages like capacity building of local electricians, involves households thereby increasing awareness about power consumption

Unlike world largest solar capacity countries i.e. China and Germany, which have a high share of RSPV,[4] India has witnessed growth in solar capacity through LUS plants. While total solar capacity is 12.289 GW (as on 31st May 2017), RSPV capacity was only 0.740 GW (as on 31st March 2017). Thus, RSPV remains negligible expanse in country's total solar capacity.

The estimated technical and market potential for RSPV is about 352 and 124 GW respectively[5] (Martin and Ryor 2016). Acknowledging this potential, there is sufficient policy thrust from the new central government, under JNNSM which targets to set up 40 GW RSPV capacity by the year 2022. Annual targets have been prescribed for all states to achieve it. However, experiences of previous implementation programs as reflected in RSPV installed capacity data (Fig. 1) shows that RSPV segment has lagged and yet to take off. The literature on technology adoption explains such phenomenon by the existence of barriers, which hinder adoption of technology (Sorrell et al. 2004; Fleiter 2007; Cagno et al. 2013). Since Indian RSVP industry is at its nascent stage, identification and addressal of these barriers is crucial for achieving these targets of RSPV. Thus, the study firstly collates these barriers. Thereafter, acknowledging these formidable national targets that have to be executed in tandem with the states having varying RSPV policy outlook the study investigates whether state wide policy measures support this scaling up? Paper analyses, if (and how) existing solar/RSPV policies and regulations are addressing identified barriers. This is necessary as it has been experienced that policies and regulations, especially state-level policies are the key factors in enhancing technology adoption in infrastructure sector (Pollitt et al. 2012; Schmid 2011). The study analyses aspects under various solar policies of RSPV across top three Indian states categorized as having the highest installed capacity, both scholarly and project implementation literature.

Following this introduction, in the next section, a short literature review of Indian and global policy experience for promotion of solar is discussed. Section 2.1 describes the methodology of for identification of barriers and policy analysis. Section 2.2 discusses various barriers or challenges based on literature review and

[4]For Germany over 98% of solar capacity consists of RSPV.

[5]While market potential accounts for economic considerations and consumer acceptance, technical potential does not.

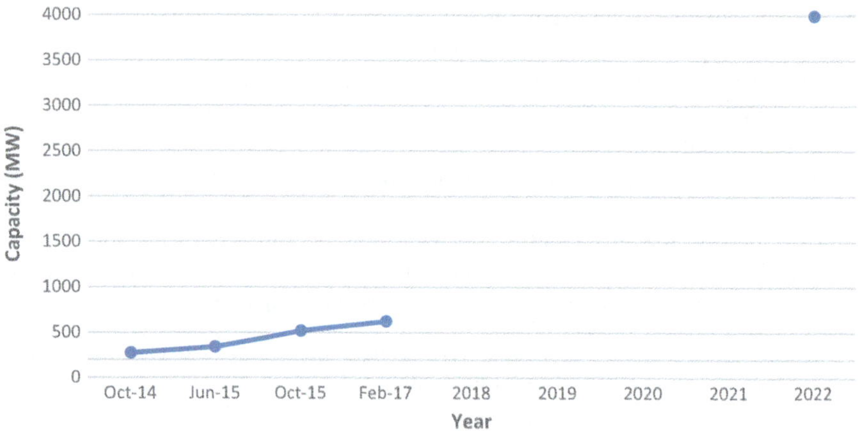

Fig. 1 Installed capacity of RSPV and targets. *Source* Martin and Ryor (2016)

interviews with key stakeholders. Section 5 analyses how policies at state level address the challenges identified. Section 6 concludes and comprehends scalability factors that the Indian states will engage in order to achieve the 40 GW target of RSPV.

2 Literature Review—Indian and Global Overview of Solar Policy Framework

Globally policies have a major role in the promotion of solar energy. A brief review of studies addressing policies for solar and RSPV usage in India and internationally is presented below.

2.1 Review of Solar and RSPV Sector in India: Stimuli for Growth

Although India is among the leading countries with high solar potential, in policy and planning, solar was first discussed as a technology for electricity generation in 3rd five-year plan[6] (1961–66) only. However, in the next two five-year plans, focus remained on hydro and thermal. Only in the 6th plan (1980–85), solar was

[6]Five-year plan are integrated nationalized economic programs which lay foundation plans for economic growth of a country (Kapoor et al. 2014). Five-year plan are developed, executed and monitored by the erstwhile planning commission (now NITI Aayog) of India.

identified as an energy source for decentralized rural areas and potential industrial uses. The Department of Non-Conventional Energy Sources was established during this plan period, with an objective to provide funding and strengthening research and demonstration of various renewable technologies including solar (Kapoor et al. 2014). Subsequent five-year plans allocated funding support for research on solar PV module technologies and setting up programs for electrifying remote villages through solar photovoltaic. A separate Ministry of New and Renewable Energy (MNRE) was established in 2006. In the 11th plan (2007–2012), primarily for climate change and energy independence reasons, country's first major policy initiative to promote solar energy—JNNSM; was undertaken as one of the missions under its National Action Plan on Climate Change (NAPCC).

Before discussing features and outcomes of JNNSM, a brief overview of key legislative changes in the electricity sector is necessary since it provides the institutional structure for policy implementation, including JNNSM. Private participation in Indian power sector began with liberalization in 1990s. Liberalisation involved a series of legislative changes, most significantly enactment of the Electricity Act, 2003 (EA 2003). EA 2003 provided several enabling provisions for the promotion of renewable energy and entrusted independent regulatory commissions at central and state level to implement these provisions. For example, Section 86 (1)(e), mandates that state regulatory commission shall specify minimum purchase obligations from renewable sources; Section 61(h), mandates appropriate commission shall specify the terms and conditions for determination of tariffs from renewable sources. Thus, having a backdrop of a good regulatory regime is a key factor for implementation of various laws for the promotion of renewables. EA 2003 also specifies that central government should bring two policies namely, National Electricity Policy and Tariff Policy and amend them suitably if needed. Accordingly, government enacted the National Electricity Policy in 2005, which promotes renewables and emphasises that the share of renewable electricity needs to be increased progressively. In 2006, the Tariff Policy was first enacted, and a revised Tariff Policy in 2016, which has a provision of renewable generation obligation (RGO) for conventional power generators. Thus, showcasing a conducive legislative environment for renewables.

Specifically for the promotion of solar the JNNSM was launched in January 2010 that targeted to install 22 GW solar capacity by 2022, in three phases; phase I (2010–12), phase II (2013–17) and phase III (2017–22). Using both grid connected and decentralized off-grid technologies, JNNSM has different targets, guidelines and mechanism/tools for different solar technologies in different phases as shown in Table 1.

Since launch of JNNSM solar capacity in India grew in a steadfast manner. In 2010, total installed capacity of solar power in India was 0.0178 GW; it reached 0.5069 GW by early 2012 and to 2.75 GW by mid-2014 (Fig. 1). This increase in four years is largely attributed to the launch of the JNNSM. Before introduction of JNNSM deployment of renewable capacity including solar was largely driven by individual subsidies approved by MNRE (Yenneti 2016). Supported by enabling legislative framework, JNNSM introduced market mechanisms with various tools

Table 1 JNNSM targets and achievement

S. No.	Phase/Period	Segment					
		Grid connected—LUS and RSPV		Off-grid		Solar collector	
		Target (MW)	Achievement (MW)	Target (MW)	Achievement (MW)	Target (million m²)	Achievement (million m²)
1	Phase I (2010–13)	1000–2000	1466	200	223	7	6.92
2	Phase II (2013–2017)	4000–10,000	–	1000	–	15	–
3	Phase III (2017–2022)	20,000	–	2000	–	20	–

like renewable/solar purchase obligations, renewable energy certificates, viability gap funding, feed-in-tariff, accelerated depreciation etc. to promote and increase private participation in solar capacity addition (Rohankar et al. 2016). As described earlier, with a new government at the central (national) level, targets of JNNSM were revised to 100 GW in 2015, including 40 GW of RSPV by 2022. To achieve this, state wise targets were set by the government. Targets for Northern, Western, Southern and Eastern regions are about 31, 28.4, 26.5 and 12.2 GW respectively (Kar et al. 2016).

Besides JNNSM, the central government has initiated various other schemes for promotion of solar energy. For instance, government of India has created a Solar City Programme under MNRE to support 60 Indian cites in development of renewable energy and energy efficiency projects aimed to reduce conventional energy demand by 10% (MNRE undated).

Alongside above efforts at central level, several states are encouraging solar/renewable through independent state solar/renewable policies. In 2009, beginning with Gujarat, presently 25 Indian States have formulated their own policies and regulations for promotion of RSPV. States are at different development stages of policies and there is considerable variation in their framing.

To an extent, the JNNSM was largely successful to promote LUS as RSPV capacity still remains low. Under JNNSM phase I, the "Rooftop PV and Small Solar Power Generation Programme" was also initiated however, installed RSPV capacity reached only 0.285 GW[7] in October 2014 and 0.525 GW in October 2015 (Martin and Ryor 2016). It is argued that despite having many advantages, RSPV is

[7]Exact data of RSPV installed capacity is not available for various limitations including a large number of consumers (as revealed by MNRE officials—interview dated March 2017). These figures are from a report by Bridge to India.

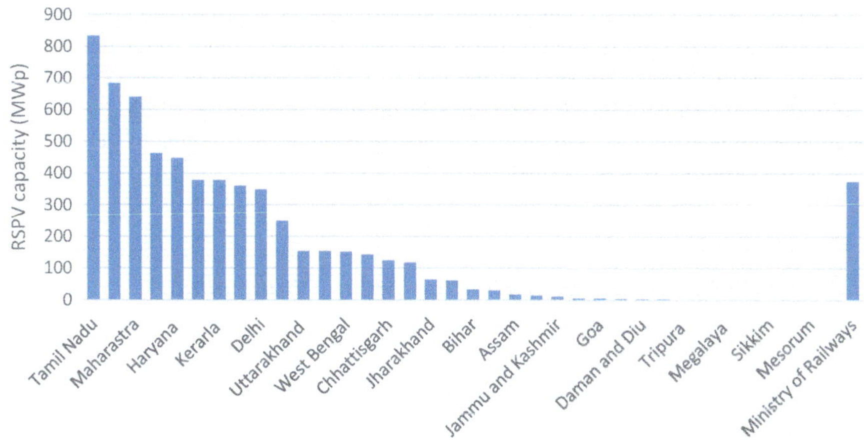

Fig. 2 Total RSPV installed capacity (MWp) across different states and central departments. *Source* MNRE (2017)

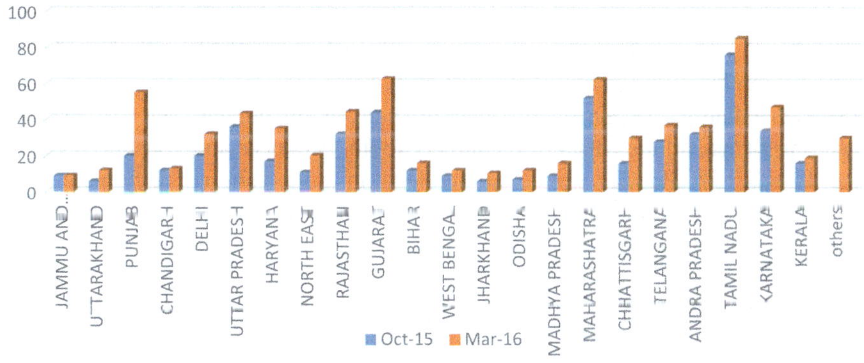

Fig. 3 State wise installed capacity for solar rooftop. *Source* Bridge to India (2015, 2016)

3–5 years behind LUS. As on 31 March 2016, India has an installed solar RSPV capacity of 0.740 GW,[8] which constitutes only 8% of total solar capacity (9.235 GW[9]). This accounts for merely 3% of total installed capacity of India.

Figure 2 shows total installed RSPV capacity across different states and central government departments like ministry of railways. Figure 3 shows, variation in RSPV installed capacity of different states between January 2017 and December 2016. As Fig. 2 shows, Tamil Nadu, Punjab, Maharashtra, Karnataka, and Haryana are states with highest RSPV capacity. In terms of capacity addition during these six month period, it was highest in the state of Punjab as shown in Fig. 3.

[8]Source: Bridge to India 2016.

[9]Data as on 31 January 2017, as per CEA and MNRE.

Despite having same central level subsidies, RSPV installed capacity vary widely across states. This reflects that RSPV faces different types of challenges that may be needed to be addressed by the policy and regulations in states concerned. Section 2.1 presents different challenges faced in promotion of RSPV but, before that an attempt to review global experience in promotion of solar in general and RSPV in particular has been made.

2.2 Global Landscape of Rooftop Solar Photovoltaic (RSPV) and Policy Levers

China leads in cumulative installed RSPV capacities at 43.5 GW, Germany, Japan, USA and Italy follow at 39.7, 34.4, 25.6 and 18.9 GW respectively. Remaining countries of the world lag behind and are below the 10 GW mark. India stands at 2 GW installed capacity in 2015 (IEA 2016; MNRE 2017).

Policy instruments have been hailed as one of the key attributes that could lead to rapid expansion of solar energy market. These fiscal mechanisms showcase a wide application variety from feed-in-tariffs (FiT), to subsidies or grants to specific credits or other financial investment routes (Timilsina et al. 2011). Self-consumption is becoming a major driver of a change from traditional energy to RSPV installations in the energy market, buoyed by policies such as FiT, net metering or net billing, placing Prosumers (both, producer and consumer) on the rise (IEA 2015). Globally and at national level in India, various policy methods or schemes that have augmented or promoted creation of sustainable infrastructure through residential RSPV are showcased as checkmark ticks against given countries in Table 2.

In Germany over 98% of solar capacity consists of RSPV. Germany released its policy on Energy Efficiency Strategy for Buildings, developed under National Action Plan on Energy Efficiency (NAPE) in 2015. Herein, 35% of the total energy consumption and one third of greenhouse gas emissions were taking place in the building sector, covering residential and non-residential buildings. Before this policy came into place, Germany's Energiewende had presented a fundamental transition by moving to a decarbonised energy system mainly based on renewable energy like solar and wind energy (Matschoss 2013). In Japan, the RSPV model show cases favourability for large-scale centralised RSPV systems. As tariffs are above the retail electricity prices, self-consumption is not incentivized. FiT program has been used to remunerate excess RSPV electricity not self-consumed for systems below 10 kW. USA offers tax credits granted by federal US government that the market finds very positive. In 2014, few states in USA have offered Value-of-Solar Tariff (VOST) as an option to net-metering. Financing is largely through indirect

Table 2 Policy specific support for RSPVs in 6 countries

Schemes[a]	Australia	Germany	India	Italy	Japan	USA
Direct capital subsidy		√	√	√	√	√
Green electricity schemes (GES)	√	√		√		√
PV specific GES						√
Renewable Portfolio standards (RPS)	√	√	√		√	√
Solar set aside RPS			√			√
Financing schemes	√	√	√		√	√
Tax credits/tax benefits	√	√	√	√	√	√
Net metering/Net billing/ Self-consumption incentives	√	√	√	√	√	√
Sustainable building requirements					√	√

[a]Schemes narrated in Table 2 are explained as follows: D*irect capital subsidy* to hold up-front cost barriers, for specific PV equipments or total installed PV system cost, *Green electricity scheme* that allows consumers to purchase green electricity usually at a premium price, Solar set aside *RPs*, that mandate utilities to source a portion of electricity through solar electricity supplies, *Investment funds (one of the financing schemes along with subsidies, generation based incentives* etc.*)* that focus on wealth creation using PV, T*ax credits*, that allows some or all expense to be deducted from taxable income streams for PV installations, *Tax benefits,* allowing accelerated depreciation on investments in RE devices, *Sustainable building requirements*: means changing norms for development of new buildings, across function/type to include PV and in order to lessen the energy requirements

Source Sundaray et al. (2014)

public funding and/or absorbed by utilities along with third party financing alternatives. Californian Solar Initiative has been financed similarly for around 60% of the installed residential systems, and third parties are allowed to benefit from tax breaks. Thus, absorbing high up-front investment through solar leases (IEA 2015).

In China, rooftop PV initiative is driven by a committed FiT scheme. Financing mechanism is through a surcharge paid by electricity users. German markets are driven by self consumption rather than supply to the grid, consumers receive a high premium, are incentivised by more than 30%, and are paid above the retail electricity price.

International Solar Alliance (ISA) is supported by 120 countries and is having a collective commitment of mobilising more than 1000 billion US Dollar of investments for affordable solar energy by 2030 (COP21 2015). Main premise of ISA is that solar energy technologies have made significant progress and can be considered for meeting a sustainable energy supply demand (Goel 2016). Declining technology costs, in particular, solar photovoltaic (PV) module prices halving between 2010 and 2012, accounting for a fifth of the increase of modern renewable energy consumption (GTF 2015). In 2015, total grid connected installed capacity globally amounted to 225.6 GW (IEA 2016). In summary of the above backdrop, it can be inferred that suitable policy measures addressing the local needs, markets driven by self-consumption, incentivising financing mechanism to adopt RSPV, wide application solar PV technology and its declining costs may lead to significant progress in meeting the solar energy supply side.

3 Research Methodology

Research methodology has been guided keeping in mind collation of policy features that may help in scaling up the activities and adoption of RSPV and limited to Indian context. The barriers to adoption of RSPV were identified in two steps. First, based on literature review a list of key issues were identified. Literature review was undertaken at global and India level in order to understand features, broadly categorised such as regulatory provisions, commercial aspects, financial mechanism and technology requirements that may help support the solar sector. Thereafter, semi-structured interviews were conducted with key stakeholders. These stakeholder were chosen from higher levels in bureaucracy from the concerned ministry, i.e. MNRE, scientists from MNRE, high level bureaucrat in the state of Telangana (which is one of latest states to roll out the solar energy policy), consultants who are working in this sector and supporting various state governments in policy formulation, seniors officials from Rural Electrification Corporation under Ministry of Power and one developer who has been implementing roof top solar PV technology in residences across north Indian states. Consequent to identification of the list, it was discussed with these stakeholders and finalized. Finally, existing policies for top three states chosen based on highest share of RSPV installed capacities were reviewed. Review was undertaken to identify, if and how these challenges are addressed in the policy. Thus, making an attempt to understand how solar policies and regulations were addressing barriers to RSPV adoption, through comparative analysis of states policies and regulations.

4 Challenges for Scaling Up RSPV

Challenges faced for scaling up of RSPV can be broadly classified into regulatory, commercial and implementation, financial and technical. This is based on a review of various implementation reports, scientific literature and discussions with key stakeholders (Annexure 1). Table 3 summarizes key challenges faced in promotion of RSPV.

Table 3 Key challenges faced in promotion of RSVP

Challenges for promotion of RSPV	Source
Regulatory issues	
Net metering or gross metering but its operationalization	Gambhir et al. (2012), Thakur and Chakraborty (2016), Hairat and Gosh (2017)
Lack of market/commitment for green energy	UK AID, Gosh et al (2015), Rohankar et al. (2016), Hairat and Gosh (2017)
Skewed consumer tariffs	Verma et al. (2016)
Commercial and implementation issues	
Lack of awareness amongst consumers/ stakeholders	Goel (2016), Verma et al. (2016), Martin and Ryor (2016)
Lack of skilled manpower	Martin and Ryor (2016)
Insufficient political involvement and support	Kapoor and Dwivedi (2017), UKAID (2016)
Rooftop availability	WB IFC 2014 Gandinagar: 81
Financial issues	
High upfront installation cost and need of appropriate business models	Narula and Reddy (2015), Gambhir et al. (2012), Kapoor and Dwivedi (2017), Kappagantu et al. (2015), Sukh and Mandavilli (2016)
Convoluted subsidies and poorly managed financing options	Kapoor and Dwivedi (2017)
Financial health of distribution utilities	Ghosh et al. (2015)
Technical issues	
Quality and efficiency of solar PVs	Sharda (2015), Manju and Sagar (2017)
Connectivity issues at low voltage and poor distribution grid/infrastructure	Chattopadhyay (2014), Magal et al. (2014), Sukh and Mandavilli (2016)
Forecasting and grid balancing issues	Chattopadhyay (2014), Magal et al. (2014), Richts et al. (2015)

4.1 Regulatory Issues

Regulators can greatly enhance RSPV adoption by prescribing and enforcing a set of standard rules and regulations, which can reduce transaction costs and gestation period associated with installation of RSPV. Regulators approve tariff trajectory for consumers, specify interconnection standards, details of energy accounting, billing, monitoring and verification, banking facilities from utilities, dispute redressal mechanism etc. All these factors influence RSPV adoption. Clear and simple institutional structure along with defined timelines would help to bring about greater transparency and accountability in the programme. Following are some of the key developmental issues faced in these areas;

a. *Net metering versus gross metering, but its operationalization*

The grid connected RSPV work either on net-metering or gross metering basis. In a net-metering system, surplus power generated after consumers own consumption is transferred to the grid and consumer is billed only for 'net' electricity (total consumption minus own production). In a Gross metering system, entire solar power generated is supplied to the grid at government offered tariffs known as FiT. Though the debate on comparative advantages and disadvantages between gross metering and net-metering approach continues, many studies conclude that net-metering approach (to encourage in situ generation primarily for self-consumption) better suits Indian conditions where distribution sector is weak (Gambhir et al. 2012; Thakur and Chakraborty 2016; Hairat and Gosh 2017; CARE 2015). Different states are at different stages of development of metering regulations and majority of state regulators have adopted net metering system. Whilst regulations can be improved, the real challenge lies in their operationalization. Lack of enforcement results in poor clarity of process at utility level and involves multiple permissions causing delays. It is suggested that regulators require utilities to collect and publicly make available data on connection requirement, new interconnections etc. and should penalize for non-compliance. In many cases, consumers reported confusion and uncertainty about net-metering specifications like maximum RSPV system size and compared to load, transformer loading, length of contract signed with utility etc. (Martin and Ryor 2016). Effectiveness in net metering covers aspects of meter locations, accuracy of reading, types of meters, sealing etc. needing classification and mitigation. Customer confidence would need to be built along with setting up of proper energy accounting processes.

b. *Lack of market/commitment for green energy*

Regulatory commissions need to ensure that the mechanisms for promotion of green power like renewable purchase obligation (RPO) or solar purchase obligations (SPO) function adequately. Since, per unit cost of electricity generated from RSPV will be higher compared to LUS projects, utilities might not prefer power purchase from RSPV for fulfilling their RPO obligations[10] (UKAID, Gosh et al. 2015; Rohankar et al. 2016; Hairat and Gosh 2017).

c. *Consumer tariffs*

Consumer tariff is the most important driver for adoption of RSPV. Tariffs vary state wise, consumer category wise and connection type. Highest tariffs are paid by industrial and commercial consumers followed by residential and agricultural consumer. Artificially suppressed tariff because of deep structural and political reasons, weakens the economic case for adoption of RSPV for some consumers e.g. residential and agricultural (Verma et al. 2016). For example, higher/increasing power tariff in Tamil Naidu and Maharashtra enhanced RSPV adoption (UKAID 2016). Financial support may be required for installations of rooftop systems in

[10]To promote RSPV, it is suggested that regulators can adjust RPO benefits from RSPV for utilities as it has a comparative advantage of avoided transmission and distribution losses (UKAID).

residential and institutional premises, where utility tariffs are still not at par with solar rooftop levelised cost of energy (Ayog 2015).

4.2 Commercial and Implementation Issues

Key commercial and implementation issues are:

a. *Lack of awareness amongst consumers/stakeholders*

Consumers as well as developers face problem of lack of information and asymmetric information regarding the government policies, available incentives and advanced PV technologies (Verma et al. 2016; Martin and Ryor 2016; Goel 2016). People still believe that solar power is very expensive and pose apprehension about technical feasibility of RSPV. This often results lack of thrust, poor conceptual understanding among stakeholder as well as spillover effects like high cost of capital as lending agencies perceive RSPV as a high risk asset class (Verma et al. 2016). Higher level of public awareness in Tamil Naidu is reported a key factor for higher adoption rates of RSPV (UKAID 2016). Hence, need for high level of public awareness and information penetration mechanism as to rightly educate consumers on economic and technical aspects associated with usage of RSPV.

b. *Lack of skilled manpower to manage and drive RSPV industry*

For its smaller system sizes, RSPV industry is more labour intensive than LUS, hence the availability of skilled workforce to design, install and maintain RSPV is reported a big challenge[11] (UKAID 2016). Besides this, capacity need to be built for key stakeholders like utilities, banks and regulators for greater understanding of functioning, technical issues related to integration with grid and its impact thereon. Limited institutional capacity of utilities (in terms of lack of capacity for synchronization and interconnection, lack of capacity among meter readers etc.) was a key barrier reported in a pilot study with residential consumers in south India (Martin and Ryor 2016).

c. *Insufficient political involvement and support*

Distribution utilities are a key stakeholder in all RSPV business models. In long term, RSPV installation might lead to utility's loss of consumers, hence, reduction in revenue. Some utilities might not support RSPV for these reasons. As most of the utilities are government owned, firm political signals or anchoring also matters. Ministers at central and state level should emphasize on utilities to actively support RSPV (UKAID 2016; Kapoor and Dwivedi 2017).

[11]Although this has positive side as well since it will generate more employment. It is estimated that 40 GW will generate 500,000 short term jobs and 140,000 long term jobs across country (UKAID).

4.3 Financial Issues

Key financial issues are:

a. *High installation cost and need of appropriate business models*

Even though the cost of PV modules have shown a declining trend, upfront installation costs remain high (minimum of approx. INR 75,000/kW). The payback period is long, varying from 6 to 10 years (depending on consumer tariff). Hence, remain a key barrier to RSPV adoption. Addressing this would require innovative financing mechanisms in order to provide finances through debt arrangement or other third party arrangement. Narula and Reddy (2015) examine financial viability of RSPV for current cost and electricity tariff in Delhi, for two business models (self-owned and third party owned).[12] They found that for smaller size (2.5 and 5 kWp), only self-owned model is financially viable along with 30% subsidy. Whereas, for large size (greater than or equal to 10 kWp) both self-owned and third party owned are financially viable without subsidy. At prevailing tariff structure, Gambhir et al. (2012) shows that it makes financial sense for those consumers paying higher tariffs (like commercial) to install RSPV. State would need to promote and demonstrate cases of sustainable business models in residential, commercial and industrial sector to attract private investments (Kapoor and Dwivedi 2017; Kappagantu et al. 2015).

b. *Convoluted subsidies and poorly managed financing options*

Subsidies were introduced to enhance solar power adoption. 30% subsidy on capital cost (for residential customers) is provided by central government. However, not releasing subsidies on time, poor advocacy such as media reports claiming cancellation of these subsides, had a large negative impact on RSPV adoption as experienced in the state of Tamil Nadu (Kapoor and Dwivedi 2017).

c. *Financial health of distribution utilities*

Ghosh et al. (2015) assessed financial feasibility of RSPV for industrial, residential and off-grid rooftops in the state of Karnataka. Study found that RSPV is not being adopted because distribution utility has imposed a condition of RSPV size not to exceed 75% of the consumer's connected load owing to its poor financial health. Because of this cap, RSPV generation can not exceed monthly consumption and consumers will have less incentives to install RSPV. The study concluded that the

[12]Under self-owned business model (also known as CAPEX model), roof owner invests and owns RSPV system whereas in third party model (also known as OPEX model), a third party (separate from roof owner and utility) invests and owns RSPV system. Only 10% of existing RSPV capacity is built on third party business model. Though, it offers many benefits like access to low cost financing, greater ability to take and mitigate risks, economies of scale and greater effectiveness of utilizing government incentives/tax benefits, it is yet to take up because of high contract default arising from rapid declining cost of solar energy (Goel 2016).

cap is the stumbling block and unless it is removed RSPV adoption can never gain momentum.

4.4 Technical

Key technical issues are:

a. *Quality and efficiency of solar PVs*

According to Sharda (2015) poor quality of solar module is an issue faced by many RSPV consumers. Since, most of the RSPV components are imported, performance of these technologies in Indian conditions are not known fully as they are in the developing stages (Manju and Sagar 2017).

b. *Connectivity issues at low voltage and poor distribution grid/infrastructure*

Technical issues like connectivity issues at low voltage or clarifying appropriate voltage has been an issue and needs to be aligned (Chattopadhyay 2014; Magal et al. 2014; Sukh and Mandavilli 2016). Thus, containing erratic behaviour of low voltage through appropriate protection system would need to be commissioned.

c. *Forecasting and grid balancing issues*

Energy security and grid stability is another aspect that may be impacted due to multiple supply sources (Chattopadhyay 2014; Magal et al. 2014; Richts et al. 2015).

5 Are Policies Addressing Identified Challenges/Barriers for Scaling Up RSPV Initiatives

With a view to understand whether the policies rolled out by the top three Indian states, namely, Tamil Nadu, Punjab and Maharashtra have addressed few of the above mentioned challenges, Table 4 provides a comparative summary context of such policy measures adopted.[13]

From the above it is evident that States may still need to undertake many measure to scale up the RSPV implementation. While net metering has been provided by many States, processes or systems for effective implementation and monitoring of net metering remain unclear. Aspects of innovative financing for tackling high installation cost will need to be introduced, though incentives are limited to grant of subsidies. A political anchor appears missing in all the three states. It is interesting to note that all the top three states recognise RSPV potential

[13]Annexure 2 summarizes key measures of solar policies of some states.

Table 4 Policy measures to address barriers and support provisioning of RSPV

Issue category	Issues identified	State	Policy/Regulation to address the issues	How issue is being addressed in policy/regulation	Remarks
Regulatory	Net metering or gross metering but its operationalization	TN	TN net metering regulation (2013) preceded by TN solar policy (2012)	– State's regulatory commission allows net metering through two different meters, one for measuring solar power generation and other for import/export measurement – Overall distribution capacity restricted to 30% of the distribution transformer capacity	– Leading states have a solar policy followed by net metering regulation. However, Maharashtra had directly adopted the net metering policy. These are likely policy enablers for adopting RSPV systems
		Punjab	Policy on net metering since 2014 preceded by New and Renewable Sources of Energy Policy (NRSE) 2012	State's regulatory commission allows two meter configuration, solar meter and consumer meter, with storage. – Overall distribution capacity restricted to 30% of the rated capacity of the distribution transformer	
		Maharashtra	Net metering for rooftop systems regulations adopted in 2015	– Overall distribution capacity restricted to 15% of the rated capacity of the distribution transformer	
	Lack of market/ commitment for green energy	TN	Policy provide solar purchase obligation	6% SPO	Lesser than central level prescribed limit at 8%
		Punjab	Renewable Purchase Obligation (RPO) and its	Minimum RPO percentages of the total generation from renewable sources defined up to	Lesser than central level prescribed limit at 8%

(continued)

Table 4 (continued)

<table>
<tr><th>Issue category</th><th>Issues identified</th><th>State</th><th>Policy/Regulation to address the issues</th><th>How issue is being addressed in policy/regulation</th><th>Remarks</th></tr>
<tr><td></td><td></td><td></td><td>compliance) Regulations, 2011</td><td>2015. From total 4%, 0.19% in 2014–15, 2.5% in 2019–20 to come from solar</td><td></td></tr>
<tr><td></td><td></td><td>Maharashtra</td><td>Notified RPO regulations for 2016–2020</td><td>RPO % in FY 2016–17 is 11% in total (10% non-solar and 1% solar). This will increase to 15% by FY 2019-20 (11.5% non-solar and 3.5% solar)</td><td>Lesser than central level prescribed limit at 8%</td></tr>
<tr><td></td><td rowspan="3">Skewed consumer tariffs</td><td>TN</td><td rowspan="3">Based on consumers tariffs for residential/domestic, low tension category obtained from respective distributions companies in respective years</td><td><table><tr><th>TN</th><th>% increase from 2010 to 2016</th></tr><tr><td>For units less than 50</td><td>172.73</td></tr><tr><td>For units less than 100</td><td>150.00</td></tr><tr><td>For units less than 250</td><td>34.62</td></tr></table></td><td rowspan="3">Tariffs have been increased over years and all three states display variation in tariff increase. However, a review of the regulatory commission's tariff note, reports average cost of supply as higher than the consumer tariffs across the three states. Thus, indicating a subsidised energy supply component by the State</td></tr>
<tr><td></td><td>Punjab</td><td><table><tr><th>Punjab</th><th>Slab (low–high)</th><th>% increase from 2010 to 2016</th></tr><tr><td rowspan="3">Load upto 50 kW</td><td>1–100</td><td>45.3376</td></tr><tr><td>101–300</td><td>35.8407</td></tr><tr><td>301+</td><td>37.2385</td></tr></table></td></tr>
<tr><td></td><td>Maharashtra</td><td><table><tr><th>Maharashtra</th><th>Slab (low–high)</th><th>% increase from 2010 to 2016</th></tr><tr><td rowspan="3">Consumption slab KWh</td><td>1–100</td><td>68.4211</td></tr><tr><td>101–300</td><td>81.0069</td></tr><tr><td>301–500</td><td>73.92</td></tr></table></td></tr>
</table>

(continued)

Table 4 (continued)

Issue category	Issues identified	State	Policy/Regulation to address the issues	How issue is being addressed in policy/regulation	Remarks
Commercial and implementation issues	Lack of awareness amongst consumers / stakeholders Lack of skilled manpower to manage and drive RSPV industry Insufficient political involvement and support	TN Punjab Maharashtra	TN net metering regulation (2013) preceded by TN solar policy (2012) Punjab Policy on net metering since 2014 preceded by New and Renewable Sources of Energy Policy (NRSE) 2012 Maharashtra Net metering for rooftop systems regulations adopted in 2015	TN introduced a consumer guide for awareness along with 2012 policy, net metering, application steps, technical feasibility and safety requirements described. Facilitation from Nodal agency stated. Separate organisation called, TEDA created for implementing the policy and capacity building. Political support not addressed in policy, however, has mandated government department to install solar rooftops Punjab does not have a separate consumer guide for awareness. Policy provides 220 day timeline from application to site installation, and information on net metering arrangement. Skilled manpower and political support not addressed in policy Maharashtra has issued a general guideline for consumers. Political support not addressed in policy	Consumer awareness appears crucial for spearheading the policy initiatives
Financial issues	High installation cost and need of	TN Punjab Maharashtra	Same as above	TN solar policy provides generation based incentives of Rs. 2 per unit for first 2 years,	Various types of financial mechanisms have been introduced by states. Thus,

(continued)

Table 4 (continued)

Issue category	Issues identified	State	Policy/Regulation to address the issues	How issue is being addressed in policy/regulation	Remarks
	appropriate business models Convoluted subsidies and poorly managed financing options Financial health of distribution utilities			Rs. 1 per unit for next 2 years and Rs. 0.5 for subsequent 2 years TN has provided, apart of GBI mentioned above, exemption from payment of electricity tax for first 5 years, tax concessions, exemption from demand cut, apart from single window clearance by TEDA, nodal agency. Tax incentives to manufactures has been outlined. Financial health of distribution utilities not addressed in policy, but, overall distribution capacity has been defined Punjab has identified PEDA as nodal agency for processing of subsidies/incentives from central/state level. Financial health of distribution utilities not addressed in policy, but, overall distribution capacity has been defined Maharashtra has identified MEDA as nodal agency for processing of subsidies/incentives from central/state	indicating financial issues that may need resolution in the long run too Dedicated agency to deal with financial incentives has also been created

(continued)

Table 4 (continued)

Issue category	Issues identified	State	Policy/Regulation to address the issues	How issue is being addressed in policy/regulation	Remarks
				level. Financial health of distribution utilities not addressed in policy, but, overall distribution capacity has been defined	
Technical issues	Quality and efficiency of solar PVs Connectivity issues at low voltage and poor distribution grid/infrastructure Forecasting and grid balancing issues	TN Punjab Maharashtra	Same as above	TN has provided a backdrop of many solar technologies in the policy. It promotes new plant and machinery approved by MNRE or as per plant load factor approved by test centres. Net metering solar PV system size as per grid has been enlisted. Forecasting not addressed in policy Punjab has enlisted technical and interconnection requirements have been identified separately for solar PV. Forecasting not addressed in policy Maharashtra has identified rooftop solar PV system, no specific efficiency defined. Threshold limit of Rooftop Solar PV system for given voltage level. Forecasting not addressed in policy	Roof top solar PV accepted across the three states. Few states have identified vendors for eligible customers to choose from. Though, these three states are not amongst them

Source Authors compilation based on policy reviews of the three states and more for self-consumption, than as a method to sale electricity by consumers. Policies have already capped or restricted the overall RSPV capacity to the rated capacity of the distribution transformer of utilities. Thus, acknowledging high solar future potential and in cognisance of helping tackle the financial health of distribution companies. Lack of roof top availability has not been addressed by these states, thus, technical issues like forecasting and grid balancing has also remained untouched.

A review of few other policies, for example for the state of Haryana, new aspects pertaining to residential RSPV segment and its scalability has also come for forefront. Haryana has prescribed minimum area limits of 500 square yards for mandatory RSPV installation while giving incentive to existing building. This may bring forth the aspects of minimum roof top requirements in area terms as well. It may be inferred that RSPV installation cannot be looked at single lens for all residential category and that policies must specifically address rooftop area needs. These aspects can be included at the national level in the mid-term corrections requirements of JNNSM too and also for various states to adopt. Though the study is limited in covering these area based aspects and planning standards that may be covered at state and municipal levels and other related technological and construction parameters. These maybe examined in detail in future research.

Thus, based on above analysis it can be concluded that while solar/RSPV policies and regulations address some barriers, other challenges to a greater extent depend on efficient functioning of entire power sector and requires cohesive interventions at sector level. For example, charging tariff that is reflective of actual cost of supply, which would make economics of RSPV more sustainable. However, this is another area that requires investigation at implementation level, and remains a research area to be explored in future.

6 Limitations of the Study

The study is largely based on secondary sources looking at comparative policy review of top three states and primary discussions with few key stakeholders. One of the key stakeholders are the consumers who have installed RSPV systems. Challenges being faced at the consumer end for implementation of solar at residential sector, especially towards installation, commercial and financial issues, and regulatory provisions remain as yet to be identified. Thus, in future these aspects may be covered as a research area.

7 Conclusions and Way Forward

From the above analysis it is evident that while many measures are being taken to sustain the RSPV power provisioning across states, and a policy path has been laid, many issues still need to be addressed. For scaling up RSPV initiatives, and to reach to a national goal of 40 GW for RSPV sector, the national and state machineries in tandem may need to spearhead many measures covering policy aspects on implementation and monitoring of these schemes. Majority states have adopted self-consumption through net metering and largely based on PV technologies, making it a tried and tested model. Excess power is either not paid or is paid at average power procurement cost or can be banked for a limited period. States have adopted measures to reduce the negative influence of sale of electricity though RSPV that may affect the businesses of distribution utilities, and consequently may be resisted by them. The top most three states studied have identified the overall distribution capacity limit, in words means promotion of self-consumption over sale for the residential sector.

Planning aspects, such as planning standards at national, state and municipal levels may have to be introduced as mid-term correction to JNNSM. Also, other challenges in power sector that impact the solar capacity development like the average cost of supply being higher than tariff charged to consumer may have to relooked in the long run for making the economics of RSPV more sustainable. Challenges being faced at the consumer end where RSPV has been installed need to be identified through primary research.

Given the nascent stage of RSPV industry, this analysis may be utilised by various states, policy makers, manufactures and distributors for consistent provisioning of RSPV. In all, the ending thought is a positive and sustainable growth model, largely self-consumption driven for the residential sector that may lead to creation of sustainable infrastructure.

Annexure 1

See Table 5.

Table 5 List of semi-structured interviews conducted

Organization type	Persons interviewed
Government/Ministry	5
Project consultant	1
Developer	1
Total	7

Annexure 2

See Table 6.

Table 6 Summary of solar policies of some key states

Tamil Nadu(TN), stands at 1st position, has 14% of all India rooftop installed capacity
1. Proposed to generate 3000 MW of solar energy by 2015—1500 MW utility scale; 350 MW solar rooftops and 1150 MW under REC mechanism 2. Solar purchase obligation (SPO) along with net metering: For certain categories of consumers (like HT and LT commercial) it mandates SPO, starting with 3% till December 2013 and 6% from January 2014 3. Domestic rooftop generation based incentives (GBI): GBI of INR 2 per unit for first two years, INR 1 per unit for next two years and INR 0.5 per unit for subsequent 2 years will be provided for all solar or solar-wind hybrid rooftops being installed before 31 March 2014 4. Tariff based competitive bidding for procurement, as well as REC and carbon credit benefits 5. Exemption from payment of electricity tax to 100% on electricity generated, for 5 years
Punjab, 2nd position, 12% 1. Targets 1000 MW of solar power generation by 2022, net metering allowed, cover 3rd party 2. Technical assistance: Technical assistance provided for grid interfacing and power wheeling 3. PV projects exempted from obtaining any NOC/consent under pollution control laws 4. For consumer, maximum capacity of RSPV to be 80% of sanctioned connected load 5. Technical specification for equipment's, energy meters specified, with o&m procedures 6. Solar Renewable Purchase Obligation (SRPO): Quantum of electricity consumed from RSPV system by consumers shall qualify towards compliance of SRPO for the distribution licensee
Maharashtra, 3rd position, 11% 1. **SRPO**: Regulation allows, quantum of electricity consumed from RSPV by consumers shall qualify towards compliance of SRPO for the distribution licensee 2. Detailed technical specification for net-metering arrangement with energy accounting and settlement mechanism 3. Regulation lays procedure for application and registration for installing RSPV system, with renewable energy certificate mechanism
Karnataka, 4th position, 8% 1. Targets 3% of projected power consumption i.e. minimum 2000 MW, solar power capacity by 2021. It constitutes 1600 MW large utility scale and 400 MW rooftop solar 2. Both net-metering as well as gross-metering arrangements are promoted 3. State government promotes energy-efficient design standards with options like grid tied building integrated PV (BIPV) based building architecture 4. State government contemplates to amend building by laws in respect of floor area ratio (FAR) in co-ordination with local government and urban development department to exempt FAR
Haryana, 4th position, 8% 1. **Mandatory installation of Solar Power Plant (SPP)** of 3–5% of connected load, else provision of penalty. Extended to select categories of connected load, an extract below: • New Residential buildings having 500 square yards and above • Existing residential buildings to be provided with financial incentives to promote RSPV • All government and private educational institutes with 30 kW and above • All private hospitals, commercial and industrial establishments with 50 kW and above

(continued)

Table 6 (continued)

Tamil Nadu(TN), stands at 1st position, has 14% of all India rooftop installed capacity
• All housing complexes with plot size of 0.5 acre and above 2. **Net metering for captive use** has been incentivised by Regulator, adjusted at 90% of the electricity consumption, during 1 year, with financial incentive of 25 paisa per unit in their bills on the solar power generated 3. **Use of rooftop of Government buildings and vacant Land on lease/rent** basis to IPPs/ RESCO[a] for setting up solar power projects 4. **Expenses for power evacuation** shall be borne by the State transmission utility for Extra High Voltage (EHV)/High Voltage (HV) transmission line up to a distance of 10 km from inter connection point 5. **Waiving off electricity related charges**: All electricity taxes and cess, electricity duty, wheeling charges, cross subsidy charges, Transmission and distribution charges and surcharges will be totally waived off for roof top solar power projects **6. Price preference shall be given to IPPs** **7. Banking of solar power generated by eligible producers,** utilised within a year from date of banking power, and not to be utilised during peak hour loadings **8. Exemptions for solar power projects from many departments on:** • Land use approvals, external development charges scrutiny fee and infrastructure development charges from TCPO[b] till valid purchase power agreement • Forest dept. and environment clearances from Haryana Pollution Control Board • 100% exemption from payment of fee and stamp duty charges for registration of rent/lease deed for the land required
Gujarat, 5th position, 6% 1. **State capital subsidy** of Rs. 10,000 per kW, given to private residential consumer, after successful installation, up to Rs. 20,000 per consumer, first 100,000 applicants to be considered initially. Installation limited up to 2 kW for scheme purposes, while sanctioned load/capacity of plant can go up to 100% of contracted load 2. **Enabling provision for public awareness by listing Empanelled Agencies (EA)** for residential rooftop solar PV systems. EA's role to design the system with all accessories and equipment, supply, Install, Commission and maintain it for 5 years with free replacement of guaranteed parts against manufacturing defects. EA will obtain all necessary approvals
Uttar Pradesh, 5th Position, 6% 1. **Promoting Self consumption/captive use**: Installation of Grid connected Rooftop Solar Photovoltaic will be based on Net Energy Metering /Net Energy Billing method 2. **Mandatory are reservation for Government institutions**: By Government/public institutions at least 25% area of the chair area of its office building for RSPV 3. **Nodal Agency earmarked**, to facilitate in eligible entities in availing subsidy or benefit from Central/State Government: Uttar Pradesh New and Renewable Energy Development Agency shall be the Nodal Agency 4. **Empanelment of system integrators**: Nodal Agency shall empanel system integrator(s) for the implementation of target capacity
Kerala, 5th Position, 6% 1. **Solar Procurement Obligation (SPO)**: Mandatory for all Commercial consumers having above 20 kVA connected load, Industrial users with above 50 kVA connected load • All HT/EHT consumers and high consuming domestic consumers with more than 500 units per month shall have to procure 0.25% of their energy consumed through SPO and clause with 10% increase per annum

(continued)

Table 6 (continued)

Tamil Nadu(TN), stands at 1st position, has 14% of all India rooftop installed capacity
2. **Floor Area considerations:** Mandatory for all: • New domestic buildings with floor area in between 2000 and 3000 ft^2 to install 500 W solar PV system • Buildings above 3000 ft^2 to install at least 1000 W solar PV system 3. **Mandatory solar power for common amenities in residences**: All residential flats/ apartments shall use 5% of the energy for common amenities 4. **No open access charges:** for wheeling the power within the State 5. **Exemption of electricity Duty**: Full exemption from the Electricity duty 6. **Banking facility**: Conditional Banking facility shall be available to captive generators after considering system constraints 7. **Facilitating for subsides from MNRE**: Nodal agency, i.e. ANERT, shall act as a facilitator for the developer for making available the subsidy from MNRE or any other central agency and act as single window 8. **System provider/integrator**: Tri-partite agreement involving also the facility owner of the roof top solar plant and KSEB, ensuring continued technical support to the plant, conduct periodical maintenance to the plant as per the standards[c]
NCT Delhi, 5th Position, 6%
1. **Mandatory solar PV** installation in govt. buildings for area above 500 m^2 2. **Virtual net metering** encouraged for consumers who do not access to suitable roof or are in group housing or living in apartments. This is in addition to net metering for residential consumers and group net metering for consumers who are unable to utilise all energy produced on their rooftops 3. **Generation based incentives (GBI)**: GBI is encouraged to reduce payback time, rate of INR 2 per unit of gross solar energy generated 4. **Other exemptions**: exemption from payment of electricity tax and cess, open access charges, conversion charges, wheeling banking, transmission charges, cross subsidy charges 5. **Building byelaw amendments**: height of module structure shall be deducted from total height of building under planning byelaws • Approval exemptions from Municipalities and other urban development bodies 6. **Facilitation from State's Nodal agency**: includes help in getting clearances/approvals/ availing subsidies from different State agencies, maintaining a green fund to disburse GBI, capacity building and building awareness

Sources Authors compilation based on solar policies of respective states
[a]IPP – Independent Power Producers, RESCO = Renewable Energy Service Company
[b]TCPO = Town and Country Planning Department
[c]Kerala was one of the early State to release a solar related policy. Thus, it recognised that system integrators were required as technology was new, building small investors' confidence and adopting its technology is required during initial stages

References

Ayog, N. (2015). *Report of the Expert Group on 175 GW RE by 2022*. Available at: http://niti.gov.in/writereaddata/files/writereaddata/files/document_publication/report-175-GW-RE.pdf. Last accessed April 7, 2017.

BMWi. (2015). *Energy Efficiency strategy for buildings*. Federal Ministry of Economic Affairs and Energy. Available at: https://www.bmwi.de/Redaktion/EN/Artikel/Energy/energy-efficiency-strategy-forbuildings.html. Last accessed April 7, 2017.

Bridge to India. (2015). India Solar Handbook 2015.

Bridge to India. (2016). India Solar Rooftop Map 2016.

Care Ratings. (2015). *Solar rooftop: Opportunities and challenges*. Available at: http://www.careratings.com/upload/NewsFiles/SplAnalysis/Solar%20rooftop%20opportunities%20and%20challenges.pdf. Last accessed April 7, 2017.

Chattopadhyay, D. (2014). Modelling renewable energy impact on the electricity market in India. *Renewable and Sustainable Energy Reviews, 31,* 9–22.

Cogno, E., Worrell, E., Trianni, A., & Pugliese, G. (2013). A novel approach for barriers to industrial energy efficiency. *Renewable and Sustainable Energy Reviews, 19*(2013), 290–308.

COP21. (2015). *India and France Launch International Solar Energy Alliance at COP21*. Available At: http://newsroom.unfccc.int/clean-energy/international-solar-energy-alliance-launched-at-cop21/. Last accessed April 7, 2017.

COP21. *IEA at COP21*. Available at: http://www.iea.org/cop21. Last accessed April 7, 2017.

Fleiter, T., Schleich, J., & Ravivanpong, P. (2007). Adoption of energy-efficiency measures in SMEs—An empirical analysis based on energy audit data from Germany. *Energy Policy, 51* (2012), 863–875.

Gambhir, A., Dixit, S., Toro, V., & Singh, V. (2012). *Solar Rooftop PV in India. Prayas Energy Group*. Available at: http://www.prayaspune.org/peg/media/k2/attachments/Solar_Rooftop_PV_in_India.pdf. Last accessed April 7, 2017.

Goel, M. (2016). *Solar rooftop in India: Policies, challenges and outlook.*

Gosh, S., Nair, A., & Krishnan, S. S. (2015). Techno-economic review of rooftop photovoltaic systems: Case studies of industrial, residential and off-grid rooftops in Bangalore, Karnataka. *Renewable and Sustainable Energy Reviews, 42*(2015), 1132–1142.

GTF. (2015). *Progress towards Sustainable Energy 2015*. Global Tracking Framework Report. Available at: http://www.se4all.org/sites/default/files/l/2013/09/GTF-2015-Full-Report.pdf.

Hairat, M. K., & Gosh, S. (2017). 100 GW solar power in India by 2022—A critical review. *Renewable and Sustainable Energy Reviews, 73*(2017), 1041–1050.

IEA. (2014). Technology roadmap: Solar photovoltaic energy 2014 International Energy Agency (IEA).

IEA. (2015). Trends 2015 in photovoltaic applications Survey Report of Selected IEA Countries between 1992 and 2014, Report IEA-PVPS T1-27:2015.

IEA. (2016). Snapshot of global photovoltaic markets, report IEA PVPS T1-29:2016.

Kapoor K. K., Dwivedi Y. K. 2017 A take on solar power in India. *Economies and Political Weekly* Feb. 18, 2017 Vol LII No. 7 21-24

Kapoor, K., Pandey, K. K., Jain, A. K., & Nandan, A. (2014). Evolution of solar energy in India: A review. *Renewable and Sustainable Energy Reviews, 40*(2014), 475–487.

Kappagantu, R., Daniel, S. A., & Venkatesh, M. (2015). Analysis of rooftop solar PV system implementation barriers in Puducherry smart grid pilot project. *Procedia Technology, 21* (2015), 490–497.

Kar, S. K., Sharma, A., & Roy, B. (2016). Solar energy market development in India. *Renewable and Sustainable Energy Reviews, 62*(2016), 121–133.

Magal, A., Engelmeier, T., Mathew, G., Gambhir, A., Dixit, S., Kulkarni, A., et al. (2014). Grid integration of distributed solar photovoltaic (PV) in India A review of technical aspects, best practices and the way forward. A Prayas (Energy Group) Report 2016.

Manju, S., & Sagar, N. (2017). Progressing towards the development of sustainable energy: A Critical Review of the current status. Applications, developmental barriers and prospects of solar photovoltaic systems in India. *Renewable and Sustainable Energy Reviews, 70*(2017), 298–313.

Martin, S., & Ryor, J. N. (2016). *Prosumers in Bengaluru: Lessons for scaling rooftop solar PV.* World Resource Institute Working Paper May 2016.

Matschoss, P. (2013). *The German energy transition, status, challenges, and the finnish perspective*. Available at: www.fiia.fi/assets/publications/bp128.pdf. Last accessed April 7, 2017.

MNRE (2012). *Jawaharlal Nehru National Solar Mission, Phase-ii, Policy Document*. Available at: http://mnre.gov.in/file-manager/UserFiles/draft-jnnsmpd-2.pdf. Last accessed April 7, 2017.

MNRE. (2017). Based on data received from MNRE official.

Narula, K., & Reddy, B. S. (2015). *Will net metering model for residential rooftop solar PV projects work in Delhi? A financial analysis*. http://www.ijrer.org/ijrer/index.php/ijrer/article/viewFile/1970/6586. Last accessed April 7, 2017.

Pollitt, M. (2012). The role of policy in energy transitions: Lessons from the energy liberalisation era. *Energy Policy, 50*(11), 128–137.

REN 21. (2016a). *Renewables 2016, Global Status report*. Available at: http://www.ren21.net/wp-content/uploads/2016/06/GSR_2016_Full_Report.pdf. Last accessed April 7, 2017.

REN 21. (2016b). *Renewables 2016, Global status report, key findings*. Available at: http://www.ren21.net/wp-content/uploads/2016/06/GSR_2016_Key_Findings.pdf. Last accessed April 7, 2017.

Richts, C., Strauss, P., & Heinemann, D. (2015). Report on forecasting, concept of renewable energy management centres and grid balancing. Indo-German Energy Programme—Green Energy Corridors, GIZ GmbH.

Rohankar, N., Jain, A. K., Nangia, O. P., & Dwivedi, P. (2016). A study of existing solar power policy framework in India for viability of the solar projects perspective. *Renewable and Sustainable Energy Reviews, 56*(2016), 510–518.

Schmid, G. (2011). *The development of renewable energy power in India: Which policies have been effective?* Available at: http://www.unige.ch/ses/dsec/repec/files/11103.pdf. Last accessed April 7, 2017.

Sharda, J. (2015). *India's solar ambitions—Challenges and options*. Available at: www.equitorials.com.

Sorrell, S., O'Malley, E., Schleich, J., & Scott, S. (2004). *The economics of energy efficiency*. Cheltenham, UK: Edward Elgar Publishing.

Sukh, N., & Mandavilli, R. (2016). Working Paper on financing the solar photovoltaic (PV) rooftop revolution in India. A report by YES Bank Ltd.

Sundaray, S., Mann, L., Bhattacharjee, U., Garud, S., & Tripathi, A. K. (2014). *Reaching the sun with rooftop solar*. Available at: http://mnre.gov.in/file-manager/UserFiles/Rooftop-SPV-White-Paper-low.pdf. Last accessed April 7, 2017.

Thakur, J., & Chakraborty, B. (2016). Sustainable net metering model for diversified India. *Energy Procedia, 88*(2016), 336–340.

Timilsina, G. R., Kurdgelashvili, L., & Narbel, P. A. (2011). *A review of solar energy*. Markets, Economics and Policies. Available at: http://documents.worldbank.org/curated/en/546091468178728029/pdf/WPS5845.pdf. Last accessed April 7, 2017.

UKAID. (2016). Solar rooftop policy coalition unleashing private investment in rooftop solar in India.

UNNFCC. (2017). *India and France Launch International Solar Energy Alliance at COP21*. Available at: http://newsroom.unfccc.int/clean-energy/international-solar-energy-alliance-launched-at-cop21/.

Verma, R., Hernandez, D. D., Sivaram, V., & Rai, V. (2016). A national certification scheme to enhance trust and quality in residential solar PV market. *The Electricity Journal, 29*(2016), 11–14.

Yenneti, K. (2016). The grid connected solar energy in India: Structures and challenges. *Energy Strategy Reviews, 11–12*(2016), 41–51.

Sustainable Subdivision Design and Energy Consumption of Households in the Hot and Humid Tropical Climate of Darwin

S. Safarova, S. T. Garnett, E. Halawa, J. B. Trombley, L. Law and J. van Hoof

Abstract This paper examines the relationship between sustainable subdivision design principals, local microclimate, and household electricity consumption. The energy-efficient subdivision design principals, energy consumption, and adjustment behaviour of 36 households were investigated in two suburbs of the city of Darwin. The participating households completed a questionnaire on family structure, thermal preferences, and behavioural adjustment to the indoor environment. Electricity consumption of households was recorded at 30-min intervals from Nov 2015 to Aug 2016. The Muirhead suburb, designed with energy-efficiency and climate-responsive design principals, has 16.9 lots per ha in dense areas and a minimum lot area of 450 m^2. Another suburb, Lyons, has 14.4 lots per ha and a minimum lot area of 525 m^2. Households were divided by lot areas into three categories: category 1 (450–610 m^2), category 2 (611–710 m^2), and category 3 (>710 m^2). In Muirhead, the average daily consumption of category one household in the warmer wet season was 98 Wh/m^2 per person compared to 154 Wh/m^2pp in Lyons. In the cooler, dry season 48 Wh/m^2pp in Muirhead and 87 Wh/m^2pp in Lyons. The cooling load calculated using multiple regression analysis showed that the main difference in energy consumption between two suburbs was in the base load. Thus, the energy-efficiency and climate-responsive design principals, applied to the subdivision, mitigate the impact of urban heat on cooling energy consumption in the hot and humid climate of Darwin. These are preliminary results and further investigation of the factors that have an impact on energy consumption of participating households is continuing.

S. Safarova (✉) · S. T. Garnett · E. Halawa · J. B. Trombley
Centre for Renewable Energy, Research Institute for the Environment and Livelihoods, Charles Darwin University, Ellengovan Drive, Casuarina, DARWIN 0909, Australia
e-mail: shokhida.safarova@cdu.edu.au

L. Law
Environment Geography and Sustainability Group, College of Marine and Environmental Sciences, James Cook University, 6811, Cairns, QLD 4870, Australia

J. van Hoof
Fontys University of Applied Sciences, Dominee Theodor Fliednestraat 2, 5631 BN Eindhoven, The Netherlands

W. Leal Filho et al. (eds.), *Sustainable Development Research in the Asia-Pacific Region*, World Sustainability Series, https://doi.org/10.1007/978-3-319-73293-0_24

Keywords Sustainable design · Energy efficient building · Subdivision design

1 Introduction: Energy Efficiency of Households and Urban Density

One of the objectives of the Australia's National Strategy for Ecologically Sustainable Development is to achieve more sustainable use of energy in urban areas through the efficient subdivision and building design and promotion of research aimed at improvement of the residential development standards related to energy efficient sustainable urban form (Department of Environment and Energy 1992). One of the design concepts of sustainable urban form is density, the ratio of people or dwelling units to the land area. High density can minimize energy consumption and costs of resource transportation (Jabareen 2006). An increase of urban density and reduction of energy demand are two of the more challenging tasks that modern urban planners have to deal with in hot and humid tropical cities. An increase in air temperature due to the urban heat island effect can result in an increase in demand for cooling energy. There are many factors that have to be considered by urban planners in order to mitigate the effect of urban heat (Crawford et al. 2009). Previous research showed that the important mitigation factors are shading of urban structures and provision of unobstructed wind flow (Lin et al. 2010; Lin 2016; Ng 2016). The wind plays important role in the dissipation of heat in the urban environment (Ng 2016). Therefore, tropical urban cities should be designed with parallel streets aligned to the prevailing breezes, with parks and open spaces connected by breezeways that allow the unobstructed flow from the windward to the leeward side of a city. The wind flow also provides natural ventilation in buildings designed for passive cooling and therefore might influence the cooling energy demand (Baker 1987; Givoni 1994).

The goal of achieving energy efficient development should be supported by the implementation mechanisms outlined in planning and building regulations. There are two views how the energy-efficiency of developments can be achieved through planning regulations. One highlights the necessity of an introduction of development assessment tools, while the other suggests the introduction of guidance on the climate-responsive design of subdivisions into planning policies (Colia and March 2012; Clark 2001; Berke and Conroy 2000). Both approaches require feedback from case studies that provide a better understanding of the relationship between subdivision design, climatic conditions, and household energy consumption. The real life data on household energy consumption is becoming vital in the wake of global warming and widening of the tropical belt (Seidel and Rendel 2007). The study of the relationship between energy-efficient subdivision design, thermal preferences of occupants, and their preferred methods to adjust the indoor conditions can provide the basis for guidelines and forecasting tools (Feriadi and Wong 2004; Hwang et al. 2009). The aim of this study is to provide a case study of subdivision design aimed to employ prevailing breezes and mitigate the effect of urban heat in the hot and humid tropical climate and develop recommendations for tropical subdivision design guidelines. The main objective of this paper is to investigate the impact of hot and humid climatic conditions on cooling energy consumption of households living in suburbs that differ in density and subdivision design.

2 Case Study

The impact of increased density and subdivision design on household energy consumption was investigated in the tropical city of Darwin, Australia. The Northern Territory Planning Scheme has minimum lot size requirements outlined for different residential zones. The requirement for lot size in single dwelling (SD) zone is not less than 800 m^2, while for SD in greenfield areas it is no less than 450 m^2. The planning scheme outlines that the lot configuration and orientation should allow a dwelling to take advantage of prevailing breezes. The minimum side and rear setback for buildings in single dwelling zones is 1.5 m from the wall to the lot boundary. However, the subdivision design guidelines do not include examples of lot orientation or configuration, street orientation or the layout of street blocks that would provide better wind flow.

The research was undertaken in the build-up, the wet and the dry seasons. The build-up is pre-monsoonal period followed by the wet season and characterised by light winds, isolated showers, and thunderstorms. In total 36 households from two neighbouring suburbs (Muirhead and Lyons) participated in this study. Muirhead and Lyons are both greenfield developments. Muirhead was designed as a higher density neighbourhood with 16.9 lots per ha and a minimum lot area of 450 m^2. The subdivision has parallel streets aligned to the prevailing breezes in north-wet/south-east or north-east/south/west directions (Fig. 1). The majority of lots have greater depth than width. The guidelines for Muirhead require a 4.5 m side setback between lot boundaries and buildings, 80% permeable fencing, and deciduous vegetation to provide unobstructed flow for breezes. Another requirement outlined in guidelines is minimum 6 star house energy rating. Parallel roads, local parks, and natural areas of open space are planned to provide breeze paths. Muirhead was still in the development stage when this research has started and had ongoing construction works. Lyons was designed with the continuous line of houses built along the main road on the east side in (Fig. 1). Lyons has 14.4 lots per hectare with a 525 m^2 minimum lot area. The majority of construction works have been completed in Lyons by 2011.

Participating households were divided into three categories by lot area: category 1 (450–610 m^2), category 2 (611–710 m^2), and category 3 (>710 m^2). The characteristics of participating households are presented in Table 1.

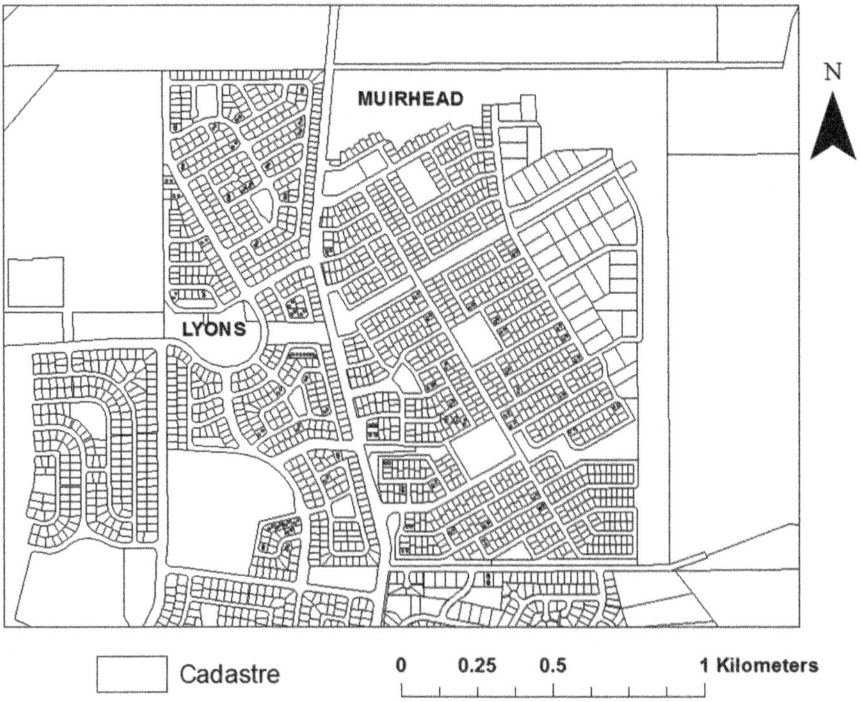

Fig. 1 Cadastre Map of Muirhead and Lyons suburbs

Table 1 Characteristics of participating households

Lot category	Number of households		Average conditioned area (m^2)		Average number of household members	
	M	L	M	L	M	L
I (450–610 m^2)	10	2	114	156	3	2
II (611–710 m^2)	7	9	137	149	4	3
III (>710 m^2)	1	7	149	134	4	4

M—Muirhead, L—Lyons

In Muirhead, 16 participating households lived in a single storey, heavyweight construction houses and two households lived in lightweight construction houses built on concrete slabs on the ground. The majority of lots orientated with the long axis in north-northwest direction. In Lyons, 16 participating households lived in a single storey, heavyweight construction houses, one household in a two-storey house with a ground floor of heavyweight construction and a lightweight construction for the upper level and one household lived in an elevated house of lightweight construction. All habitable rooms in houses had air-conditioning installed. The majority of houses had split system air-conditioners installed. All participating households had solar water heaters installed. In Muirhead, only two of the participating households had swimming pools compared to five households with pools in Lyons.

3 Methods

A survey questionnaire was developed to investigate the occupants' preferred method of climate control of the indoor environment. Surveys of participants in the 'build-up', wet and dry seasons were conducted in living rooms of houses in Muirhead and Lyons. Participants were asked to choose the preferred adjustment from the list, such as operation frequency of air-conditioners and fans or opening windows and doors.

A weather station was installed in Muirhead to collect half-hourly air temperature, humidity, wind speed, and wind direction data from January 2016 to January 2017. The wind monitor was installed 7 m above the ground. Half-hourly energy consumption data for participating households was collected from the Power and Water Corporation from December 2015 to January 2017. The average energy consumption of participating households in each suburb was used for comparative analysis. The half-hourly data was aggregated to calculate the daily average household energy consumption in each suburb. The average daily energy consumption of households per conditioned square meter and per household member was used for comparative analysis within the different lot categories. A multivariate regression analysis was applied to determine the base load and cooling energy consumption of the average household in each suburb.

4 Results and Discussion

4.1 Weather Data

The weather data collected from the weather station in Muirhead showed that monthly mean temperature was between 28–30 °C all year except the period from June to August when mean temperature was 24.7–27.3 °C with the minimum monthly mean temperature of 24.7 °C recorded in July 2016 (Fig. 2a). The maximum monthly mean temperature was recorded in March and November 2016 at 29.9 °C. The monthly maximum temperatures were around 35 °C from January to April and from October to December. These results correlate with the Bureau of Meteorology report for Darwin stating that record warm temperatures were in several individual months including January, February, March, and April in 2016. The monthly mean relative humidity reached its maximum 80% in the wet season,

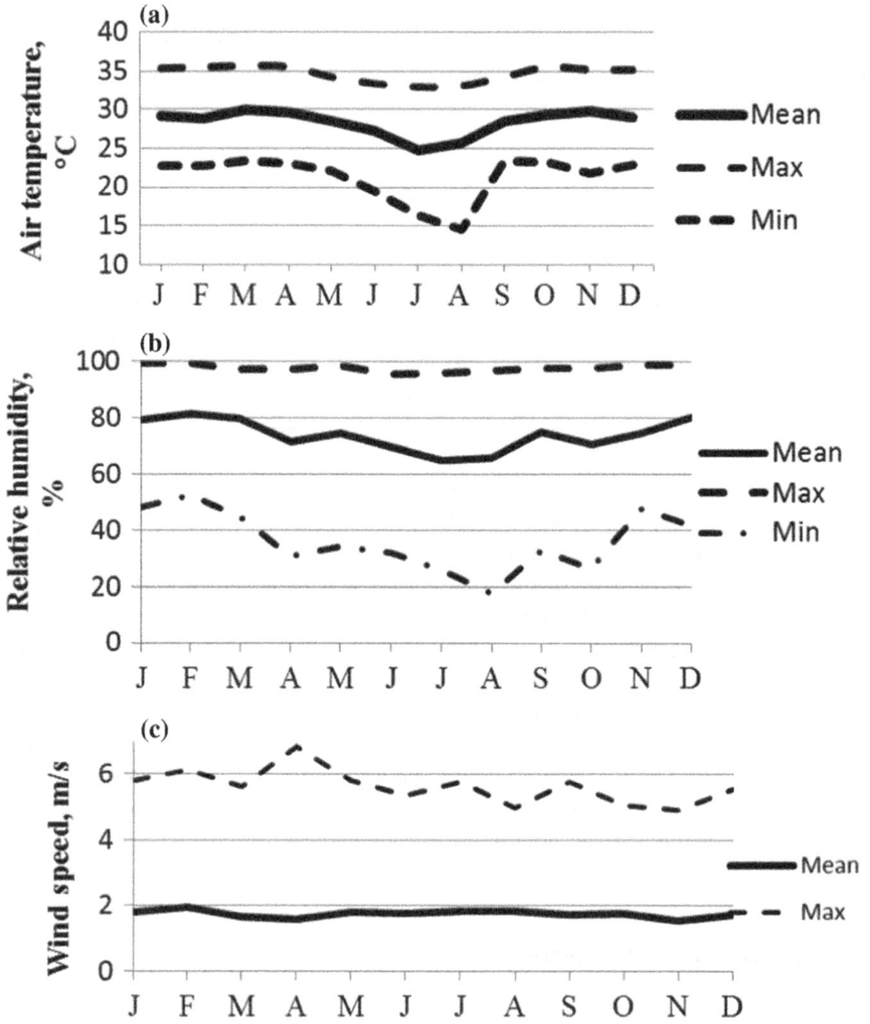

Fig. 2 Weather data collected from the reference station in Muirhead: **a** air temperature, **b** relative humidity, **c** wind speed

February and December 2016 (Fig. 2b). However, the monthly mean relative humidity stayed above 60% throughout the year. The monthly mean wind speed remained below 2 m/s during the whole year. The maximum wind speed was 7 m/s (Fig. 2c). The direction of much of the wind within the 2–4 m/s speed range was between north-northwest and north-northeast. The high percentage of breezes from north-east and south-east was recorded in the dry season.

4.2 Survey

The survey analysis showed that the total percentage of votes for 'always' and 'often' open windows was the highest in the dry season (94% votes in Lyons and 88% votes in Muirhead). In Muirhead, the percentage of respondents who preferred to open windows 'always' decreased from 35% in the build-up to 21% in the wet, and increased to 44% in the dry season. In Lyons, the percentage of votes for 'always' open windows decreased from 50% in the build-up to 45% in the wet season and increased to 61% in the dry season. Overall the percentage of respondents who preferred to open doors 'always' or 'often' increased from 60% in the build-up to 100% in the dry season in both suburbs. The higher percentage of participants using windows and doors in the dry season can be explained by lower air temperature and humidity compared to the wet season and the build-up (Fig. 2a), and north-easterly and south-easterly prevailing breezes. The air temperature in July 2016 (dry season) was 20.3–30 °C compared to 26.8–33 °C in November 2016 (build-up). The lower percentage of respondents that preferred to have open windows and doors during the build-up and the wet season in Muirhead was due to several reasons. First, the respondents complained about the amount of dust in the air from construction works in the suburb. Second, the majority of respondents in Muirhead were 25–44 years old and were dissatisfied with air temperature above 28 °C compared to respondents in Lyons who in the majority were over 45 years old and were dissatisfied with temperatures above 30 °C. The participants in Lyons, who preferred not to open windows but used open doors, complained about the short distance between the neighbouring houses and mentioned that most of the time in the build-up and the wet season the air movement through windows was not enough for thermal comfort. Some of the participants preferred to have closed windows and doors in the wet season to avoid humid air entering the room and instead run fans.

The fans were observed in use in the majority of houses in both suburbs during the build-up and the wet season because respondents were not able to attain thermal comfort through natural ventilation alone. It correlated well with the survey results that showed the operation of fans as the most preferred adjustment method in the build-up and the wet season. In Muirhead 84% and in Lyons 90% of respondents preferred to 'always' operate a fan/fans in the build-up and during the wet season, compared to 50 and 61% in the dry season.

The percentage of Muirhead respondents who preferred to operate air-conditioners 'always' or 'often' increased from 70% in the build-up to 83% in the wet, and decreased to 28% in the dry season. The percentage of Lyons respondents who preferred to operate air-conditioners 'always' or 'often' reduced from 40% in the wet to 17% in the dry season. None of the respondents had the air-conditioner in use at the time of the survey in the dry season. While in the wet season, there were 67% of Muirhead respondents and 40% of Lyons respondents filling out a questionnaire in an air-conditioned room.

4.3 Energy Consumption

The half-hourly energy consumption of households was investigated in both suburbs. The average half-hourly energy consumption was 0.6 kWh in Muirhead and 0.7 kWh in Lyons. The average energy consumption of Muirhead households was below 0.6 kWh from 24:00 to 15:30. After 15:30 energy consumption increased steadily from 0.6 to 0.9 kWh at 18:00, remained at 0.9 kWh until 20:30 then reduced to 0.6 kWh at 23:00 (Fig. 3). The average energy consumption of Lyons households was around 0.7 kWh between 24:00 and 14:00, steadily increased from 0.7 to 1.1 kWh between 16:30 and 19:00, remained at 1.1 kWh until 20:00, and steadily reduced to 0.7 kWh at 23:30. The average energy consumption across all hours in Muirhead was lower than in Lyons except in the afternoon (13.00–13.30) when the consumptions were equal. All months had a similar trend, except February 2016 when Muirhead households consumed in average by 0.04–0.3 kWh more than Lyons households between 9:30 and 17:30 with the maximum difference of 0.3 kWh observed at 13:00.

Weekly profile showed that average energy consumption increased on weekends in Muirhead and Lyons (Fig. 4). The average daily consumption of Muirhead households was by about 10 kWh higher on weekends compared to weekdays. On weekends the average energy consumption of Lyons households increased by 4 kWh. Muirhead households in average consumed by 6–7 kWh less than Lyons households every day of the week. A similar trend was observed across all months.

The annual profile of the average daily energy consumption shows that consumption was lower during June–August 2016 (dry season) than the rest of the year (Fig. 5). The average daily consumption in Muirhead was lower than in Lyons across all months and ranged from 16 kWh in July 2016 to 35 kWh in February 2016. In Lyons, the average daily consumption ranged from 23 kWh in July 2016 to 41 kWh in November 2016. The annual profile shows the connection between survey results for Muirhead participants (83% preferred to always operate air-conditioners in wet season) and an increase in energy consumption in February–March 2016. While survey results in Lyons did not have the connection with energy consumption in the build-up or wet seasons. There were only 40% of participants in

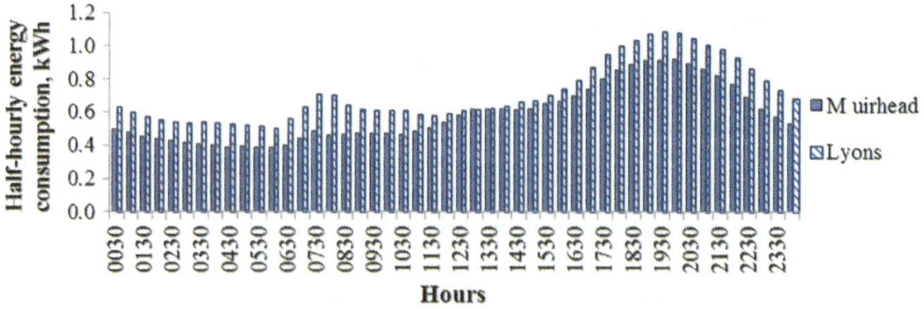

Fig. 3 Half-hourly profile of the average energy consumption

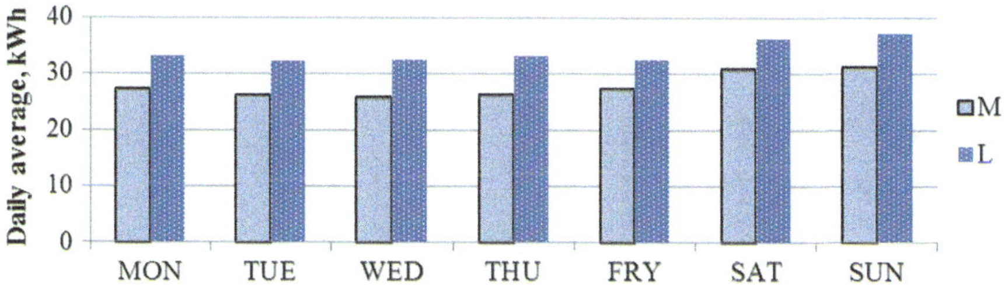

Fig. 4 The weekly profile of the average energy consumption. M—Muirhead, L—Lyons

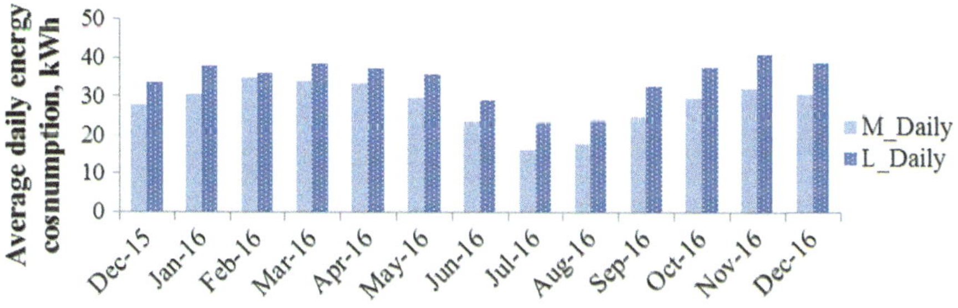

Fig. 5 Annual profile of average daily energy consumption: M—Muirhead, L—Lyons

Lyons who preferred to operate air-conditioners 'always' or 'often' compared to 83% in Muirhead, but average daily energy consumption ranged between 38–41 kWh in Lyons compared to 30–35 kWh in Muirhead during the build-up and the wet seasons. The average daily energy consumption in Muirhead in July was by 7 kWh less than in Lyons.

The annual profile of energy consumption of households in different lot area categories is presented in Fig. 6. The daily energy consumption was investigated per square meter of conditioned area and per household member. The average energy consumption of Muirhead households in the first category was higher than average consumption of Lyons households in the same category in January, February, and April 2016, but lower for the rest of the year. The highest average daily consumption of 98 Wh/m^{2}pp was in April 2016 and the lowest consumption of 48 Wh/m^{2}pp was in July 2016 (Fig. 6a). The average energy consumption of Lyons households ranged from 59 Wh/m^{2}pp in February 2016 to 154 Wh/m^{2}pp in December 2016.

Muirhead households in the second lot category consistently consumed less than Lyons households in the same category. In Muirhead, the highest average daily consumption of 74 Wh/m^{2}pp was in November 2016. In Lyons, highest average energy consumption of 123 Wh/m^{2}pp was in March 2016 (Fig. 6b). The high consumption in March and November 2016 correlates with the maximum mean temperature recorded by the weather station in the same months (Fig. 2a).

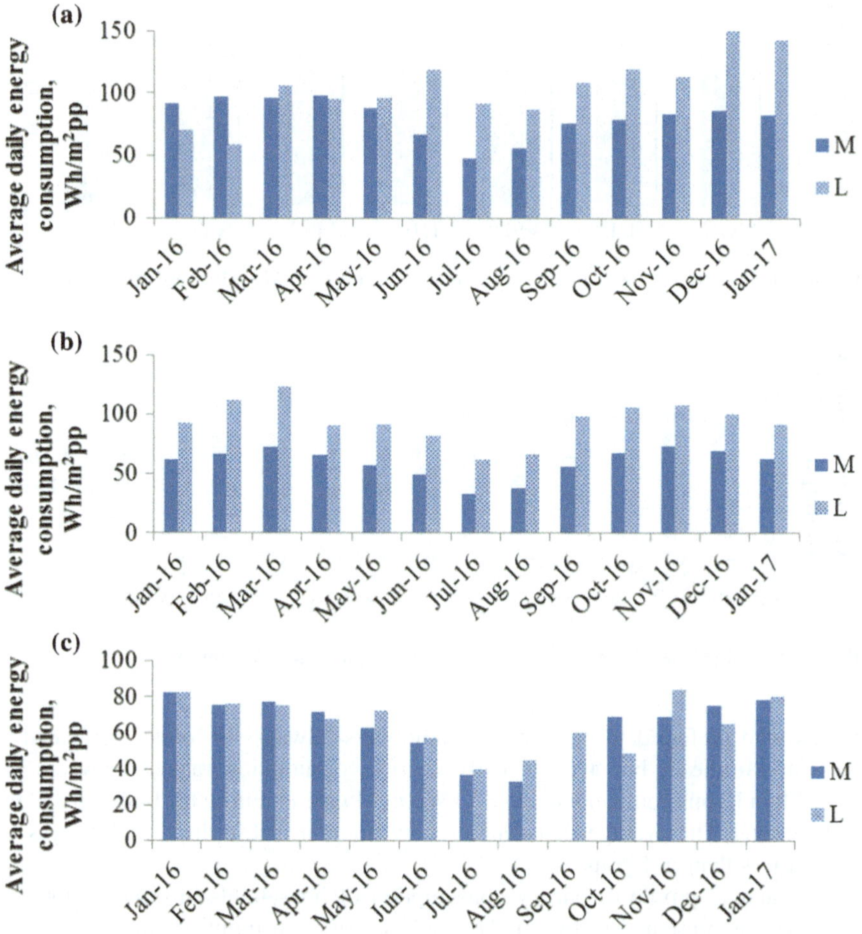

Fig. 6 Annual profile of daily energy consumption of households: **a** lot category 1; **b** lot category 2 and **c** lot category 3. M—Muirhead, L—Lyons

The average energy consumption of the Muirhead household in the third lot category was higher than the average consumption of Lyons households in the same category in March, April, October and December 2016. In Muirhead, the highest average consumption of 82 Wh/m^{2}pp was recorded in January 2016 and the lowest consumption of 33 Wh/m^{2}pp in August 2016. Data for September 2016 consumption of the Muirhead household in this category was not available therefore is not presented in the graph. Lyons seven households in the third category had highest average consumption of 85 Wh/m^{2}pp in November 2016 and lowest consumption of 40 Wh/m^{2}pp in July 2016.

4.4 Cooling Energy Consumption

The difference between annual and daily profiles of household energy consumption in two suburbs showed the necessity to investigate cooling energy alone. The main weather variables that impact energy consumption in the tropical climate are temperature and humidity. Therefore, the cooling energy should include sensible and latent loads. Previous studies on cooling energy are based on cooling degree days (CDD) or cooling degree hours (CDH) that quantify how outside temperature is warmer than the indoor base temperature and the duration of the cooling required (Durmayaz et al. 2000). Total cooling degree hours for a day can be expressed as:

$$CDH = \sum_{i=1}^{24} (T_{out} - T_b)^+ \quad (1)$$

where T_{out} is outdoor temperature (dry bulb temperature) and T_b is the base temperature indoors. In this study, the decision was made to use wet bulb temperature (WBT) as the base temperature to take account for both humidity and temperature (Krese et al. 2012). The correlation analysis of daily average energy consumption and CDH_w was performed for WBT base values of 15, 17, 18, 18.5, 19, and 19.5 °C. The best correlation with the daily energy consumption data for both suburbs was found to be with a CDH_w base of 18 °C, which had a Pearson coefficient 0.812 for Muirhead and 0.802 for Lyons with $p < 0.001$. Therefore CDH_w with the base 18 °C WBT, daily mean wind speed and mean wind direction were included as predictor parameters in multivariate regression analysis of daily average consumption. The results of regression analysis are presented in Tables 2, 3 and 4.

According to regression models, the average daily base load is 14.9 kWh for Muirhead households and 22.1 kWh for Lyons household. Wind speed and wind direction both have a negative relationship with energy consumption at significant p values for both suburbs. The cooling load calculated based on regression models is shown in Table 5. The cooling load of the average household in Muirhead is higher

Table 2 Regression model for daily average energy consumption of Muirhead households

Model	Unstandardized coefficients		Standardized coefficients	t	p	95.0% confidence interval for B	
	B	Std. error	Beta			Lower bound	Upper bound
Base load (kWh)	14.9	0.8		19.062	0.000	13.390	16.472
Ws (m/s)	−0.5	0.2	−0.093	−2.870	0.004	−0.864	−0.161
Wdr (rad. deg.)	−0.7	0.2	−0.111	−2.996	0.003	−1.117	−0.231
CDH_w	0.108	0.004	0.875	24.030	0.000	0.099	0.117

Table 3 Regression model for daily average energy consumption of Lyons households

Model	Unstandardized coefficients		Standardized coefficients	t	p	95.0% confidence interval for B	
	B	Std. error	Beta			Lower bound	Upper bound
Base load (kWh)	22.1	0.7		31.770	0.000	20.761	23.503
Ws (m/s)	−0.33	0.2	−0.070	−2.093	0.037	−0.645	−0.020
Wdr (rad. deg.)	−0.620	0.2	−0.118	−3.101	0.002	−1.014	−0.227
CDH_w	0.093	0.004	0.868	23.329	0.000	0.085	0.101

Table 4 Regression model summary

Model	R	R^2	Adjusted R^2	Std. error of the estimate
Muirhead	0.82	0.67	0.66	3.6
Lyons	0.83	0.68	0.68	4.1

Table 5 Cooling load based on models

Months	Daily cooling load (kWh)	
	Muirhead ±3.6 kWh std. error	Lyons ±4.1 kWh std. error
January	16.1	14
February	17.2	14.8
March	18.3	15.8
April	15.3	13.3
May	13.7	11.9
June	9	7.8
July	2.3	2.1
August	3.9	3.5
September	13.2	11.4
October	13.6	11.8
November	15.8	13.6
December	16.5	14.3

by 2 kWh in most months but this difference is within the standard error of both models.

The Muirhead model describes 67% of the variation in daily energy consumption of the average household. The Lyons model describes 68%. The outputs of the models and observed energy consumption are presented in Fig. 7. Based on data analysis and model output the days with high residuals are mainly weekends, public and school holidays. The model showed that the main difference in household energy consumption in Muirhead and Lyons is due to the base load.

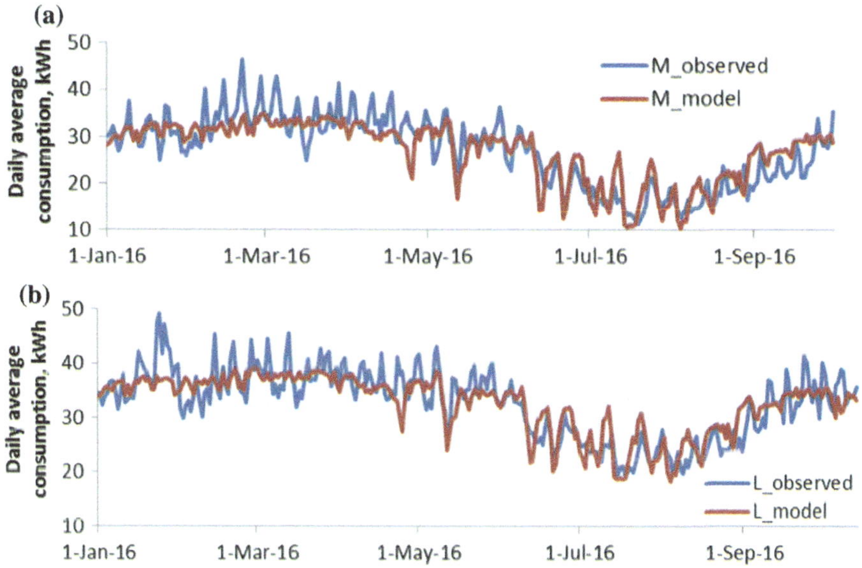

Fig. 7 Daily energy consumption of the average household in Muirhead (**a**) and Lyons (**b**)

5 Conclusion

This study investigated the energy consumption of households in two suburbs of Darwin. The results of this study showed that most of the participating households in both suburbs preferred to use open windows and doors during the dry season. The survey responses correlate well with low energy consumption in July and August 2016. During the build-up and wet seasons, the majority of households in Muirhead preferred to use air-conditioners, which correlates well with an increase in energy consumption during these seasons. However, there was no correlation of the preferences of Lyons households for the build-up and wet seasons with energy consumption for those seasons. The energy consumption of the average household in Lyons was consistently higher than in Muirhead. The regression analysis showed that the difference in energy consumption was due to the base load. Thus, the participating households in two suburbs used in average the same amount of energy on cooling. In conclusion, the subdivision designed with consideration of energy-efficiency and climate–responsive design principals mitigates the impact of urban heat on cooling energy consumption in the hot and humid tropical climate.

This study results showed that the performance criteria for subdivision design and lot configuration and orientation in the local planning scheme should be reviewed. The subdivision guidelines should include climate-responsive design principals that are crucial for the future sustainability of the development in hot and humid tropical cities. The guidelines should include recommendations for connectivity of breezeways, the maximum size of the street block, and the minimum ratio of the floor area to lot area. The guidelines should present the best examples of climate-responsive design in the region. The research and audits of the actual energy performance of urban developments conducted at a larger scale on a regular basis will provide the case studies for guidelines and planning policies and data for energy consumption forecasting tools. Considering the high preference of occupants to use air-conditioners for thermal comfort in the build-up and the wet season further research aimed to improve the thermal performance of the tropical houses is recommended.

This study is subject to several limitations. First, there was limited number of households willing to participate in this study and the survey results are only applicable to the research area. Second, because of limited time, data, and resources the study focused only on the cooling energy consumption and did not research other household energy uses. These are preliminary results and further investigation of the factors that have an impact on energy consumption of participating households is continuing.

References

Baker, N. (1987). *Passive and low energy building design for tropical island climates*. London, UK: Commonwealth Secretariat.

Berke, P. R., & Conroy, M. (2000). Are we planning for sustainable development? *Journal of the American Planning Association, 66*(1), 21–33. https://doi.org/10.1080/01944360008976081.

Clark, M. (2001). Domestic futures and sustainable residential development. *Futures, 33*(10), 817–836.

Collia, C., & March, A. (2012). Urban planning regulations for ecologically sustainable development (ESD) in Victoria: Beyond building controls. *Urban Policy and Research, 30*(2), 105–126.

Crawford, J., Davoudi, S., & Mehmood, A. (Eds.). (2009). *Planning for climate change: Strategies for mitigation and adaptation for spatial planners*. London, UK: Routledge.

Department of Environment and Energy. (1992). National Strategy for Ecologically Sustainable Development. Australian Government 1992.

Durmayaz, A., Kadıoğlu, M., & Şen, Z. (2000). An application of the degree-hours method to estimate the residential heating energy requirement and fuel consumption in Istanbul. *Energy, 25*(12), 1245–1256.

Feriadi, H., & Wong, N. (2004). Thermal comfort for naturally ventilated houses in Indonesia. *Energy and Buildings, 36*(7), 614–626.

Givoni, B. (1994). *Passive low energy cooling of buildings*. Toronto, Canada: Wiley.

Hwang, R., Cheng, M., Lin, T., & Ho, M. (2009). Thermal perceptions, general adaptation methods and occupant's idea about the trade-off between thermal comfort and energy saving in hot–humid regions. *Build and Environment, 44*(6), 1128–1134.

Jabareen, Y. R. (2006). Sustainable urban forms: Their typologies, models, and concepts. *Journal of Planning Education and Research, 26*(1), 38–52.

Krese, G., Prek, M., & Butala, V. (2012). Analysis of building electric energy consumption data using an improved cooling degree day method. *Strojniški Vestnik-Journal of Mechanical Engineering, 58*(2), 107–114.

Lin, T. P. (2016, March). Management of shading and public places. In *Urban Climate Challenges in the Tropics: Rethinking Planning and Design Opportunities* (pp. 49–77).

Lin, T. P., Matzarakis, A., & Hwang, R. L. (2010). Shading effect on long-term outdoor thermal comfort. *Building and Environment, 45*(1), 213–221.
Ng, E. (2016). Urban air ventilation in high-density cities in the tropics. In *Urban climate challenges in the tropics: Rethinking planning and design opportunities* (pp. 79–110).
Seidel, D. J., Randel, W. J. (2007, October). Recent widening of the tropical belt: Evidence from tropopause observations. *Journal of Geophysical Research: Atmospheres, 112*(D20).

Author Biography

Shokhida Safarova Shokhida Safarova completed a Masters of Applied Science majoring in Tropical Urban and Regional Planning, at James Cook University in 2014. Her main research goal is to investigate an impact of modern urban design on the energy performance of residential dwellings. In her master's thesis, she assessed the passive design of residential detached dwellings in the tropical climate and energy efficiency strategy in an Australian policy context. After completing her degree, she was involved in the Tropical Sustainable Design Case Studies Project, which aimed to capture and share the knowledge and best tropical expertise in the built environment. She then continued her studies as a Ph.D. candidate at Charles Darwin University. Her research is focused on energy efficiency assessment of houses and urban design in Darwin.

The Role of Biochar in Sustainable Agriculture, and Climate Change Mitigation for Sustainable Cities

Sandra Rodrigues and Edmund Horan

Abstract Sustainable development is facing a transformation. One solution finding its place as a key strategy in a number of sustainable development goal areas is biochar. Biochar is a geo-engineered bi-product that is similar in appearance to charcoal which differs from charcoal in that the organic biomass waste from which it is created is heated in an enclosed chamber to low temperatures and deprived of oxygen. This paper investigates the latest studies in crop productivity, commercial viabilities, and the socio-economic options of biochar production at a local and international level in the pursuit of sustainability and climate change mitigation. In terms of sustainability the uptake of biochar as a model for sustainable agriculture is steadily gaining momentum due to its agronomic potential for crop yields, organic soil carbon, carbon sequestration and climate change mitigation. The economic viability of biochar production varies markedly around the world. The most promising opportunities for soil carbon sequestration have derived from voluntary carbon markets and government initiatives such as Australia's Emissions Reduction Fund. The scheme promoted the integration of biochar in land use management practices at a farm scale. If biochar is to have a more profound impact on a global scale, then countries need to adopt a sustainable biochar strategy which is universal along with standardised sustainability protocols.

Keywords Sustainable cities · Biochar · Soil productivity · Climate change mitigation

S. Rodrigues (✉) · E. Horan
School of Engineering, RMIT University, 124 La Trobe Street, Melbourne, VIC 3000, Australia
e-mail: s9776089@student.rmit.edu.au

E. Horan
e-mail: edmund.horan@rmit.edu.au

W. Leal Filho et al. (eds.), *Sustainable Development Research in the Asia-Pacific Region*, World Sustainability Series, https://doi.org/10.1007/978-3-319-73293-0_25

1 Introduction: Origins from Terra Preta de Indio to Terra Preta Nova

International standard guidelines define biochar as "a solid material obtained from thermochemical conversion of biomass in an oxygen limited environment" (IBI 2015). Although biochar is similar to charcoal it does differ from charcoal in that biomass such as wood, manure, leaves and so on is heated in an enclosed chamber to regulated low temperatures (<700 °C) and deprived of oxygen (Lehmann and Joseph 2009).

Biochar is being developed worldwide via a range of pyrolysis technologies with two functions in mind. Firstly, as a soil amendment to improve soil productivity, increase carbon sinks, and improve soil water filtration. Secondly as an environmental management strategy to mitigate greenhouse gas emissions through the conversion of biomass into carbon sequestration products such as biochar, bio-oil and syngas (IBI 2015; Lehmann and Joseph 2009).

Similar to Australia, the Amazon has very poor soils of low fertility with the exception of the highly fertile terra preta soils which are high in soil organic matter and nutrients such as nitrogen, phosphorus, potassium and calcium including a high char content resulting from forest fires and in-field burning practices (Sohi et al. 2009).

The terra preta nova geo-engineered soils (biochar) originated from a concept developed by Wim Sombroek who "promoted the idea of developing new dark earths as carbon stores and sinks for intensive cultivation" (Woods and Denevan 2009). The main drive for this new technology is that terra preta soils are gradually losing soil fertility through the loss of essential micro-nutrients; therefore, we need to invest in developing geo-engineered solutions that actually retain nutrient capacity in terra preta soils and other poor soils worldwide (Sombroek et al. 2002).

Anthropogenic black soils are quite predominant in the Amazon region with localized estimates of approximately 50,000 ha of terra preta between the Tapajos and Curua Una rivers, south of Santarem (Woods and Denevan 2009), with some patches known to vary in size from 5 to 300 ha in many places (Sombroek et al. 2002). Terra mulata soils share a synergy with terra preta soils influenced by daily activities and land use from pre-Columbian tribes with settlement in the region ranging from 3000 to 8000 years old (Denevan 1996; Glaser and Birk 2012). Land use practices have also undergone tremendous change following post-Columbian settlement by the Portuguese about five centuries ago beginning with the landing of Pedro Alvares Cabral on Brazilian soil in the year 1500 (Nowell 1936).

What remains of these pre-Columbian people and settlements is still a mystery, and so is their ancient methods of agriculture. What is clear though is that there is a strong association between the formation of terra preta soils and kitchen middens (garbage mounds). Terra preta soils are a result of discarded waste (organic and inorganic) in waste pits situated adjacent to living areas within the settlement (Myers et al. 2003). Over the centuries settlements have been used and reclaimed by different indigenous groups and the waste disposal has accumulated and broken down into soil organic matter.

1.1 Contemporary Indigenous Practices

The Kayapo are a native Amazonian group which practices sustainable agriculture involving the use of a biochar process. Their village of Gorotire is located in the southern area of Para and Northern Mato Grosso in Brazil on the margins of the Rio Fresco, a tributary of the Xingu River (Hecht 2003). The village of Gorotire was occupied by the Kayapo around the 1930s, following displacement from their traditional territory away from European sorcery i.e. disease. According to Hecht (2003, p. 358) oral accounts suggest that "the ancestors of the Gorotire group, lived in one ancestral village until the end of the nineteenth century".

The Kayapo use low intensity "cool fires" to control weeds, reduce biomass in grasslands and forest understories, create fire breaks, stimulate new plant growth, stimulate seed germination, and enhance soil fertility in certain vegetation types (Hecht 2003). The practice of low combustion burning creates one of the main components of soil stability in Amazonian Dark Earths—charcoal of different shapes and sizes. Nutrient additions involve the mulching of palm material in the fields which is afterwards burned; in field cooking hearths: sweet potato, manioc and corn with the ashes spread over the fields when cooled; and crumbled ant and termite mounds as a soil amendment (Hecht 2003).

The Kayapo use a form of mosaic burning, selectively choosing patches to be burned. Determining the right time to burn is chosen by the elders and chiefs of the tribe. However, it has to occur before the 'birth' of the August moon and before the buds of the Piqui tree reach blooming maturity (Posey 1985).

The Kayapo have been creating 'forest islands' for as long as they can remember. These forest patches are a key aspect to the long-term sustainability of the Amazonian environment, its wildlife and its people. Posey (1985, p. 142) describes the creation of forest islands as "an interesting process. Compost heaps are prepared in existing forest patches from sticks, woody limbs, and leaves. These are allowed to rot, then beaten with sticks to produce a mulch. This mulch is subsequently taken to a selected spot in the campo and piled onto the ground. Slight depressions in the surface are usually sought out because they are more likely to retain moisture".

The author further describes how "these depressions are filled with the mulch, which is mixed with soil from mounds of a termite (*Nasutitermes*) called *rorote*, and smashed up bits of the nest of an ant called *mrum kudja* (*Azteca* Sp.). Living ants and termites are included in the mixture. The resulting mounds of earth, called *apete-nu* (newly formed vegetation clumps), are generally one to two metres in diameter and 50–60 cm deep. The *apete-nu* are usually formed in August and September during the first rains of the wet season and then nurtured by the Kayapo as they pass along the savanna trails to their gardens (*puru-nu*). Over the years, the *apete-nu* 'grow' into large *ape-ti* (large forest islands with many tall trees about 1 hectare in size over a 10-year period)" (Posey 1985, p. 142).

The Kayapo have demonstrated that land management practices that integrate and benefit the whole ecosystem including its people can be sustainable. Through a process of adding mulch, ash, charcoal, ant and termite mounds, and a systematic mosaic burning management of selective areas, there is opportunity to build resilience and associated benefits in plant communities for food production and medicines, to encourage wildlife, attract game and birds, firewood, materials for daily living (i.e. weaving material for baskets and wood for bows and arrows), and other flow on effects such as soil fertility (Posey 1985).

2 International Contemporary Case Studies

There is a certain urgency to establish biochar's credibility not only to reduce greenhouse gas emissions and sequester carbon but to also have agronomic benefit for food production and soil health. In this section, we examine three case studies and evaluate their benefits and constraints.

2.1 Case Study: Growing Cacao in the Toledo District of Belize

In order to combat the common fungal disease monilia, a frequent management practice in cacao orchards involves the trimming back of extra tree foliage to increase the air flow between cacao trees and shade trees. This creates an amount of pruning waste which in the past was left to decompose or burnt in situ (slash and burn) thus releasing carbon dioxide into the atmosphere (CG 2014).

By providing farmers with financial incentive in the form of carbon payments to stimulate interest and uptake in the cacao orchards of Toledo, biochar was produced using the biomass pruning waste as a feedstock resulting in a more sustainable solution to waste management and a reduction in atmospheric emissions from previous slash and burn practices (CG 2014).

In the space of twelve months of applying locally made biochar from pruning waste from cacao and shade trees, to a mixture of age stands there was noticeably more growth, higher yields of cacao pods, and an increased resilience to the fungal disease monilia. The use of biochar for enhancing productivity has also instigated a fundamental change in how farmers produce their cocoa seedlings (CG 2014).

Traditionally these once grafted would have been planted directly into the soil and left to natural variability to ascertain its survival rate. In the new approach the cocoa seedlings are grown in the nursery for a longer period of time ($\sim$8 months) using biochar as a soil amendment. The biochar's water retention capacity reduced water usage by 50%, and within three months a 100% growth rate was observed in the number of seedlings that attained an 8-mm diameter (CG 2014).

One of the key constraints seeing this farming initiative into the future is that its current levels of biomass waste are not sustainable to produce the quantity and quality of biochar needed to support the local cocoa industry. Additional sources of sustainable feedstock such as sawmill residues and rice husks were investigated and found to be suitable (CG 2014).

The use of these additional feedstocks will change the biochar's chemical characteristics and nutrient availability, and therefore will have a differing effect on the growth and establishment of seedlings in the nursery and as a soil amendment of cocoa trees in the orchards. Looking into the future, the introduction of biochar in Toledo's cocoa orchards will support the propagation and growth of at least 45,000 cacao plants per year and social, economic and environmental benefits to farmers and their community (CG 2014).

2.2 *Case Study: Growing Brussels Sprouts in San Mateo County, California*

Between the spring of 2012 and the fall of 2014, a field trial was conducted on two test plots involving three treatments, raw biochar, poultry manure compost, and a biochar-compost mix to test the effect on brussels sprout response to crop yield, soil health, nitrate leaching, and soil carbon sequestration (RCD 2016).

The field trial suggested that the raw biochar and the biochar-compost mix treatments resulted in a neutral or negative effect on brussels sprout yield. The exception was the poultry manure compost which resulted in a neutral or positive effect on brussels sprout yield. In fact, during the second harvest in the fall of 2013 all the treatments including the control treatment experienced the lowest brussels sprout yields in the course of the three-year trial although no conclusive explanations were determined (RCD 2016).

In previous biochar studies liming and pH have been suggested as important contributing factors in stimulating plant growth (Macdonald et al. 2014) especially as brussels sprouts need a pH greater than 6.5 for optimum growth (RCD 2016). In this instance the application of liming was not found to have any significant effect on brussels sprouts yield. Furthermore, it was noted that one of the plots had exhibited significantly higher soil organic matter (SOM). It was suggested that SOM in addition to water and nutrient retention perhaps provided the stimulus for the increase in the plot's crop yield in contrast to the other plot where the treatments had no effect on SOM levels (RCD 2016).

Weather and site-specific factors may have played an integral part in the outcome of the crop yields in the two plots. Extreme weather variations went from hot summers with little rainfall to minimum rainfall late in the season which may have had some influence on the lower crop yields experienced in the second harvest

(RCD 2016). In addition, plot specific differences such as using different brussels sprout varieties, the amount of site shade, and variation in soil conditions may have had some influence on the one treatment (poultry manure compost) that produced a positive crop yield result in the first harvest (RCD 2016).

This project indicates that we have much to learn about the use of biochar in agriculture and the nuances between the varying nutritional aspects and the growth requirements of different crops.

2.3 Case Study: Growing Soybeans and Forage Biomass in Quebec, Canada

One of the largest biochar field trials currently underway pertains to the growing of soybeans and forage biomass in Quebec, Canada. Established in May 2008, the field trial used a commercial biochar (CQuest™) made from wood waste utilizing the fast pyrolysis process and applied to a clay loam soil. Due to a loss of 30% of biochar material in the initial application, the biochar plot only received an application rate of 3.9 t/ha instead of the anticipated 5.6 t/ha rate. Liming of the clay loam soil was also concurrently incorporated with the biochar amendment to a depth of 20 cm and lime was also applied to the control plot (Husk and Major 2009).

The first crop of soybean seeds (*Glycine max* (L.) Merr.) was planted in the soil early June 2008 and harvested four months later. The second crop of forage biomass was a mixture of rye grass (*Lolium multiflorum* Lam.), red clover (*Trifolium pretense* L.), timothy (*Phleum pretense* L.), and oats (*Avena sativa* L.) was planted in early June 2009 and harvested two months later due to an extremely wet season. Biological soil parameters are good indicators of soil health. During the second year of rotational cropping the earthworm density and endomycorrhizal fungi colonization was on average greater in the biochar amended plot (Husk and Major 2009).

Morphological parameters indicated that plant height for soybean was greater in the biochar amended plot and the number of root nodules and seed pods was lower with biochar. For the oats, the most significant difference occurred in plant height, root and total plant length in the biochar amended plot before the harvest. After harvesting the oats only root length remained significant whilst the number of oat seeds per plant increased, however, it was not significant (Husk and Major 2009).

In terms of yield and plant density, the first harvest of soybean produced a 20% yield increase and notable increases in soybean plant density ranging from 11% (one 4 m^2 quadrat) to 68% (50 plants on 2 adjacent rows) in the biochar amended soil. In the second harvest, soya yield increased by 17% and the remaining forage biomass yield increased by 99% with biochar whilst forage plant density exceeded its expectations at 102% (Husk and Major 2009).

In the third year of the trial, forage seeding included a mixture of rye grass (*Lolium multiflorum* Lam.), red clover (*Trifolium pretense* L.), and timothy (*Phleum pretense* L.) with no re-seeding of oats. Above ground plant biomass, moisture content, total dry matter yield, and forage nutritional quality all increased in the biochar amended plot. The notable exception was the plant fibre content showing low levels in the biochar amended soil which is ideal for this forage crop (Husk and Major 2011).

In terms of plant variety distribution, the biochar amended plot had a higher percentage of clover (51%) followed by ryegrass (44%) in comparison with the control plot which was inversely dominated by 60% ryegrass and 35% clover. In assessing the forage biomass benefits for milk production, it was estimated that the forage grown in the biochar amended soil had an overall increase in projected milk production ranging from 16.4% increase in milk per metric ton to 43.6% increase in milk per day whilst milk per hectare resulted in a 19.6% increase (Husk and Major 2011).

Of all the case studies examined in this section only the Quebec study had an appropriate field trial design that included replication and statistical analysis based on the International Biochar Initiative's biochar standards aimed at standardizing testing and measurement of biochar internationally (IBI 2015).

3 Economic Assessment of Biochar Application

Van Zwieten et al. (2009) argues that in order for biochar to become viable and sustainable it has to have an economic value and an associated carbon market price. In their study, commercially available poultry litter, green waste, and paper mill biochar were evaluated using field derived data and found to have an economic value of $329, $85, and $72/tonne respectively which included a carbon trading value of $38/ton for poultry litter, $73/ton for green waste, and $40/ton for paper mill. In addition, the authors also estimated that poultry litter biochar applied at two application rates of 10 and 50 t/ha to sweet corn generated an economic benefit of $465 and $1100/t respectively (Van Zwieten et al. 2009).

In contrast in evaluating biochar application to soil based on two methods, broadcast-and-disk (BAD) and trench-and-fill (TAF), Williams and Arnott (2010) applied commercially available hardwood waste biochar (CQuest™) utilizing the BAD method at rates of 2.5 t/acre to 50 t/acre, and found that costs ranged from $29 to $300/acre whilst for TAF application the rates ranged from 5 to 75 t/acre with costs ranging from $26 to $1280/acre. This potential high cost of application via the two methods would need to be offset by tariffs and a market mechanism to become viable including development of innovative application method solutions that are both cost effective and sustainable (Williams and Arnott 2010).

The compendium of available published literature (Collison et al. 2009; Joseph 2009; McCarl et al. 2009; Shackley et al. 2011) merely focuses on short term economic scenarios whilst what is actually needed is a cost benefit analysis that goes beyond the next 20 years and which incorporates all aspects of production and soil amendment application of biochar (Dickinson et al. 2015). This is an important determinant in establishing whether the adoption of biochar by farmers will have an economic benefit or an economic deficit for their investment in soil carbon land management practices presently and into the future.

4 Agriculture and Soil Carbon Sequestration Potential in Australia

There has been considerable interest in storing and retaining carbon in soils for a number of decades since it was determined that healthy soils are productive soils. As a potential soil amendment biochar has a number of functions including the capacity to sequester greenhouse gas emissions as a mitigation strategy to ameliorate climate change. It is argued that "soils hold the largest terrestrial store of organic carbon. Globally, estimated stores of organic C are approximately 684–724 Gt (1 Gt = 1015 g = 1 Pg) in the top 30 cm, 1200–1550 Gt in the uppermost metre of soil and around 2300–2450 Gt in the upper 2 or 3 m" (Viscarra Rossel et al. 2014, p. 2954).

The current accumulated carbon stocks have been heavily influenced by zonal climate, lithology, soil type, biota, and Australia's 200-year-old agricultural land use practices. It is noted that the combination of low temperatures and waterlogging tends to slow down the process of decomposition and mineralization of organic carbon which leads to carbon persisting in the soil for much more extended periods of time. This is particularly relevant to southern Australia states i.e. Victoria and Tasmania (Viscarra Rossel et al. 2014).

Downie et al. (2011) argues that there is potential to further increase soil carbon stocks and soil fertility through anthropogenic derived amendments such as biochar and charcoal, as evidenced by the highly fertile soils of the Amazon (terra preta de indio), North-West Europe (plaggen soils), and the ancient agricultural terraced soils of the Andes. Australia shares similarities with the Amazon, as its pre-European anthropogenic soils were also created as a result of traditional cooking earths and refuse mounds over generational timelines (Downie et al. 2011).

To date very little is known about Australia's dark earths or Terra Preta Australis (TPA) soils but by their very existence it is possible that Australian soils generally ascribed as being quite nutrient poor have the potential to play a bigger role in climate change amelioration, food security and agricultural sustainability (Downie et al. 2011). In the study conducted by Downie et al. (2011) around 30 oven mounds were identified above the flood zone of the Murray River along the New South Wales-Victoria border and classified as Cumulic Anthroposol soils.

Radiocarbon dating places the soils between 650 and 1609 years BP which is consistent with earlier discoveries in the 1970s dated between 600 and 3500 years BP. These dates therefore support the theory of continuous organic black carbon persistence in Australian soils for hundreds to thousands of years (Downie et al. 2011). Soil sampling estimated the average increase of the total carbon mean within the top 30 cm to be 3.35%. If this level of increase in soil carbon could be achieved through the integration of biochar to the top 30 cm layers across the 42 million ha of cropped soils, Australia would be able to sequester approximately 23 Gt CO_{2-e} (Downie et al. 2011).

4.1 *Australia's Emissions Reduction Fund*

The Emissions Reduction Fund (ERF) is a voluntary scheme set up by the Australian Government with the functional objective of firstly meeting our climate change international commitments under the Kyoto Protocol, and secondly to achieve Australia's emission reduction target of 5% below 2000 levels by the year 2020 by stimulating economic incentives to organizations and individuals to take up emission reduction mitigation strategies and technologies (DEE 2016).

The ERF was enacted through the *Carbon Credits (Carbon Farming Initiative) Act 2011*, the Carbon Credits (Carbon Farming Initiative) Regulations 2011, and the Carbon Credits (Carbon Farming Initiative) Rule 2015 (CER 2016). The ERF scheme offers eight project types which includes soil carbon and agriculture and rewards businesses and community through crediting and purchasing of carbon credits. For businesses, a safeguard mechanism has also been included to deter double dipping (DE 2016).

The soil carbon projects are aimed specifically at building soil carbon through a change in farming practices and hopes to achieve a reduction of 7.8 Mt CO_{2-e} through existing projects that include biochar whilst agricultural based projects have generated emission reductions so far of 1.0 Mt CO_{2-e} stemming from beef cattle, milking cows, methane from effluent waste at piggeries, and nitrous oxide emissions in irrigated cotton farms through improved fertilizer efficiency (DE 2016).

Crediting occurs when registered ERF businesses that reduce emissions through a range of approved activities are credited with one Australian Carbon Credit Unit (ACCU) for every tonne of reduced emissions or every tonne of sequestered soil carbon. Once the credited ACCU's have been issued by the Clean Energy Regulator (CER), businesses can then sell their ACCU credits at auction to the Australian Government via the CER or through voluntary carbon markets (DE 2016).

5 Conclusion: Future Options for Biochar in Supporting Sustainable Development Goals

Biochar has a greater global potential future, one that supports amelioration of climate change and greenhouse gas emissions in addition to soil carbon sequestration. This will underpin agricultural production serving our future sustainable cities. While some evidence demonstrates that Australian indigenous people practised a form of biochar enriched land management, 200 years of colonial agricultural processes have significantly diminished this process. We have much to learn from sustainable land management practised by indigenous people in Australia and across the Pacific to the indigenous tribes in South America.

As an environmental management strategy for farmers in order for biochar to become viable and sustainable it has to have an economic value and an associated carbon market price that ultimately provides incentives for uptake. Even with incentives the market prices need to be at an economic value that provides economic returns not economic losses.

Biochar's credibility is its premise founded on sustainability principles. It is estimated that the global maximum sustainable technical potential of biochar to mitigate climate change is about 12% of anthropogenic related emissions. However, there is still a lot of uncertainty and sustainability challenges surrounding sustainable sourcing of feedstock quality and quantity.

The introduction of international guidelines and standards across the biochar industry should see biochar's profile raised across all levels of government and industry. To date the global biochar industry has grown to twice the size reported in 2013s state of the industry report so the future for biochar looks promising if the reported barriers and constraints can be overcome.

References

CER. (2016). About the emissions reduction fund. The Clean Energy Regulator. Australian Government http://www.cleanenergyregulator.gov.au/ERF/About-the-Emissions-Reduction-Fund. Last accessed 6/10/2016.

CG. (2014). Biochar: A sustainable waste management solution and effective soil amendment for cacao growers in Central America. *International Biochar Initiative* www.biochar-international.org/sites/default/files/Belize_case_study.pdf. Last accessed 10/09/2016.

Collison, M., Collison, L., Sakrabani, R., Tofield, B., & Wallage, Z. (2009). *Biochar and carbon sequestration: A regional perspective*. Low Carbon Innovation Centre: University of East Anglia, United Kingdom.

DE. (2016). The emissions reduction fund–what it means for you: How Australian businesses and the community can benefit from the Emissions Reduction Fund. Canberra, ACT: Department of the Environment.

DEE. (2016). *About the Emissions Reduction Fund*. Canberra, ACT: Department of the Environment and Energy.

Denevan, W. (1996). A bluff model of riverine settlement in prehistoric Amazonia. *Annals of the Association of American Geographers, 86,* 654–681.

Dickinson, D., Balduccio, L., Buysse, J., Ronsse, F., Van Huylenbroeck, G., & Prins, W. (2015). Cost-benefit analysis of using biochar to improve cereals agriculture. *Global Change Biology Bioenergy, 7,* 850–864.

Downie, A., Van Zwieten, L., Smernik, R., Morris, S., & Munroe, P. (2011). Terra preta australis: Reassessing the carbon storage capacity of temperate soils. *Agriculture Ecosystems & Environment, 140,* 137–147.

Glaser, B., & Birk, J. (2012). State of the scientific knowledge on properties and genesis of anthropogenic dark earths in Central Amazonia (terra preta de indio). *Geochimica et Cosmochimica Acta, 82,* 39–51.

Hecht, S. (2003). Indigenous soil management and the creation of Amazonian dark earths: Implications of Kayapo practices. In J. Lehmann, D. Kern, B. Glaser, & W. Woods (Eds.), *Amazonian Dark Earths: Origin – properties – management*. Dordrecht: Kluwer Academic Publishers.

Husk, B., & Major, J. (2009). *Commercial scale agricultural biochar field trial in Quebec, Canada over two years: Effects of biochar on soil fertility, biology and crop productivity and quality*. Drummondville, Quebec: BlueLeaf Inc.

Husk, B., & Major, J. (2011). *Biochar commercial agriculture field trial in Quebec, Canada – year three: Effects of biochar on forage plant biomass quantity, quality and milk production*. Drummondville, Quebec: BlueLeaf Inc.

IBI. (2015). Standardized product definition and product testing guidelines for biochar that is used in soil. *International Biochar Initiative, Version, 2,* 1.

Joseph, S. (2009). Socio-economic assessment and implementation for small-scale biochar projects. In J. Lehmann & S. Joseph (Eds.), *Biochar for environmental management: Science and technology*. London: Earthscan.

Lehmann, J., & Joseph, S. (2009). Biochar for environmental management: An introduction. In J. Lehmann & S. Joseph (Eds.), *Biochar for environmental management: Science and technology*. London: Earthscan.

Macdonald, L., Farrell, M., Van Zwieten, L., & Krull, E. (2014). Plant growth responses to biochar addition: An Australian soils perspective. *Biol Fertil Soils, 50,* 1035–1045.

McCarl, B., Peacocke, C., Chrisman, R., Kung, C., & Sands, R. (2009). Economics of biochar production, utilization and greenhouse gas offsets. In J. Lehmann & S. Joseph (Eds.), *Biochar for environmental management: Science and technology*. London: Earthscan.

Myers, T., Denevan, W., Winklerprins, A., & Porro, A. (2003). Historical perspectives on Amazonian Dark Earths. In J. Lehmann, D. Kern, B. Glaser, & W. Woods (Eds.), *Amazonian Dark Earths: Origin – properties – management*. Dordrecht: Kluwer Academic Publishers.

Nowell, C. (1936). The discovery of Brazil – accidental or intentional? *The Hispanic American Historical Review, 16,* 311–338.

Posey, D. (1985). Indigenous management of tropical forest ecosystems: The case of the Kayapo Indians of the Brazilian Amazon. *Agroforestry Systems, 3,* 139–158.

RCD. (2016). Biochar field trials in San Mateo County, CA: Final project report. United States Department of Agriculture-Natural Resources Conservation Service, California Department of Conservation and San Mateo Resource Conservation District. Half Moon Bay, California.

Shackley, S., Hammond, J., Gaunt, J., & Ibarrola, R. (2011). The feasibility and costs of biochar deployment in the UK. *Carbon Management, 2,* 335–356.

Sohi, S., Lopez-Capel, E., Krull, E., & Bol, R. (2009). Biochar, climate change and soil: A review to guide future research. CSIRO Land and Water Science Report 05/09, p. 64.

Sombroek, W., Kern, D., Rodrigues, T., Cravo, M., Jarbas, T., Woods, W., & Glaser B. (2002). Terra preta and terra mulata: Pre-columbian amazon kitchen middens and agricultural fields, their sustainability and their replication. In R. Dudal (Ed.) *Symposium 18. Anthropogenic factors of soil formation*. 17th World Congress of Soil Science, Bangkok, Thailand.

Van Zwieten, L., Kimber, S., Orr, L., Morris, S., Downie, A., Sinclair, K., Joseph, S., & Chan K. (2009). Agro-economic valuation of biochar using field derived data. In *Proceedings of the 1st Asia Pacific Biochar Conference*, May 17, 2009. Gold Coast, Queensland, Australia.

Viscarra, Rossel R., Webster, R., Bui, E., & Baldock, J. (2014). Baseline map of organic carbon in Australian soil to support national carbon accounting and monitoring under climate change. *Global Change Biology, 20,* 2953–2970.

Williams, M., & Arnott, J. (2010). A comparison of variable economic costs associated with two proposed biochar application methods. *Annals of Environmental Science, 4,* 23–30.

Woods, W., & Denevan, W. (2009). Amazonian dark Earths: The first century of reports. In W. Woods, W. Teixeira, J. Lehmann, C. Steiner, A. Winklerprins, & L. Rebellato (Eds.) *Amazonian Dark Earths: Wim Sombroek's Vision* (p. 502). Springer Science.

Synchronization of Sustainable Development with Land Development

Kristopher Adam Orlowski

Abstract Sustainable development, particularly the word sustainable has been overused and misused to the extent that its true meaning is no longer thought of especially in the field of land development. There are intrinsic conflicts between the notions surrounding the development of land and sustainability which is heightened by society and its values. These are the fundaments for the knowledge gap existing between uniting true sustainability with that of land development that is accustomed modern society in terms of definition and notion, let alone a model. This has led to three objectives of this research project which include; (1) Scrutinization of how development of land since the Industrial Age has conferred to our notions of sustainability; (2) Exploring possible synergies and tradeoffs between land development and sustainable development; (3) Identifying the possible ways and implications of the change towards a sustainable society. A strong methodology evolved around an integrated conceptual frame work is used which involves multi-scale perspectives in aiming to achieve comprehensive insight on this complex issue. Significant outcomes are expected which are not already prevalent in literature, principally, exploration of synergies and tradeoffs between sustainable development and land development that can be exploited and handled to put forward the ideals of true sustainable land development.

Keywords Sustainable development · Land development · Sustainability Synchronization

K. A. Orlowski (✉)
Department of Infrastructure Engineering, University of Melbourne,
Parkville, Melbourne, VIC 3010, Australia
e-mail: kristopher.orlowski@unimelb.edu.au

W. Leal Filho et al. (eds.), *Sustainable Development Research in the Asia-Pacific Region*, World Sustainability Series, https://doi.org/10.1007/978-3-319-73293-0_26

1 Introduction

The word sustainable is overly misused in the context of land development to the extent that its core meaning is forgotten or rather ignored (DuFault and Khol 2015). Land development and management is now becoming a concern due to the growth in population and rapid urbanization. The scarcity of land will affect everybody in the most dramatic ways as it is core to our survival on numerous levels from food production to livelihood.

Sustainability and land development in their own right have an innate conflict due to the underlying competing principles. This complex issue revolves around the notion that growth is not sustainable and thus current practices of land development are also not sustainable. Values of accumulation and consumption are deep rooted in society and are amongst the prime root sources leading to land degradation. Therefore it is clear that a profound change, right down to our individual character is needed to ensure new norms to be established in society so that conduct on a national and global scale will pursue sustainability resulting in appropriate outcomes from both sustainability and land development perspectives.

The Project Scope of this area of study is restricted to the topics of sustainable development and land development which will be looked into on individual, societal and global scales. The timeframe in question is from the beginning of the Industrial Revolution to the present and futurity.

The definition of Sustainable development for the purposes of this study is that Sustainable Development is the progression towards sustainability. In its richest sense, sustainability is the ability to continue to act or behave a certain way indefinitely.

As Jonathon Porritt, Chairman of the British Government Sustainable Development Commission eloquently put it "If something is sustainable, it means we can go on doing it indefinitely. If it isn't, we can't" (Poole 2005).

The definition of Land development for the purposes of this study is that the development of land simply refers to the human impact on a parcel of land or in the general sense ground area. It may involve:

(a) anthropogenic altering of the landforms from a natural or current state to a desired state for ease of use or other benefits
(b) building and constructing on top or within landforms for the benefit (usually economic) of the owner or users of the land
    Such purposes of land development and its uses are those evolved around housing, agriculture, infrastructure or industry.

Combining sustainability and land development we get the term Sustainable land development which is the term that relates the development of land to the ideals of sustainability. Essentially it's sustainable development in an applied manner. In this sense we can discreetly list a number of key goals as shown in Table 1 that sustainable land development would achieve.

Table 1 Sustainable land development desirables

Item	Goal
Water and food	Sufficient quantity and quality to meet the demand on the dependent population Furthermore a healthy and stable seasonal ground water level and cycle is to be maintained
Greenhouse gas emissions	Not more than that of the natural environment, which the land is over can absorb
Air	Free from toxic pollutants
Energy	Derived only from renewable sources that are part of closed lifecycles
Natural resources	No depletion over time, full recycling with no unusable waste
Biodiversity	Promoted and preserved
Equity	Across society in all forms (educational, intergenerational, privileges, rights, legal and health)

Ideally resources are to be shared and communities equally consume and produce in the same manner. That is the equity principle that overrides cadastral boundaries.

2 Literature Review

There have been numerous studies into perceived forms of sustainable development of land. Perceived in the sense that none are self-sufficient or otherwise meet all the goals listed in Table 1 (p. 2). One of the more recent and 'greener' models of land development is that of CSD (Conservation Subdivision), also known as Conservation Development. The former term was made popular by Arendt et al. (1996) (Arendt 1999a, b, 2004). In essence CSD maximizes open space, which act as the organizing structure of the development which is in the form of a series of protected lands which encompasses the area and supremely interconnect with neighbouring communities as a conserved natural system. This does not necessarily mean the development must have fewer buildings nor does it mean that there will always be less profit for a developer (Soule 2000). Houses are closer together and lots are generally smaller, they are arranged to preserve as much land as possible. The cost benefits for developers and utility service providers are primarily in the form of savings from not having to construct as much infrastructure such as those evolved around roads, pathways, drainage, sewerage, water, gas and electricity (Mohamed 2006).

Fiscal benefits are plentiful as there are decrease costs in: maintenance of public infrastructure, providing fire and ambulance services, and police protection to even collecting garbage. However lack of incentives to compensate for reduced profits from lots has been identified as a barrier by numerous authors claiming this to be a primary economic factor against CSD (Allen et al. 2012; Bosworth 2007; Bowman and Thompson 2009; Carter 2009). However this is not without challenge, it's been contended that in fact in conservation development houses not only sell for more, they sell faster and also save on construction costs when compared to houses built in a conventional development (Bowman et al. 2009). This demonstrates how uncertain and undeveloped the field of conservation subdivision is and that there is every opportunity to capitalise on this type of development that can profit industry and the environment.

Considerations around environment are inherently prioritised in CSD design due to the development revolving around open space and protected lands. Therefore there are benefits to biodiversity and wildlife when good CSD incorporates sensitive protected land such as riparian areas (Lenth et al. 2006). This can be extended to claim conservation of scenic views, historic and archaeological features, recreational provision and perhaps most significantly farmland conservation. Conservation of farming land is a pressing issue as shown in developing countries where there are rural households that participate in agricultural activities but do not have enough land to even produce food for their family, and this problem will only increase with rising population (Timmer 2015).

It must be noted that CSD does not encompass anything more than allocation of space to residential area and open areas. There is so much more that needs to be taken into consideration when building for the future. It has been established that for human society to be sustainable, we must not confine ourselves only with our relationship with the environment, rather we should be concerned with the nature of society and the way it is organised (Espinosa Salazar and Walker 2011). Therefore a primarily hard sustainable view, encompassing holistic considerations including issues of population, growth and values which can lend together with the principles of conservation development in order to better resolve sustainable land development.

There is a limitation in that even when we develop a theoretical sustainable land development model, land management is powerfully governed by laws and regulations, as well as public expenditure priorities (Bank 2006). How to best alter these laws and regulations to promote sustainable land development is outside the scope of this project, however, it is an opportunity for further research.

Conservational design of subdivision coupled with current best practices in home design, newest technologies in energy and efficiency and values akin to preservation is where current industry and society is up to in terms of progression towards sustainability in a land development context.

Currently, works in the fields of sustainability and land development are arguably miss-focused, in that there is a seemingly overwhelming consensus in that we shall adapt our model of land development to fit into popular sustainability principles rather than starting with fundamental sustainability principles and evolve land development around them. That is, the notion of sustainability itself must be part of the core to any application of it so as to keep the meaning of the word prevalent and meaningful.

The knowledge gap is in that of the uniting of true sustainability (that which can continue indefinitely) with that of land development accustomed modern society in terms of definition and notions, let alone a model. Formed from the literature review, the intention of addressing these particular research questions outlined below is to provide new insight into the issue.

How does the development of land since the Industrial Age confer to our notions of sustainability?

What are possible synergies and tradeoffs that can be drawn from land development and sustainable development?

In what ways and with what implications can the change towards a sustainable society be brought about?

The multidimensional issues of sustainable development and land development ought to be looked at on different scales. This is due to complexity of the problem as this issue changes in terms of its own composition when looked at from different scales. Simply put multiple scale problems hold different properties on different levels, the problem with this is that the "existence of distinguishable complete mechanisms" acting on different scales (Brackbill and Cohen 2014). Therefore using a multi-scale approach such as that proposed for this project (individual, societal and global) in applying the methodology gives possibility to insight with a comprehensive system perspective.

A conceptual framework is used as the basis of outlining methodology. This exploits the nature of the issue and the method of applying perspective of different scales as it has been demonstrated and concluded that conceptual frameworks are a qualified and proficient way to tackle issues covering small and large scales (Maxwell 2009; Ravitch and Riggan 2012). Since the topic of synchronization of sustainable development with land development encompasses diverse ideas (many conflicting), it is necessary to use an open methodology which can accommodate this need. The use of a conceptual framework meets this need as it is a principally a method to organize ideas to achieve the purpose of a research project (Shields, Patricia and Rangarjan 2013).

3 Methodology

The research in question will be directed in a three tiered manner which is evolved around the scale or boundary of consideration. The three levels of scales that are being referred to are:

1. Individual
2. Societal
3. Global

An overarching progression of contemplation and analysis is recognized and followed with a mindset of any of the former scales or boundaries of consideration.

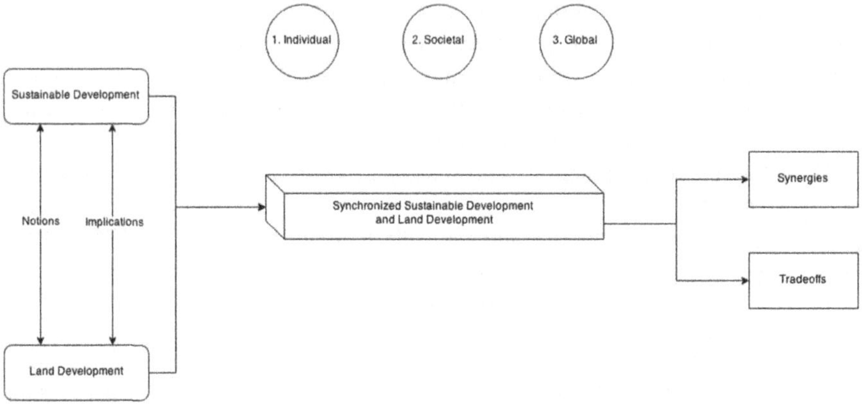

Fig. 1 Conceptual framework Orlowski (2017)

This is referred to as the conceptual framework and is simplified graphically in Fig. 1.

4 Method

Keeping in mind the three levels of scales throughout each stage, the conceptual framework in use begins with characterization of sustainable development and land development in their own respected lights, which is as they are perceived currently as an individual entity.

Stemming from these characteristic description are links of association between the two. They are revolved around the:

(a) notions one may comprehend that are associated between sustainable development and land development.
(b) implications of pursuing either land development or sustainable development on the feasibility and state of the other (Fig. 2).

The butterfly diagram shows that the understanding sustainable development and land development as their own sense and detailing the notions and implications between them permits an informed finding of a synchronized sustainable development and land development schema. The conception and constitution of such a schema will entail thought experiments and conversant and well-versed conjecture.

Analysis of synchronization between sustainable development and land development will yield distinct synergies and tradeoffs which may be exploited and handled.

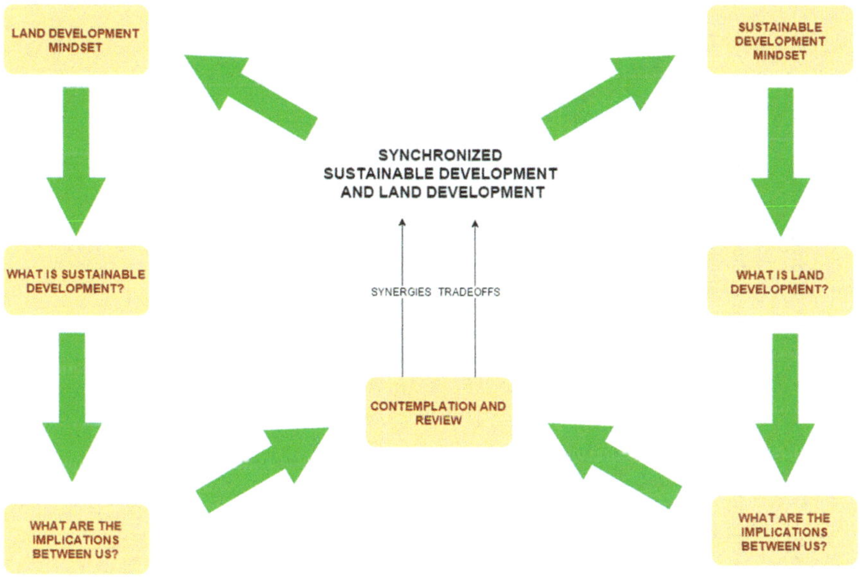

Fig. 2 Butterfly diagram Orlowski (2017)

5 Discussion

Tables 2 and 3 have been created with the aim to aid in comparison and under standing a rounded and holistic view of the issues in question and further develop them in the context of synchronizing sustainable development with land development. In fact, it is shown that all current prominent notions and views are com patible with this research and together offer a balanced view.

The conundrum of synchronized Land development and Sustainable development is shown in Fig. 3. There are clear differences in the core principles and concepts between the two on all scales; individual, societal and global. Based upon current notions and the implications of these notions on the item itself one can source what these notions must have their own implications of affecting the item itself in order to transform them into a state where they are in synchronization. That is, that the idea and reality of land development is itself sustainable development.

The individual, societal and global nature of this issue as depicted in Fig. 3, draws the need for understanding of all people and their belief and attitude system in order to drive change towards sustainable land development. Ecological Economics 'Policy assessment and simulation of actor orientation for sustainable development' excellently rationalises this view:

Table 2 Coordination and organisation table of definitive studies

Citation	Journal/ Book name	Paper title/case study	Strengths/bias	Supports/advocates	Authors progression
Rokeach (1973)	The Nature of Human Values	Rokeach Value Survey	• Extensively used by physiologists and sociologist • Served as a basis of further research	Moderately few 'Terminal Human Values' exist which are defined as desirable end-states. 1. True friendship 2. Mature love 3. Self-respect 4. Happiness 5. Inner Harmony 6. Equality 7. Freedom 8. Pleasure 9. Social recognition 10. Wisdom 11. Salvation 12. Family security 13. National security 14. A Sense of accomplishment 15. A World of beauty 16. A World at peace 17. A Comfortable Life 18. An exciting life Results indicate that individualism—achievement and collectivism—affiliation are the underlying dimensions of the RVS (Johnston 1995)	1. Treat the synchronization of sustainable development and land development on different scales, individual, societal and global 2. Identify the terminal values associated with each 3. Explore the influencing drivers behind these values 4. Priorities these values

(continued)

Table 2 (continued)

Citation	Journal/ Book name	Paper title/case study	Strengths/bias	Supports/advocates	Authors progression
Bossel (2000)	Ecological Economics	Policy assessment and simulation of actor orientation for sustainable development	• Environmental systems analysis • Decision support systems • Simulating complex systems and Social change	Sustainability is gained from ensuing those who adequately care for their own interests and respect the interests of those which they depend Understanding the interests of actors and systems is key If we can identify the fundamental interest of actors and systems then we can predict the behavior and outcomes	1. Let us define the outcome of sustainable development and land development and the behaviors necessary to achieve this 2. Create achievable goals and targets to get to this goal 3. Through the use of hortatory policy and education of particularly the next generation; establish the fundamental interest which will be the key driver of change
Robinson (2004)	Ecological Economics	Squaring the circle?	• Assesses the theory and the practice of sustainable development • Concerned about the development of industrialised countries	There is no singular method or approach that shall be seen as correct. Sustainability is a political act not a scientific concept Neither and end state or a process, "rather an emergent property of a conversation about what kind of world we collectively would like to live in now and in the future" (Robinson 2004)	Living sustainably and sustainable development is what we make of it. Any initiative or idea that is to be classed as sustainable must be supported collectively. Therefore emphasis should be placed on how we would *like* land to be used now and into the future rather than how land *needs* to be used now and into the future

Table 3 Coordination and organisation table of prominent notions and views

Notion/view	Authors progression
Sustainability as an end state	Finite amount of notions and implications of these notions exist between sustainable development and land development These can be identified and characterized We can synchronize the use and development of land with sustainability based on these through exploited the synergies and overcoming the tradeoffs through policy
Sustainability development as an process	Define the steps and short term goals needed to implement change at multiple scales. Evaluation at the end of each term on what the steps and goals should be for the following term
Three pillar approach • Social • Ecological • Economic	Synchronization of Land development with Sustainable development would bring the social and ecological aspects to the forefront of decision making as compared to the current economic approval for land development
Dualistic Relationship between humanity and nature	Use of this dualistic principle to establish balance and trade-offs between undeveloped natural land and developed land
Dematerialization Reducing the amount of matter and the energy throughput per economic activity unit Resocialization Increase well-being per economic activity unit	If adherence to resocialization and dematerialization became a norm, then the compounding the reduction in the need to economic activity with the reduction of consequence of economic activity would provide the framework for sustainable use of land

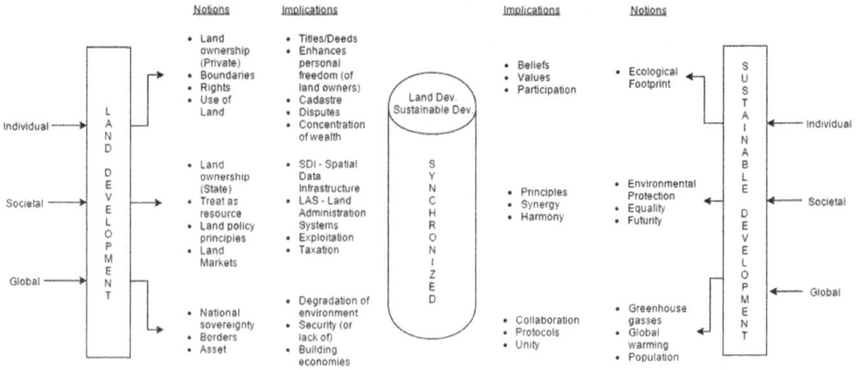

Fig. 3 The conundrum of synchronized land development and sustainable development Orlowski (2017)

> Sustainable development is not arbitrary. It has to remain within the strict boundaries of an accessibility space that is defined by physical constraints (laws of nature, causal relationships, physical environment, solar energy flow, material resource stocks, carrying capacity), by constraints of time and system laws (delays, inertia, permissible rates of change, feedback, and self-organization), and by the constraints of human actors (intellectual and organizational ability, culture, ethics and values, technology, social and political system) (Bossel 2000). Behaviour is shaped by the perceptions of these constraints by the human actors, not by their actual state. These perceptions—the cognitions of actors—are therefore powerful determinants of future development, and must be properly accounted for in policy assessments and development studies. (Bossel 2000)

Thus value systems and perceptions are the underlying resistors but also prospective drivers for the shift to a sustainable society.

As value systems and perceptions take time to change, particular additional measures can be implemented in order to bring about this desired state more promptly. One of such is described below in the context of synchronizing sustainable development and land development through the understanding of synergies, tradeoffs and their implications to better shape the measure to be emplaced.

Idea: Ensure everyone has a right to use of land. Take away the 'right' to purchase land/property that you directly will not use. Modify land tenure for companies to only use land for 'best use' based upon geographic/inherit suitability for current needs/demands from society. Meaning take away the privilege to privatise monopoly of land and exploitation of land.

Synergies

- Increase liquidity in the market which further allows for rapid development and change
- Equity principle is met at a basic level
- Encourages higher density living as population increases
- Tackles the core primer to the threat of humanity, overconsumption. Creates a foundation for rapid change in other aspects of consumption, such as food, water, resources and goods
- Technology advancements, especially in the 3D cadastre
- Concentration of wealth limited in not only in itself occurring but its impact on the environment
- Affordability of living is significantly increased
- Transparency in the cadastre and relevant information can be made open
- Institutes favourable outlook from travellers to host nations/regions
- Will draw out a sense of belonging for each individual
- No homelessness and lower poverty levels

Tradeoffs

- Traditional economics of land abolished
- New norms needs to be established
- Travelling overseas or to regions where you do not know anyone may be more difficult

- Each property must have a "permanent" registered "user", nobody can be a "user" of two properties simultaneously
- Nations/states have more control of foreigners but locals have more freedom
- Conformity in the way people travel and stay in a nation other than their own
- Lower immigration and increase in 'us and you' mentality between nations

Implications

- Cooperation on a large scale necessary to ensure best use of land
- Cooperation on a societal scale to create this norm
- Cooperation on an individual scale to live this way and ideally advocate

Such thought experiments which utilises the conceptual framework as presented in Fig. 1 may lead to new understandings, holistic views and potential steps to work towards true sustainability, one of which can be implemented today for a planned and controlled indefinite future.

This process of thought experiment can be carried out on current land systems such as CSD (conservation sub division) in which are clearly not sustainable due to it being built around growth. A hard sustainable view, encompassing holistic considerations including issues of population, growth and values which can lend together with the principles of conservation development in order to better resolve sustainable land development enough so that the notion will change from land development to land redevelopment.

6 Conclusion

The methodology used is that of a conceptual framework which is geared around the foci of the set of research questions presented in this paper. The method outlined encompasses this conceptual framework and procedurally outlines the relationships and associations between elements. In doing so, it addresses in detail each research question and hence achieves the project outcomes.

An integrated conceptual frame work has been used involving multi-scale perspectives which has achieved its aim in providing comprehensive insight on this complex issue of the synchronization of sustainable development with land development.

The project outcomes are that of answering the proposed research questions. The proposed research questions are formed from the basis of a detailed literature review into the topic of sustainable development and land development. The intention of addressing these particular research questions is to provide new insight into the issue.

They are as follows:

How does the development of land since the Industrial Age confer to our notions of sustainability?

The rapid increase of use of land for resources to bring us here today has drawn us to question our current trend and motivation with the concern for the future. The word sustainable is often termed to describe this awareness in many aspects to the level in which it has been overused and misused to the extent that its true meaning is no longer thought of especially in the field of land development. The intrinsic conflicts between the notions surrounding the development of land and sustainability is heightened by society and its values. Thus there is a need to unite true sustainability; that of which ensures the ability to continue to act or behave a certain way indefinitely with that of land development.

What are possible synergies and tradeoffs that can be drawn from land development and sustainable development?

There are numerous potential synergies in synchronizing sustainable development with land development, some of the more prevalent ones are as follows:

- Increase liquidity in the market which further allows for rapid development and change
- Technology advancements. Especially in the 3D cadastre
- Concentration of wealth is limited
- Housing affordability is increased
- Transparency in the cadastre and open information can be established simply
- Institutes favourable outlook from travellers to host nations/regions
- Tackles the core primer to the threat of humanity, overconsumption. Creates a foundation for rapid change in other aspects of consumption, such as food, water, resources and goods
- Will draw out a sense of belonging for each individual

Tradeoffs are required to in order to facilitate this synchronization. Primarily there are two:

1. Traditional economics of land abolished
2. New norms needs to be established

In what ways and with what implication can the change towards a sustainable society be brought about?

Value systems and perceptions are the underlying resistors but also prospective drivers for the shift to sustainable society.

As value systems and perceptions take time to change particular additional measures can be implemented in order to bring about this desired state more promptly. One of such is described in the context of synchronizing sustainable development and land development through the understanding of synergies, tradeoffs and their implications to better shape the measure to be emplaced. That is: Ensure everyone has a right to use of land. Take away the 'right' to purchase land/property that you directly will not use. Modify land tenure for companies to use land for best use based upon geographic/inherit suitability current needs/demands from society. Meaning take away the privilege to privatised monopoly of land and exploitation of land.

7 Recommendation

How to best alter laws which govern regulation of development of land in order that they align and promote sustainable land development and how best to transform their current norms voluntarily is an outstanding opportunity for further research and development.

Acknowledgements An expression of gratitude is made towards Assoc. Prof. Graham Moore for cooperative and available support of this project. A show of appreciation is duly righted to Dr. Yongping Wei for her generous and comprehensive guidance and advice.

References

Allen, S. C., Moorman, C. E., Peterson, M. N., Hess, G. R., & Moore, S. E. (2012). Overcoming socio-economic barriers to conservation subdivisions: A case study of four successful communities. *Landscape and Urban Planning, 106*(3), 244–252. https://doi.org/10.1016/j.landurbplan.2012.03.012.

Arendt, R. (1999a). *Growing greener: Putting conservation into local plans and ordinances.* Washington, DC: Island Press.

Arendt (1999b). Growing green: Conservation subdivision design. *Planning Commissioners Journal, 33*. Washington, DC: Island Press. Accessed at http://plannersweb.com/wp-content/uploads/1999/01/155.pdf.

Arendt, R., et al. (1996). *Conservation design for subdivisions: A practical guide to creating open space networks*. Washington, DC: Island Press.

Arendt, R. (2004). Linked landscapes: Creating greenway corridors through conservation subdivision design strategies in the northeastern and central United States. *Landscape and Urban Planning*, *68*, 241–269.

Bank, W. (2006). Sustainable land management: Challenges, opportunities, and trade-offs. Washington, DC: World Bank.

Bossel, H. (2000). Policy assessment and simulation of actor orientation for sustainable development. *Ecological Economics, 35*(3), 337–355. https://doi.org/10.1016/S0921-8009(00)00218-4.

Bosworth, K. (2007). Conservation subdivision design: Perceptions and reality. MS Thesis, University of Michigan.

Bowman, T., & Thompson, J. (2009). Barriers to implementation of low-impact and conservation subdivision design: Developer perceptions and resident demand. *Landscape and Urban Planning, 92*(2): 96–105, Waveland Press Inc.

Bowman, T., Thompson, J., & Colletti, J. (2009). Valuation of open space and conservation features in residential. Waveland Press Inc.

Brackbill, J. U., & Cohen, B. I. (2014). *Multiple time scales*. Burlington: Elsevier Science, Academic Press Inc.

Carter, T. (2009). Developing conservation subdivisions: Ecological constraints, regulatory barriers, and market incentives. *Landscape and Urban Planning, 92*(2), 117–124.

DuFault & Khol. (2015). 'Sustainability': Is it a dirty word? While consumers endure the semantics, what words can enable real discussion, and not just greenwashing - or eye rolling? *theguardian.com*. Accessed at http://www.theguardian.com/sustainable-business/2015/mar/25/sustainability-eco-green-natural-buzzwords-greenwashing.

Espinosa Salazar, A. M., & Walker, J. (2011). *A complexity approach to sustainability [electronic resource]: Theory and application*. London: Imperial College Press; Singapore: Distributed by World Scientific Pub. Co., c2011.
Johnston, C. S. (1995). The rokeach value survey: Underlying structure and multidimensional scaling. *Journal of Psychology, 129*(5), 583. https://doi.org/10.1080/00223980.1995.9914930.
Lenth, B. A., Knight, R. L. & Gilbert, W. C. (2006). Conservation value of clustered housing developments. Island Press.
Maxwell, J. (2009). Designing a qualitative study. In L. Bickmam & D. Rog (Eds.), *The sate handbook for applied social science research* (p. 222). Thousand Oaks, CA: Sage.
Mohamed, R. (2006). The economics of conservation subdivisions. *Urban Affairs Review, 41*(3): 376–399. Sage Publications.
Orlowski, K. A. (2017). *Synchronization of sustainable development with land development*. World Sustainability Series, Springer: Sustainable Development Research at Universities in the Asia-Pacific.
Poole, M. (2005). *Touchstones: Ethics for the survival of our species*. Goldpanner Books.
Ravitch, S. M., & Riggan, M. (2012). *Reason and rigor: How conceptual frameworks guide research* (p. xiii). Thousand Oaks CA: Sage Publications.
Robinson, J. (2004). Squaring the circle? Some thoughts on the idea of sustainable development. *Ecological Economics, 48*(4), 369–384. https://doi.org/10.1016/j.ecolecon.2003.10.017.
Rokeach, M. (1973). *The nature of human values*. Free Press.
Shields, P., & Rangarjan, N. (2013). A playbook for research methods: *Integrating conceptual frameworks and project management* (p. 24.). Stillwater, OK: New Forums Press.
Soule, C. (2000). *The conservation subdivision design project: Booklet for developing a local bylaw*. Metropolitan Area Planning Council. Accessed at http://www.greenneighborhoods.org/OSRD.pdf.
Timmer, C. P. (2015). *Food security and scarcity: Why ending hunger is so hard*. Philadelphia: University of Pennsylvania Press.

Closed Loop Food Production and the 'Greening' of Corrections Facilities: Using Composted Kitchen Waste for Fresh Food Production

Wes Death and Edmund Horan

Abstract Food represents a large proportion of putrescible waste in landfill in Australia, with very little food waste processed to recover resources. This has environmental impacts in the form of greenhouse gas generation and other externalities. Prisons typically waste high levels of food compared to the general community and, with the exception of a few small-scale and isolated composting examples in Australian prisons, the majority of prison food waste is sent to landfill. Internationally, especially in the U.S.A., the so-called 'greening of corrections' has gained pace over the past few decades with the movement espousing similar objectives to certain United Nations (UN) Sustainable Development Goals. The literature suggests that prisoner participation in sustainability measures, environmental stewardship programs, landscape gardening and or agricultural/horticultural production decreases the chance of recidivism and is likely to improve the chance of post-release employment in the sustainability industry. As part of the greening of corrections, there are a number of examples of large-scale food waste composting operations used as the basis for significant, and often lucrative, fresh food production in prisons. Various U.S.A. prisons have demonstrated substantial waste disposal savings and further savings on fertilizers and fresh produce as a result of food waste composting processes. The types of food waste composting technologies and techniques utilised in these U.S.A. prisons, the savings in waste disposal fees and fertilizer costs, and the consequent food production could have beneficial impacts if modified for Australian conditions.

Keywords Food waste · Corrective services · Food production

W. Death (✉)
Master of Sustainable Practice, RMIT University, Melbourne, Australia
e-mail: s3045001@student.rmit.edu.au

E. Horan
Program Director—Master of Sustainable Practice, RMIT University, Melbourne, Australia

W. Leal Filho et al. (eds.), *Sustainable Development Research in the Asia-Pacific Region*, World Sustainability Series, https://doi.org/10.1007/978-3-319-73293-0_27

1 Introduction: Food Waste in Australia

Sustainability Victoria's (SV) Statewide Waste and Resource Recovery Infrastructure Plan (SWRRIP) found that in 2011–12 roughly 1,461,000 tonnes of waste was sent to landfill (Sustainability Victoria 2015). Of the waste sent to landfill, 929,000 tonnes was food waste and of this approximately 31,000 tonnes were recovered for processing (Sustainability Victoria 2015). The Victorian Organics Resource Recovery Strategy (VORRS) found that only 4% of food waste across all sectors was recovered (SV 2015).

There are various environmental impacts of food waste to landfill. A 2009 report by the Department of Environment, Water, Heritage and the Arts ('The Full Cost of Landfill Disposal in Australia') discusses "externalities" that include impacts on the environment, human health or social amenity of food waste to landfill (Department of Environment and Energy 2009).

The Intergovernmental Panel on Climate Change found that food waste is a significant contributor to greenhouse gas (GHG) emissions and accountable for approximately 5% of the global greenhouse budget. Bogner et al. (2008) found that this 5% consists of methane (CH_4) emission from anaerobic decomposition of solid waste and carbon dioxide (CO_2) from wastewater decomposition.

A significant proportion of prison waste comes from food. Lou and Nair's (2009) study at Hakea prison in Western Australia (WA) found that 68.2% of the prison's daily waste stream sent to landfill was putrescible waste and that 89.9% of that was food waste. While it is unknown exactly how much food waste in Australia comes from prisons, a community waste stream analysis conducted by Griffin et al. (2009) in the United States of America (U.S.A.) found that prison food waste per person was around three times as high as household food waste.

While there is little published information about large-scale composting or food waste processing in Australian prisons, there are some examples in New South Wales (N.S.W.) and Queensland where prisoners have been involved in small-scale processing or composting food waste for use in prison gardens (Department of Attorney General and Justice 2012; Department of Community Safety 2010). One particular success story is the Alexander Maconochie Centre, a minimum-maximum security prison in the Australian Capital Territory (A.C.T.), which uses vermiculture to compost food waste destined for the prison's gardens. A. Penders (personal communication, 5 April 2017) reported that between 560–700 l of food waste and additional shredded paper is put into their worm farms each week. With this composted food waste being used to enrich the soil, A. Penders (Email correspondence, 5 April 2017) advised that the prison produced around 5800 kg of vegetables during 2015–16. However, there was no information available in the literature relating to state-wide corrections organics recovery strategies, suggesting that the use of kitchen waste to support food production in prisons is not done systemically.

Davis et al. (2016) define Closed Loop systems as: "systems that are designed to incorporate waste back into one or more stages of production". There is little published information regarding Closed Loop food production in Australian prisons. While there is information about food production from kitchen waste in prisons internationally, mainly in the U.S.A., there is little published about the *process* of establishing such closed loop food production systems. There is a dearth of published case studies where prisons sought to close the production loop and described the entire process, including successes, failures or lessons learned. Therefore, there is very little information available that might guide the process for new programs that is based on past trial and error.

In the U.S.A., there is a body of research regarding food waste processing/composting in prisons, mostly in Washington State and New York State but with examples in many other states. It appears that U.S.A. prisons processing food waste to compost utilise the product on-site, either with landscaping or in food production, with some prisons donating their compost locally or selling it to prison employees. Mendrey (2013) provides Washington State's Sustainability in Prisons Project (SPP) as an example of a prison network that has implemented on-site food waste processing and food production and through this enjoyed substantial financial savings, prisoner job training and improved sustainability performance.

The reason as to why Australian States and Territories have not followed suit with network-wide strategies such as those employed in U.S.A. prisons is not clear and may vary across the States and Territories. However, there are indications that policies are taking shape and that the practical implementation of composting initiatives should follow.

There is recognition at the South Australian state level that correctional facilities are significant generators of waste (Zero Waste SA 2010). Such recognition has led to the generation of waste reduction targets at the Victorian state level. For example, the Victorian Department of Justice and Regulation specified in their 2014–15 Annual Report that they aim to reduce food waste to landfill by 10% compared to 2014–15 levels (Department of Justice and Regulation 2014–15). The Victorian State government has expressed interest in using food waste as the fuel for compost to increase on-site food production. Specifically, their 2013–14 Annual Report mentioned the installation of waste digesters to convert food and green waste into fertilizer for on-site horticultural production (Department of Justice and Regulation 2013–14).

While there is little experience with food waste to compost programs in Australian prisons, there is ample information from the U.S.A. corrections setting to guide both implementation and sustainability locally.

2 Composting in Prisons

A number of U.S.A. correctional facilities have documented both implementation of composting programs and their successes. With regards to program set-up, there is literature detailing composting techniques and technologies used to process food waste in prisons (Smith et al. 2006; WRAP, Bush et al. 2015). There are also some key case studies that discuss these different techniques and technologies and their suitability in different settings. These studies address some of the important factors that affected the efficiency of composting systems, including why they were suited to their particular corrections environment, and discuss other issues like food waste separation techniques and skills training (Emerson 2014; Goldstein 2003; Marion 2000; Ulrich and Nadkarni 2009).

Similarly, there is a convincing body of research in the U.S.A. justice system confirming that food waste composting programs have significantly reduced organic waste to landfill. For example, Ulrich and Nadkarni (2009) found that Cedar Creek Corrections Centre (C.C.C.C.) in Washington State removed around 30.2 kg of organic waste from landfill per prisoner per annum due to their composting programs (2009). Flammer (2014) found that a correctional facility with 3500 prisoners in San Diego County California uses large vermicomposting beds to process 15 tonnes of food waste per annum. As a consequence of the diversion of food waste from landfill and food waste processing on-site, there is evidence that composting has delivered significant financial benefits to participating prisons. For example, Emerson (2014) provided the example of Waymart Correctional Facility in Pennsylvania housing 1400 prisoners that saves approximately USD $30,000 per annum on waste disposal costs. Marion (2000) stated that Gowanda Correctional Facility in New York realised annual waste disposal savings of up to USD $89,959 by composting their food and yard waste in a covered windrow composting facility.

3 Agricultural and Horticultural Production in Prisons

Growing food in prisons can be a lucrative exercise and there is a significant body of published literature showing its high monetary value. In the U.S.A., there is a great deal of information confirming that in-house agricultural and horticultural production can significantly decrease the prison network's reliance on external suppliers, and deliver significant food mile and carbon emission reductions alongside triple bottom line benefits. As a surrogate measure of this benefit, MacCready (2014) found that some states in the U.S.A. have passed legislation requiring the establishment of prison gardens, mainly on the basis that the measures save the state money.

Bush et al. (2015) advised that across Washington State, approximately 77,110 kg of fresh produce is grown in correctional facilities. In Florida, Sherman (2014) estimated that during 2013–14, the value of produce grown in Florida correctional facilities was USD $4.5 million with a budget of USD $1 million (weight of crop was 4.03 million kg). Sherman (2014) estimated that this was equal to savings of USD $3.5 million. In the Australian context, N.S.W. and W.A. appear to have published the most information regarding successful food production in prisons. Information sourced from the Department of Corrective Services W.A. Annual Reports indicate 100% self-sufficiency in the supply of milk, eggs, and red meat, while internal production of fruit and vegetables meets 70% of their requirements (largely through three prison farms). Corrections W.A. reported that three prison farms supplied a total of 1,122,189 l of milk, 572,094 kg of meat and 110,337 dozens of eggs, which represented a total value of approximately $7 million (WA Department of Corrective Services 2013). In addition, market gardens in public prisons collectively produced vegetables to a value of $815,703 (WA Department of Corrective Services 2013).

However, there is no published information in the Australian context showing how much, if any, of our prison's agricultural/ horticultural production has utilised composted food waste as fertilizer or feed. Such data is available in the U.S.A. context. In Washington State at the CCCC, Ulrich and Nadkarni (2009) found that compost made on-site met nearly all the facility's fertilizer requirements and contributed to an annual production of 5900 kg of fresh produce for use in the prison. Similarly, Mendrey (2013) provides an example of a corrections centre in Shelton (Washington State) that uses their compost to produce 6803 kg in food from food waste, half of which is donated to local food banks. There is nothing demonstrably unique about these U.S.A. facilities, and the evident horticultural skills of W.A. prisons would clearly translate well to the processing of food waste to compost. If this was realised, it would provide not only an alternate activity for prisoners but also a substantial cost saving in terms of reduced waste removal and lower expenditure on fertilizer.

4 Sustainability Programs in U.S.A. Prisons, Green Jobs Training and Post-release Employment

Closed Loop food production is just one element of the 'greening' of Corrections—a phenomenon that has gained significant momentum in the U.S.A., especially over the past 20 years. The most widely publicised network of such prisons is the SPP in Washington State.

The SPP was established in 2003 in partnership with Evergreen State College. Broadly, the SPP seeks to increase the resource efficiency of prisons and to deliver a range of environmental, personal or therapeutic benefits, alongside training prisoners in green industries. LeRoy et al. (2012) found that the SPP has delivered on its overarching aim of achieving significant economic savings.

Other states in the U.S.A. have implemented similar measures to increase the sustainability credentials of corrections. For example, in 2012 the Illinois Department of Corrections unveiled a plan to create more efficient and sustainable facilities, minimise resource use and maximise recycling, develop workplace skills in sustainable industries, and develop links to green industries for post-release employment in order to reduce recidivism (Illinois Department of Corrections 2012). Similarly, in 2012 the Ohio Department of Rehabilitation and Corrections developed a 3-year strategic sustainability plan to reduce consumption of water and electricity and decrease waste disposal costs. The program seeks to provide skills training to offenders on environmental issues, green jobs skills and to develop links to green industries and employment opportunities. The program's goal is to save money, preserve the environment, and prepare inmates to lead sustainable lives post release (Ohio Department of Rehabilitation and Correction 2012).

As a result of the UN Sustainable Development Summit in September 2015, world leaders pledged to adopt 17 Sustainable Development Goals (SDGs) including 169 targets effective from 1 January 2016. These goals, with the overarching aims of fighting inequality and injustice, fixing climate change and ending extreme poverty, are designed to frame national policy agendas over the following 15 years. The prison sustainability reforms already discussed can achieve aspects of the SDGs, namely the fourth and twelfth goals relating to "inclusive and quality education for all (to) promote lifelong learning" and "sustainable consumption and production patterns", respectively (United Nations 2016). The priorities set out in the aforementioned prison sustainability plans in Washington State and Ohio sought to satisfy similar goals as articulated in the SDGs in regards to equal access to affordable and quality technical education and the promotion of peace and non-violence. Also, such measures also satisfy the twelfth SDG as they seek to "substantially reduce waste generation through prevention, reduction, recycling and reuse" (United Nations 2016).

While it is acknowledged that recidivism levels are hard to track due to a lack of resources for monitoring, there is some evidence that prisoner participation in sustainability/prison garden programs in the U.S.A. has reduced recidivism. One example results from collaboration between the Insight Garden Program (IGP) and the Planning Justice organisation at San Quentin prison in Marion County California.

An element of Planning Justice's work is to address injustices in the food movement by promoting economic, social and food justice. In the IGP, prisoners are trained in organic food production and permaculture and involved in the planning, designing, building and maintaining of prison gardens and, most importantly, attempts are made to link the participants with jobs in the industry upon release. A 2011 survey conducted by IGP in MacCready (2014) found that 117 men employed post-release from 2003 to 09 re-offended 10% of the time compared to 70% for non-program participants. MacCready (2014) also discussed another program in a 2014 dissertation reporting similar levels of re-offending (10%) for those involved in the program.

A similar study by Van Der Linden (2015) found that recidivism rates for graduates of green programs were between 10–24% depending on the specific program. Despite these positive findings indicating that participation in prison sustainability programs decreases recidivism, Van Der Linden (2015) acknowledges that data on the topic comes from small and self-selected samples.

While there appears to be a positive correlation between participants in Green prison initiatives and decreased recidivism, there also seem to be positive impacts upon post-release employment prospects. The U.S. Bureau of Labor Statistics found that between 2010 and 2011 those employed in "green" jobs increased from 3.1 to 3.4 million (Elliott and Lindley 2017) and Morgan (2011) advised that some within the not-for-profit sector believe that new green jobs should present increasing opportunities for ex-offenders in the U.S.A.; however, there is a lack of data confirming that prisoners have been able to obtain sustainability-focused jobs. Unfortunately, due to the difficulty and expense involved in tracking ex-prisoners, there are no reliable figures about green program participant post-release employment in the sustainability sector, environmental stewardship sector, landscape gardening and/or the agricultural/horticultural industry. However, one study discussed by Clarke (2011) asking prisoners and prison staff about the benefits of prisoner participation in SPP programs found that an increase in job skills and employment prospects was the third highest perceived benefit behind interest in environmental topics and positive social interactions. Furthermore, there are indications that participants were more resilient, had more life skills and were generally more employment ready. According to various studies, horticulture as a form of indirect therapy has a number of psychological benefits and healing outcomes that impact upon employment prospects (Krasny 2015; MacCready 2014; Sevin 2015). Rice and Remy (1998) refer to garden programs that cultivate "healthy self development" with McGuinn and Relf proposing that juvenile offenders relish the opportunity to prove that they can be successful in horticultural work and use such success to develop a positive self image (Elings 2006).

5 Conclusion

It is clear that prisons produce significant waste, of which food represents a high percentage. This paper provides evidence that food waste to landfill can have associated externalities and represents a significant missed opportunity both in terms of monetary savings and prisoner rehabilitation.

Despite Australia lagging behind countries like the U.S.A. in the institution of Closed Loop programs, there appears to be a growing awareness amidst policy makers that there are social, environmental and financial benefits of composting prison food waste. There is some evidence in Australia that isolated measures have been implemented where compost from food waste is being used to produce fresh produce on-site. International examples demonstrate how composting food waste on-site can deliver significant waste disposal savings and provide inputs for meaningful on-site fresh fruit and vegetable production. That there are established horticulture and agriculture programs already operating in Australian prisons suggests that translating the findings from the U.S.A. correctional context is feasible.

Importantly, food waste to compost initiatives have formed a part of broader prison sustainability reforms which include training in horticulture, landscaping, environmental stewardship and gardening. Evidence is also available showing that participating in such programs can be a form of indirect therapy for prisoners, which in turn can reduce recidivism and improve prisoner post-release employment prospects.

In sum, the evidence base from our overseas counterparts suggests this myriad of benefits is also available to Australian prisons, and creates a strong argument for further research into Closed Loop food production in Australian prisons.

References

Bogner, J., Pipatti, R., Hashimoto, S., Diaz, C., Mareckova, K., Diaz, L., Kjeldsen, P., Monni, S., Faaij, A., Qingxian, G., Tianzhu, Z., Mohammed Abdelrafie, A., Sutamihardja, R. T. M. & Gregory, R. (2008). Mitigation of global greenhouse gas emissions from waste: Conclusions and strategies from the Intergovernmental Panel on Climate Change (IPCC) Fourth assessment report. Working Group III (Mitigation). *Waste Management & Research, 26*, 11–32.

Bush, K., Vanneste, J., Pacholke, D., Trivett, J., Sinclair, S., & Heinitz, E. (2015). Department of corrections walks the walk on sustainability. *BioCycle, 56*(65–68), 71.

Clarke, S. E. (2011). Assessing the rehabilitative potential of science and sustainability education in prisons: A study of the Sustainable Prisons Project. Master's thesis, Environmental Studies, The Evergreen State College, Olympia, Washington.

Davis, S. C., Kauneckis, D., Kruse, N. A., Miller, K. E., Zimmer, M., & Dabelko, G. D. (2016). Closing the loop: Integrative systems management of waste in food, energy, and water systems. *Journal of Environmental Studies and Sciences, 6*, 11–24.

Department of Attorney General and Justice. (2012). *Annual Report 2011–12* [Online]. Department of Attorney General and Justice Available: http://www.justice.nsw.gov.au/Documents/AnnualReports/AGJ_AR_2011-12_Complete.pdf. Accessed September 30, 2016.

Department of Community Safety. (2010). *Corrections News* [Online]. Department of Community Safety - Queensland Government. Available: http://www.correctiveservices.qld.gov.au/Publications/Corrections_News/2010/February/CN_February_Full.pdf. Accessed February 7, 2017.

Department of Environment and Energy. (2009). *The full cost of landfill disposal in Australia* [Online]. Department of Environment and Energy. Available: http://www.environment.gov.au/protection/national-waste-policy/publications/full-cost-landfill-disposal-australia. Accessed September 30, 2016.

Department of Justice and Regulation. (2013–14). *Department of Justice Annual Report 2013–14* [Online]. 1 Treasury Place, Melbourne: Victorian Government. Available: http://www.parliament.vic.gov.au/file_uploads/Department_Of_Justice_Annual_Report_2013-14_bpXQJ5G1.pdf. Accessed May 28, 2017.

Department of Justice and Regulation. (2014–15). *Annual Report* [Online]. Victorian Department of Justice and Regulation Available: http://assets.justice.vic.gov.au/justice/resources/d2a58cac-e746-4942-a6be-9e4edc28db85/djr_annualreport2015.pdf. Accessed February 2, 2017.

Elings, M. (2006). *People-plant interaction: the physiological, psychological and sociological effects of plants on people*. Farming for Health: Springer.

Elliott, R. J. R., & Lindley, J. K. (2017). Environmental jobs and growth in the United States. *Ecol Econ, 132*, 232–244.

Emerson, D. (2014). Prison food waste composting a long-term success. *BioCycle, 55*, 36.

Flammer, R. (2014). Worms compost food scraps at correctional facility. *BioCycle, 55*, 34–36.

Goldstein, N. (2003). Comparing composting technologies at correctional facilities. *BioCycle, 44*, 28–32.

Griffin, M., Sobal, J., & Lyson, T. A. (2009). An analysis of a community food waste stream. *Agriculture and Human Values, 26*, 67–81.

Illinois Department of Corrections. (2012). *Sustainability plan 2012* [Online]. Illinois Department of Corrections Available: https://www.illinois.gov/idoc/reportsandstatistics/Documents/2012IDOCSustainabilityPlan.pdf. Accessed September 30, 2016.

Krasny, M. E. (2015). Greening programs to facilitate prisoner reentry.
Leroy, C. J., Bush, K., Trivet, J., Gallagher, B. (2012). *The sustainability in prisons project - An overview (2004–2012)* [Online]. Available: http://sustainabilityinprisons.org/wp-content/uploads/2016/02/Overview-cover-text-reduced-size.pdf. Accessed September 30, 2016.
Lou, X. F. & Nair, J. (2009). Evaluating alternative waste management strategies for Hakea prison.
Maccready, S. D. (2014). *Food, farming, and our justice system: Horticulture programs in correctional settings.* 3648372 D.P.A., University of La Verne.
Marion, J. (2000). Composting 12,000 tons of food residuals a year. *BioCycle, 41,* 30–35.
Mendrey, K. (2013). Correctional facility composting in Washington State. *BioCycle, 54,* 32–35.
Morgan, L. (2011). Green job training in prisons benefits everyone. *Corrections Today, 73,* 34–37.
Ohio Department of Rehabilitation and Correction. (2012). *Three year sustainability plan - 2012* [Online]. Ohio Department of Rehabilitation and Correction Available: http://www.drc.state.oh.us/web/Sustainability_Plan.pdf [Accessed].
Rice, R. (1998). Impact of horticultural therapy on psychosocial functioning among urban jail inmates. *Journal of Offender Rehabilitation, 26,* 169–191.
Sevin, A. (2015). GREEN LIFE: Helping rebuild lives post-incarceration. *Earth Island Journal.* San Francisco: Earth Island Institute.
Sherman, A. (2014). *Edible crops program grew under scott.* Florida. Politifact.
Smith, D. R., Cawthon, D. L., Sloan, J. J., & Freeman, T. M. (2006). In-vessel, mechanical rotating drum composting of institutional food residuals. *Compost Science & Utilization, 14,* 155–161.
Sustainability Victoria. (2015). *The statewide waste and resource recovery infrastructure plan* [Online]. Level 28, Urban Workshop 50 Lonsdale Street Melbourne: Sustainability Victoria (SV). Available: http://www.sustainability.vic.gov.au/our-priorities/statewide-waste-planning/2015-2020-priorities/statewide-waste-and-resource-recovery-infrastructure-plan. Accessed 14 January 2017.
SV. (2015). *Victorian organics resource recovery strategy* [Online]. Level 28, Urban Workshop 50 Lonsdale Street Melbourne: Sustainability Victoria. Available: http://www.sustainability.vic.gov.au/organicsstrategy. Accessed 14 January 2017.
Ulrich, C., & Nadkarni, N. M. (2009). Sustainability research and practices in enforced residential institutions: Collaborations of ecologists and prisoners. *Environment, Development and Sustainability, 11,* 815–832.
United Nations. (2016). *Sustainable development goals* [Online]. United Nations. Available: http://www.un.org/sustainabledevelopment/sustainable-development-goals/. Accessed 27 May 2017.
van der Linden, S. (2015). Green prison programmes, recidivism and mental health: A primer. *Criminal Behaviour and Mental Health, 25,* 338–342.
WA Department of Corrective Services. (2013). *Annual Report 2012–13.* Department of Corrective Services.
WRAP. *Her Majesty's prisons case study - UK* [Online]. Waste and Resource Action Program Available: http://www.wrap.org.uk/sites/files/wrap/HerMajesty'sPrisons.pptx. Accessed February 1, 2017.
Zero Waste SA. (2010). *Zero waste SA: Business plan 2010–11 and future directions 2011–13* [Online]. Government of South Australia. Available: http://www.zerowaste.sa.gov.au/upload/2010ZWSABUSINESSPLAN20.1.11.pdf. Accessed May 28, 2017.

Part IV
Case Studies: Sustainable Buildings

Manifestos for Sustainable Development: Sustainable Modular Steel-Precast Concrete Building Construction System for Dwellings in Singapore

Kian Heng Liew

Abstract Existing and ageing landed residential dwelling properties in Singapore are continuously undergoing reconstruction and improvements. Due to the many faceted architectural designs that are unique to each and every individual owner's brief, most of such constructions are non-repetitive and use of reinforced in situ concrete with brickworks is commonly adopted. With recent drastic manpower shortage, such traditional construction, which normally takes 6–12 months to complete, is becoming very challenging. The use of sustainable modular steel-precast concrete building construction system paves a path for future construction of small-scale one-off design dwellings. This paper explores the implementation, implications, innovations and achievements of using this modular system in one such project. This is likely the first project in Singapore to test-bed such system for landed residential dwelling addition and alteration. It was conceived by a professional engineer, the project expert with the full support of the owner right from the onset. The paper offers professionals, builders, developers and other stakeholders in sustainable development to meet their goals in such building constructions.

Keywords Sustainability · Efficacies · Safety · Productivity · Steel-precast concrete · Dwellings · Green

1 Introduction

Singapore has been Southeast Asia's most modern city in building constructions and developments. The city blends Asian and colonial architectures and cultures. Its unique ethnic tapestry exudes the construction of diverse and faceted architectural designs. Its rapid rate of development due to economic progress and population

K. H. Liew (✉)
SIM-RMIT, Institute for Engineering Leadership, NUS, Liew Strategics,
25 Bukit Batok Crescent #04-04, Singapore 658066, Singapore
e-mail: liew@strategics.com.sg

W. Leal Filho et al. (eds.), *Sustainable Development Research in the Asia-Pacific Region*, World Sustainability Series, https://doi.org/10.1007/978-3-319-73293-0_28

shift signify the need for continual upgrade, reconstruction and improvement of infrastructure and housing needs, landed residential dwellings in Singapore have become a staple for unique designs. Due to the diverse society, each dwelling has its own unique flavour, originating from the owner's personal preference. This introduces the demand for more complicated designs. Because of these diverse architectural projects, and very limited compact land constraints, most of such construction are non-repetitive and traditionally use reinforced in situ concrete with brickworks. These offer a huge scope in the planning and designs for improvements and redecoration works which is called as additions and alterations (A&A) in Singapore. A&A is defined as minor works to an existing property. If the proposed work is less than 50% of the existing landed house's gross floor area (GFA), external walls and structural elements, the proposal will be considered under A&A[1] by the building authorities. Should there be an increase in GFA in the A&A, and especially with multitude of varied designs, traditional formwork concreting is normally adopted.

The construction sector in Singapore is seen as a low productivity sector because of its "dangerous, dirty and demeaning"[2] and evident employment of a sizable number of foreign workers. This hiring comes from the circumstance that Singapore nationally has a manpower shortage. Acute shortage is probably the biggest challenge for Singapore on a macro and micro scale, especially in the building and construction industry. With these recent drastic manpower shortages, such traditional construction methods will take an even more extensive and lengthy time to complete. These traditional constructions can take up to 6–12 months for such A&A works. Furthermore, complications and problems such as site constraints, noise and dust complaints, and potential damages to neighboring properties can pose a gargantuan setback to the construction process. Thus, prolonging the time and cost for the construction to be completed.

Pioneering the way forward, Singapore has done exceptionally well in public housing, and should be looking at new ways to tackle modern construction problems in A&A to landed residential dwellings. With new technology available, these problems may be mitigated in the immediate future. Sustainable modular steel-cum-precast concrete construction method is possibly the next step forward by using the best in steel and concrete.

This paper explores the implementation, implications, innovations and achievements of using this modular system in one such A&A project. This is likely the first project in Singapore to test-bed such system for landed residential dwelling addition and alteration. It was conceived by a professional engineer, the project expert with the full support of the owner right from the onset.

[1]https://www.ura.gov.sg/uol/circulars/2005/sep/dc05_21.

[2]http://www.nas.gov.sg/archivesonline/speeches/view-html?filename=2000052503.htm.

2 Design

The design of this project used structural steel and precast concrete to meet the owner's requirement of a simple and cost-effective solution to blend in with the existing semi-detached house. After much deliberation, modular structural steel and pre-cast construction was considered the most suitable with the use of hollow steel sections and pre-cast concrete slabs and walls. Aesthetically, the hollow steel sections and pre-finished pre-cast walls and slabs combination was acceptably appealing by the owner.

In addition to complying with all the government statutory requirements and by-laws, the design of the project had to incorporate detailing of the structural and architectural elements. How the elements were modularized into manageable portions was meticulously deliberated. The design was to ensure that each module could be fabricated, transported and installed to enable robust and stable conditions through out construction as a permanent building.

Therefore, a raft foundation was proposed to be of cast in situ to receive the structural steel modules and precast walls instead of several separate individual pad footings. The structural steel was divided into seven modules. All steel was welded in the workshop with the largest hollow section of 200 × 100 × 6 mm and all joints between modules were innovatively bolted on site without any welding (Figs. 1, 2, 3 and 4).

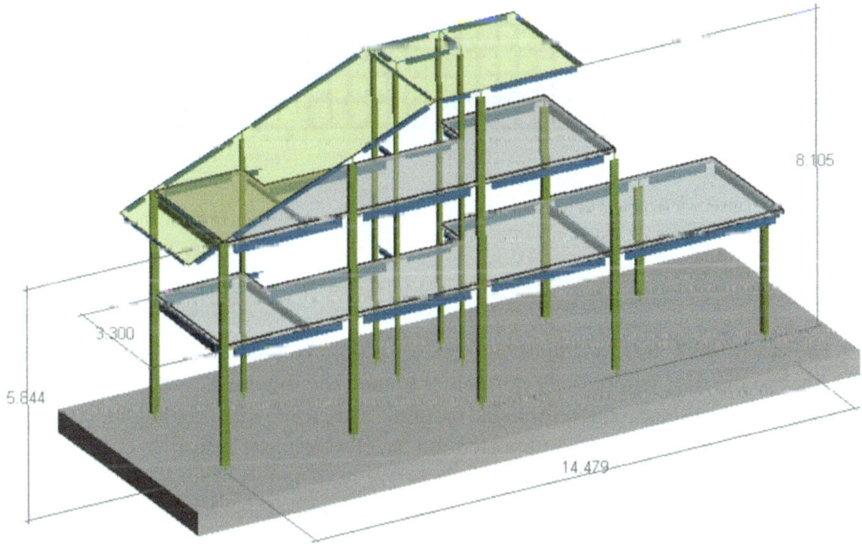

Fig. 1 3-D perspective

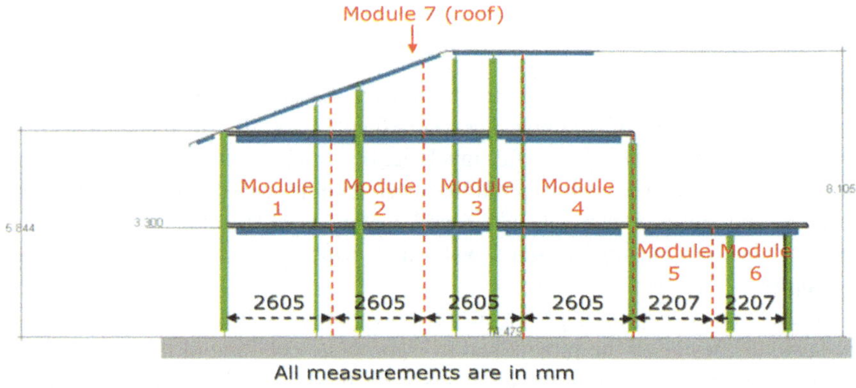

Fig. 2 Structural steel modules

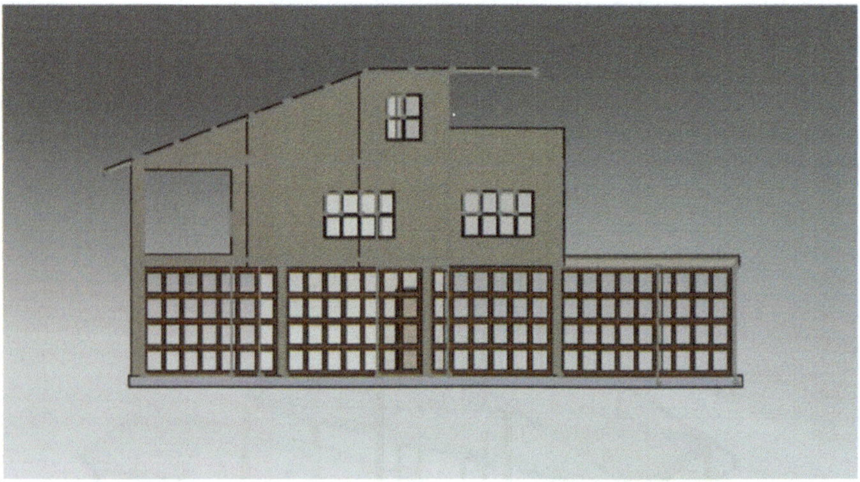

Fig. 3 Side elevation

The walls were 100 mm thick pre-cast in one long section, longest being from the floor to the roof without any horizontal joints. There were eleven pre-cast panels and eighteen pre-cast concrete slabs in total. The roof was constructed of light-weight metal roofing sheet supported on steel C channel purlins connected to $100 \times 100 \times 6$ mm hollow section rafters.

The major focus for the project was based around efficiency, productivity and sustainability. Each step of the pre-construction, construction and post-construction process was methodically thought-out to ensure that the most proficient way of handling the project was possible. Unlike traditional construction, the use of a modular construction is more labour-saving, cost-saving and time-saving.

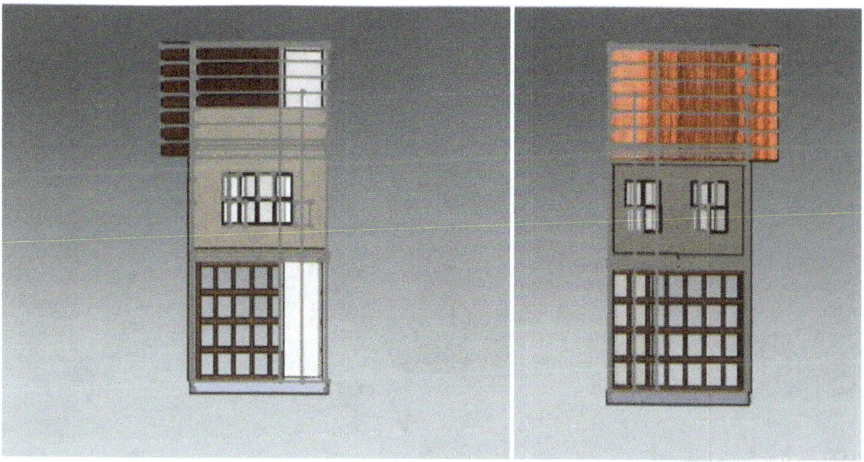

Fig. 4 Front and back elevation

3 Labour-Saving

Traditional construction is a process whose primary structural elements are constructed entirely or largely on site. There are many problems and difficulties that come with traditional construction method. Firstly, traditional construction is a linear process that means that each construction step has to be completed before the next can begin. This means that time is wasted and not utilized as efficiently as possible. Secondly, weather and labour can greatly affect the construction process. More specifically, weather damage and inconsistent labour yield can greatly affect quality and productivity. In Singapore, there are monsoon periods that cause extended periods of rainfall. The rain can potentially cause damage to construction materials, machines and processes. Unlike machines, where productivity is constant and expected, manual human labour is inconsistent and varied. Hence, the amount of productive labour yield fluctuates constantly. Thirdly, since all construction occurs on-site, the schedule is dependent on weather conditions. This significantly interrupts the amount of on-site construction that can be done as well as under-utilizes labour as they cannot work under inclement conditions.

The estimated man-days for traditional construction method compared with modular construction method actually completed for the project is as shown below. The labour savings of (1221 − 719 = 502 man-days) is estimated to be about 40% (Table 1).

Another measure of labour saving is by comparison with the building authority prevalent industry practices and Code on buildability. The buildability score was created by the Building Construction Authority (BCA) of Singapore for all buildings to raise productivity in the built sector and reduce the reliance on foreign

Fig. 5 Lifting and installing onto pre-fixed sleeves of first module on site

Table 1 Traditional versus modular construction man days

Work	Traditional (Man × Days)	Modular (Man × Days)
Raft foundation	6 × 7	6 × 7
Structure (RC or Steel)	6 × 30	6 × 14
Concrete slab	10 × 30	8 × 14
Concrete walls	10 × 30	8 × 14
Roofing and skylight	4 × 14	4 × 14
Lift installation and commissioning	2 × 10	2 × 10
Electrical power, lighting and lightning system	3 × 21	3 × 18
Sanitary and water installation	3 × 7	3 × 5
Aluminum and glass works for windows, doors and lift	5 × 21	5 × 18
External works and drainage	5 × 10	5 × 10
Other finishing works	4 × 21	4 × 21
Total	**1221**	**719**

workers.[3] BCA calculates Buildable Design Score using the Code of Practice and also Buildable Design Appraisal System (BDAS).

This Code sets out the requirements of minimum Buildable Design Score and minimum Constructability Score for buildings, the method for determining the Buildable Design Score and the Constructability Score as well as their submission procedures. It also sets the requirements for the specific productive technologies

[3]https://www.bca.gov.sg/BuildableDesign/others/copbdnov2015.pdf.

Table 2 Labour saving index

Description	Labour saving index (points)
Pre-cast concrete system: full pre-cast	1.00
Structural steel system: steel beam and column (without concrete encasement)	1.00
Bonus points: innovative structural steel connections	2.00
Bonus points: recommended pre-cast joint types ($\geq$65%)	3.00
Mechanical connection for vertical precast wall joints ($\geq$65%)	2.00
Precast concrete wall: off-form precast concrete external walls and columns	1.00
Bonus points: design without complex form	3.00

that need to be complied with by developments built on Government Land Sales sites for which the technologies have been stipulated as land sales conditions.[4]

BDAS was developed as a means to measure the potential impact of building design on the usage of labour.[5] It consists of three components: Structural System, Wall System and Other Buildable Design Features.

Table 2 shows the maximum points possible as scored by this project. Even though BDAS buildability score only applies to buildings with a Gross Floor Area (GFA) of 2000 m^2 and above, and the project building covers less than 2000 m^2, this addition and alteration project still grades highly in the various Structural and Wall system requirements (see Footnote 3).

The Labour Saving Index (LSI) is defined as: a value given to a particular building system which reflects the relative difference in site labour productivity associated with the various structural and wall systems. In certain instances, the LSI could be further lowered to discourage the use of labour intensive elements or components. A LSI is also given for the use of prefabricated reinforcement/cages in cast in situ components (see Footnote 3).

As such, the points acquired by this project show that it complies optimally and favourably with the requirements of the BCA for buildability. Thus, furthering the project's initiative to similarly push for more such productive and efficient construction methods.

4 Time-Saving

As mentioned, traditional construction is a linear process where each step has to be completed before the next can start. Looking an illustrated timeframe for this modular construction method verses traditional construction method, the vast

[4]Ibid.

[5]https://www.bca.gov.sg/BuildableDesign/buildable_design_bdas.html.

Table 3 Schedule of traditional construction method versus modular construction method

Day schedule (traditional)	Traditional method	Modular method	Day schedule (modular)
1–2	Grading and site preparations	Grading and site preparations	1–2
3–5	Foundation excavation, compaction, services and rebars	Foundation excavation, compaction, services and rebars	3–5
		Fabrication modules and pre-cast concrete	3–14
6–9	Foundation concreting and complete	Foundation concreting and completion	6–9
10–14	Concrete curing and prep for 2nd storey R.C. columns	Concrete curing and prep for 1st storey steel sleeves	10–14
15–25	Form work for 2nd storey	Start shipping modules	15
		Installation setting modules and shipping of pre-cast concrete	16–19
		Installation of pre-cast concrete slabs and walls	18–23
26–46	Rebar and Concrete 2nd storey	Roofing works	25–39
47–57	Form work for 3rd storey	Electrical, lift and sanitary works	40–58
57–77	Rebar and Concrete 3rd storey	Aluminum and glass works	50–68
78–119	Wall and roofing works	External works and drainage	60–70
110–150	Electrical, lift and sanitary works	Finishes	60–81
140–161	Aluminum and glass works		
150–160	External works and drainage		
150–171	Finishes		

Note Traditional method would take about around 180 days (6 months) while the modular method actually took about 90 days (3 months) to complete this project

differences in construction and time efficiency can be realized. Table 3 illustrates the comparative differences between both methods for this project.

As seen in the table above, the modular method takes about half the time to complete, 90 days against 180 days, than a traditional would have taken. This is 50% time saving on completion. Upon reviewing, the reason for this drastic difference in time is a result of the simultaneous construction of foundations, modules and pre-cast concrete. Unlike the traditional method (linear) that has to wait for the foundations to be completed before on-site construction can begin level-by-level.

For this project, all columns, beams and walls were fabricated in modules in a workshop before setting up at the construction site. Each steel member was designed beforehand to fit specific site-verified dimensions. The productivity and quality of the works were not affected by weather or machineries limitations.

Furthermore, inspection and welding tests were all done before the transportation to the construction site.

The walls and slabs were also made in another workshop casting yard. The walls and slabs were pre-finished with all recesses and holes and thus did not require any further hacking works on site. Also, plastering work or skim coat did not need to be applied on-site. Similarly, these were inspected and materials tested for quality check.

With all steel welded, walls and slabs pre-cast, these just had to be installed. The installation of the steel members was done by lifting the pre-fabricated steel modules by a 100T crane, slotted into pre-installed sleeves on the foundation. The pre-cast slabs were placed onto the structural steel to serve as a platform without the need of erecting any scaffolds.

Consecutively pre-cast walls were clad onto the structural steel modules and slotted into pre-installed dowel bars anchored into the raft foundation. All the pre-cast slabs and walls were secured by innovative steel-concrete connections using chemical anchor bolts without any welding on site.

5 Cost-Saving

Construction costs for traditional methods is normally misconceived to be cheaper than modular methods. However, this trend may not be true as modular construction becomes more advanced and widespread.

For this project, the breakdown costs are as follows (Table 4).

The Gross Floor Area for this project is 90 m^2. So the cost per GFA is S$2411 per m^2. In Rider Levett Bucknall's Singapore Construction Market Quarterly report, the cost per CFA for landed residential semi-detached house is S$2450–S $3300 per m^2.[6] CFA is defined as the area of all building enclosed covered spaces measured to the outside face of the external walls including covered basement and above ground car park areas (see Footnote 6). The Gross Floor Area of 90 m^2 for this project includes car porch and small canopies which is synonymous to CFA.

This project cost per GFA (hence CFA) is almost similar to Singapore's residential landed semi-detached houses—similar to prevalent national construction prices generally based on traditional method of construction. Hence, even for such a small project, the unit cost is indeed competitive.

This proves that using modular construction is not the more expensive option. Using innovative designs, such as without any external walls supported on beams but instead directly onto the raft foundation in one full panel, innovative connections as well as adoption of optimum materials for modular construction can substantially lower the cost of the project.

[6]http://assets.rlb.com/production/2016/06/21003826/RLB-Construction-Market-Quarterly-Update-Singapore-March-2016-1.pdf.

Table 4 Breakdown costs of project

Works	Costs (S$)
Raft foundation	10,000
Structural steel	35,000
Pre-cast concrete supply	10,000
Pre-cast slab installation with self leveling screed/topping	9000
Pre-cast wall installation	8000
External works and drainage	5000
Roofing and skylight	17,000
Lift including installation and commissioning	50,000
Electrical power, lighting and lightning system	10,000
Sanitary hard wares supply	2000
Sanitary and water installation	6000
Aluminum and glass works for windows	20,000
Other finishing works	5000
Fees and other admin costs	30,000
Total	217,000

Should the cost of the lift be taken out, then the cost per GFA for building works would be S$1856 per m^2. This would be 23% lower than the lowest national construction figure.

6 Sustainability

In Singapore, every building has to follow the Code for Environmental Sustainability of Buildings. The code was put in place to establish environmentally friendly practices for the planning, design and construction of buildings. The code sets out the minimum environmental sustainability standard for buildings and administrative requirements. It adopts the BCA's Green Mark criteria as the compliance method in assessing the environmental performance of a building development. The minimum Green Mark score for building works related to a residential building is 50 points and the maximum Green Mark score is 100 points.

Based on the GFA criteria, this project was not required to comply with the code. However, even if it is assessed, this project strives to be as efficient and environmentally sustainable as possible. Thus, using the code as a benchmark for comparison, one of the areas that was put into consideration for sustainability is the structure's Concrete Usage Index (CUI). CUI is an indicator of the amount of concrete required to construct a superstructure that includes structural and

Table 5 Building concrete area, volume and concrete usage index

Member	Area (m^2)	Volume (m^3)	Total floor area (m^2)	Total volume (m^3)	Concrete usage index (m^3/m^2)
Front Walls	5.37	0.537	73.0[a]	15.5	0.212
Side Walls	59.2	5.92			
Back Walls	6.93	0.693			
Interior Walls	4.97	0.497			
Slabs	73.0	7.30			
Parapet	5.33	0.533			

[a]Same as floor slab area since total constructed floor is fully covered by precast slabs
The calculation of CUI is just a part of the Green Mark Scheme. It is under "Part 3—Environmental Protection". This part focuses on design, practice, and selection of materials and resources that would reduce the environmental impact of built structures

non-structural elements. CUI is defined as volume of concrete in cubic metres (m^3) to cast a square metre (m^2) of constructed floor area.[7] The calculation of CUI does not include concrete used for external works and sub-structural works such as basements and foundations.[8] Table 5 illustrates the total area, volume and CUI of the concrete structure for this project.

In this part, the use of sustainable construction, recycling and the adoption of building designs, construction practices and materials that are environmentally friendly and sustainable are encouraged (see Footnote 7). There are prerequisite requirements to achieve Green Mark Gold^{Plus} and Green Mark Platinum. Attain 3 points for Green Mark Gold^{Plus} and 5 points for Green Mark Platinum. Points are calculated by CUI. Table 6 shows the breakdown for the point system.

As seen from Table 5, the CUI of the concrete structure achieved is 0.212. This means that the points allocated is 5—Green Mark Platinum prerequisite is optimally attained.

7 Safety

The Ministry of Manpower (MOM) of Singapore has put in place Workplace Safety and Health Act (WSH). WSH covers safety, health and welfare of persons at work in a workplace. It requires stakeholders to take reasonably practical steps for the safety and health of workers and others affected by work.[9] WSH is a vital part of a

[7]https://www.bca.gov.sg/SustainableConstruction/others/sc_cui_final.pdf.

[8]Ibid.

[9]http://www.mom.gov.sg/workplace-safety-and-health/workplace-safety-and-health-act/what-it-covers.

structure to develop good safety habits in all individuals, so as to build a strong safety culture in workplaces. There are three guiding principles of WSH: reduce risk at the source by requiring all stakeholders to remove or minimise the risk they create, encourage industries to adopt greater ownership of safety and health outcomes and impose higher penalties for poor safety management and outcomes (see Footnote 9).

Safety is number one priority when it comes to construction. This project offers no different opinion and planning on safety was from the onset. All pre-construction, construction and post-construction methods were meticulously thought out to mitigate risks and increase safety. This modular method simplifies the construction process and, in so doing, reduces workplace health and safety risks by avoiding additional and unnecessary construction processes.

Scaffolding platforms introduce many safety risks that could be easily nullified by using this innovative modular construction method. The greatest risk presented by scaffolding is falling from a high elevation. This is the most common scaffolding hazard when working at heights.

The people under the scaffolds are also at risk due to falling debris, tools or materials. Such were totally eradicated for this project.

The pre-cast walls, modular steel members and slabs were fitted together, like a jigsaw puzzle prefinished without the need of additional building procedures such as to external facade. This negates the need for any temporary works. Eliminating scaffold platforms as there were no in situ casting, no concrete encasement, no plastering and no skim coats ensured deliberate major reductions of risks on site. It was deliberate eliminating risk at source by design from the onset.

This modular construction method only required a crane to construct. In this project, a 100T crane was used to install pre-fabricated steel modules, pre-cast slabs, pre-cast walls and roof.

Figure 6 shows workers guiding the crane operator to place concrete slabs that are used for staging the construction processes. Thus, eliminating the need for scaffold platforms. These slabs just have to be installed and slotted onto the steel members. Concrete slabs are sturdier and provide a safer platform to work on than scaffolds and staging. This is possibly the best, optimum and safest use of pre cast slabs working at heights (Figs. 5 and 7).

Table 6 Breakdown of CUI point allocation

Project CUI (m^3/m^2)	Points allocation
≤ 0.70	1
≤ 0.60	2
≤ 0.50	3
≤ 0.40	4
≤ 0.35	5

Fig. 6 Lifting and placing of pre-cast slabs onto modules

Fig. 7 Lifting of pre-cast wall and installing and clad onto modules at side elevation

8 Implementation

Implementation was as stated previously, where pre-cast walls, pre-fabricated slabs and steel members are installed in modules. Every project on site has its challenges. For such a unique project, there were site issues anticipated, identified early and attended to quickly. Some implications and improvements could be made and are specified below.

Site measurements had to be adjusted and retaken constantly to tally with the design dimensions. On-going fabrication of steel members, pre-cast slabs and pre-cast walls had to be adjusted to meet actual site conditions. Further requirements, such as owner's final detail information, like toilet pipe openings, and clashes with existing obstruction of services also had to be taken into consideration.

Delivery of the long span slabs and pre-cast walls required long trailers and multiple sturdy supports to ensure that no cracking or excessive deflection materialized in the slabs and pre-cast walls due to vibrations created during transportation. Lifting of slim 100 mm thick 9 m walls need special cranage skills to avoid snapping. Road clearances had to be taken into account during transportation. Different routes had to be planned beforehand to ensure that no obstructions could appear along the trip.

On site, the large and long trailers and crane required traffic control throughout the whole lifting and construction process. There were road restrictions when the crane and trailer were parked side by side along the road.

Tolerances must be in good order. Otherwise, jacking would be needed to be carried out to force fit the joints on-site. Furthermore, small gaps between slabs were required to be filled due to slight site modifications. Special innovative connections using chemical anchors had to be adjusted due to the presence of rebar in concrete.

The roofing sheet could have been installed onto the roof rafters and purlin entirely at ground level. However, due to site constraints, this could not be implemented. Instead, the roofing sheets were placed after the rafters-purlins were assembled. Mechanical and electrical services work started after structural work was completed with the roof covered much earlier than traditional method. Pre-form power point openings were made on the walls. All cables were laid on the floor and buried with the topping. Water pipes were all placed above ceilings and were concealed along existing walls of the house. Only waste water pipes were exposed below the soffit of the slabs. The lift was installed concurrently with all the mechanical and electrical works directly after the roof was totally completed. Architectural works such as the aluminum windows, doors and glass for the toilets and lifts were then installed followed by painting works. These were all completed without the need for scaffolding platforms. Only movable ladders were used throughout the whole duration of the project. Water proofing was done to all wet areas in the building. Using special cementitious polymers with high strength self-leveling flowable grout to level the floor slabs.

Final painting work was using low volatile organic compounds to improve air quality, health and wellbeing of occupants. Thus complying with BCA's Smart and Healthy Building requirements.[10] Such simple green concepts were initiated.

[10]https://www.bca.gov.sg/greenmark/others/GM_RB_2016_criteria_pilot.pdf.

The Fire Safety Act in Singapore has no requirements for the structural steel to be fire protected for such landed residential dwellings. This is a further advantage in terms of efficacies and sustainability in addition to have an aesthetically appealing structure to owner's satisfaction.

9 Conclusion

This unique modular steel-precast concrete system of construction is safe, green and sustainable and beneficial to all stakeholders.

This project has certainly opened a new door for construction in Singapore on A&A projects for landed residential dwellings. It shows how modular construction method can be an alternative to traditional construction method for small scale addition and alteration works.

Modular construction is component based and pre-fabricated method that has its merits. This project illustrates how well it scores on government authorities' building guidelines. Green Mark, Constructability Score and Buildable Design Appraisal System were just some of the standards and Code of practices that the project scored well in—even though it did not have to comply. These BCA standards and strategic priorities are useful in sharing with global initiatives in sustainability developments to meet their sustainable development goals. Housing policy and sustainable urban development strategies in Australia too share common priorities in this area of research.[11]

The project also minimized wastages, and observed good cleanliness and clearing practices on site—no carpentry formwork was used and no laitance concrete or steel was left over. Lintels and stiffeners were also not required. In the balance of steel and concrete construction design, this project illustrates that optimum steel and optimum pre-cast concrete can give efficacies in productivity and savings as well. Other examples of mitigating potential risks and faults were the absence of welding on site and horizontal joinery. By removing the need for welding on site, fire safety is improved. Without the need of horizontal joints, possibilities of leakages were avoided.

This is an innovative modular model project that could be emulated and implemented on all such residential dwelling developments in Singapore. If used for other bigger developments, it would help to achieve Green Mark score of Platinum. As a professional engineer, qualified person, acting as expert witness as well as adjudicator on building construction disputes, the adoption of such optimum modular system could increase the sustainable economy of Singapore and certainly reduce many building disputes on noise, dust and damages widely prevalent in the industry.

[11]https://www.ahuri.edu.au/__data/assets/pdf_file/0012/2028/AHURI_Final_Report_No39_Housing_policy_and_sustainable_urban_development.pdf.

It also benefits society as it helps to mitigate conflicts due to leakages, because of its innovative leak proof technology. Overall, the owner was satisfied with the outcome on safety, cost, quality and delivery.

Though this project is small, it had a big impact on the construction industry in Singapore. It was successfully test-bedded and was a proof of concept beyond compliance requirements. Hence, more such innovative and sustainable building construction could be adopted in sync with the authorities' push for a safe and green industry for Singapore and countries regionally as well as globally undertaking similar works.

References

Authority, B. a. (2012, June 15). *Sustainable Construction*. Retrieved April 3, 2017, from Building and Construction Authority of Singapore: https://www.bca.gov.sg/SustainableConstruction/others/sc_cui_final.pdf.

Authority, B. a. (2015, December 15). *Code of Practice on Buildability*. Retrieved March 29, 2017, from Building and Construction Authority of Singapore Web Site: https://www.bca.gov.sg/BuildableDesign/others/copbdnov2015.pdf.

Authority, B. a. (2016). *Green Mark Criteria Pilot*. Retrieved April 1, 2017, from Building and Construction Authority of Singapore Web Site: https://www.bca.gov.sg/greenmark/others/GM_RB_2016_criteria_pilot.pdf.

Authority, B. a. (n.d.). *Buildable Design BDAS*. Retrieved April 3, 2017, from Building and Construction Authority of Singapore Web Site: https://www.bca.gov.sg/BuildableDesign/buildable_design_bdas.html.

Bucknall, R. L. (2016, June). *RLB Construction Market Quarterly Update Singapore*. Retrieved April 2, 2017, from Rider Levett Bucknall Web Site: http://assets.rlb.com/production/2016/06/21003826/RLB-Construction-Market-Quarterly-Update-Singapore-March-2016-1.pdf.

Buildings, D. (2016, June 29). *Modular vs Traditional Construction*. Retrieved April 6, 2017, from Designing Buildings Web Site: https://www.designingbuildings.co.uk/wiki/Modular_vs_traditional_construction.

Manpower, M. o. (2017). *Workplace Safety and Health Act*. Retrieved March 29, 2017, from Ministry of Manpower (Singapore) Web Site: http://www.mom.gov.sg/workplace-safety-and-health/workplace-safety-and-health-act/what-it-covers.

Nicole, G. (2003, June). *Housing policy and sustainable urban development: Evaluating the use of local housing strategies in Queensland, New South Wales and Victoria*, Australian Housing Urban Research Institute, From AHURI Final Report No, 39, Web Site: www.ahuri.edu.au/__data/assets/pdf_file/0012/2028/AHURI_Final_Report_No39_Housing_policy_and_sustainable_urban_development.pdf.

See, F. C. (2005, September 5). *Clarifications on the Distinction between Additions and Alterations and Reconstruction Works for Landed Housing for the Purpose of Planning Approval*. Retrieved April 2, 2017, from Urban Redevelopment Authority of Singapore Web Site: https://www.ura.gov.sg/uol/circulars/2005/sep/dc05_21.

Singapore, N. A. (2000, May 25). *Archives Online, Speeches*. Retrieved April 4, 2017, from National Archives of Singapore Web Site: http://www.nas.gov.sg/archivesonline/speeches/view-html?filename=2000052503.htm.

Review on Sustainable Building Design and Construction in the Rural Context: The Case of Building Ampara, Sri Lanka

R. Palliyaguru, G. Karunasena and Susan Ang

Abstract The hands-on experience from the recently completed Deakin's iDiDe (Intercultural Dialogue Through Design) program in 2016 concluded that there is a clear and critical need to empower rural communities in the Ampara region of Sri Lanka to achieve United Nation's Sustainable Development Goals 2015–2030. Although the contribution of building design and construction processes for sustainable community development has been frequently discussed in other research, it has been mostly with reference to urban areas, urban population, urban problems and/or one or more of selected "sustainability" philosophies in the building domain among many, demonstrating substantial disintegration. Taking the above into consideration, as part of a much larger research initiative originated through the Deakin's iDiDe program, this paper presents the need for an integrated framework for sustainable building design and construction in the rural context with a specific focus on Ampara region of Sri Lanka with an ultimate goal of empowering its communities. This was achieved through the hands-on experience gained through the iDiDe study program conducted by Deakin University partnered with a number of institutions and organizations and rural communities in Ampara region, Sri Lanka; and a critical literature synthesis. Study findings concluded that adoption of the Integral Sustainable Design and Construction (ISDC) framework for rural contexts is vital to address prevailing physical, social and cultural issues.

Keywords Integrated framework · Sustainable building design and construction Rural community development · Empower community

R. Palliyaguru (✉) · G. Karunasena · S. Ang
School of Architecture and Built Environment, Deakin University,
Gheringhap St., Geelong, VIC 3220, Australia
e-mail: r.palliyaguru@deakin.edu.au

W. Leal Filho et al. (eds.), *Sustainable Development Research in the Asia-Pacific Region*, World Sustainability Series, https://doi.org/10.1007/978-3-319-73293-0_29

1 Research Background and Introduction

Deakin University's iDiDe (Intercultural Dialogue Through Design) is a student study tour program for research and designing buildings for sustainable community development and capacity building in rural communities. This mobility model has made sustained contributions towards strengthening established and new partnerships through ongoing development of new projects, and mutual and long-term commitments to the United Nations Sustainable Development Goals within the Indo-Pacific region. As part of the iDiDe program, five multidisciplinary student groups from Deakin have travelled to the Ampara region in Sri Lanka and initiated the "Building Ampara" Project which is an integrated sustainable rural community project concerned with long term sustainable development. It is a structured program offering intercultural immersive learning experiences which utilizes a multidisciplinary and integrated perspective in sustainable design, eco-tourism, cultural preservation, and rural community infrastructure development in the conceptualisation (feasibility and design) and project development for realisation of prototype buildings.

The hands-on experience from the iDiDe programme at Ampara in 2016 depicted that there is a clear and critical need to empower rural communities in this part of the country to achieve United Nation's Sustainable Development Goals 2015–2030. Rural communities in Ampara confront long term struggles where progress in women empowerment, children education and income generation are slow due to disparities between real community needs and top down policy driven development initiatives in combination with sporadic volunteer and funding efforts by both local and international "Not-for profit organisations". Further, the other critical findings are heavy disintegration in the building design and construction processes, lack of application of sustainability principles and disintegration of stakeholder groups and professionals from essential disciplines during building design and construction processes, resulting in non-sustainable and poor building outcomes. On the other hand, although the sustainability of building design and construction processes has been frequently discussed in other research, it has been mostly with reference to urban areas, urban population, urban problems or one or more of the selected "sustainability" philosophies in the building domain among many, again demonstrating a substantial disintegration. Accordingly a new research challenge emerged through iDiDe: "Is there a need for an integrated framework for sustainable building design and construction in the rural context"? Hence, the outcomes of the iDiDe programme included a proposal for an integrated sustainable vision for community development for Ampara, and a vision to implement an integrated approach to sustainable building that could facilitate capacity building. It basically offered a holistic community driven vision developed with rural community involvement that embraced co-design and co-building methods for three prototype designs (Adaptable Classroom, Community Centre and Low Cost House), with the identification of the following four key principles:

- Participatory community based feasibility studies and concept design development.
- Co-design and co-building techniques with community input and involvement.
- Use of local materials, local techniques, references of vernacular with new vision and modern applications as a first option.
- Introduce the use of bamboo as a building material with longer term intentions of instigating higher levels of government agency for a bamboo industry. (Currently, the use of bamboo in of Sri Lanka was limited for crafting purposes.)

Accordingly, this paper aims to investigate the need to create an integrated framework for sustainable building design and construction in the rural context with a specific focus on the Ampara region of Sri Lanka.

2 Research Methodology

The findings from the recently completed iDiDe study program in Ampara provided the initial underpinning and justification for the research gap to be addressed through this paper: what are the needs to create an integrated framework for sustainable building design and construction in the rural context? In order to address this gap, initially identified critical community problems of rural communities in Ampara through the iDiDe program, were further elaborated and compared with the current situation in other developing countries through a critical literature synthesis. Secondly, this literature synthesis was further expanded to review the principles/philosophies of sustainable development in the built environment with an aim to identify how and why such principles/philosophies could be adopted and improved to suit the sustainable community development needs in the rural context, specifically for the Ampara region. Context analysis, intercultural dialogue, interpersonal communication and cross-cultural and multi-leveled collaboration alongside supervised participation in community engagement activities were used as research techniques when gaining hands-on experience through the iDiDe study program.

3 Critical Problems Confronted by Rural Communities: "Ampara" as a Rural Region

It is worth distinguishing between urban and rural areas: an area which is not an urban area is considered as a rural area. On the other hand, rural areas can be defined as an areas where the population density is low. However, the existing approaches to define the rural areas do only rely on population density. For example, in Australia, there are three methods to define remoteness and rurality. These three classifications are named: (1) Rural Remote and Metropolitan Areas classification (RRMA); (2) Accessibility/Remoteness Index of Australia (ARIA);

and (3) Australian Standard Geographical Classification (ASGC-RA) which are respectively based on the size of a community, distance from population centres, and access to services. In fact, the definition for rural places is different in different countries for different periods and in different regions. For example, the Australian government decided to use the ASGC system to replace the RRMA system for administering remote and rural regions in 2009. In early Europe, a rural area was defined as a place with less than 2000 people, now the OECD methodology classifies the areas with a population density below 150 inhabitants per km^2 as rural (European Commission 2017). In Sri Lanka, "*rural sector is geographically demarcated as the area outside the boundaries of local administrative authorities of municipal councils and urban councils*" (Wickramasinghe 2010). These local authorities are named as such authorities based on the availability of developed infrastructure that defines urban way of living by the Urban Development Authority (Wickramasinghe 2010).

Although the first Millennium Development Goal to halve the number living in poverty in 1990 has been achieved to a certain extent, much still remains to be done around the World (United Nations 2015). In 2012, nearly 900 million people still lived on less than \$2 per day. Across 114 developing countries close to one billion people live in poverty in rural areas, which translates into 70% of the total rural population lives in poverty (Dixon 2015; Jazairy et al. 1992; World Bank 2016). Moreover, there is a growing inequality of income in many developing countries (OECD 2015). These differences are most evident in the rural areas. There are other disparities in opportunities available for those living in rural areas; for example, in health care, secondary education and employment opportunities.

In Sri Lanka, nine out of ten poor people live in rural areas. The poverty levels in Sri Lanka rose up due to the 30-year civil conflict in the north and east of the country which resulted in about 100,000 being killed and about 800,000 people being displaced from their homes and sources of livelihood; and also due to the 2004 tsunami. Both these major catastrophes affected the east of Sri Lanka where "Ampara" is located. Ampara is the main town of Ampara district which is one out of the twenty five districts in Sri Lanka. Districts are the second-level administrative divisions, and are included in a province. Ampara district is located in the Eastern Province of Sri Lanka about 360 km from the capital Colombo—the capital city of Sri Lanka. Ampara is situated in a lowland region experiencing a hot, humid, tropical climate, with maximum temperatures between 25 and 30 °C year-long and Relative Humidity a muggy 85%, creating uncomfortable conditions in the Sun for much of the day (Johansson and Emmanuel 2006). Ampara district being the geographically fourth largest district in Sri Lanka, spans a wider area providing home to people from different ethnicities, religions, languages, income levels and so on. While people living in and around the main towns of Ampara enjoy a certain level of prosperity, there are plenty of remote and rural communities where people do not even enjoy the bare minimum living standards.

The majority of the rural community of Ampara are small grade farmers. Apart from poor people in rural areas in Ampara being affected by the 30-year conflict and the tsunami disaster, agricultural growth in these areas has been sluggish.

Small-scale farmers produce most of the agricultural output, but their production systems are hampered by neglect, poor economies of scale, low investment levels resulting from poor financial services, inappropriate or limited technology, fragmented landholding, post-harvest losses, inconsistent produce pricing and trade policies, and market constraints. Accordingly, growing nature of poverty, inequality of income and disparities in opportunities available for health care, secondary education (especially for girls), gender equality, clean water and sanitation are the most evident concerns in most of the rural areas of Ampara. There is a significant lack of infrastructure such as roads, electricity, irrigation, communication facilities, safe drinking water which limits people's ability and opportunity to earn income through off-farm activities.

Protection issues of women and children is another major concern in Ampara. Due to the 30-year conflict and the devastating 2004 tsunami, there was an increase in the number of households headed by women, which exposed economic hardship. Sexual and domestic violence, poor capability of women to response to domestic violence issues, lack of information regarding rights and opportunities, under representation in family and local decision making opportunities, high child mortality rates—especially at childbirth, lack of infrastructure that focuses on preventative medicine, malnutrition due to lack of education and/or food security etc. are some of the other persistent issues that justify the lack of women and children protection at large. Moreover, women suffer from lack of financial control, independence and responsibility due to lack of employment, skills, training and leadership opportunities; lack of culture that facilitates employment and training for young women (i.e.: their most obvious opportunity for security is to start a family); no culture of financial responsibility and accountability for women in families; and family responsibilities that hinder employment, financial and training opportunities for women. There exists unevenly distributed family responsibilities between men and women. Patriarchal social structures and hierarchy disadvantage women and expose them to physical, social and financial vulnerability such as dowries, the custom of women not being seen as a part of her husbands' family, divorce rights, marital rape etc. In summary, women in rural villages are unable to attain further training or remain in education longer or gain financial independence due to perceived or real lack of opportunities in these area. This leads women to focus solely on family building activities. All these issues prevail in remote and rural parts of Ampara, but extent of it is a matter that requires further research.

Accordingly, there is a critical requirement to address the above described specific issues effecting the rural through appropriate rural-specific sustainable development programmes. Hence, the next section is focused on how and why the principles for rural sustainable development should vary from urban sustainable development to further understand how the "sustainability" principles, framework and theories in the building domain must be re-shaped to suit the rural contexts.

4 Sustainable Development in Rural Context

Sustainable development had become an environmental catchphrase since the recent past decades. It was embraced as the new paradigm of development by a wide range of non-governmental as well as governmental organizations. There are many definitions for sustainable development, but the most widely accepted definition is from Our Common Future, also known as the Brundtland Report, from the United Nations World Commission on Environment and Development (WCED) which was published in 1987: "*Sustainable development is development that meets the needs of the present without compromising the ability of future generations to meet their own needs*" (Du Pisani 2006, p. 89). This concept was embraced widely due to fear that economic growth may threaten the health of the people and the survival of the planet after the World War two. Until now, the action plan for sustainable development, Agenda 21, has covered social and economic development, the conservation and management of resources for development, strengthening the role of major groups involved in achieving sustainable development and means of implantation; it tries to provide treatment for fundamental issues, including the inequities between the rich and the poor, wasteful consumption and the population explosion and integration of environment and development (Reid 2013).

There are also some definitions about sustainable development in rural areas, for example, it is pointed out that sustainable development in rural areas means "*a stable socio-economic development of rural areas, volume increasing of agricultural output, improvement of agricultural effectiveness, achievement of full-employment of rural population and increase in their level of living, rational land use*" (Belyaeva et al. 2016, p. 6890). The sustainable development in rural areas aims to provide the balance between the preservation of the material and spiritual needs of the countryside and trend of modernization (Nistreanu et al. 2009). Belyaeva et al. (2016) also point out that the key aims of sustainable development in rural areas is to make condition for achieving welfare for people, to create capacities of self-developing, to protect cultural values, to preserve natural resources for reproducing and long-term using in economic activities of tourism, crafts, agriculture, trades, recreation, and other areas. Accordingly, it is evident that, sustainable development in non-rural and rural areas both involves providing balances between the environment and economic growth and improvement of welfare for population. However rural area is a particular area with much specificity (Iagaru and Pompilica 2014), therefore it needs more comprehensive consideration according to economic, cultural, social and environmental conditions.

The next section discusses the principles of sustainable development in the built environment with an aim to identify the loopholes that diminishes the contribution of building design and construction processes to sustainable rural community development.

5 Sustainable Development in the Building Domain

As we see it, the construction industry is responsible for shaping the world, whether it be through urban or rural based buildings or infrastructure. While the construction industry is a universal trade, different practices can be adopted amongst different cultures, whether it, be due to legislation or current trends in a particular country. With the built environment around the world being responsible for 30% of raw materials used, 42% of energy use, 25% of water use, 12% of land use, 40% of atmospheric emissions, 20% of water effluents and 25% of solid waste (DSEWPC 2012), there is a progressive attentiveness to the effect that the construction industry has on the environment. Kibert et al. (2000) further reiterated this, stating that "*construction and operation of the built environment in countries apart of the Organization for Economic Cooperation and Development (OECD) account for the greatest consumption of material and energy resources of all economic sectors*". Moreover, in both urban and rural areas, rapid development brings environmental problems such as water pollution, air pollution and social issues such as rural to urban migration and left-behind children. Somewhere over the years, this realisation has been adopted by industry professionals, resulting in today's industry being heavily weighted towards 'green', "sustainable" and "regenerative" solutions.

5.1 Sustainable Development

Sustainable development includes three broad components; social, environmental and economic often known as the 'triple bottom line', which brings environmental responsibility, social awareness and economic profitability objectives to the fore in the built environment and facilitates a wider community (Ali and Nsairat 2009). It may be perceived as a long-term exercise carried out by various industries to achieve sustainable development within the parameters of economic, social and environmental cornerstones. Thus, construction industry which traditionally focused on time, cost, and quality was shifted to a new paradigm by focusing economic, environmental and social aspects. Such construction would bring environmental responsibility, social awareness, and economic profitability to the fore in the built environment and facilities to a wider community. Sustainability in construction offers first-rate response to the present environmental and socio-economic problems as it is an application of the principles of sustainable development to the comprehensive construction cycle from the extraction of raw materials, through planning, design and construction of buildings and infrastructure, until their final deconstruction and management of the resultant waste (Yunus and Yang 2011). The main challenge for the industry is to play an integral part in reducing the impacts of its activities on the environment and local communities. Simply, how construction development process can be aligned with community development for long-term sustainability or to enhance the local resilience.

A sustainable community uses its resources to meet current needs while ensuring that adequate resources are available for future generations. It seeks a better quality of life for all its residents while maintaining nature's ability to function over time by minimizing waste, preventing pollution, promoting efficiency and developing local resources to revitalize the local economy. Among many initiatives, designing energy efficient and people friendly buildings is one of the important aspects of the sustainable communities.

5.2 Green Development

In response to the serious and irretrievable climatic changes, the green revolution has taken place in the building sector too. It proposes to basically alter the built environment by creating energy efficient, healthy and productive buildings that cut back the significant impacts of buildings on urban life and global environment (United States Environmental Protection Agency (USEPA) 2009; United States Green Building Council (USGBC) 2009 cited Gou et al. 2013). A paper on sustainable development (2006) mentioned that in 1950–60s the "green" thought began to be accepted worldwide as the "green movement" was lifted and performed among western countries. The "green" thought intended to protect the natural resources, alter human behavior, convene the ecological virtuous cycle of nature, and make sure the safety of human existence. It finally has resulted in the modern build green movement. By the end of 1980s, "sustainable development" had become the worldwide program of action, and at the same time, ecology, sociology, and other subjects extended to the architecture domain, and then the "green architecture" concept came out naturally. The global sustainability goals have led to the development of the green building movement. Further, green building is the status of the effort in achieving sustainability in construction practices (Sinha et al. 2013). Hence, green building has become one of the greatest and emergent concepts to date. Architects, Designers, and home owners are being guaranteed with the cost saving potential, prominence of energy saving, contemporary look, and the symbiotic relationship with environment that green buildings possess (Isnin et al. 2012). Thus, construction activities may not only include new building projects or infrastructure and utilities alone, there is an emergent demand for converting buildings towards green (Douglas 2006 cited in Isnin et al. 2012). A report by McGraw Hill highlights the growth and demand 'green' design has placed on the industry, with industry professionals seeing an increase in market share, from 13% in 2009 to 60% at the time the report was released in 2013 (McGraw-Hill 2013). An interesting development that came about from this survey was that the growth was not just localized to common economized countries, it was a globally recognized evolution. It appears that 'green' thinking has progressed a step further over the years, with the thought process behind going green changing from the placebo of 'doing the right thing', to a practice required for businesses in order to obtain consistent work.

With such a heavy focus on being 'green' and having as little impact on the environment as possible, the term sustainability has been thrown around more times than that can be imagined when it comes to today's construction industry. The concept of being sustainable, as opposed to 'green' construction, sits with a heavy presumption of there being a strong relationship between the two. However, it is essential to distinguish between the two terms in order to understand why we should be implementing these practices in urban and rural construction in the same way. Green design is the process of developing a project with the intention of environmental performance being as sufficient as that can be—this associated with green buildings, are those structures that excel others in relation to their environmental performance. Sustainability is a different thinking process opposed to green construction. It is not about making a building more environmentally friendly, it is about making the two elements co-exist without compromising the ability of future generations to meet their own needs (World Commission on Environment and Development 1987). In a way, green construction is just one of many directions one can take in order to be more sustainable, but a building as such cannot be directly sustainable (Gibberd 2001). It can be concluded that sustainability is more about how a building can contribute to the 'social, ecological and economic health of where it functions', whereas green design is primarily directed at 'doing less harm' or, more generally, reducing the degenerative consequences of human activity on the health and the integrity of ecological systems' (Cole 2012a).

5.3 *Benchmarking: Green Development*

Green design was the original answer to how we can make the construction industry more environmentally friendly through the introduction of multiple 'green' initiatives, transitioning what was an unconventional exercise into a mainstream practice (Kibert 2008). The success of green buildings depends on the quality and efficiency of the green systems introduced. If the building installed with less quality system, it will neither reach the environmental goals nor create the estimated benefits. Therefore, the market demands a direction to differentiate green buildings from traditional buildings through the use of standard, transparent, objective, and verifiable measures of green, which guarantees that the minimum green requirements have been reached (Lacouture et al. 2008). Hence, a range of green building assessment tools have emerged in the past twenty years, which are used to assess and benchmark the levels of achievement in the green revolution and to establish a common language and standards of measurement to delineate green buildings differentiating from traditional buildings (Yudelson 2008, cited in Gou et al. 2013). Green assessment tools were primarily introduced to evaluate specific aspects of a building, relating to sustainability goals (McKay 2007). Once measured, buildings could be more easily compared with current and past building practices and other green buildings. Wallhagen (2010) further verifies that the green assessment tools could also be employed to create guidelines, benchmarks and ratings for building construction practices with low environmental impact, and also promote green building practices around the world, acting as a stimulus for the green building movement.

The first assessment tool introduced was the Building Research Establishment Environmental Assessment Method (BREEAM) (Baldwin 1998 cited Lacouture et al. 2008) and, the most representative and widely used green assessment tools are Leadership in Energy and Environmental Design (LEED), Comprehensive Assessment System for Building Environmental Efficiency (CASBEE), Green Star, Green Building Index (GBI)—Malaysia, Green Mark—Singapore, Hong Kong Building Environmental Assessment Method (HK-BEAM) and the Pearl Rating System for Estidama (Sustainability) (Boonstra and Pettersen 2003; McKay 2007). These assessment tools are designed to engage the minds of builders and architects alike. Certifications are awarded based on the performance of the building measured against explicitly declared criteria (Cole and Howard 2005), some examples being indoor air quality, energy consumption or storm water management. While different country may differ in their assessment tools, the baseline principles essentially are the same across the board.

5.4 Adaptable and Resilient Development

As part of the *1992 Rio Declaration on Environment and Development*, sustainability should be integral across Sri Lankan development, as worldwide building life-cycles account for 40% of CO_2 emissions (Abeysundara et al. 2009). However, as Sri Lankan rural buildings already consume minimal resources, only emitting 0.72t of CO_2 per capita in 2011 against the world average of 4.9t, it instead faces challenges to maintain sustainability as economic development occurs, already rising to much as 226.4% from 1990 to 2011. The Asian Development Bank estimates it must cut emissions growth by 50% by 2100 as part of a global effort to stabilise climatic changes, which if continued unabated by 2050 sea average temperatures will rise by 3 °C and sea levels will rise by 0.2–0.6 m devastating the low-lying Ampara district (Bedford and Cook 2016). Similarly, extreme weather events would increase in severity, such as the 2004 Tsunami which killed and displaced thousands of people. Therefore, long-term focused integration of adaptable and resilient design features is also critical. Yet recommendations proposed must consider Sri Lanka's developing economic context, with only $2290 per capita in 2010 against the world average of $9097. Despite this, it has a relatively high Human Development index of 0.686 in 2011, which has potential to rapidly and sustainably increase through these solutions (Kishnani 2012).

5.5 Regenerative Development

After many years of substantial attention on the policy and practice of sustainability, a question has been raised about the ability of "sustainable development" of being really sustainable (Conte and Monno 2016; Blowers et al. 2012). Of course, sustainability works well in the three broad pillars of social, environmental and economic, but the ability of this original concept to "*reestablish cooperation between the natural and the human worlds for a mutual beneficial development*" is negative (Conte and Monno 2016). Therefore, what if "development" is "regenerative"? Regenerative development aims to create a thriving, healing and vital outcome for all stakeholders. The term "regenerate" is generally concerned with three things: a radical change for the better, the creation of a new spirit and return energy to the source. The concept of regeneration attempts to recover the weakened relationship between "human" and "planet"—in other words it addresses the "root cause of human imbalance with the planet". It is the process of cultivating the capacity and capability of people, communities and other natural systems to renew, sustain and thrive. While sustainable development approach is concerned with minimization of depletion of natural resources and impact on the environment, regenerative development approach is concerned with enabling natural resources supply systems continually, self-renewing or regenerative in their operation (Cole 2012b; Hes and Plessis 2015). John Tillman Lyle introduced the term 'regenerate design' as "*an approach to the design of urban landscapes which enables them to regenerate lost ecosystems*" (Hes and Plessis 2015). In fact, "sustainable development" will not be fully sustainable or attained without that development being "regenerative". In other words, regenerative development is not different from sustainable development, but it is an essential constituent of sustainable development, which is so far missing in theory and practice.

While green assessment evaluation tools such as BREEAM have become quite popular as they provide benchmarks for measuring the rate of sustainability of buildings and act as the guidelines of a development process, they have substantial limitations in evaluating the economic, social and institutional aspects of sustainability (Gou and Xie 2016; Conte and Monno 2016). Most of these tools are focused excessively on environmental sustainability (Gou and Xie 2016). Therefore, the need for a more integrated sustainability assessment approach has been recognized by many researchers in the recent past (e.g. Gou and Xie 2016; Conte and Monno 2016; Cole 2012b). They predominantly call for directing sustainability assessment towards a regenerative approach.

There are a number of framework such as LENSES, Arup ArupSpeAR, REGEN, Peter and Wills framework that offer unique and structured process for this regenerative development process where as application of these framework largely depends on the context in which they are used and the designers who use them (Gou and Xie 2016). These framework do not actually provide new sustainability evaluation systems or levels of certification (Gou and Xie 2016). Thus there is a crucial research need to establish a framework that allows assessment of sustainability from a "regenerative development" angle.

6 Potential Integrated Framework for Sustainable Building Design and Construction, in the Rural Context: Why, How and What?

Across Asia, even when sustainable systems are employed, the fragmented design process leads to implementation focused only on under-performing short-term solutions (Kishnani 2012). The resultant lack of long-term sustainable planning leads to issues, for example such as Sri Lankans desires for modern buildings to feature 'western' Air Conditioning with short term (1–5 years) energy-efficiency and low operational costs, but with increased long-term risks as availability and cost of energy fluctuates (Kumara et al. 2016). The unintegrated design and construction culture can be corrected via an Integrated Design and Construction Process. Integration in the design process is "*a methodological framework for correcting the prevailing culture of short-termism and fragmentation*" (Kishnani 2012). "Continuity" feature built-in in the concept of regenerative development is of course a good beginning to envisage an integrated approach for building design and construction.

Currently in Sri Lanka, the high-level sustainability-related concepts in the built-environment are mostly targeted at urban buildings, with minimal attention to rural buildings, poor communities and their critical problems. However, the rural built-environment is also in need of engrossing the essential sustainability principles to produce socially, environmentally, economically responsible design and construction that will also enable co-evolution of human and natural world. The purpose will be to provide reasonable solutions to the existing critical problems in these rural areas on their specific issues e.g. poverty, women and children protection and education which are often interlinked (Sect. 2 for further information on context-specific community problems in Ampara). The current call for redirecting sustainability assessment towards the regenerative approach further justifies the need for a framework that encapsulates the requirements posed by different "sustainability" philosophies into a single integrated framework. Such a framework will essentially address how poor outcome and processes used to design and construct buildings in rural areas can be negated. But given the diverse nature of existing rural community problems, the framework shall be something that will incorporate these rural community problems with a view to optimize the contribution of the building design and construction processes to sustainable rural community development.

Considering the context specific community issues in Ampara, such as social and cultural barriers for women and children, poverty levels due to post-war and livelihood instability, vulnerability to natural disasters and damages encountered from previous major natural disasters etc. certainly highlight the need for following elements in a potential conceptual framework, in order to ensure that design and construction is "truly" sustainable and integrated:

- Consider building performance beyond environmental impact and economical returns;
- Incorporate quantitative metrics to map the physical and social context as well as spatial relationships related to design and construction;
- Incorporate qualitative metrics to measure performance against how people sense and feel in or around a building;
- Incorporate qualitative metrics to find out about cultural values, world views, symbols, and social meanings associated with buildings to refine and improve the process for integrated sustainable design, construction, building operation and maintenance. This is particularly an opportunity to consider social and cultural implications of design and construction processes by embracing community needs (including wellbeing, combating poverty, women and children protection etc.) and establishing cooperation between the natural and human worlds for mutual benefit and development;
- Consider establishing resilience in the built environment;

7 Conclusion and Way Forward

As an outcome, the paper presents a conceptual idea of an integrated building design and framework for sustainable rural community development in Ampara region of Sri Lanka with an ultimate goal of empowering its communities. It demonstrates that expansion and modifications are possible for "sustainability" philosophies when they are adopted and integrated for rural community development with an eye to solving the pressing needs of these communities while utilizing their strengths and opportunities. The findings specifically highlight the requirement for considering "qualitative" aspects of sustainability such as how people sense and feel in or around a building and cultural values, world views, symbols, and social meanings associated with buildings. In addition, the importance of "quantitative" aspects of the sustainability, such as set targets for environmental performance to refine and improve the process for integrated sustainable designs, construction, building operation and maintenance has been highlighted. Accordingly, adoption of the integral sustainable design and construction (ISDC) theory and investigating its appropriateness for rural contexts has been recognized as a direction for the way forward of this research.

References

Abeysundara, U. G. Y., Babel, S., & Gheewala, S. (2009). A Matrix in life cycle perspective for selecting sustainable materials for buildings in Sri Lanka. *Building and Environment, 44*(5), 997–1004.

Ali, H. H., & Nsairat, F. A. (2009). Developing a green building assessment tool for developing countries—Case of Jordan. *Building and Environment, 44*(5), 1053–1064.

Bedford, D., & Cook, J. (2016). Chapter 3—The likely impacts. In *Climate change, examining the facts* (pp. 103–149). Santa Barbara: ABC-CLIO.

Belyaeva, G. I., Ermoshkina, E. N., Kosyakova, I. V., Pankratova, L. E., & Zotova, A. S. (2016). Strategic analysis of sustainable socioeconomic situation of rural areas in the Samara Region of the Russian Federation. *International Journal of Environmental and Science Education, 11*(14), 6889–6897.

Blowers, A., Boersem, J., & Martin, A. (2012). Is sustainable development sustainable? *Journal of Integrative Environmental Sciences, 9*(1), 1–8.

Boonstra, C., & Pettersen, T. D. (2003). *Tools for environmental assessment of existing buildings*. UNEP Industry and Environment. Available at: http://www.bvsde.paho.org/bvsaia/fulltext/tools.pdf.

Cole, R. J. (2012a). Transitioning from green to regenerative design. *Building Research & Information, 40*(1), 39–53.

Cole, R. J. (2012b). Regenerative design and development: Current theory and practice. *Building Research & Information, 40*(1), 1–6.

Cole, R. J., & Howard, N. (2005). Building environmental assessment tools: Current and future roles. In *The 2005 World Sustainable Building Conference*, Tokyo.

Conte, E., & Monno, V. (2016). The regenerative approach to model an integrated urban-building evaluation method. *International Journal of Sustainable Built Environment, 5*(1), 12–22.

Department of Sustainable, Environment, Water Pollution and Communities. (2012). *Construction and demolition waste guide—Recycling and re-use across the supply chain*. DSEWPC, Australia. Available at https://www.environment.gov.au/system/files/resources/b0ac5ce4-4253-4d2b-b001-0becf84b52b8/files/case-studies.pdf.

Dixon, C. (2015). *Rural development in the third world*. Oxon, UK: Routledge Revivals.

Du Pisani, J. A. (2006). Sustainable development–historical roots of the concept. *Environmental Sciences, 3*(2), 83–96.

European Commission. (2017). *Urban-rural typology*. Available at: http://ec.europa.eu/eurostat/statistics-explained/index.php/Urban-rural_typology. Retrieved January 6, 2017.

Gibberd, J. (2001). The opinion of Gibberd. *Sustainable Building, 3*, 41.

Gou, Z., Prasad, D., & Lau, S. S. (2013). Are green buildings more satisfactory and comfortable? *Habitat International, 39*, 156–161.

Gou, Z., & Xie, X. (2016). Evolving green building: Triple bottom line or regenerative design. *Journal of Cleaner Production, 153*, 600–607.

Hes, D., & Plessis, C. D. (2015). *Designing for hope: Pathways to regenerative sustainability*. Oxon: Routledge.

Iagaru, R., & Pompilica, I. (2014). The impact of the current situation on the development of non-agricultural activities in rural areas situated at the border of Sibiu. *Agricultural Management/Lucrari Stiintifice Seria I, Management Agricol, 16*(1), 225–229.

Isnin, Z., Ahamad, S. S., & Yahya, Z. (2012). Challenges of the unknown building material substances for greener adaptation projects. *Procedia—Social and Behavioral Sciences, 68*(2012), 53–62.

Jazairy, I., Alamgir, M., & Panuccio, T. (1992). *The state of world rural poverty: An inquiry into its causes and consequences*. New York: International Fund for Agricultural Development of United Nations (IFAD).

Johansson, E., & Emmanuel, R. (2006). The influence of urban design on outdoor thermal comfort in the hot, humid city of Colombo, Sri Lanka. *International Journal of Biometeorology, 51*(2), 119–133.
Kibert, C. J. (2008). *Sustainable construction: Green building design and delivery* (4th ed.). Hoboken, New Jersey: Wiley.
Kibert, C. J., Sendzimir, J., & Guy, B. (2000). Construction ecology and metabolism: Natural system analogues for a sustainable built environment. *Construction Management and Economics, 18*(8), 903–916.
Kishnani, N. (2012). Chapter 10—Integration. In *Greening Asia: Emerging principles for sustainable architecture* (pp. 100–108). Singapore: BCI Asia.
Kumara, W. H. C. D., Waidyasekara, K. G. A. S., & Weerasinghe, R. P. N. P. (2016). Building management system for sustainable built environment in SriLanka. *Built Environment Project and Asset Management, 6*(3), 302–316.
Lacouture, C., Sefair, J., Florez, L., & Medaglia, A. L. (2008). Optimization model for the selection of materials using a LEED-based green building rating system in Colombia. *Building and Environment, 44*(6), 1162–1170.
McGraw-Hill Construction. (2013). *World green building trends: Business benefits driving new and retrofit market opportunities in 60 countries*. McGraw-Hill Construction. Available at: http://www.gbcsa.org.za/wp-content/uploads/2013/06/WGBC-Trends-Report_2013.pdf. Retrieved March 5, 2017.
McKay, J. (2007). Green assessment tools: The integration of building envelope durability. In *11th Canadian Conference on Building Science and Technology*, Canada, Alberta.
Nistreanu, M., Maracineanu, F., & Constantin, E. (2009). *Sustainable regional development—Policies and strategies* (p. 6). Bucureşti: Editura Nouă.
OECD. (2015). *In it together: Why less inequality benefits all*. Paris: OECD Publishing. https://doi.org/10.1787/9789264235120-en.
Reid, D. (2013). *Sustainable development: An introductory guide*. London: Routledge.
Sinha, A., Gupta, R., & Kutnar, A. (2013). Sustainable development and green buildings. *DrvnaIndustrija, 64*(1), 45–53.
The World Bank. (2016). *Rural population*. World Bank Group. Available at: http://data.worldbank.org/indicator/SP.RUR.TOTL. Retrieved December 1, 2016.
United Nations. (2015). The Millennium Development Goals Report 2015, United Nations: NewYork.
United States Environmental Protection Agency—USEPA. (2009). *Definition of green building*. U.S. Environmental Protection Agency. Available at: https://archive.epa.gov/greenbuilding/web/html/about.html. Retrieved February 4, 2017.
Wallhagen, M. (2010). *Environmental assessment of buildings and the influence on architectural design (Master's thesis)*. Stockholm, Sweden: Royal Institute of Technology.
Wickramasinghe, W. (2010). Rural development measures: Indicators and indices for Sri Lanka. *Sri Lanka Journal of Agrarian Studies, 14*(1&2), 23–44.
World Commission on Environment and Development-WCED. (1987). *Report of the world commission on environment and development: Our common future*. WCED. Available at: http://www.un-documents.net/our-common-future.pdf. Retrieved February 7, 2017.
Yunus, R., & Yang, J. (2011). Sustainability criteria for industrialised building systems (IBS) in Malaysia. *Procedia Engineering, 14*, 1590–1598.

Green Modular Concept of Sustainable Kampong Cityblock in Indonesia

Budi Prayitno

Abstract The change in housing form from landed houses to more than 20-story high-rise apartments that prioritizes the aspects of a high efficiency often no longer considers the aspects of habitation culture. Kampongs as a form of urban vernacular settlement that has a habitation culture of close-knit neighboring culture are very vulnerable against high-rise urban living culture. This paper aims to analyze the process of the habitation culture transformation through green-modular approach. The green approach used in this model involves sustainable design principles of 'co-habitation architecture concept'. The neighboring behavior of kampongs with a very close neighboring culture needs to be approached by using sustainability principles in the context of high rise kampong cityblock architecture. The concept of co-working responds to the needs of social interaction; co-benefit concept responds to the needs of accommodation to support informal economic activities; and co-exist concept responds to the needs of harmony for tropical architecture that utilizes a combination of natural thermal comfort system and artificial thermal comfort system. Meanwhile, the modular approach used in this model involves the principle of space efficiency coordinated in modular system and space effectiveness principle that can respond to the needs of interaction space. The variations of organic living units connected through a corridor system with pocket space are hoped to be able to accommodate neighboring culture and informal economic activities in a modular space coordination system.

Keywords Green modular · Kampong cityblock · Neighboring culture Pocket space · Living unit

B. Prayitno (✉)
Department of Architecture and Planning, Faculty of Engineering,
Gadjah Mada University, Jl Grafika 2 FT UGM, Yogyakarta 55281, Indonesia
e-mail: budiprayitno_ugm@yahoo.com

W. Leal Filho et al. (eds.), *Sustainable Development Research in the Asia-Pacific Region*, World Sustainability Series, https://doi.org/10.1007/978-3-319-73293-0_30

1 Introduction

Understanding sustainable design principles cannot be done only by fulfilling normative standards from social, economic, and environment aspects. What is more important is how we can meet the demand of local values. By taking an observation case of sustainable building in urban area, this research attempts to dig into the local values in social, economic, and environment aspects of a settlement transformation phenomenon from landed settlement to high-rise settlement in an urban center area. The transformations from horizontally landed living space in providing neighboring activities, horizontally landed livelihoods/informal economic activities, and from horizontally tropical environment to vertical high-rise settlement have different local values that should be worked on in the context of sustainable green and smart building.

Kampong, which is a typical landed settlement in Indonesia with its alley architecture in an organic form, dominant neighboring/social interaction values, and veranda and shady alleys, is experiencing a massive and intensive transformation process into a vertical high-rise settlement. Relatively, the transformation of landed settlement into low-rise settlement does not drastically change the social system, livelihoods, and built environment system. This is very different from the transformation of kampong settlement into high-rise settlement where the neighboring system is highly regulated according to space utilization, efficiency, mobility, the rigid function of the settlement, and the restricted usage of the passive thermal comfort system.

Consequently, there are unsuitable living cultures, space isolations, monotony built environment, environmental pressures, and the absence of urban democracy. Rejuvenation projects in landed settlements and low-rise settlements that turn them into high-rise settlements raise a lot of design problems relating to local identity loss in social interaction or neighboring activities and livelihoods values (i.e. home-based informal economic activities) and tropical thermal comfort values that need to be adapted and applied in high-rise living culture and environment.

1.1 Sustainable Living Space

Kampong life, which has its own identity, must be maintained as one of the common cores or an important defining concept that is very important. Identity of space is the basic concept in creating the quality of place's public environment as a production of space (Perry 1945; Habraken 1972; Habraken et al. 1976; Cuttler 1976). Some designing theories of ideal city as a network/connection between varied multi levels of settlements by considering modern urban function that qualifies smart and green spirit become one of the references in settlement rejuvenation urban regeneration process. Sustainability approaches have existed in the concept of 'garden city' to today's green and smart urbanism, which are then adopted and adapted into many concepts of ideal city in order to harmonize urban living style with natural and cultural local values (Eaton 2002).

Kampong living style that has a strong neighboring value is extremely vulnerable to disappearing due to the demand of space utilization efficiency and limitation in the use of tropical passive system principle in fulfilling thermal comfort. Kampong alleys that function as a shared space to fulfill the needs of neighboring culture have to be continuously adaptive to sustainable kampong cityblock design (Prayitno 2013). The adaptive transformation can be done through the implementation of kampong alley into the corridor system configuration in high-rise kampong cityblock construction.

Corridor configuration serves as a support system with varied housing areal unit for the infill. This concept refers to a structural transformation theory in mass building to support efficiency through modular coordination principles (Habraken 1972), which is then developed into 'self-customized' system from each of the occupant (Habraken et al. 1976) and into the system of built environment entity (Habrake 1983) that should be analyzed in a transformation system of ordinary structure (Habraken 1998).

The variety of settlement transformation must be applied in a vertical settlement system (Prayitno 1994). The transformation of kampong alley into vertical settlement corridor must be able to provide the function of pocket shared space for neighboring. The concept of co-habitation space in high-rise kampong cityblock is a sustainability approach in an architectural engineering habitat.

1.2 Sustainable Livelihoods

The understanding of social sustainability in the context of kampong living culture, other than its relation to neighboring culture, is its relation to economic sustainability. This is due to the fact that most people in kampong society are of low income people, who depend much on informal economic activities (livelihoods). In this case, the process of kampong transformation into urban high rise settlement needs an approach based on livelihood functions that can generate home-based income (Clarke 1992). These informal economic activities are usually performed to fulfill the needs of market segmentation that is proportionate with its level of service scope (Clarke 1995) and for livelihoods and survival strategies (Chambers 1995).

The space for informal economic activities is closely related to a series of space and territory conflicts between public and private realm (Madanipour 1999). A territorial agreement in performing home-based economic activities in landed kampong settlement is very flexible. This will be very different from the activities in high rise settlement where the rules and regulations of space utility are so rigid that the repressive control by the occupants of high rise settlement often creates social conflicts. This leads to a loss of local values where the society tries to fulfill their domestic needs in their closest neighborhood, such as in landed kampong settlement. In other words, the self-sustainability of kampong living culture from the aspect of informal economic activities does not occur.

1.3 Sustainable Built Environment

The life in a landed kampong settlement in the tropics is characterized by verandah tropical architecture with the principle of natural lighting and ventilation, also with the principles of flat, shaded, and porous/ventilated walls. The transformation of landed settlement into high rise settlement creates a very different type of space. Limited construction type has to take into account effects of turbulence should they use cross ventilation. The same goes with position, shape, and ventilation areas, in which the direction and the speed of the wind count. This is why climate condition and tropical ecology should be a foundation in developing these high rise multilevel settlements. Wind circulation system for corridor ventilation or for settlement unit becomes a challenge in designing sustainability in built environment habitat. Vertical verandah settlement design principle can also implement shaded and ventilated building façade, rooftop ecological corridor, and ecocell as the vertical linkage integration (Yeang 2006). Installations of green pocket space in some sides of the corridor and planting wall on some sides of building facade can give an answer to built environment sustainability in the tropics.

Besides ventilation, natural light, shaded elements, and greenery utilizations, extremely high rainfall in the tropics should also be taken into consideration in sustainable built environment. Zero run off and rain water harvesting can be two of the solutions to recycle natural energy. Psycho-social, bio-physical, and technological environment approaches are to recycle urban environmental energy (Cuttler 1976). The implementation of high-rise kampong cityblock in the tropics can give an answer to transformation of horizontally built environment into high-rise vertically built environment by applying eco spirit in creating sustainable built environment.

2 Research Area and Methods

By selecting a highly densely populated area in urban center in East Jakarta for the research area, this research attempts to observe the transformation of their habitation kampong living culture. This research involves 3 types of observation area, namely landed kampong settlement area, walk-up/low rise kampong settlement, and high rise kampong settlement. The number of samples taken from each area is 90 households. The observation areas are highly densely populated in urban center in East Jakarta. Kampung Pulo area and Kampung Melayu are selected for the landed kampong settlements. Pulo Gadung area is selected for the walk up kampong settlement, and Jatinegara Barat is selected for the high rise kampong settlement.

Kampong Pulo is a settlement area with high density, with a population of 9597 and an area of 11.93 ha, making it 804 people/ha. Almost 8 ha is used for built-up area. In addition, the landed kampong settlement in Kampong Pulo consists mostly of second-story buildings because of the high risk of flood. The residents of Kampong Pulo work in informal economic sector because of the limited access to funding source and career fields. They have jobs such as shopkeepers, traders, valets, porters, etc.

The daily routines done by the residents of Kampong Pulo are as follow. They work in the morning and interact with their neighbors out in the afternoon and evening. Activities with neighbors usually take place in public and semi-public places such as on the streets, in communal rooms, street markets, *pos ronda* (a post/gathering place for neighborhood night-watch), etc. Activities in semi-private rooms are usually religious gatherings, *arisan* (regular neighborhood gathering), soccer night (watching soccer games together), or simply hanging out and chatting with one another. The occupants of low rise kampong settlement in Pulo Gadung and high rise kampong settlement in Jatinegara Barat are occupants from settlement relocation program from the flood plain of Ciliwung River near Kampong Pulo. The program is a part of a river normalization program. Low-rise kampong settlement Pulo Gadung and high-rise kampong settlement Jatinegara Barat is a low cost apartment project in East Jakarta. This kampong aims to accommodate people who have been relocated from the flood plain of Ciliwung River since June 2015 (Figs. 1, 2 and 3).

The data are collected through a questioner, interview, and observation. These methods are conducted on the three locations by using five sustainability performance aspects: neighboring, availability of verandah, availability of area for trading, availability of greenery, and housing type. The data, perceptions towards sustainability performance obtained from the questioner and interview with residents. The results of the analysis are used as the basis for a design simulation using Space Syntax to assess the level of interactions among neighbors and participative interviews with the residents.

By using Space Syntax as the framework, the method uses visual integration approach. In terms of scale, the author uses global and local approach. These two scales are used to see the effect of the intangibility of space layout to be accessed in

Fig. 1 Sustainable alley in landed kampong settlement in the form of economic function for informal business activities (left), social function for neighboring (middle) and environmental function for greenery (right). (*Source* Author)

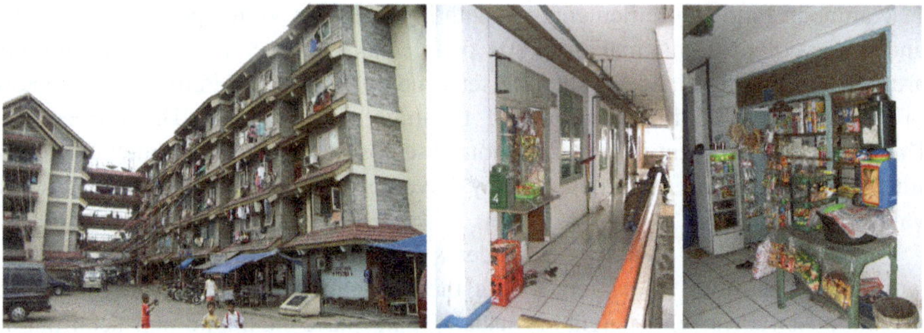

Fig. 2 Livable livelihoods activities in low-rise kampong settlement in the form of front-porch business activities in ground floor (left), in balcony (middle) and in corridor (right). (*Source* Author)

Fig. 3 Low sustainability performance values in high-rise kampong city-block settlement. There are no space for informal economic activities (left), for neighboring activities (middle) and for greenery (right). (*Source* Author)

a corridor form system with developed function. Space system is composed of two main components, which are layout and configuration that affect movements and access. Therefore, interactions in a space layout are affected (Hillier 1984).

Visibility that allows easy control and access will cause a pocket space on a certain floor to have a better and stronger visual integration. A corridor form that functions as a common space will be easily recognized if it has a high visibility value. The pattern for placing pocket space in the right corridor will determine the 'liveliness' of the corridor space. One of the methods in Space Syntax that is used to see the strength of a room is visual integration approach. Visual integration is direct and indirect relationship among spaces in a system. A space with a high visual integration value means that the space has a high accessibility value that it is easily accessed by its users. By identifying high and low values of visual integration and intangibility, it will be easy to plan an effective corridor form to be used for a public space.

3 Sustainability Performance of Built Environment Transformation

The results of the surveys and interview show that the sustainability performance values of the landed kampong settlement are the highest among the three areas. These values are: 94.4% for neighboring, 86.7% for verandah, 42.2% for trading area, 45.6% for greenery, and 71.1% for housing type. The values for walk-up/low rise kampong settlement are: 84.4% for neighboring, 75.6% for verandah, 32.2% for trading area, 21.1% for greenery, and 36.7% for housing type. And the values for high rise kampong settlement are: 18.9% for neighboring, 0% for verandah, 5.6% for trading area, 5.6% for greenery, and 6.7% for housing type (Fig. 4).

The observations on occurring activities and the configuration of the alleys in landed kampong settlements, the corridors in walk up kampong settlement, and the corridors in the high rise kampong settlement suggest that high level of neighbor interactions that usually occur in pocket spaces in kampong alleys will get based on the number of floors. The higher the settlement, the lower the level of interaction among neighbors. This is due to the much higher efficiency demand on high rise settlement buildings that must eliminate pocket spaces usually used for neighbor interactions. Similarly, the housing types in landed kampong settlement are very diverse, whereas the higher the building/apartment, the more uniform housing type will be. This is based on the consideration of the space usage efficiency and the construction cost. Neighboring culture, which is a utilization of a shared space of kampong alleys that are constructed out of organic configuration of housing unit with varied areal spaces and shapes in landed kampong settlement in Kampong Pulo, shows a high level of interaction.

By looking at the precedents living activities through space syntax analysis in the low rise and high rise kampong settlement, the interaction patterns can be typologized into five patterns.

The following patterns are the results of the analyses on several models and unit samples that are integrated with pocket spaces.

1. The opening factor of each unit towards pocket space determines its strength as a space that is more public in nature. On the other hand, if the opening of a residential unit moves away from the pocket space, it can be said that the space has low space integration, and it tends to be private in nature (Fig 5).

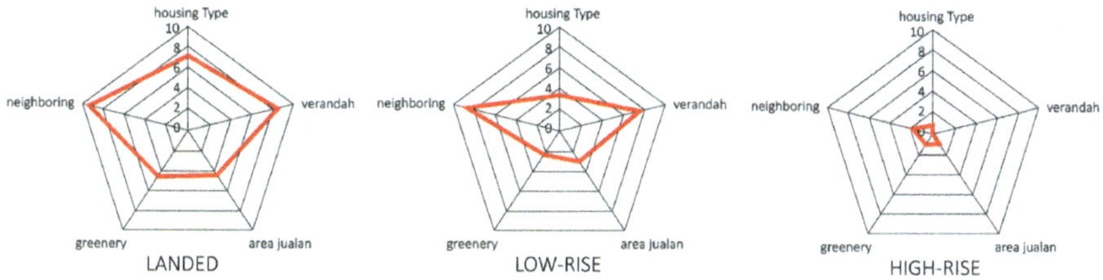

Fig. 4 Sustainability performance of built environment transformation. (*Source* Author)

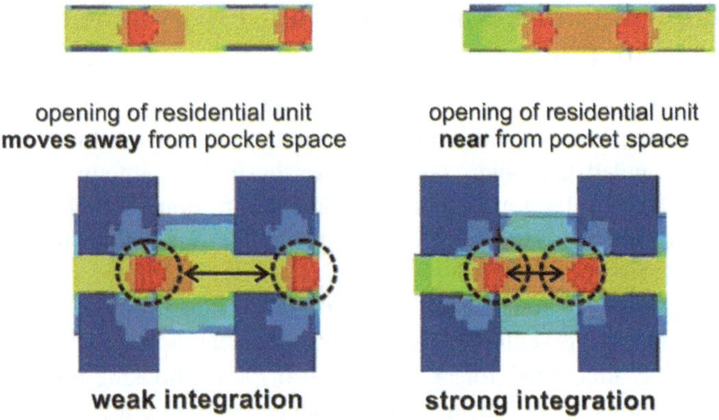

Fig. 5 Two integration values of two pocket spaces that are face to face in a double loaded corridor. (*Source* Author)

2. The placement of a pocket space with a face-to-face pattern is stronger than either crisscross or single pattern. The face-to-face pocket spaces relating to an opening (door) can create a strong space integration (Fig. 6).
3. The longer the corridor, the stronger the corridor's space integration, and the weaker the space integration of pocket space in residential units (Fig. 7).

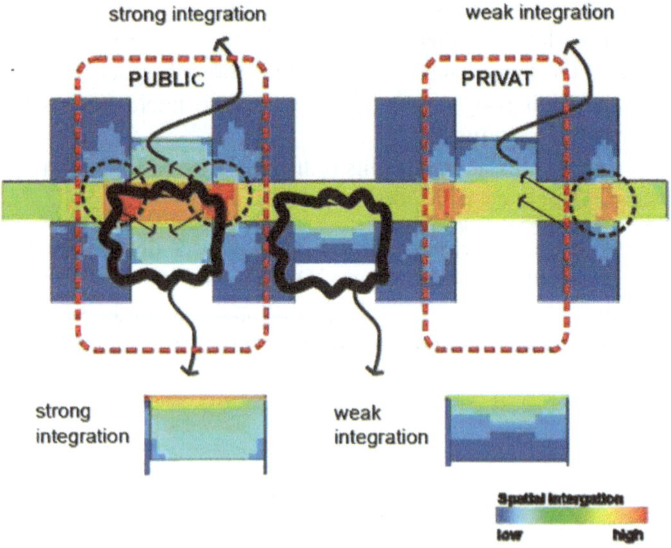

Fig. 6 An integration value in a combined pocket space in a double loaded corridor. (*Source* Author)

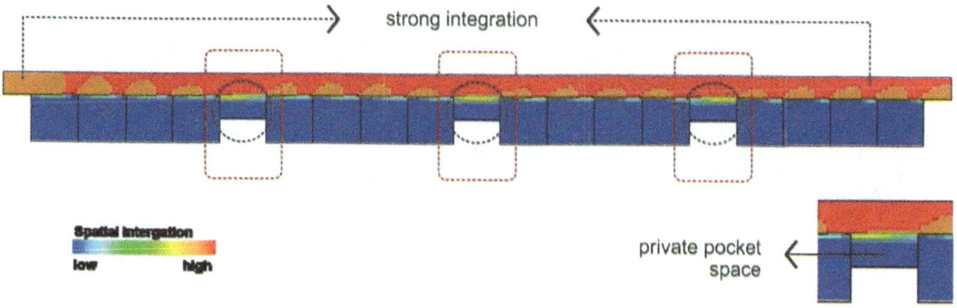

Fig. 7 Integration value in a pocket space in a single loaded corridor. (*Source* Author)

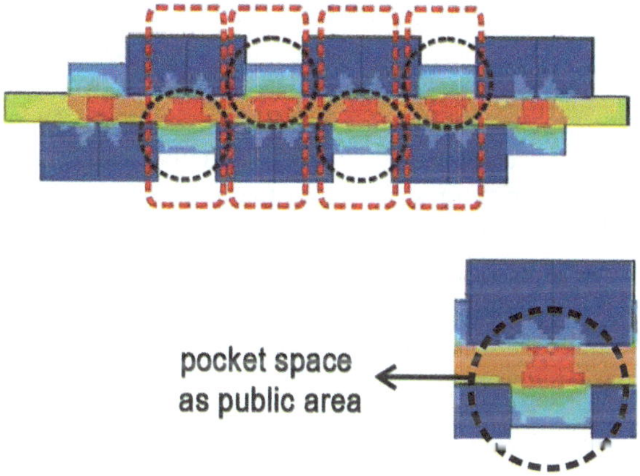

Fig. 8 Integration value in a crisscross pocket space in a double loaded corridor. (*Source* Author)

4. The crisscross form of residential units and pocket spaces is an optimum composition that increases the space integration of a pocket space in a linier residential unit (Fig. 8).
5. The form of residential units and pocket space that is crisscrossed with a long corridor does not affect the strength of the pocket space of residential units (Fig. 9).

Several comparisons between alley type with and without pocket space through space syntax simulation in landed kampong settlement shows a significant difference. There is no neighboring interaction in alley without pocket space because the alley is only for circulation (through), while alley with pocket space has stopping points where neighboring interaction frequently occur. Space use behavior in type alley with pocket space can be transformed into s corridor design in low rise (walk up) kampong settlement.

However, this does not happen in high rise kampong settlement because the shared space in the corridor does not create social interaction. This is due to the space condition where there is no furniture that supports social interaction because of the strict regulations on public space use. Moreover, there is a rule that does not

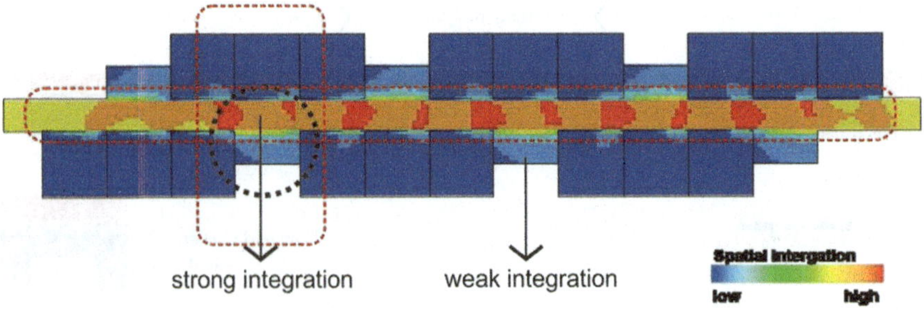

Fig. 9 Integration value in a crisscross pocket spaces in a double loaded corridor. (*Source* Author)

allow the occupants to use the corridor at all, either as a semi private space or for a semi public space, whereas people would usually use the corridor for their activity extension, from private activity to semi private (where the corridor is used as a guest room or from semi private activity to semi public in the form of home-based informal economic activities, usually called front-porch business activities). Because of the restriction on the corridor use, activities of semi private and semi public territories become private space territory activities. In other words, they internalize activities which they would usually "extend". Consequently, social and economic interactions do not occur in high rise kampong settlement.

This can be seen in the space syntax simulation result, which shows that high space integration and connectivity values only in front of vertical circulation facility (lift).

The same goes with green pocket space, which is usually constructed in landed kampong settlement alley configuration that is adapted into the corridor system in low rise kampong settlement. However, it does not exist in the corridor configuration in high rise kampong cityblock settlement. This is because there is a strict regulation on areal space use based on the quantity of space affectivity and efficiency, without considering the thermal comfort capacity. Ventilation where air should naturally circulate using passive natural cross ventilation principle was not taken into account in building façade and wall designs in the housing unit. In addition, photovoltaic as solar energy utilization is not applied in high rise kampong settlement. Other energy utilization such as rain water harvesting component in high rise kampong settlement is not applied. Planting wall or vertical plantation to reduce sun radiation is not applied. Consequently, the sustainability value in the transformation process from landed kampong settlement to high rise kampong settlement is very low in terms of social (living space), economic (livelihoods), and environmental aspect (built environment).

4 Design Recommendation for Sustainable High Rise Kampong Cityblock

Referring to the best practices from the past experience of transforming landed kampong settlement into low rise walk-up kampong settlement and the lessons learned from transformation failure in high rise kampong settlement, there are some recommended design principles involving considerations related to theories with housing type variations, ecology, and modular coordination spirit (Figs. 10, 11, 12, 13, 14 and 15).

Design principle 1 is a multi-level living space, which suggests independent neighboring clusters that still consider vertical mobility efficiency in order that self-sufficiency of the living space community can be achieved, with pocket spaces as shared space that enable the occupants to have social interactions and to generate income (livelihoods).

Design principle 2 is ecology spirit, which suggests green architecture elements such as passive lighting system, natural air circulation, photovoltaic application, and rain water harvesting to reduce energy use.

Design principle 3 is smart porta modular coordination, which suggests support system such as the design of configuration system for building types based on corridor system, which provides pocket space to support the flexibility (porta) of areal variety and the type of the unit that are smart and modularly coordinated.

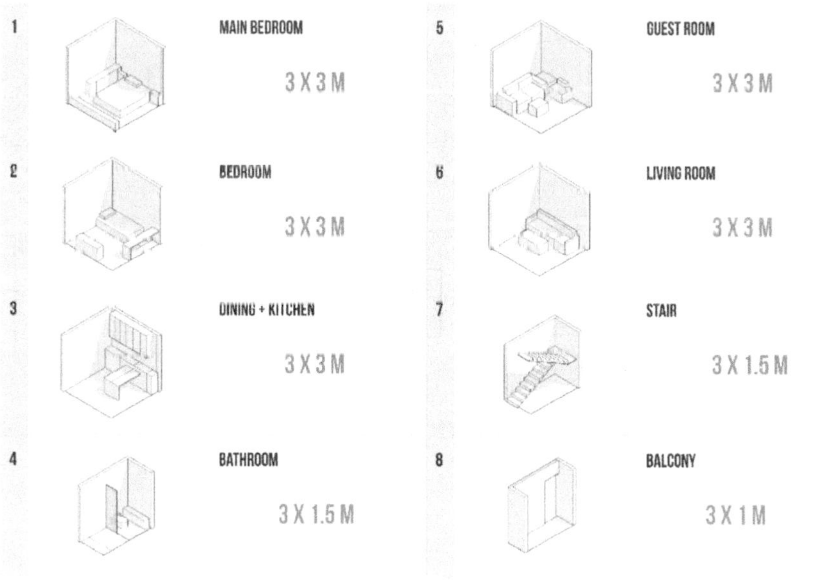

Fig. 10 Space modular system. (*Source* Author)

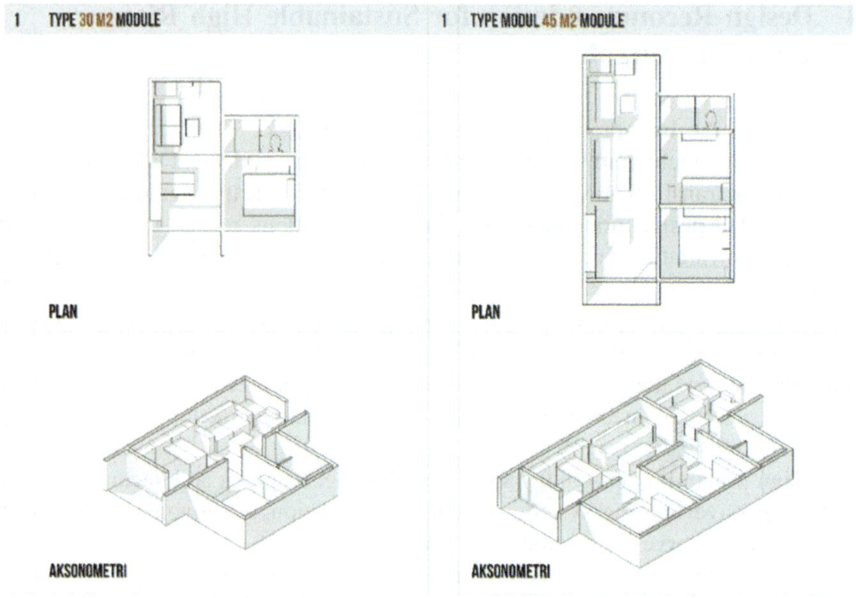

Fig. 11 The variety of prototype module. (*Source* Author)

Fig. 12 Housing type modular system. (*Source* Author)

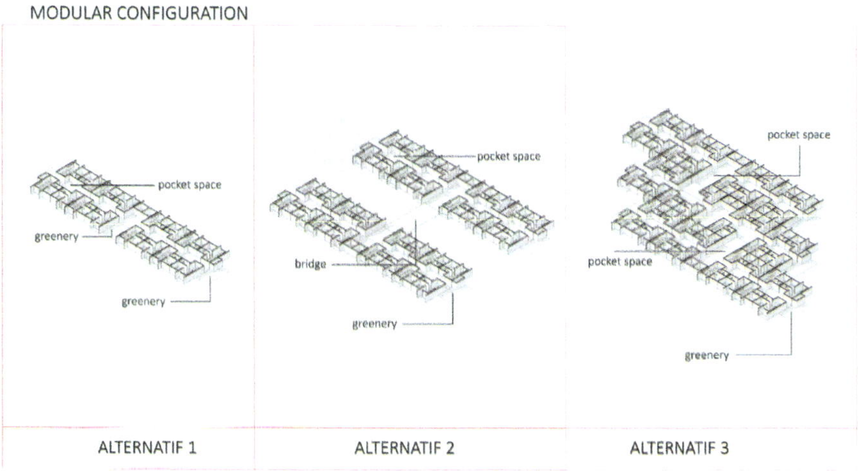

Fig. 13 Variety of building layout. (*Source* Author)

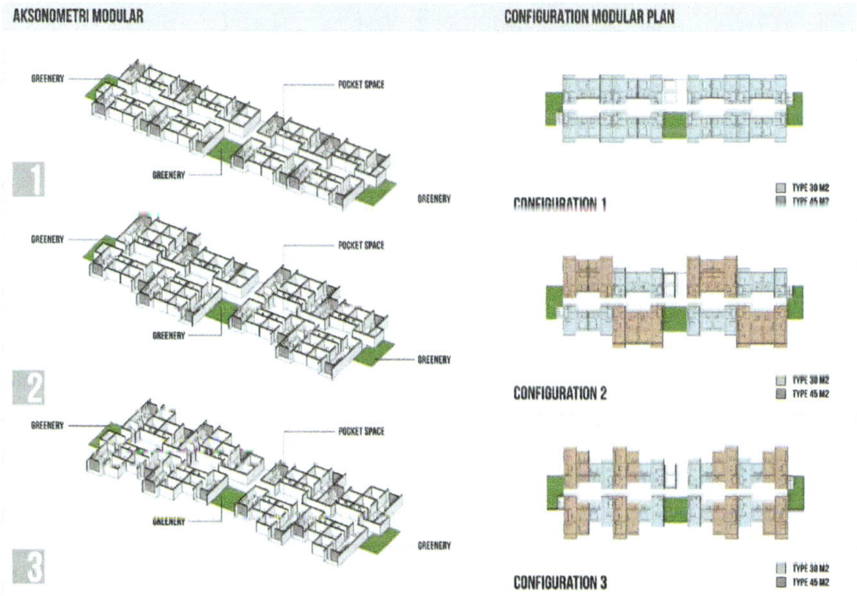

Fig. 14 The variety of organic corridor-based living unit modular configuration. (*Source* Author)

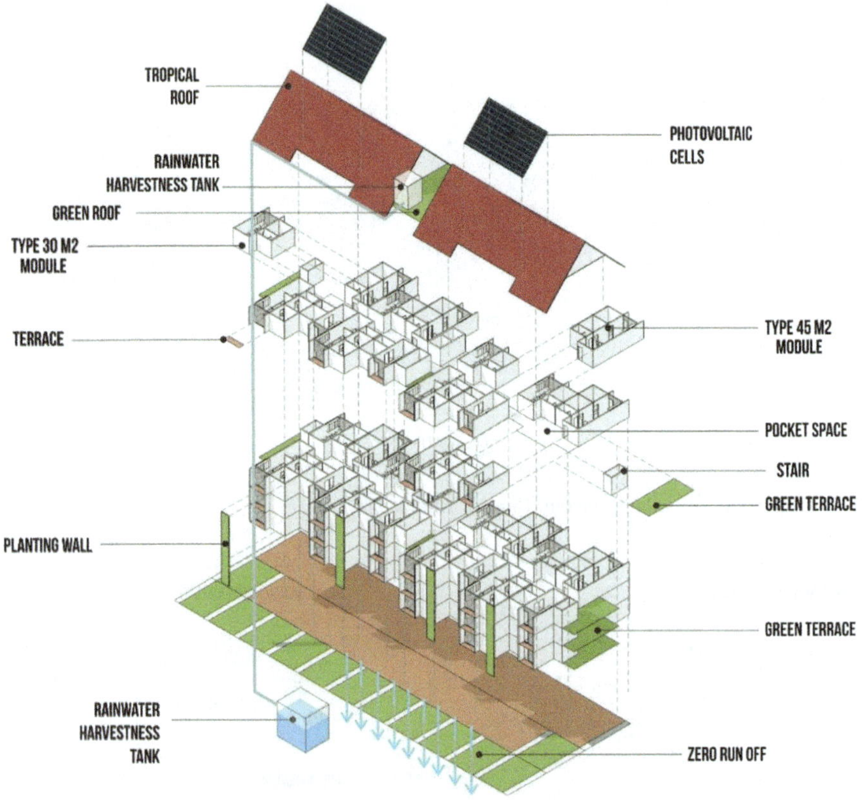

Fig. 15 Schematic model of sustainable high rise kampong city block. (*Source* Author)

5 Conclusion

Based on the survey, interviews with the residents, and observations on of neighbor-related activities, availability of trading space for low income residents who are the majority in the kampong settlement, availability of verandah and greenery (vegetation), it can be concluded that the transformation of landed settlement village into low rise/walk up kampong settlement still has a good sustainable performance value. In contrast, the transformation of the settlement into high rise kampong settlement has a very low sustainability performance value. From the results of this analysis the study carried out a participatory design simulation involving residents from the three settlement types.

References

Chambers, R. (1995). Poverty and livelihoods: Whose reality counts? *Environment and Urbanization, 7*(1), 173–204.

Clarke, G. (1992). Towards appropriate forms of urban spatial planning. *Habitat International, 16*, 149–165.

Clarke, G. (1995). Megacity management: Trends and issues. In J. Stubbs & G. Clarke (Eds.), *Megacity management in the Asian and Pacific region* (p. 180). Manila: Asian Development Bank and United Nations/World Bank Urban Management Programme for Asia and the Pacific Countries.

Cuttler, L. S. (1976). *Recycling cities for people: The urban design process* (p. 241). Boston: Cahner Books International Inc.

Eaton, R. (2002). *Ideal cities, utopianism and the (un)built environment* (p. 212). London: Themes and Hudson Ltd.

Habraken, N. J. (1972). *Support: An alternative to mass housing* (87 p). Cambridge: MIT Press.

Habraken, N. J. (1983). *Transformation of the site* (p. 70). Cambridge, MA: A Water Press.

Habraken, N. J. (1998). *The structure of the ordinary* (p. 116). Cambridge, MA: MIT Press.

Habraken, N. J., Boekholt, J., Thyssen, A., & Donjens, P. (1976). *Variations, the systematic design of support* (210 p). Cambridge: MIT Press.

Hillier, B., & Hanson, J. (1984). *The social logic of space* (p. 66). Cambridge: Cambridge University Press.

Madanipour, A. (1999). Why are the design and development of public space significant for cities? *Environment and Planning B: Planning and Design, 26*, 879–891.

Perry, D. C. (1945). *Making space: Planning as a model of though* (212 p). Thousand Oaks: Sage Publication Inc.

Prayitno, B. (2013). An analysis of consolidation patterns of Kampung Alley living space in Yogyakarta, Indonesia. *Journal of Habitat Engineering and Design, 6*(1), 99–112.

Prayitno, B., et al. (1994). Low cost rental flat development in Indonesian large cities. *Journal of Architecture Institute of Japan, 517*, 243–250.

Yeang, K. (2006). *Ecodesign, a manual for ecological design* (190 p). Britain: Willey Academy.

Author Biography

Assoc. Prof. Budi Prayitno is a senior lecturer and researcher in the fields of architecture and urbanism at Gadjah Mada University, Yogyakarta, Indonesia, where he is the director of Center of Sustainable Habitat Engineering. Currently, he is responsible for coordinating the urban regeneration projects in Indonesia. He is also received the Regional Center of Expertise on Education for Sustainable Development (RCE Recognition Award 2012) for his publication of "Kampong Upgrading and Greening: Enabling and Learning Processes for Consolidation-based Urban Settlement Redevelopment Project". He has spent 20 years as an architect, urban planner, and advisor to the Ministry of Public Works and Housing and Ministry of National Development Planning.

The Unintended Consequence of Building Sustainably in Australia

Tim Law and Mark Dewsbury

Abstract What makes a sustainable house? One might suggest it should be energy-efficient, resilient to climate change and still comfortable. Indeed in Australia, we see aspects of these three priorities being exercised: energy-efficiency standards being introduced into residential requirements of the National Construction Code in 2003, bushfire requirements expressed as a national standard in 2009, and the constant demand for more efficient and round-the-clock climate control. All these actions relate to Sustainable Development Goal (SDG) 13: Climate Action. One might assume that these trends mark progress for both the environment and the home owners. However there is a dark side to the story, because in the very effort of reducing greenhouse gas emissions (also one of the functional objective of the national construction code), the construction industry has inadvertently implemented practices that have led to entrapment of moisture in buildings, thus compromising their habitability. Using data from Tasmania, this chapter shows how common mistakes in building science, design and construction have led to a widespread increase of condensation in buildings located in cool climates. Condensation has further led to other problems with mould and health (SDG 3: Good Health and Well-being), making new code-compliant houses potentially uninhabitable after experiencing their first winter. These challenges need to be in the wider discussion of architecture, construction, indoor microbiology and public health when sustainable housing standards are being promoted.

Keywords Energy-efficiency · Australia · Building code · Bushfire Legislation · Mould

T. Law (✉) · M. Dewsbury
School of Architecture and Design, University of Tasmania,
Launceston, Australia
e-mail: tim.law@archsciences.com.au

W. Leal Filho et al. (eds.), *Sustainable Development Research in the Asia-Pacific Region*, World Sustainability Series, https://doi.org/10.1007/978-3-319-73293-0_31

1 Introduction

In Australia the building codes ensure that there are minimum standards for residential dwellings. Indeed when the codes were first written in 1988, structural sufficiency, fire safety, health and amenity were the listed objectives. Home occupants could have a legitimate expectation that any house bought should be safe, healthy and have sufficient amenity. Subsequently, energy efficiency provisions were introduced into the Building Code of Australia in 2003, starting as a requirement for housing (ABCB 2010). Various states within Australia would introduce different stated values of star ratings developed by the Nationwide House Energy Rating Scheme (NatHERS) for residences. Australian states independently and incrementally moved from the initial 4-star to the current 6-star requirements for energy efficiency. Table 1 illustrates the introduction of the requirements over time. The star ratings ascertain the amount of energy required to condition a house (and to a lesser extent artificially light or heat pools) in MJ/m^2 per annum and is determined at the design and specification stage of a house.

National Construction Code (NCC) energy efficiency requirements timeline for various building types (ABCB 2016).

After a major bushfire in the state of Victoria in 2009, also called "Black Saturday" a new national standard (AS3959, 2009) was introduced to specify how new buildings were to be designed around bushfire prone areas. Currently in Victoria, without exemption all houses need to have some measure of bushfire attack resistance. Part of the strategy invoked preventing embers from entering roof spaces and a common construction method to deal with that is to lay glasswool insulation blankets under the ridge and eaves of a metal roof to fill up the voids under a profiled metal roof, reducing the chance of ember entry. However, this approach also eliminates the option of naturally ventilating the roof space.

Table 1 National Construction Code (NCC) energy efficiency requirements timeline for various building types ABCB (2016)

	Housing	Multi-residential	Non-residential
2000	*Announcement to mandate minimum energy efficiency requirements in the BCA*		
2003	Housing provisions introduced		
2005		Multi-residential buildings provisions introduced	
2006	Housing stringency increased (5 star minimum)		Non-residential buildings provisions introduced
2009	*Announcement to increase stringency for all buildings*		
2010	Housing stringency increased (6 star minimum)	Multi-residential buildings stringency increased	Non-residential buildings stringency increased

In the 2010 update to the Building Code of Australia (BCA), besides increasing the stringency of energy efficiency provisions, the new standard was included for construction in bushfire-prone areas. The industry, in response to making buildings more thermally efficient, had to also make them tighter. Air tightness of buildings not only contained heat better, it was also better at keeping cinders from blowing into the buildings. All these actions relate to Sustainable Development Goals (SDG) 13: Climate Action. One might assume that these trends mark progress for both the environment and the homeowners.

However with increased air tightness, the industry started noticing a disturbing new trend: many new residential buildings were encountering copious amounts of condensation. The persistent damp from condensation has led to other problems with mould and its deleterious effects on human health which compromise another goal, SDG 3, that of 'Good Health and Well-being'. Using data from Tasmania, this chapter will demonstrate how common mistakes in building science, design and construction have led to a widespread increase of condensation in buildings located in cool climates.

Condensation is a complex phenomenon and requires as a background some understanding of the psychrometry of air, vapour permeability and the temperature of the condensing surface.

2 Psychrometry of Air

Where air moves freely it will carry moisture with it. Where air is stopped by a barrier, moisture may still move through, assuming the air barrier was vapour permeable (Babbitt 1939; Barre 1938). To understand the moisture content and transport through building elements we turn to psychrometry, the study of moist air. With any combination of two variables (typically being any combination of dry bulb temperature, wet bulb temperature or relative humidity) one can locate the coordinates on the psychrometric chart, and from there determine the dew point temperature, the point at which air reaches saturation and any further cooling will result in precipitation.

Historically, vapour pressure management and condensation risk analysis were taught within applicable technical and tertiary courses (Szokolay 2008), which utilised Australian and international guidelines (CSIRO 1991a, b; van Straaten 1967). However, it appears that due to the general leakiness of historical Australian construction practises and the general 'de-sciencing' of many technical and tertiary training courses, which have often blamed pressures from funding arrangements, this critical component of built fabric design and construction has been removed from key learning outcomes and regulatory requirements. This combination of a simplified professional education, the aspiration for more complex building envelopes and the lack of guidance and regulation has established a significant knowledge gap between the technical considerations required to safely achieve more thermally efficient, structurally safe and healthy buildings and the current industry capacity of construction trades, building surveyors, environmental health officers engineers and the building design professions.

Psychrometry is one of the principle tools building mechanical engineers use to size and select air-conditioning equipment, and balance the loads between sensible heat, where heating or refrigeration causes temperature change, and latent heat, where energy is accounted for by a phase change like evaporation or dehumidification without a change in temperature. However other fields have very limited knowledge of psychrometry, if any at all. In the 2015 ABCB Condensation Survey, less than a third of the respondents were aware of how to apply a hygrothermal (condensation) analysis at the design stage.

Without a common understanding in psychrometrics it would be difficult to appreciate the significance of air tightness, vapour permeability and thermal bridging. Building scientist Lstiburek (2010) explains how a building envelope has four principal functions which, in order of priority will be: rain control, air control, vapour control and thermal control. It would seem self-evident that an occupant in a cool climate would rather a house that was cold but dry, then one which was warm but damp. Unfortunately the contrary is largely observed which could derive from the emphasis on energy efficiency instead of vapour management in many Australian houses. These three of Lstiburek's four earlier principles are often misunderstood or ignored. In contrast, whilst the industry well understands the importance of rain control, the irony is that without understanding psychrometry, we have created an environment for the production of 'indoor rain' through condensation.

To illustrate this, in Fig. 1 we see psychrometric measurements during early spring of a house in Tasmania, the southernmost state of Australia, and here coded House-BN. The heat pump is run 24 h a day at a thermostat setting of 18 °C, a typical heating system and setting for most houses in the temperate climates of Australia. Measurements in the roof are coincident in time but presented separately

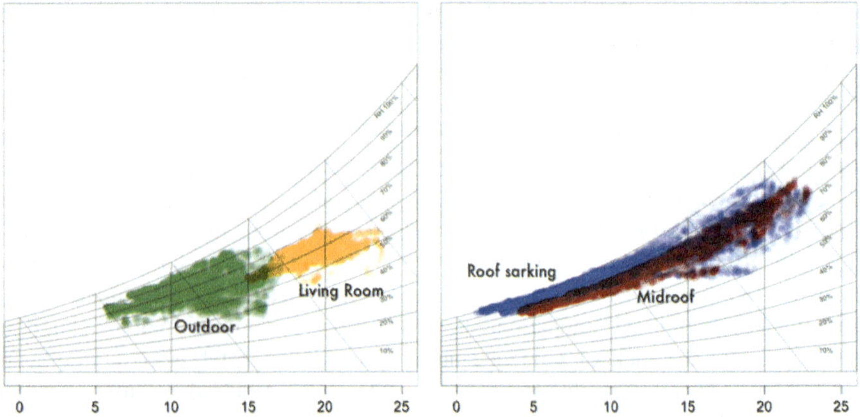

Fig. 1 Psychrometric charts of House-BN in Tasmania measured during late winter

Fig. 2 Selection of time-lapse photography to show persistence of droplets throughout the day

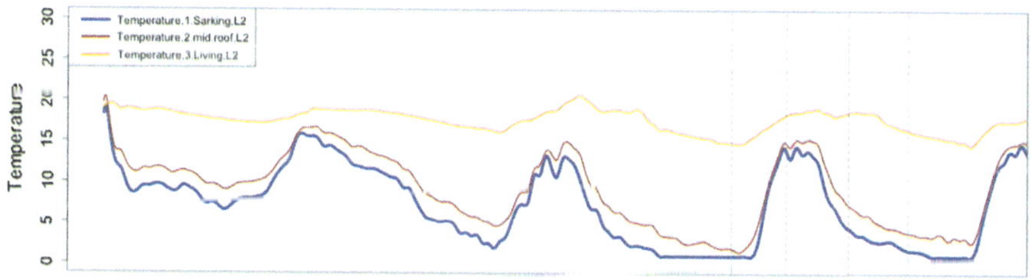

Fig. 3 Temperature measurements in House-BN during same period, grey lines indicate timing corresponding to above time-lapse photographs

for clarity. Outdoor air, after being heated, increases in temperature and absolute humidity before entering the roof space. In this space, the absolute humidity increases further, indicative of moisture accumulation, and condenses against the sarking, resulting in precipitation under the roof. A snippet of time-lapse photography (Fig. 2) cross referenced against the temperature logs (Fig. 3) provided a visual confirmation of the persistency of condensation, showing that even though condensation was not continuously occurring, once the droplets formed they did not dry out during the day. This persistence of moisture is observed throughout the months of winter on time-lapse photography.

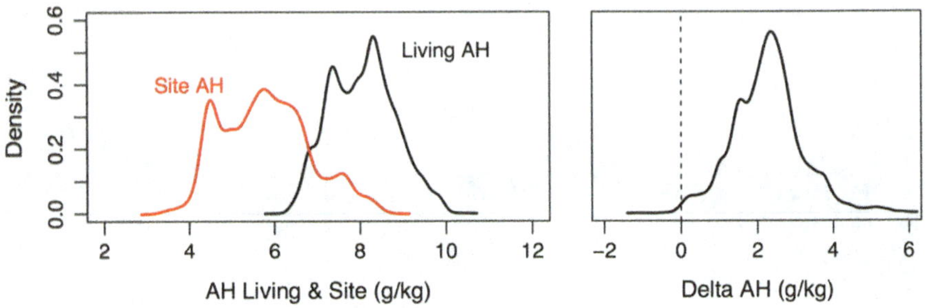

Fig. 4 Frequency distribution of absolute humidity of site and living room air, and the difference between the two values for House-BN

3 Vapour Diffusion, Air-Tightness and Permeability

When still air has different amounts of moisture, indicated by its Absolute Humidity (AH), diffusion occurs naturally from higher to lower concentration. AH is calculated psychrometrically from measurements of temperature and RH and the values in the living room compared to the site AH is illustrated in Fig. 4. The delta AH shows that living AH is almost always higher than site AH with the mean difference about 2.5 g of vapour per kg of dry air. This difference in AH highlights a need for ventilation and a building fabric designed to manage vapour.

In House-BN the house occupant, typical of most Australians during winter, does not ventilate the house at anytime throughout the day. We have corroborated our measurements with an interview with the owner. Besides security and privacy, the common conception is that it is most energy-efficient for the heat pump to be run continuously instead of only when occupants are present. This also corresponds to many occupants expectation that the house should already be comfortable before they return home, without having to wait for the air and surface temperatures to be comfortable. In Australia although there are no stipulations for space heating/cooling temperatures, there is an indication of expectation of a range between 18–26 °C, in the context of how energy consumption is to be calculated for a performance-based verification method (Section JV3, ABCB 2016). To meet this expectation of round the clock comfort, the most energy-efficient heating system are heat pumps. Hence a cycle of self-reinforcement is produced, with the impression that continuos space conditioning with heat pumps is the most energy-efficient and cost-savvy approach to modern comfort.

Whilst this thinking is arguably reasonable, it leads to an important caveat, that in a continuously conditioned space without ventilation, the entire burden of vapour management is transferred to the building fabric of walls, floors and ceilings. Here, the vapour can be managed by either infiltration at junctions and penetrations, or allowed to permeate through the fabric. On a national average, Australian houses reportedly achieve 15.4 ACH@50 Pa (air changes per hour at a pressure difference of 50 Pa) with houses in Hobart (Tasmania's capital city) significantly more airtight, with an average 7.9 ACH@50 Pa (Ambrose 2015). Uncontrolled infiltration is clearly not a viable way of managing moisture in cool climates. So instead of having indoor air replaced by the relatively drier outdoor air, the vapour from indoor air has to be allowed to diffuse through the building fabric of plasterboards, insulation, building wrap and then exhausted by a ventilated cavity.

The most commonly used building wraps in Australia are foil-based, the cheapest wrap to install. When punched with holes they are often advertised as 'breathable', a purely marketing term that carries no definition or control under the relevant standard (AS4200, 1994). However when the space between studs is filled with insulation batts, there is no free movement of air, so vapour does not 'find' its way to those openings. Instead when vapour reach the foil wrap, it encounters an impermeable surface and condenses against the wrap as it cools. In many cases, like the common brick veneer construction, once the house is completed it is prohibitively costly to replace a foil wrap with a vapour permeable one.

Ventilated cavities are not always required by code or by manufacturer's details. Thus even when a selected wall cladding is vapour impermeable, like sheet metal or exterior insulation finishing systems (EIFS), many builders opt not to include wall battening to create a ventilated cavity, fearing that the additional work will price them out of a competitive tender. They in turn blame the architect or building designer for not requiring battens in the drawings, who in turn claim they are not required to know about the phenomenon of condensation, nor is there any legislative stipulation to ventilate the wall cavity. Most owners are not cognisant of this problem as the mould develops in the interstitial spaces. Even a borescope is of limited use as the walls are packed with insulation. The cladding has to be peeled back (Fig. 5) to reveal the extent of damage at House-MW2, which was newly built and occupied for less than 2 months. To increase ventilation in roofs, walls and sub-floors, the authors have developed design solutions to common architectural

Fig. 5 Often the true effect of condensation happens inside the wall framing. A builder helps dismantle the sheet metal cladding that is hard fixed against the studs (no battens) causing the wrap to be thermally bridged with the cold exterior. Indoor vapour, unable to pass through the foil wrap condenses on the insulation which acts as a reservoir, keeping the studs and bottom plate a moist breeding ground for mould

details for the Tasmanian government (BSOL 2014) disseminated to the building industry as a non-mandatory design guide. The intention was to offer designers with alternative options that increased airflow in the interstitial spaces as a starting point, and on this awareness to develop an appreciation for the more complex topics that follow.

4 Condensing Surfaces in Buildings

Whilst psychrometry analyses the condition of moist air in a space, condensation frequently occurs on surfaces with a temperature difference. The warm space (like a heated living area) supplies the energy for air to hold more moisture in vapour state. When this vapour laden air encounters a cold surface below its dew point it will condense on it. In Fig. 6 the mid-roof and site (verandah) temperatures go House-BN are compared against the dew point of the living room. The frequency distribution show that for almost half the time, site temperature is below living dew point so building components in thermal communication between these two spaces, like aluminium window frames, will result in condensation (Fig. 6). Similarly, the uninsulated part of the ceilings will allow cold roof space to cause condensation on any uninsulated part of the ceiling. Australia's experience with the poorly managed Home Insulation Program (Parker 2014) has raised awareness to the importance of leaving a clear 200 mm around down lights to avoid fires (Standards Australia, 2007). This however leaves an uninsulated area around the down light where mid roof temperatures were below living dew point, which for House-BN was 76% of the time (Fig. 7). Two of these dripping down lights were directly above the master bed (Fig. 8).

Roofs have been observed to be a particularly problematic area as the ceiling plenum covers many sources of moisture ingress and is itself one of the coldest surfaces, especially with the radiative losses on a clear night. Figure 9 illustrates the monthly average amount of radiation and daily measurements. However when

Fig. 6 An example of what happens when there is thermal bridge between site temperature and living dew point: condensation on the aluminium frame of a south facing window

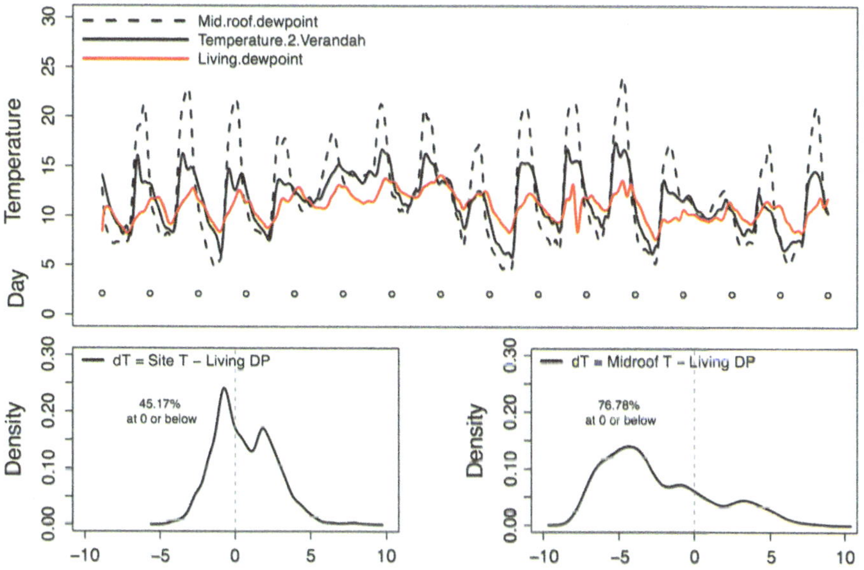

Fig. 7 Top: time series plot of midroof temperature, site temperature and living room dew point. Bottom: frequency distribution of difference in temperature, where a negative value indicates condensation: site temperature versus living dew point, and midroof temperature versus dew point

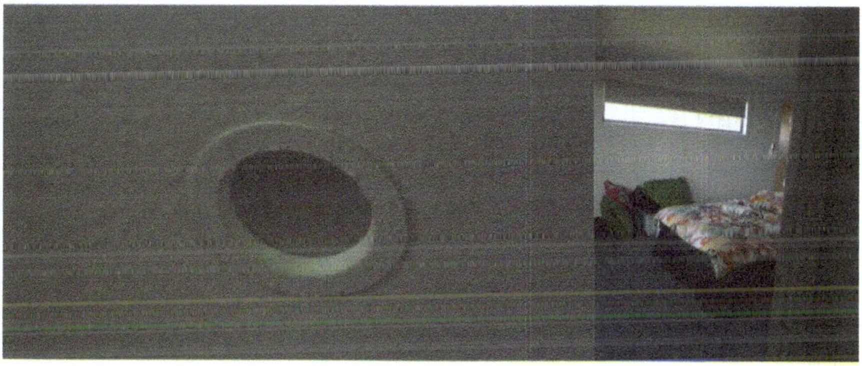

Fig. 8 An example of what happens when there is thermal bridge between mid-roof temperature and living dew point: condensation around the uninsulated downlight

overlaid with temperature of the roof sarking there is little correlation between radiation and roof temperature, i.e. a day with higher than average radiation does not appear to result in a higher than average roof temperature. Throughout winter, the roof is colder than the living room, even on sunny days.

With a roof so cold, even the conventional thinking of ventilating the roof space needs to be critically reconsidered. Figure 10 shows how the sarking is below site

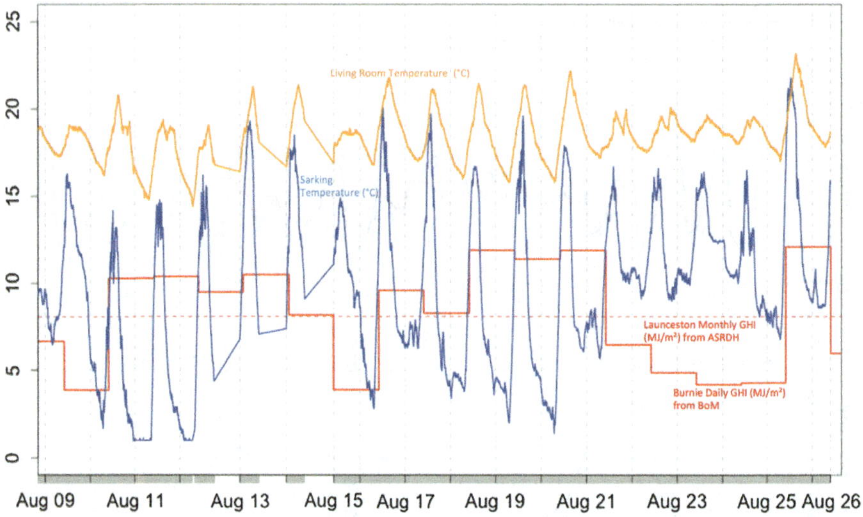

Fig. 9 Time series of solar radiation: dotted red line shows monthly average from the Australian Solar Radiation Data Handbook (2006) of the nearest city Launceston, solid red line are measurements by the Bureau of Meteorology, Australia (in MJ/m^2). Temperatures measurements in House-BN at roof sarking and inside living room

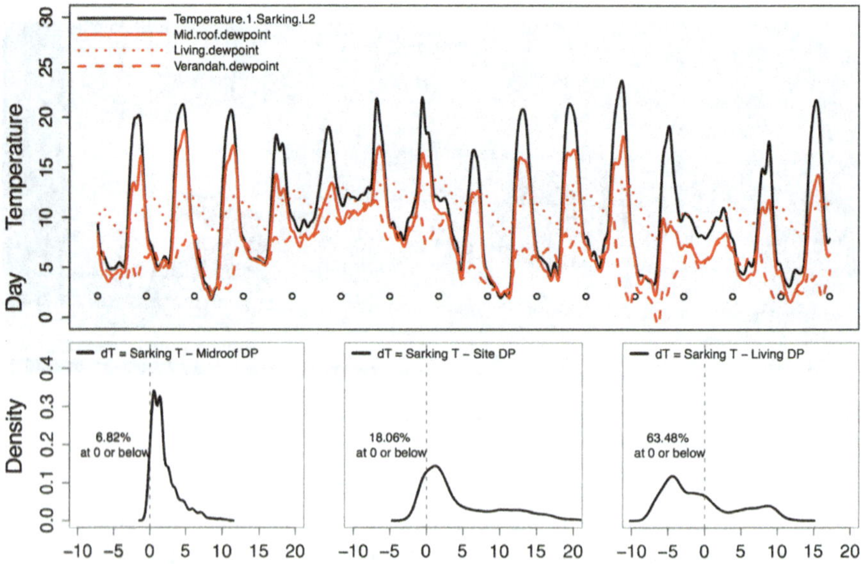

Fig. 10 Top: time series plot of sarking temperature, against the dew points of midroof, site, and living. Bottom: frequency distribution of difference in temperature, where a negative value indicates condensation: sarking temperature versus living dew point, sarking temperature versus site dew point, and site temperature versus living dew point

Fig. 11 Blower door apparatus modified for testing roof cavity

Fig. 12 Installation of roof ventilation, in total 4 of these vents were installed

dew point more frequently (18% of the time) than it is below the roof dew point (7% of the time). In other words if more ventilation were to be encouraged, this particular roof would suffer even more condensation. Of particular concern is that any air from the living room leaking into the roof cavity would most likely meet a condensing surface (63% of the time). When the blower door apparatus was modified to pressure test the roof (Fig. 11), we found major internal air leakages in powerpoints and pocket doors. When the roof cavity was tested at 50 Pa it measured 71.2 ACH before 4 roof vents were installed and 72.7 ACH after (2.1% increase). The roof vents, an example seen in Fig. 12, thus had little effect on the leakiness of the roof. We can surmise that the source of moisture was primarily from the habitable areas travelling through service and construction penetrations into the roof cavity and condensing on the underside of the cool roof (Fig. 13).

Figure 14 compares the BN roof (white colour, sarking foil) with MW2 (green colour, insulated foil blanket). Both roofs experience unshaded sun. It is apparent the effect of roof colour where MW2 is able to benefit from solar absorptivity, even

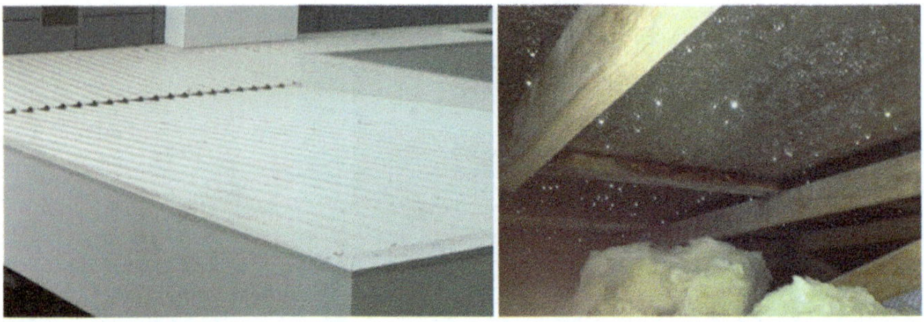

Fig. 13 The constantly wet, and often dripping underside of the roof

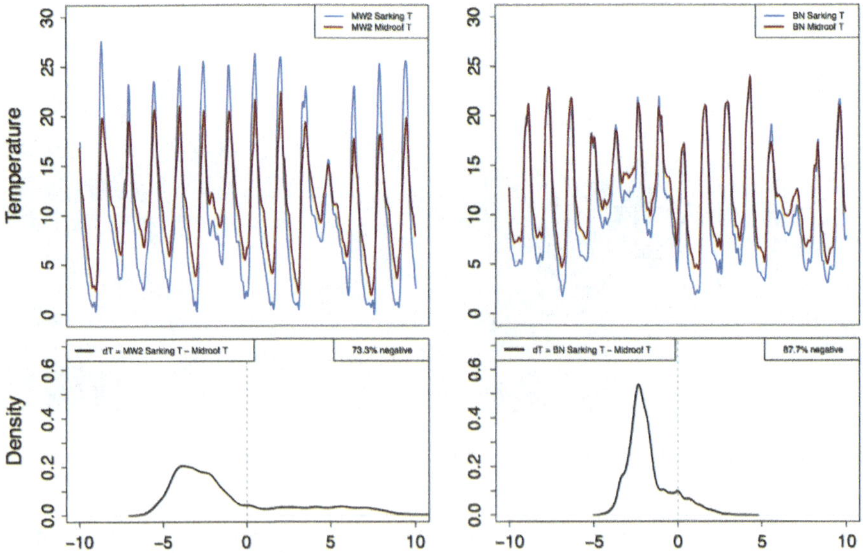

Fig. 14 Time-series and frequency distribution of (left) MW2 mid-tone coloued roof and (right) BN white roof

with a foil blanket. Without this insulation, the peaks would be even higher during daytime to assist in drying the roof cavity.

The role of the roof to minimise condensation is not well understood, and attempts to increase energy efficiency by adding insulation to ceilings result in a variety of problems that exacerbate condensation. A number of common roofs are illustrated in Fig. 15.

For the double pitched roofs like gable or hipped, orientation of the roof to the winter sun is immaterial as there will always be part of the roof exposed to the sun and this will increase the drying potential of the roof space, as long as it is well ventilated. The standard pitch of 22.5° allows moisture to run along the underside

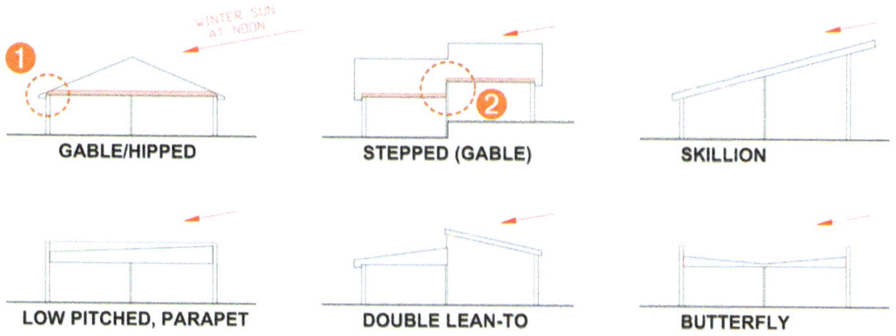

Fig. 15 Common roof types in houses, each can be potentially problematic with condensation in winter

of the roof and drain at the gutter, assuming it is not blocked by ceiling insulation. When the roof is fitted and insulation installed after that, the installers are not able to determine where the edge of the walls are and typically push the insulation hard against the underside of the roof, creating a path for moisture to run into the ceiling. Particular care also needs to be given to instances where vertical surfaces appear in a roof, like when the ceiling is stepped in line with the site. Insulation installers often forget to insulate the vertical surface as they should the entire ceiling with the same level of insulation. Single pitched roofs like skillion roof are often used to allow higher walls on the sunny side, but this orients the roof away from the sun and limits its drying potential. Double lean-to roofs separate the two roofs and, for the shaded side, limits both the solar exposure and ventilation opportunities. This similarly affects the low pitched roof and butterfly roof, which has the same disadvantages and the additional problem of creating suspended droplets that will not flow down into the gutter.

5 The Problem with Condensation

Dampness can affect up an estimated 10–50% of buildings in Australia, Europe, India, Japan, and North America (WHO 2009; Andersen et al. 2011). The ABCB Survey in Fig. 16 shows that respondents estimate that an overall one third of new Australian buildings suffering from condensation problems, with minor variations across climate types.

In many ways, condensation inside a building can be as bad as water ingress arising from a building defect or faulty plumbing. However, as opposed to such problems that typically occur in an ageing stock of buildings, most home owners would not expect their brand new houses to have a moisture problem of this nature. Unless something catastrophic happens, many will not even be aware they are experiencing a moisture problem every bit as serious as a leak.

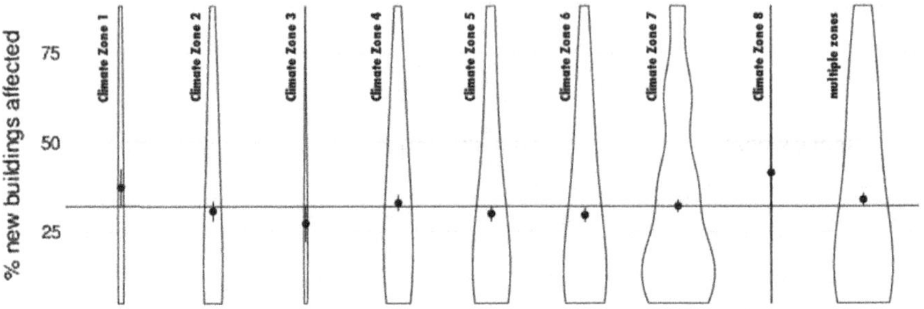

Fig. 16 Distribution of responses by states in Australia to the question, "What do you believe is the overall proportion of new residential buildings (both houses and apartments) affected by condensation?" Sorted by climate zone as follows: Climate zone 1—High humidity summer, warm winter; Climate zone 2—Warm humid summer, mild winter; Climate zone 3—Hot dry summer, warm winter; Climate zone 4—Hot dry summer, cool winter; Climate zone 5—Warm temperate o Climate zone; 6—Mild temperate; Climate zone 7—Cool temperate; Climate zone 8—Alpine; Multiple—Respondents that work in more than one climate zone. Each "violin" shows the distribution of state estimates, similarly to a smoothed histogram. The breadths are proportional to the subsample sizes; the central dot and bars show the estimated mean and its standard error. The horizontal line indicates the overall mean

In terms of structural safety, the moisture build up has been a cause for collapsed ceilings when the ceiling insulation is overburdened with so much moisture the furring channels could not support the additional weight. Additionally, constant wetting and drying of timber trusses has the possibility of compromising the nail plates, an alleged cause for the collapsing of a roof in 2002 at the Riverside Golf Club, South Australia (James 2006; Palmer 2012). Moisture also accelerates fungal and borer decay of untreated timber frames. In New Zealand the 'wet building syndrome' that emerged from monolithic construction (a form of construction that gives the building a seamless appearance) has resulted in many timber framed buildings succumbing to structural damage. There an ongoing class action lawsuit is filed by about 1000 claimants representing over 300 buildings for $250 million against a plaster cladding manufacturer for the moisture related problems that are common with this method of construction (Thorn 2016).

Although the structural damage from moisture can be serious, it is usually a prolonged matter of a decade or more before a timber structure deteriorates to a point it is deemed unsafe. In contrast, a mouldy dwelling can be deemed unhealthy within its first winter. To understand the impact of mould better we have to another field, that of mycology, with a focus on the health implications of mould.

6 Mould Growth

Mycology is the study of fungi, a kingdom of which moulds are classified under. Mould is a general term to describe microscopic filamentous fungi. Due to the ubiquity of cellulose-based nutrient sources in timber-framed buildings lined with paper-faced plasterboards, moulds can grow anywhere in a building with enough moisture, and especially in the dark areas. By the time an occupant sees the mould appearing on the painted walls it is most likely there would much more to be found in the dark interstitial spaces behind the wall and ceiling lining. Mould is always an indicator of damp, and if it cannot be traced to a cladding or plumbing leak then condensation should be the seen as the most likely culprit.

The conditions that affect the rate of growth of mould is a combination of available moisture and available nutrient. Figure 17 shows isopleths for the germination and growth of mould from the Australian Institute of Refrigeration, Air-Conditioning and Heating (AIRAH 2016) and the World Health Organisation (WHO 2009). Whilst in both cases, relative humidity (RH) has been presented as a determiner of growth rate, it should be noted that the RH of air is an indirect predictor of mould growth, microbiologists recognising that 'the effects of relative humidity on airborne microbes are complex and involve phase changes at the molecular level' (Kowalski 2005, p. 157). With so many aspects of aerobiological behaviour not fully understood it is no wonder that in the *ASHRAE (American Society for Heating, Refrigeration and Air-conditioning Engineers) Position Document on Airborne Infectious Diseases*, the only mention of humidity is that "humidity affects survival of the infectious agent although not always in predictable ways." (ASHRAE 2009, p. 5) Hence merely controlling RH in a space is a rather blunt and energetically costly way of controlling microbes like moulds (Law 2013).

Rather that RH of air, fungal growth is predominantly controlled by the moisture content of a substrate and is likely if the water activity (aw) of a material exceeds 0.76–0–96, depending on fungal species, temperature, time and composition of a

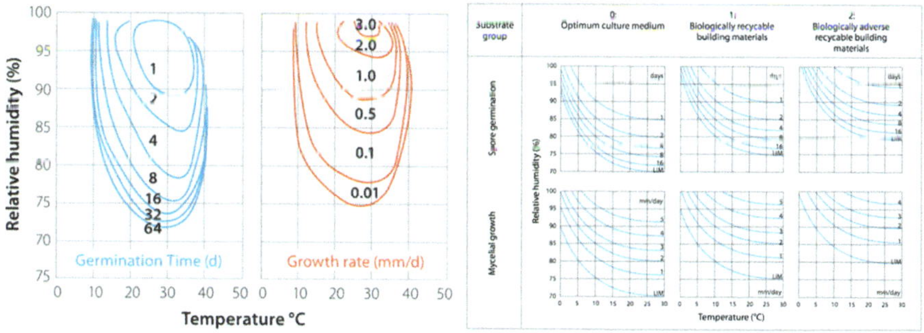

Fig. 17 Typical conditions that affect germination and growth rates for moulds as established by AIRAH (2016) in its application manual "DA20 Humid Tropical Air Conditioning" and the WHO (2009) "Guidelines for indoor air quality: dampness and mould"

material (Pasanen et al. 1992). Barbosa-Cánovas et al. (2008) provide the following formulation to determine the water activity based on substrate temperature and the dew point temperature of the air around the sample.

$$a_w = \exp\left(\frac{bT_d}{c+T_d} - \frac{bT_s}{c+T_s}\right)$$

where

a_w water activity
T_d dew point temperature
T_s AS3959
a 0.611 kPa
b 17.502 C−1
c 240.97 C

In Fig. 18, measurements were made of a House-MW2 in Tasmania. Based on the living room temperature and dew point there would have been no instance when the RH would have exceeded 80% (or 0.8 equivalent RH), let alone experience condensation. However since the sheet metal cladding was not constructed with a cavity, i.e. it was hard-fixed against the timber studs with a layer of vapour impermeable wrap in-between, it would be valid to assume there was certain amount of thermal bridging occurring around the timber frame, especially the top and bottom plates of the wall structure. Using the outdoor temperature as T_s we see

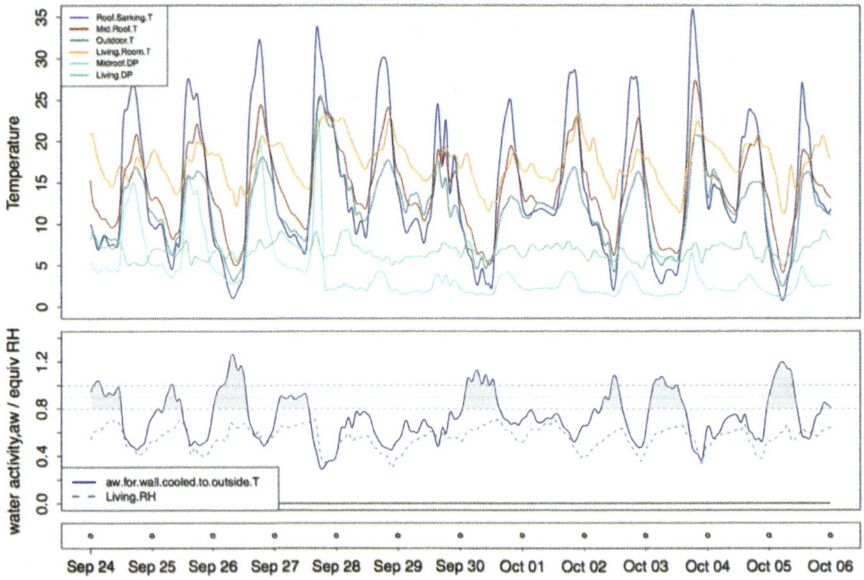

Fig. 18 House-MW2 in Tasmania, measured in Sept 2014

Fig. 19 Condensation forming on the kitchen flooring due to poorly installed insulation and a vapour impermeable vinyl finish. Around the window reveals the thermal bridging of the structure results in condensation

Fig. 20 Under the bed, the wall and carpet have limited exposure to the heating and experience thermal bridging to the external walls

a very different profile of the aw for parts of the wall that were uninsulated, with frequent occurrences of a_w exceeding 0.8; and even exceeding 1.0 indicating the availability of 'free water'. If this water was not evaporated by the next day it will accumulate, and over time provide a continuous period of dampness to permit mould growth. In this house, condensation was visible around the windows and on the kitchen vinyl flooring (see Fig. 19). Mould was observed in many instances on walls (especially behind furniture (Fig. 20) and inside the walls and ceiling (Fig. 21). Spore samples were collected on settle plates and the following genera were identified: *penicillium*, *zygomycete*, *cladosporium* and *aspergillus*. After occupying this house for about a month after construction, occupants of this house were unwell with varying symptoms which were not present prior to occupation: one parent had laryngitis and a child had constant headaches.

Fig. 21 Under the bed, the wall and carpet have limited exposure to the heating and experience thermal bridging to the external walls

7 Mould and Health

Mould can affect human health by an infectious route, typically through respiration, and produce diseases like *Aspergillosis*, *Histoplasmosis*, *Coccidioidomycosis* (Kendrick 2000). Alternatively, mould can cause toxicosis through skin contact, inhalation or ingestion. Mycotoxins (mould toxins) are a secondary metabolite, so they are not used in the primary function of growth and reproduction. These mycotoxins accumulate and are then released sporadically (Kendrick 2000, p 145), possibly for competitive exclusion depending on the antagonist (Zain 2011). This is of particular concern in cool climates, as toxin production usually increases at low temperatures (Wannemacher and Wiener 1997).

Concerned for the need to establish minimum safety standards to reoccupy houses two years after Hurricane Katrina, detailed sampling and identification were carried out in houses heavily contaminated with mould (Bloom et al. 2009). Although the researchers established a case for the requirement for personal protective equipment for retuning residents, they found that the amounts of fungi in a house did not correlate with the amount of mycotoxins. The ability for moulds to create different mycotoxins at different rates under different conditions has made it very difficult to establish the health impact under the conventional understanding of toxicology: that the poison is in the dose. If one were to proceed on this trajectory, trying to establish objective measures of health effects associated with mycotoxin exposure in comparison with a control population, one would invariably come to this conclusion: no definitive or causal relationship. And we see this repeatedly (Page and Trout 2001; Kelman et al. 2004) and confirmed by Täubel and Hyvärinen (2016) who state, "Support from epidemiological studies in clarifying potential health effects upon indoor mycotoxin exposure is almost completely absent."

However, the inability to establish a toxic dose has not prevented organisations from laying unequivocal positions on the damage mould has on health. For example the World Health Organistion (WHO 2009) in its extensive review of literature concludes that, "Sufficient epidemiological evidence is available from studies conducted in different countries and under different climatic conditions to show that the occupants of damp or mouldy buildings, both houses and public buildings, are at increased risk of respiratory symptoms, respiratory infections and exacerbation of asthma." (p. 93) This is one of the most cited health impact of mould in damp buildings and used in a number of position statements, like the American Industrial Hygiene Association (AIHA 2013).

On another front, Shoemaker (2011) explains, "We now know that this idea from toxicology has little bearing on mold illness, as indeed mold illness isn't toxicological at all, but is immunological instead." As a physician, he found mould patients presenting similar symptoms to those who suffered from dinoflagellate toxin exposure (like during a toxic algal bloom) he treated elsewhere. These negatively charged toxins (mostly called ionophores), do not get excreted but are reabsorbed further down the gut. The accumulated toxics causes individuals with susceptible genetic makeup to have an over sensitised immune response that places them under chronic inflammation. Essentially, according to Shoemaker, they suffer more from the friendly fire of their defence mechanism than from the invader itself. He has created a protocol around the diagnosis and treatment of the syndrome termed CIRS-WDB, (Chronic Inflammatory Response Syndrome in Water-Damaged Buildings) and estimates that 24% of the population have the genetic disposition to develop CIRS if they were exposed to a WDB. Given that up to 40% of buildings in some countries have mould (Sivasubramani et al. 2004) there is a high likelihood that susceptible patients will encounter problematic buildings on a daily basis.

In Shoemaker's treatment protocol, cholestyramine (CSM) therapy was found to be particularly effective in eliminating biotoxins by preventing their reabsorption. His research is featured in the active an growing *Toxic Mould Support Australia* group (http://www.toxic-mould-support-australia.org). In an experiment involving 14 patients in a double-blind, placebo-controlled, clinical trial (Shoemaker and House 2006), patients were (1) exposed to a WDB, (2) treated with CSM, (3) ceased CSM, (4) re-exposed to the building that caused the sick building syndrome and (5) treated with CSM again. CSM was provenThe researchers note, "Health status continued to show marked improvement following CSM therapy while the study participants avoided re-exposure to the WDBs. However, all participants relapsed within 7 days of re-exposure to the WDBs."

In their comprehensive literature review on the deleterious effects of biocontaminants, Thrasher and Crawley (2009) reach a similar conclusion, urging that the "medical profession must recognize the importance of immediate removal of occupants from the toxic environment." Many would not see an urgency in leaving a mouldy building, and many owners would rather remediate it than to abandon it entirely. The following excerpt of a blog from a client of Thrasher is noteworthy (Adrienne 2014):

> I made contact with a leading toxicologist in the field, Dr. Jack Thrasher, and he explained that even with remediation, a systemic mold problem like ours was virtually impossible to eradicate. On October 4, 2008 we vacated our home. As advised, we treated the home like a fire—bringing nothing with us.

http://wholenewmom.com/health-concerns/black-mold-symptoms-mold-exposure-symptoms/

It is of immense consequence that a house can not only be deemed uninhabitable, but irreparable as well. How sick does an occupant need to be before a house is uninhabitable? How mouldy does a building need to be before it is considered irreparable? And who should make the call? Internationally, the legal experts in this area are the industrial hygienists (Peake 2015). In Tasmania, the Public Health Act (State of Tasmania 1997) places these two responsibilities on the Environmental Health Officers (EHOs) and Building Surveyors respectively. The EHOs in particular have "very broad powers of entry and inspection" (Workplace Standards Tasmania 2009) and the authority to issue rectification notices and, upon non-compliance from the owner, to follow up with closure orders. Despite efforts to introduce guidance in understanding and remediating mould (Kemp and Neumeister-Kemp 2010a, b) there are no mandatory standards for the determination of the severity and required remediation of mouldy premises. Mould severity guidance to EHO's (Department of Health and Human Services, Tasmanian Government 2015) is superficial, based largely on visual inspection of visible areas. Without understanding that dew point and water activity interaction is most severe in the interstitial spaces, most EHOs will not be able to look in the right places, much less require that the builder remove cladding and lining for a proper evaluation. Moreover the lack of detailed guidance places substantial responsibility on the EHO to make a subjective determination with enormous consequence to the builder, surveyor, designer and occupant. It is unsurprising then that to date, closure orders and rectification notices have not been known to be enforced for mould problems in Tasmania.

There is a significant resistance to the enforcement of the *Public Health Act* on unhealthy premises. It not only impacts the building design and construction industry, but the banking (home mortgages) and insurance industries as well. The lack of commitment to the injurious presence of mould contrasts to other countries where definitive positions are taken. In the United States, the National Association of Home Builders has stated that "growth of any type of mold in a home is never acceptable" (Small 2003). In Canada, the National Building Code (NRC CCC 2015) stresses that energy performance compliance "is not dependent on occupant interaction" (Section 9.36.5.3, clause 5), but on the building fabric and any associated automated mechanical systems. Taken together, the best practise would be to have buildings designed and constructed robustly, and fitted out with automatic ventilation systems if necessary, such that the occupants could use the house in any reasonable way without fear of mould occurrence.

8 Conclusion

In order to even understand the significance of this problem, there is a need thus to bridge three distinct disciplines and the match between three profiles: the building profile, microbiological profile, and occupant health profile. Buildings need to be understood in the building physics of psychrometry and vapour management from design, specification and construction. This physical environment needs to be studied with its impact on the biotoxin producing micro-organisms and their effect on human immunology.

Condensation is a complex and systemic process with serious structural and health repercussions. Many of the complexities cannot be understood by a single discipline and there it is timely to address this using a multidisciplinary approach. Condensation and subsequent mould growth appears to be a longstanding problem with indications that it might have been recently exacerbated in Australia by increased air tightness and thermal differentials that resulted from increased energy-efficiency and bushfire legislature, and also increased market-driven demands for thermal comfort.

The competitive building industry has led to inadequate consideration of vapour management at the design stage and improper installation during construction. Whilst there is acknowledgement of the condensation problem, there is widespread reluctance by any party from the construction or healthcare sectors to take a decisive and categorical position.

The Australian condensation experience highlights the imperative of taking a generalist and over-arching view of sustainability, where all SDGs are considered simultaneously. Any attempt to move one at the expense of the other can only lead to a short-termed solution, and cannot be a truly sustainable effort. In this instance, climate action (SDG 13) had been pursued at the expense of health and well-being (SDG 3). Until this trend is arrested, we will continue to build thermally efficient houses that compromise health and well-being.

In the iconic 1997 Australian movie *The Castle* the protagonist says, "It's not a house, it's a home. A man's home is his castle... You can't just walk in and steal our homes." There is due indignation when one is robbed of his own home. Condensation threatens to do exactly that. The irony with condensation is that this was far more infrequent before energy efficiency standards pushed for higher thermal performance. The insertion of sustainability considerations into the mission of the ABCB now appears to challenge the prior objectives for safety, health and amenity. In extreme cases, these new houses can become uninhabitable within their first winter. When one in three Australian houses are damp enough to trigger one in four people to have a chronic inflammatory response, we need to rethink if we have been too preoccupied with one sustainable development goal and completely forgotten that the house is our home, our castle. It is supposed to be our place to feel safe and healthy.

Acknowledgements of Funding Source and Student Contribution The case studies House-MW2 and House-BN were undertaken as funded research by the Building Standards and Occupational Licensing, Tasmanian Government under project RT106855, "Investigation of Destructive Condensation in Australian Cool temperate Buildings". Access to the condensation survey results was possible though the Australian Building Codes Board under a research consultancy RT108992, "Scoping study of condensation in residential buildings". The authors gratefully acknowledge the contribution of Dr. Stephen Tristram, Senior Lecturer in Medical Microbiology from the School of Health Sciences, University of Tasmania (UTAS) for identifying the moulds in the collected samples. Dr. Des FitzGerald and Dr. Bennet McComish, both from the School of Maths & Physics, UTAS, have assisted greatly in the statistical analysis of the ABCB Survey data.

References

ABCB (Australian Building Codes Board). (2010). *BCA Section J assessment and verification of an alternative solution*. Canberra: Australian Government and States and Territories of Australia.

ABCB (Australian Building Codes Board). (2016). *NCC volume one energy efficiency provisions* (4th ed.). Australian Government and States and Territories of Australia.

Adrienne. (2014, June). *Are your health problems black mold symptoms?* Retrieved from http://wholenewmom.com/health-concerns/black-mold-symptoms-mold-exposure-symptoms/.

AIRAH, Australian Institute of Refrigeration, Air Conditioning and Heating. (2016). *DA20 humid tropical air conditioning. australian institute of refrigeration, air conditioning and heating*.

Ambrose, M. D., & Syme, M. (2015). *House Energy Efficiency Inspections Project–Final Report*. CSIRO, Australia.

American Industrial Hygiene Association. (2013). *Position statement on mold and dampness in the built environment*.

Andersen, B., Frisvad, J. C., Søndergaard, I., Rasmussen, I. S., & Larsen, L. S. (2011). Associations between fungal species and water-damaged building materials. *Applied and Environmental Microbiology, 77*(12), 4180–4188. https://doi.org/10.1128/AEM.02513-10.

ASHRAE. (2009). *ASHRAE position document on airborne infectious diseases*.

Babbitt, J. D. (1939). The diffusion of water vapour through various building materials. *Canadian Journal of Research, 17a*(2), 15–32. http://doi.org/10.1139/cjr39a–002.

Barbosa-Cánovas, G. V., Fontana Jr, A. J., Schmidt, S. J., & Labuza, T. P. (2008). *Water activity in foods: Fundamentals and applications*. Wiley.

Barre, H. J. (1938). *The relation of wall construction to moisture accumulation in fill type insulation* (Doctoral). Iowa State College. Retrieved from http://lib.dr.iastate.edu/cgi/viewcontent.cgi?article=14546&context=rtd.

Bloom, E., Grimsley, L. F., Pehrson, C., Lewis, J., & Larsson, L. (2009). Molds and mycotoxins in dust from water-damaged homes in New Orleans after hurricane Katrina. *Indoor Air, 19*(2), 153–158. https://doi.org/10.1111/j.1600-0668.2008.00574.x.

BSOL, Building Standards and Occupational Licensing. (2014). *Condensation in Buildings: Tasmanian Designers' Guide*. Department of Justice, Tasmanian Government.

CSIRO. (1991a). *Condensation in houses*. Division of Building, Construction and Engineering.

CSIRO. (1991b). *Condensation in houses (Notes on the Science of Building No. NSB 61)*. Commonwealth Scientific and Industrial Research Organisation.

Department of Health and Human Services, Tasmanian Government. (2015). *Guide to assessing unhealthy premises*. Retrieved from http://www.dhhs.tas.gov.au/__data/assets/pdf_file/0004/223366/Guide_to_Assessing_Unhealthy_Premises_final_1June15_1.pdf.

James, C. (2006). Roof collapse women compensated. *The advertiser*. Retrieved from http://www.adelaidenow.com.au/news/south-australia/roof-collapse-women-compensated/story-e6frea83–1111112639825.

Kelman, B. J., Robbins, C. A., Swenson, L. J., & Hardin, B. D. (2004). Risk from inhaled mycotoxins in indoor office and residential environments. *International Journal of Toxicology, 23*(1), 3–10. https://doi.org/10.1080/10915810490265423.

Kemp, P., & Neumeister-Kemp, H. (2010a). *Australian mould guideline: The go-to guide for everything mould/by Peter Kemp and Heike Neumeister-Kemp*. Sydney: Messenger Publishing.

Kemp, P., & Neumeister-Kemp, H. (2010b). *The mould worker's handbook: A practical guide for remediation*. Osborne Park, W.A: The Enviro Trust.

Kendrick, B. (2000). *The fifth kingdom* (3rd ed.). Newburyport, MA: Focus.

Kowalski, W. (2005). *Aerobiological engineering handbook: Airborne disease and control technologies* (1st ed.). McGraw-Hill Professional.

Law, T. (2013). Comfort energetics: Thermal comfort under energy constraints. In *The future of thermal comfort in an energy-constrained world* (pp. 83–115). Springer International Publishing. Retrieved from http://link.springer.com/chapter/10.1007/978–3–319–00149–4_5.

Lstiburek, J. (2010). The perfect wall. *Insight*. Retrieved from http://buildingscience.com/documents/insights/bsi–001-the-perfect-wall.

NRC (National Research Council Canada) CCC (Canadian Codes Centre). (2015). Section 9.36. Energy efficiency. In *National building code of Canada*. Retrieved from http://www.bccodes.ca/BCBC_9%2036%20EnergyEfficiency.pdf.

Page, E. H., & Trout, D. B. (2001). The role of Stachybotrys mycotoxins in building-related illness. *AIHAJ: A Journal for the Science of Occupational and Environmental Health and Safety, 62*(5), 644–648.

Palmer, D. (2012, December). Nail plate separation from a truss member is one of the contributing factors behind the Riverside Golf Club roof collapse. Retrieved August 9, 2016 from http://reliablehomeinspections.com.au/real-estate-agent-discourages-against-having-building-inspections/.

Parker, J. (2014). Lessons to be learnt from the pink batts disaster [Text]. Retrieved August 12, 2016 from http://www.abc.net.au/news/2014–05–21/parker-lessons-to-be-learnt-from the pink-batts-disaster/5466762.

Pasanen, A.-L., Juutinen, T., Jantunen, M. J., & Kalliokoski, P. (1992). Occurrence and moisture requirements of microbial growth in building materials. *International Biodeterioration & Biodegradation, 30*(4), 273–283. https://doi.org/10.1016/0964-8305(92)90033-K.

Peake, D. (2015). Forensic engineering critique of mold expert opinions. *Environmental Claims Journal, 27*(1), 50–59. https://doi.org/10.1080/10406026.2014.998483.

Shoemaker, R. (2011). *Surviving mold : Life in the era of dangerous buildings*. Otter Bay Books.

Shoemaker, R. C., & House, D. E. (2006). Sick building syndrome (SBS) and exposure to water-damaged buildings: Time series study, clinical trial and mechanisms. *Neurotoxicol Teratology, 28*(5), 573–588. https://doi.org/10.1016/j.ntt.2006.07.003.

Sivasubramani, S. K., Niemeier, R. T., Reponen, T., & Grinshpun, S. A. (2004). Assessment of the aerosolization potential for fungal spores in moldy homes. *Indoor Air, 14*(6), 405–412. https://doi.org/10.1111/j.1600-0668.2004.00262.x.

Small, B. M. (2003). Creating mold-free buildings: A key to avoiding health effects of indoor molds. *Archives of Environmental Health: An International Journal, 58*(8), 523–527. https://doi.org/10.3200/AEOH.58.8.523-527.

State of Tasmania. (1997). Division 2 - Unhealthy premises. In *Public Health Act 1997 - An Act to protect and promote the health of communities in the State and reduce the incidence of preventable illness*. Hobart, Australia: State of Tasmania. Retrieved from http://www.austlii.edu.au/au/legis/tas/consol_act/pha1997126/.

Straaten, J. F. V. (1967). *Thermal performance of buildings*. Elsevier Publishing Co.

Szokolay, S. (2008). *Introduction to architectural science*. Architectural Press.

Taubel, M., & Hyvärinen, A. (2016). Occurrence of mycotoxins in indoor environments. In *Environmental mycology in public health* (pp. 299–323). Elsevier.

Thorn, A. (2016). Supreme Court decision "extremely good news" for claimants in action against James Hardie. Retrieved August 9, 2016 from http://goodcladding.co.nz/announcements.html.

Thrasher, J. D., & Crawley, S. (2009). The biocontaminants and complexity of damp indoor spaces: More than what meets the eyes. *Toxicology & Industrial Health, 25*(9/10), 583–615.

Wannemacher, R. W., & Wiener, S. L. (1997). Chapter 34: Trichothecene mycotoxins. In R. Zajtchuk (Ed.), *Medical aspects of chemical and biological warfare*. Office of The Surgeon General: Maryland.

Workplace Standards Tasmania. (2009). Unhealthy premises: A guide for environmental health officers. Hobart: Tasmanian Public and Environmental Health Service. Retrieved from http://trove.nla.gov.au/version/47658285.

World Health Organisation (WHO). (2009). *Guidelines for indoor air quality: Dampness and mould*. Denmark: World Health Organization Regional Office for Europe.

Zain, M. E. (2011). Impact of mycotoxins on humans and animals. *Journal of Saudi Chemical Society, 15*(2), 129–144. https://doi.org/10.1016/j.jscs.2010.06.006.

The manufacturer's authorised representative in the EU is Springer Nature Customer Service Centre GmbH, Europaplatz 3, 69115 Heidelberg, Germany. If you have any concerns regarding our products, please contact ProductSafety@springernature.com

Printed and bound by CPI Group (UK) Ltd, Croydon, CR0 4YY

30/09/2026

02225662-0018